BIOPROSPECTS OF MACROFUNGI
Recent Developments

Series: Progress in Mycological Research

BIOPROSPECTS OF
MACROFUNGI
Recent Developments

Editors

Sunil Kumar Deshmukh
Scientific Advisor
Greenvention Biotech Pvt. Ltd.
Pune, India

Kandikere R. Sridhar
Adjunct Faculty, Department of Biosciences
Mangalore University, Mangalore, India

Hesham Ali El Enshasy
Professor at Bioprocess Engineering Department
Faculty of Engineering
Universiti Teknologi Malaysia (UTM), Malaysia

CRC Press
Taylor & Francis Group
Boca Raton London New York

CRC Press is an imprint of the
Taylor & Francis Group, an **informa** business

A SCIENCE PUBLISHERS BOOK

First edition published 2024
by CRC Press
2385 NW Executive Center Drive, Suite 320, Boca Raton FL 33431

and by CRC Press
4 Park Square, Milton Park, Abingdon, Oxon, OX14 4RN

© 2024 Sunil Kumar Deshmukh, Kandikere R. Sridhar and Hesham Ali El Enshasy

CRC Press is an imprint of Taylor & Francis Group, LLC

ISBN: 978-1-032-38167-1 (hbk)
ISBN: 978-1-032-38169-5 (pbk)
ISBN: 978-1-003-34380-6 (ebk)

DOI: 10.1201/9781003343806

Typeset in Times New Roman
by Radiant Productions

Preface

There are several gaps in our knowledge of health, nutrition, energy, agriculture and the environment. Such as the significance of plants and animals, the fifth kingdom, and the filamentous fungi possess the capacity to combat several current challenges. Mycology being a mega-science expanded our knowledge in several folds since the beginning of 21 century. Due to a wide variety of fungi and fungi-like organisms distributed in different ecological niches, several biotechnological applications have been considered to combat the challenges in value-added foodstuffs, sustainable agriculture, energy production, environmental cleaning and industrially valued products. Owing to fungal diversity, their varied lifestyles, substrate preference and versatile metabolites, they entered the applied science by drawing the attention of ecologists, geneticists, biochemists, biotechnologists, fermentation technologists, experimental biologists and bioengineers. Several contributions have emerged in the recent past about the biotechnological applications of fungi (e.g., Nevalainen 2020, Deshmukh et al. 2022, Meyer 2022).

Macrofungi are non-conventional resources that meet the requirements of nutrition, health, agriculture and industrial applications. A global conservative estimation of fungi ranges from 2.2–3.8 million and macrofungal diversity ranges between 0.14 and 1.25 million (Hawksworth 2019, Hawksworth and Lucking 2017). Evaluation of macrofungal reserves in different ecosystems increases our knowledge of their applications and future progress. In the choice of subjects, efforts have been made to pool together chapters that possess extensive applications worldwide. Thus, the present book projects various benefits mainly on their cultivation and products (metabolites, nutraceuticals, and biocomposites). Contributions of researchers from Algeria, Austria, Egypt, Korea, Malaysia, India, Japan, Indonesia, Italy, Mexico, Poland, Spain, South Africa, Turkey, the United Kingdom, and the United States of America offer 19 chapters in different applied areas. This volume possesses seven sections in macrofungal biotechnology (cultivation, bioactive compounds, bio-composites, nutraceuticals, association with fauna, biofertilizers and biocontrol).

We hope that the effort on the biotechnological potential of macrofungi offered in this book will be informative and valuable for a wide group of readers, beginners in the fields of mycology, biotechnology, and microbiology. This contribution mainly focused on various facets of applied mycology of macrofungi of different disciplines (e.g., metabolites, products, cultivation, nutraceuticals, biocomposites, biofertilizers and pest control). We are grateful to the kind gestures of the contributors and reviewers towards on-time submission and evaluation. CRC Press has offered their co-operation by honest official formalities to offer this book on time to the readers.

Pune, India
Mangalore, India
Johor Bahru, Malaysia

Sunil K. Deshmukh
Kandikere R. Sridhar
Hesham Ali El Enshasy

References

Deshmukh, S.K., M.V. Deshpande and K.R. Sridhar. 2022. Fungal Biopolymers and Biocomposites: Prospects and Avenues, Springer Nature Singapore Pte Ltd., p. 421.

Hawksworth, D.L. 2019. The macrofungal resource: Extant, current utilization, future prospects and challenges. pp. 1–9. *In*: Sridhar, K.R. and Deshmukh, S.K. (eds.). Advances in Macrofungi: Diversity, Ecology and Biotechnology. CRC Press, Boca Raton, USA.

Hawksworth, D.L. and R. Lücking. 2017. Fungal diversity revisited: 2.2 to 3.8 million species. Microbiology Spectrum 5: FUNK-0052-2016.

Meyer, V. 2022. Connecting materials sciences with fungal biology: A sea of possibilities. *Fungal Biol. Biotechnol.*, 9: 5. 10.1186/s40694-022-00137-8.

Nevalainen, H. 2020. Grand Challenges in Fungal Biotechnology, Springer Nature Switzerland AG, p. 535.

Contents

Biofertilizers and Biocontrol

List of Contributors

Aguilar-Marcelino, Liliana
Centro Nacional de Investigación Disciplinaria en Salud Animal e Inocuidad, INIFAP, Jiutepec, Morelos, Mexico. Email: aguilar.liliana@inifap.gob.mx

Akchiche, Yasmine Fatima
University of Algiers, Faculty of Sciences, Algiers, Algeria.

Aksoy, Adel
Eskil Vocation of High School, Laboratory Veterinary Science, Aksaray University, Aksaray, Turkey. Email: adilaksoy@aksaray.edu.tr

Benbaibeche, Hassiba
University of Algiers, Faculty of Sciences, Algiers, Algeria.

Boumehira, Ali Zineddine
University of Algiers, Faculty of Sciences, Algiers, Algeria, Ecole Nationale Supérieure Agronomique - ENSA, El Harrach, Algiers, Algeria.
Email: ali.boumehira@edu.ensa.dz

Carrasco, Jaime
Centro Tecnológico de Investigación del Champiñón de La Rioja (CTICH), 26560 Autol, Spain.
Department of Plant Sciences, University of Oxford, South Parks Road, Oxford OX1 3RB, United Kingdom. Email: carraco.jaime@gmail.com

Castañeda-Ramírez, Gloria Sarahi
Centro Nacional de Investigación Disciplinaria en Salud Animal e Inocuidad, INIFAP, Jiutepec, Morelos, Mexico.

Chhipa, Hemraj
Agriculture University, Kota Rajasthan-324001, India.

Cruz-Arévalo, Julio
Universidad Tecnológica de la Selva. Entronque Toniná Km. 0.5 Carretera Ocosingo-Altamirano, Ocosingo, Chiapas, México.

Dailin, Danie Joe

Institute of Bioproduct Development (IBD), Universiti Teknologi Malaysia (UTM), Johor Bahru, Johor, Malaysia.
Faculty of Chemical and Energy Engineering, Universiti Teknologi Malaysia (UTM), Johor Bahru, Johor, Malaysia.

Darwish, Amira M. Galal

Department of Food Technology, Arid Lands Cultivation Research Institute (ALCRI), City of Scientific Research and Technological Applications (SRTA-City), 21934 Alexandria, Egypt. Email: amiragdarwish@yahoo.com, adarwish@srtacity.sci.eg

Deshmukh, Sunil K.

Greenvention Biotech Pvt. Ltd. Pune, India.
Email: sunil.deshmukh1958@gmail.com

Domínguez-Núñez, José Alfonso

Department of Forest and Environmental Engineering and Management. Universidad Politécnica de Madrid. C/ Jose Antonio Novais, 10. 28040, Madrid, Spain.

El-Deeb, Nehal M.

City of Scientific Research and Technological Applications (SRTA-City), New Borg El-Arab City, Alexandria, Egypt. Email: nehalmohammed83@gmail.com

El-Gharabawy, Hoda M.

Botany and Microbiology Department, Faculty of Science, Damietta University, New Damietta, 34517, Egypt.

EL-Enshasy, Hesham A.

Institute of Bioproduct Development (IBD), Universiti Teknologi Malaysia (UTM), Johor Bahru, Johor, Malaysia.
Faculty of Chemical and Energy Engineering, Universiti Teknologi Malaysia (UTM), Johor Bahru, Johor, Malaysia.
City of Scientific Research and Technology Applications, New Burg Al Arab, Alexandria, Egypt.

El Fadel Ousmaal, Mohamed

University of Algiers, Faculty of Sciences, Algiers, Algeria

Ferreira, Juliana Marques

Institute of Tropical Pathology and Health, Universidade Federal de Goiás, Goiânia, Brazil.

Fouad, Aya M.

Faculty of Pharmacy, Damanhour University, Egypt.

Freitas Soares, Filippe Elias de

Department of Chemistry, Universidade Federal de Lavras, Lavras, Minas Gerais, Brazil.

Fung, Shin Yee

Medicinal Mushroom Research Group (MMRG), Department of Molecular Medicine, Universiti Malaya, 50603 Kuala Lumpur, Malaysia. Email: syfung@um.edu.my

Furukawa, Hitoshi

Nagano Prefecture Forestry Center, Shiojiri, Nagano 399-0711, Japan.

Gupta, Manish K.

SGT College of Pharmacy, SGT University, Gurugram 122505, Haryana, India.

Idriss, Heba F.F.M.

Department of Soil, Plant and Food Sciences, Bari Aldo Moro University, Bari, Italy. Email: h.idriss1@studenti.uniba.it

Jędrejko, Karol

Department of Pharmaceutical Botany Faculty of Pharmacy, Jagiellonian University Medical College, 30-688 Kraków, Poland.

Kała, Katarzyna

Department of Pharmaceutical Botany Faculty of Pharmacy, Jagiellonian University Medical College, 30-688 Kraków, Poland.

Kenawy, Ahmed M.

City of Scientific Research and Technological Applications (SRTA-City), New Borg El-Arab City, Alexandria, Egypt. Email: aatta75@yahoo.com

Kinoshita, Akihiko

Kyushu Research Center, FFPRI, Forest Research and Management Organization, Kumamoto City, Kumamoto 860-0862, Japan.

Mautner, Andreas

Polymer & Composite Engineering (PaCE) group, Institute of Materials Chemistry, University of Vienna, Währinger Straße 42, 1090 Wien, Austria. Email: andreas.mautner@univie.ac.at

Mendoza-Ortega, Benjamín

Centro Nacional de Investigación Disciplinaria en Salud Animal e Inocuidad, INIFAP, Jiutepec, Morelos, Mexico.

Moloi, Neo

Sawubona Mycelium Co., Centurion, Gauteng, South Africa.

Muszyńska, Bożena

Department of Pharmaceutical Botany Faculty of Pharmacy, Jagiellonian University Medical College, 30-688 Kraków, Poland. Email: muchon@poczta.fm

Obase, Keisuke

FFPRI, Forest Research and Management Organization, Tsukuba, Ibaraki 305-8687, Japan.

Páez-León, Susan Yaracet

Centro Nacional de Investigación Disciplinaria en Salud Animal e Inocuidad, INIFAP, Jiutepec, Morelos, Mexico.

Peña, Raquel

Centro Tecnológico de Investigación del Champiñón de La Rioja (CTICH), 26560 Autol, Spain.

Pineda-Alegría, Jesús Antonio

Centro Nacional de Investigación Disciplinaria en Salud Animal e Inocuidad, INIFAP, Jiutepec, Morelos, Mexico.
Biotechnology Research Center, UAEM, Cuernavaca, Morelos, Mexico.

Qin, Zhao

Department of Civil and Environmental Engineering, Syracuse University, Syracuse, NY 13244, United States of America.
The BioInspired Institute, Syracuse University, Syracuse, NY 13244, United States of America.
Laboratory for Multiscale Material Modeling, Syracuse University, Syracuse, NY, 13244, United States of America. Email: zqin02@syr.edu

Rahi, Deepak Kumar

Department of Microbiology, Panjab University, Chandigarh, Punjab, India

Rahi, Sonu

Department of Botany, Government Post-Graduate Girls College, A. P. S. University, Rewa, Madhya Padesh, India.

Ramli, Solleh

Institute of Bioproduct Development (IBD), Universiti Teknologi Malaysia (UTM), Johor Bahru, Johor, Malaysia.

Rasid, Zaitul Iffa Abd

Institute of Bioproduct Development (IBD), Universiti Teknologi Malaysia (UTM), Johor Bahru, Johor, Malaysia.

Ravikrishnan, Venugopalan

Department of Biosciences, Mangalore University, Mangalagangotri, Karnataka, India.

Rebah, Amira

University of Algiers, Faculty of Sciences, Algiers, Algeria.

Serag, Mamdouh S.

Botany and Microbiology Department, Faculty of Science, Damietta University, New Damietta, 34517, Egypt.

Sharma, Devender Kumar

Department of Microbiology, Panjab University, Chandigarh, Punjab, India

Sridhar, Kandikere R.

Department of Biosciences, Mangalore University, Mangalagangotri, Karnataka, India. Email: kandikere@gmail.com

Sukmawati, Dalia

Faculty of Mathematics and Natural Science, Universitas Negeri Jakarta, Jakarta Timur 13220, Indonesia.

Sułkowska-Ziaja, Katarzyna

Department of Pharmaceutical Botany Faculty of Pharmacy, Jagiellonian University Medical College, 30-688 Kraków, Poland.

Tan, Chon-Seng

LiGNO Research Initiative, LiGNO Biotech Sdn. Bhd., Balakong Jaya 43300, Selangor, Malaysia.

Yahayu, Maizatulakmal

Institute of Bioproduct Development (IBD), Universiti Teknologi Malaysia (UTM), Johor Bahru, Johor, Malaysia.

Yamanaka, Takashi

Tohoku Research Center, Forestry and Forest Products Research Institute (FFPRI), Forest Research and Management Organization, Morioka, Iwate 020-0123, Japan. Email: yamanaka@affrc.go.jp

Yamada, Akiyoshi

Shinshu University, Minami-minowa, Nagano 399-4598, Japan.

Yang, Libin

Department of Civil and Environmental Engineering, Syracuse University, Syracuse, NY 13244, United States of America.
Laboratory for Multiscale Material Modeling, Syracuse University, Syracuse, NY, 13244, United States of America.

Cultivation

1

Advances in the Cultivation of Ectomycorrhizal Mushrooms in Japan

Takashi Yamanaka,[1],* Akiyoshi Yamada,[2]
Hitoshi Furukawa,[3] Akihiko Kinoshita[4] and
Keisuke Obase[5]

1. Introduction

A variety of wild mushrooms have been harvested in Japan. People in Japan enjoy mushrooms prepared in various ways, including grilled, fried (tempura), flavored rice, and soup. Thus, mushrooms have been a common food ingredient in autumn. Among these edible mushrooms, shiitake (*Lentinula edodes*), nameko (*Pholiota microspora*), maitake (*Grifola frondosa*), enokitake (*Flammulina velutipes*), buna-shimeji (*Hypsizygus marmoreus*), etc., are saprotrophic species produced by log cultivation or bed cultivation. On the other hand, ectomycorrhizal (EM) species such as matsutake (*Tricholoma matsutake*), shoro (*Rhizopogon roseolus*), and amitake (*Suillus bovinus*) are difficult to cultivate because their growth and fruiting depend on living plant roots (Yamada et al. 2017). Consequently, these EM mushrooms

[1] Tohoku Research Center, Forestry and Forest Products Research Institute (FFPRI), Forest Research and Management Organization, 92-25 Nabe-yashiki, Shimo-kuriyagawa Morioka, Iwate 020-0123, Japan.
[2] Shinshu University, Minami-minowa, Nagano 399-4598, Japan
[3] Nagano Prefecture Forestry Research Center, Shiojiri, Nagano 399-0711, Japan
[4] Kyushu Research Center, FFPRI, Forest Research and Management Organization, Kumamoto City, Kumamoto 860-0862, Japan
[5] FFPRI, Forest Research and Management Organization, Tsukuba, Ibaraki 305-8687, Japan
* Corresponding author: yamanaka@ffpri.affrc.go.jp

are costly. While truffles (*Tuber* spp.), girolle (*Cantharellus cibarius* and several related species), and porcini (*Boletus edulis* and several related species) are also EM mushrooms that are expensive food ingredients in western cuisines (Guerin-Laguette 2021), only recently have these species been reported in Japan (Kinoshita et al. 2011, Endo et al. 2014, Ogawa et al. 2019a). The artificial cultivation technique of EM mushrooms would lead to the development of stable new markets, and numerous attempts have been made to cultivate them. Establishing symbiotic associations between EM species and trees by cultivating trees with EM species is the cultivation system for these EM mushrooms. However, a few EM mushrooms (*Lyophyllum shimeji*, *Boletus* spp.) could be cultivated in the same manner as saprotrophic mushrooms (Ohta 1994b, Yamanaka et al. 2000, Ohta and Fujiwara 2003). The following steps are necessary for establishing the cultivation of EM mushrooms: obtaining inoculums (spores and pure cultures of EM spp.), cultivating EM plants *in vitro* or *in vivo*, and acclimatizing the EM plants in the greenhouse and then in the field. In addition, field inoculation with spores or mycelia of EM species on mature trees is another technique for EM mushroom cultivation. The research conducted on EM mushroom species for cultivation in Japan is presented in Table 1. This article focuses on the EM mushrooms (*T. matsutake*, *T. bakamatsutake*, *L. shimeji,* and *Tuber* spp.) and describes recent efforts for their cultivation in Japan.

2. *Tricholoma matsutake* (Matsutake)

Tricholoma matsutake is an EM species that produces the coveted edible mushroom known as "matsutake" (Figure 1A). This mushroom has been revered in Japan as

Figure 1: Matsutakes harvested in Toyooka, Nagano, Japan (A); Bakamatsutakes developed after field inoculation with mycelia of this species on an oak plant in Nara, Japan (B); Honshimejis at Nosegawa in Nara, Japan (C); Japanese black truffles from Hyogo, Japan (D) and Japanese white truffles from Mie, Japan (E).

Table 1. Status of the research for the cultivation of Japanese EM mushrooms.

EM species	Pure culture	EM formation in vitro	Fruiting body formation				
			Pure culture	EM plant in pot	Plantation of EM sapling	In the field	
						Spore spray	Mycelial inoculum
Tricholoma matsutake	Kawai and Abe (1976), Shimazono (1979), Yamada et al. (2001c)	Yamada et al. (1999b, 2006)	Ogawa and Hamada (1975), Kawai and Ogawa (1976)		Kareki and Kawakami (1985), Ka et al. (2018)		
T. bakamatsutake	Shimazono (1979), Yamada et al. (2001c)	Yamanaka et al. (2014)			Kawai (2019)		
T. portensosum	Yamada et al. (2001a)	Yamada et al. (2001a, b)		Yamada et al. (2001a)			
T. saponaceum	Yamada et al. (2001a)	Yamada et al. (2001a, b)		Yamada et al. (2001a)			
T. flavovirens	Yamada et al. (2001a)	Yamada et al. (2001a, b)		Yamada et al. (2001a)			
T. terreum	Yamada et al. (2007)	Yamada et al. (2007)		Yamada et al. (2007)			
Lyophyllum shimeji	Ohta (1994a)	Yamada et al. (2001a, b)	Ohta (1994b)	Kawai (1997)	Kawai (1999)		Kawai (1999), Mizutani (2005)
L. semitale	Yamada et al. (2001a)	Yamada et al. (2001a, b)		Yamada et al. (2001a)			
Japanese *Tuber* spp	Obase et al. (2021), Nakano et al. (2020a, 2022a)						

Table 1 contd. ...

...Table 1 contd.

EM species	Pure culture	EM formation *in vitro*	Fruiting body formation				
			Pure culture	EM plant in pot	Plantation of EM sapling	In the field	
						Spore spray	Mycelial inoculum
Rhizopogon rubescens	Yamada et al. (2001a), Shimomura (2019)	Yamada et al. (2001a), Shimomura (2019)		Yamada et al. (2001a), Shimomura (2019)		Tomikawa (2006)	
Suillus bovinus	Yamada et al. (2001a)	Yamada et al. (2001a, b)					
S. luteus	Yamada et al. (2001a)	Yamada et al. (2001a, b)		Yamada et al. (2001a)			
S. granulatus	Yamada et al. (2001a)	Yamada et al. (2001a, b)		Yamada et al. (2001a)			
S. grevillei	Qu et al. (2003)	Qu et al. (2003)				Shibata (1989), Katagiri et al. (2021)	
Boletus edulis	Endo et al. (2014)	Endo et al. (2014)					
Amanita caesareoides	Endo et al. (2013)	Endo et al. (2013)					
Lactarius akahatsu	Yamada et al. (2001a)	Yamada et al. (2001a, b)		Yamada et al. (2001a)			
L. hatsutake	Yamada et al. (2001a)	Yamada et al. (2001a, b)		Yamada et al. (2001a)			
Japanese golden chantarella (*Cantharellus anzutake*)	Ogawa et al. (2019a)	Ogawa et al. (2019b)		Ogawa et al. (2019b)			

"a taste of autumn" for more than a millennium. In the early 1940s, the matsutake harvest was approximately 12,000 Mg but has since decreased to less than 100 Mg annually. In consequence, the price per kilogram of this mushroom rose from 370 JPY in 1952 to 67,000 JPY in 2017. Along with a decline in the domestic harvest, it is necessary to implement the artificial cultivation of these mushrooms (Yamanaka et al. 2020).

This EM species primarily inhabits forests of Japanese red pine (*Pinus densiflora*), while *Picea glehnii*, *P. jezoensis*, *Pinus pumila*, *P. thunbergii*, *Tsuga diversifolia* and *T. sieboldii* have also hosted tree species (Imazeki and Hongo 1987). Asian, European, North American, and Central American specimens of matsutake and its close relatives have been documented.

Strains have been isolated from various geographic locations and utilized in phylogenetic studies based on genetic data, physiological properties, and EM formation on various host trees (Yamada et al. 2014, Narimatsu et al. 2019). Yamada et al. (1999a) described a typical EM formation by *T. matsutake* in the field. This was confirmed by inoculating pine seedlings *in vitro* with a strain of *T. matsutake* (Yamada et al. 1999b). Afterwards, many studies on EM formation using various combinations of strains and sterile seedlings of putative host species were conducted (Yamada et al. 2010, 2014, Nakano et al. 2020b). These studies demonstrated that was more likely to form EM with tree species *in vitro* than in the field, although the improved growth of tree seedlings may be affected by nutrient conditions, such as the soil materials used (Saito et al. 2018).

"Shiro", a mycelial aggregation of *T. matsutake*, forms under natural conditions in the field, in association with EM roots and soil particles, obtaining carbon sources from the tree roots with which it is associated. In autumn, matsutakes appear on the periphery of the shiro. Therefore, the cultivation of shiro in the field and/or in a greenhouse is required for the production of matsutake. Numerous field trials for developing shiros and fruiting matsutakes have been conducted, including inoculation with spores, mycelia, fragments of shiro, and planting pine saplings with EM of *T. matsutake* (Yamanaka et al. 2020). Among these trials, matsutake appeared in Hiroshima in 1983, five years after planting a pine sapling with EM of *T. matsutake* (Kareki and Kawakami 1985). In South Korea, it was reported that later matsutake occurred after pine saplings with EM were similarly planted in the field (Ka et al. 2018). However, it was not determined in either instance whether these matsutakes were genetically identical to the strains that were inoculated on the sapling.

Yamada et al. (2006) and Kobayashi et al. (2007) successfully produced shiros using EM seedlings in an *in vitro* system. These EM pine seedlings with shiros were planted in the field, but no shiros were subsequently developed (Kobayashi et al. 2015). There is no instance in which the development of shiro was positively identified after field planting.

Mycelia forming a shiro have long been believed to be genetically homogeneous, as a shiro develops from a dikaryotic mycelium after mating between two monokaryotic mycelia. However, Murata et al. (2005) reported that isolates from fruit bodies and spores from the same fruit bodies were genetically distinct, and

Lian et al. (2006) discovered that one to four genets were found within each shiro after analyzing fruiting bodies. These results demonstrated the natural occurrence of genetic mosaics in a shiro. Murata et al. (2005) suggested that the management of shiro should encourage colonization and fruit body production by maintaining a favorable mosaic status, which could be sustained by the dispersal and germination of basidiospores and their subsequent sexual interaction with shiro mycelia. This causes physiological variation among genets within a single shiro. These variations should be complementary between neighboring genet and linked to concurrent fruit body production within a shiro. Yamada et al. (2019) initially compared isolates established from the spores of a single fruiting body to determine the effect of mosaicism on the development of shiro and fruit body production. On media, the mycelial growth of these sibling isolates varied significantly. In addition, the isolates demonstrated commensal and amensal interactions with varying levels of nutrients. Then, sibling spore isolates were compared to ectmycorrhization ability (Horimai et al. 2020). The ratio of EM colonization varied significantly between isolates and was dependent on the nitrogen content of the soil. Mixed inoculations of three selected isolates with single pine seedlings revealed that paired inoculations of two of the three isolates and triple inoculations of these isolates resulted in significantly higher levels of ectomycorrhizal colonization than a single isolate. Moreover, Horimai et al. (2021) demonstrated that basidiospore germination in the EM system of a pine host produced new genets. These occurrences contribute to the comprehension of how maintained shiro structures, as well as the occurrence of alternating and newly produced genets. Besides, basidiospore inoculation in an already established EM system involving this fungus on a host of pine was beneficial for generating complex shiro structures that resemble the natural population structures.

Understanding the relationship between soil microorganisms and shiro (Vaario et al. 2011) is also essential for the establishment and growth of shiro. The active mycorrhizal zone of shiro possesses antimicrobial properties (Ohara and Hamada 1967). Lian et al. (2006) demonstrated that the EM fungal community underneath the active mycorrhizal zone is species-poor and that there are significant differences between those within and outside the shiros. Kataoka et al. (2012) demonstrated that certain bacteria, such as *Sphingomonas* and *Acidobacterium*, were present in the active mycorrhizal zone, although none were detected using the technique of dilution plating. These findings have been supported by a report by Ohara and Hamada (1967). In addition, in South Korea, fewer fungal communities were found beneath the fruiting zone of the shiro compared to the fruiting zone's interior and exterior (Kim et al. 2013, Oh et al. 2016). However, Kim et al. (2014) found no difference between the bacterial communities of the fruiting zone within and outside of the fruiting zone. Li et al. (2018) reported that oak roots colonized by *T. matsutake* changed the endophytic microbial community in the roots. According to Ohara and Hamada (1967), volatile monoterpenes like α- and β-pinene are antimicrobial metabolites. According to Nishino et al. (2016), the (oxalato) aluminate complex has antimicrobial properties. This is a product of the reaction of aluminum phosphate and oxalic acid secreted by EM fungi, and it releases soluble phosphorus that utilizes as a nutrient to spread the shiro in the soil. Additionally, Vaario et al. (2015) reported that *T. matsutake* got soil minerals from rock fragments. Thus, shiro development has been chemically and biologically described but not induced artificially.

It has been believed that bed cultivation of matsutake is not possible because this EM species has a limited capacity to decompose organic matter (Kusuda et al. 2006, 2008, Shimokawa et al. 2017). Ogawa and Hamada (1975) and Kawai and Ogawa (1976) showed that produced fruit body primordia from mycelia grown in sterilized soil or vermiculite containing a nutrient solution. However, the fruiting bodies did not develop from the primordia. Recently, Murata et al. (2018, 2019) obtained a stable mutant of with enhanced amylose and cellulose degradation induced by heavy-ion beam exposure. Using barley-based substrate cultivation, a mutant of *T. matsutake* generated by irradiation of γ-rays, produced protuberances composed of non-aerial hyphal tissues (Murata et al. 2020). Although these protuberances did not develop into fruiting bodies, they were observed in three independent experiments in succession.

3. *Tricholoma bakamatsutake* (Bakamatsutake)

In Japan, three mushroom species have been reported as closely related to matsutake: bakamatsutake (*T. bakamatsutake*), nisematsutake (*T. fulvocastaneum*), matsutakemodoki (*T. robustum*) (Imazeki and Hongo 1987). Among them, bakamatsutake is anticipated to be an equally valuable edible mushroom with a strong matsutake aroma. *Tricholoma bakamatsutake* is a member of the Fagaceae-associated species and has been reported from China, North Korea, and New Guinea, in addition to Japan (Otani 1976, Gyong 2011, Yamanaka et al. 2011). In the field and *in vitro*, this species produced chlamydospores, thick-walled pigmented spores, at the terminal end of vegetative hyphae (Shimazono 1979, Terashima et al. 1993). The addition of amino acids *in vitro* enhanced the formation of chlamydospores (Yamanaka et al. 2019). *Tricholoma bakamatsutake* can form EMs with pine seedlings *in vitro*, indicating that it is *in vitro* host range is broader than in nature (Yamanaka et al. 2014). Multiple fruit bodies of this species have recently formed on the floor of a *Quercus serrata* forest following field inoculation with the planting of a *Q. phillyraeoides* sapling inoculated with the fungus mycelia (Figure 1B) (Kawai et al. 2019). In contrast, bakamatsutake strains demonstrated an enzymatic capacity for lignin decomposition (Oikawa et al. 2020), and a private company reported the formation of fruiting bodies in bed cultivation.

4. *Lyophyllum shimeji* (Honshimeji)

Honshimeji is *L. shimeji*'s fruit body (Figure 1C). Japanese people have a high regard for honshimeji, remarking, "matsutake has a good aroma, while shimeji has a good flavor." *Lyophyllum shimeji* is an EM species associated with *P. densiflora*, *Q. serrata*, *Q. crispula*, *Q. phillyraeoides*, etc. It is necessary to establish mycorrhizal associations between *L. shimeji* and its host plants for its cultivation in the field. Kawai (1997) inoculated *P. densiflora* saplings obtained by air-layering with spawns of *L. shimeji* to produce EM in the pine saplings. Fruiting bodies were produced during the cultivation of these EM pine saplings in a pot with sterilized sandy soil. Moreover, after these EM pine saplings were planted in the pine forest, fruit bodies appeared on the forest floor (Kawai 1999).

This EM species is uniquely capable of decomposing organic matter and producing fruit bodies in bed culture. Ohta (1994a, b) reported that all strains of *L. shimeji* could grow on beech and pine sawdust and that some of these strains produced fruiting bodies on a medium containing barley, beech sawdust, and liquid synthetic nutrients. Then, four strains of *L. shimeji* were chosen from 60 wild strains for commercial honshimeji production (Ohta 1998).

5. *Tuber himalayense* and *T. japonicum* (Japanese Truffle)

"True truffles" are the hypogynous sporocarps of *Tuber* spp. (Ascomycota, Pezizales). Several truffle species, including the Périgord black truffle (*T. melanosporum*), the Italian white truffle (*T. magnatum*), the summer truffle (*T. aestivum*), and the Bianchetto truffle (*T. borchii*), are highly valued in European cultures due to their distinctive aroma (Hall et al. 2007). All known species are EM, allowing them to coexist with a variety of temperate forest tree species. Recently, Kinoshita et al. (2011) demonstrated that Japan is home to at least 20 species of Japanese truffles, the majority of which are distinct from European and North American species. Some of them were subsequently reported as new species (Kinoshita et al. 2016, 2018a, 2021).

Several experiments have been conducted to determine the optimal management of truffle cultivation in fields. Several truffle species are cultivated artificially using EM plants inoculated with these species (Hall et al. 2007, Iotti et al. 2016, Bach et al. 2021). In Japan, research is being conducted on the artificial cultivation of these truffles. Initially, two species of Japanese truffles (Figure 1D and IE) (*T. himalayense* and *T. japonicum*) were chosen for artificial cultivation because these species have been observed in many regions of Japan and their fruit bodies are of an appropriate size for human consumption. *Tuber japonicum* was collected from Miyagi to Okayama (Kinoshita et al. 2011, 2016), and *T. himalayense* from Hokkaido to Kyushu (Kinoshita et al. 2011, 2018a). Moreover, the analysis of odor-active volatile compounds revealed that 1-octen-3-ol and 3-methyl-2, 4-dithiapentane contributed significantly to the odor of *T. japonicum*, whereas 2,4-dithiapentane is the most important odorant of *T. magnatum* (Shimokawa et al. 2020). The chemical components of *T. japonicum* were identical to those of *T. magnatum* and *T. melanosporum*, both of which are edible. Acute oral toxicity tests revealed no abnormalities, with an LD_{50} of over 2,000 mg/kg under experimental conditions. These results indicated that *T. japonicum* may have a potentially high market value.

Based on information gathered from the collection of these truffles in their natural habitat, the vegetation and soil properties of their habitat were studied. These truffles are typically grown in oak forests, but birch and pine can also serve as hosts. *Tuber himalayense* inhabits neutral to slightly alkaline (pH 6.4–8.0) soils, whereas *T. japonicum* inhabits slightly acid (pH 5.6–6.0) soils (Furusawa et al.

2020). The pure cultures of these truffles have been obtained and utilized to elucidate their physiological properties. At pH 5.0 and 6.0, *T. japonicum* grew well and *T. himalayense* thrived at pH 7.0 (Nakano et al. 2020a). These findings suggest that the optimal pH for mycelial growth differs among species. The growth data collected in these studies could be used to determine the optimal pH conditions for the artificial cultivation of these truffles. For truffle mycelial growth, EM formation, and in situ fruit body production, soil pH is one of the most vital factors.

Strains of *T. japonicum* grew optimally on malt extract and modified Melin–Norkrans medium (20°C or 25°C) (Nakano et al. 2022a). This fungus utilized both inorganic (NH_4^+ and NO_3^-) and organic sources of nitrogen (casamino acids, glutamine, peptone, urea, and yeast extract). Moreover, this fungus utilizes numerous carbon sources, including monosaccharides (arabinose, fructose, galactose, glucose, and mannose), disaccharides (maltose, sucrose, and trehalose), polysaccharides (dextrin and soluble starch), and sugar alcohol (mannitol). However, growth-promoting nutrient sources and their effects varied considerably between strains. Interestingly, *T. japonicum* formed mitospores on the vegetative hyphae mycelium (Nakano et al. 2022b). Among 25 strains, 20 strains developed mitospores on modified Melin–Norkrans agar, demonstrating that mitospore formation is likely a trait shared by many *T. japonicum* strains. This *in vitro* nature of *T. japonicum* will be useful in the future for understanding the functions of mitospores in the genus under controlled environmental conditions.

Nakamura et al. (2020) investigated the local genotypic status of *T. himalayense* fruiting bodies utilizing 15 newly developed and 4 existing simple sequence repeat markers. The results revealed low genetic diversity in the truffle grounds, but that these truffle grounds remained productive over the sampling years, signifying that low genetic diversity does not necessarily hurt truffle production, at least not over an extended period. These data should significantly contribute to the continuous reproduction of Asian truffle species for which fundamental data on fine-scale genetic structure are lacking.

6. Conclusion

Many studies on the artificial fruit body formation of EM species have been conducted in Japan. In order to cultivate EM mushrooms in the field, it is necessary to create favorable conditions for the growth of EM species, based on the data on the ecology, physiology, and genetics of both EM fungi and their host plants. Most edible EM mushrooms are only found in the wild or with established techniques for their cultivation, the majority of which are only preliminary from a commercial perspective. Thus, EM mushrooms obtained from wild-collected samples may lead to the depletion of resources for cultivation research. The preservation of natural habitats where these fungi grow and perpetuate is a pressing concern. Strategies for the conservation of edible fungi's habitat provide information useful for the field cultivation of these species. However, it is not possible to implement cultivation practices with the expectation of immediate success. There is a need for more long-term research projects on the cultivation of EM mushrooms.

Acknowledgements

Many of our recent works described in this article were financially supported by a grant from the Ministry of Agriculture, Forestry and Fisheries of Japan entitled "Technology development for the optimal use of forest resources" Grant number 15653601.

References

Bach, C., P. Beacco, P. Cammaletti, Z. Babel-Chen, E. Levesque, F. Todesco, C. Cotton, B. Robin and C. Murat. 2021. First production of Italian white truffle (*Tuber magnatum* Pico) ascocarps in an orchard outside its natural range distribution in France. *Mycorrhiza*, 31: 383–388.

Endo, N., S. Gisusi, M. Fukuda and A. Yamada. 2013. *In vitro* mycorrhization and acclimatization of *Amanita caesareoides* and its relatives on *Pinus densiflora*. *Mycorrhiza*, 23: 303–315.

Endo, N., F. Kawamura, R. Kitahara, D. Sakuma, M. Fukuda and A. Yamada. 2014. Synthesis of Japanese *Boletus edulis* ectomycorrhizae with Japanese red pine. *Mycoscience*, 55: 405–416.

Furusawa, H., T. Yamanaka, A. Kinoshita, S. Nakano, K. Noguchi and K. Obase. 2020. Soil properties in *Tuber himalayense* and *Tuber japonicum* habitats in Japan. *Bull. FFPRI*, 19: 55–67. (In Japanese)

Guerin-Laguette, A. 2021. Successes and challenges in the sustainable cultivation of edible mycorrhizal fungi – furthering the dream. *Mycoscience*, 62: 10–28.

Gyong, J.S. 2011. Study on distribution characteristics of *Tricholoma matsutake* and its allied species and food culture. *Edible Fungi China*, 30 (Supp.): 110–111.

Hall, I.R., G.T. Brown and A. Zambonelli. 2007. Taming the Truffle. The History, Lore, and Science of the Ultimate Mushroom. Timber Press, Portland, pp. 304.

Horimai, Y., H. Misawa, K. Suzuki, M. Fukuda, H. Furukawa, K. Masuno, T. Yamanaka and A. Yamada. 2020. Sibling spore isolates of *Tricholoma matsutake* vary significantly in their ectomycorrhizal colonization abilities on pine hosts *in vitro* and form multiple intimate associations in single ectomycorrhizal roots. *Fungal Ecol.*, 43: 100874.

Horimai, Y., H. Misawa, K. Suzuki, Y. Tateishi, H. Furukawa, T. Yamanaka, S. Yamashita, T. Takayama, F. Fukuda and A. Yamada. 2021. Spore germination and ectomycorrhizae formation of *Tricholoma matsutake* on pine root systems with previously established ectomycorrhizae from a dikaryotic mycelial isolate of *T. matsutake*. *Mycorrhiza*, 31: 335–347.

Imazeki, R. and T. Hongo. 1987. Colored illustrations of mushrooms of Japan (vol. I). Osaka: Hoikusha, Osaka, pp. 325. (In Japanese)

Iotti, M., F. Piattoni, P. Leonardi, I.R. Hall and A. Zambonelli. 2016. First evidence for truffle production from plants inoculated with mycelial pure cultures. *Mycorrhiza*, 26: 793–798.

Ka, K.H., H.S. Kim, T.C. Hur, H. Park, S.M. Jeon, R. Ryoo and Y. Jang. 2018. Analysis of environment and production of *Tricholoma matsutake* in matsutake-infected pine trees. *Korean J. Mycol.*, 46: 34–42. (In Korean)

Kareki, K. and Y. Kawakami. 1985. Artificial formation of Shiro (fungus colony) by planting the pine saplings infected with *Tricholoma matsutake* (Ito et Imai) Sing. *Bull. Hiroshima Pref. For. Exp. Stat.*, 20: 13–23. (In Japanese)

Katagiri, K., K. Kato and K. Masuno. 2021. Effect of rainfall and soil temperature on the production of *Suillus grevillei*. *Bull. Nagano Pref. For. Res. Cent.*, 35: 83–92. (In Japanese)

Kataoka, R., Z.A. Siddiqui, J. Kikuchi, M. Ando, R. Sriwati, A. Nozaki and K. Futai. 2012. Detecting nonculturable bacteria in the active mycorrhizal zone of the pine mushroom *Tricholoma matsutake*. *J. Microbiol.*, 50: 199–206.

Kawai, M. 1997. Artificial ectomycorrhizal formation on roots of air-layered *Pinus densiflora* saplings by inoculation with *Lyophyllum shimeji*. *Mycologia*, 89: 228–232.

Kawai, M. 1999. Artificial infection and fruit body occurrence by inoculation of *Lyophyllum shimeji* mycelial culture in *Pinus densiflora* forest. *Bull. Nara For. Exp. Stat.*, 29: 1–7. (In Japanese)

Kawai, M. 2019. Researches on the cultivation of ectomycorrhizal mushrooms in Nara prefecture. *Shinrin Kagaku* 86: 30–33. (In Japanese)

Kawai, M. and S. Abe. 1976. Studies on the artificial reproduction of *Tricholoma matsutake* (S. Ito et Imai) Sing. I. Effects of carbon and nitrogen sources in media on the vegetative growth of *T. matsutake*. *Trans. Mycol. Soc. Jpn.*, 17: 159–167. (In Japanese)

Kawai, M. and M. Ogawa. 1976. Studies on the artificial reproduction of *Tricholoma matsutake* (S. Ito et Imai) Sing. IV. Studies on a seed culture and a trial for the cultivation on solid media. *Trans. Mycol. Soc. Jpn.*, 17: 499–505. (In Japanese)

Kim, M., H. Yoon, Y.H. You, Y.E. Kim, J.R. Woo, Y. Seo, G.M. Lee, Y.J. Kim, W.S. Kong and J.G. Kim. 2013. Metagenomic analysis of fungal communities inhabiting the fairy ring zone of *Tricholoma matsutake*. *J. Microbiol. Biotechnol.*, 23: 1347–1356.

Kim, M., H. Yoon, Y.E. Kim, Y.J. Kim, W.S. Kong and J.G. Kim. 2014. Comparative analysis of bacterial diversity and communities inhabiting the fairy ring of *Tricholoma matsutake* by barcoded pyrosequencing. *J. Appl. Microbiol.*, 117: 699–710.

Kinoshita, A., H. Sasaki and K. Nara. 2011. Phylogeny and diversity of Japanese truffles (*Tuber* spp.) inferred from sequences of four nuclear loci. *Mycologia*, 103: 779–794.

Kinoshita, A., H. Sasaki and K. Nara. 2016. Two new truffle species, *Tuber japonicum* and *Tuber flavidosporum* spp. nov. found from Japan. *Mycoscience*, 57: 366–373.

Kinoshita, A., K. Nara, H. Sasaki, B. Feng, K. Obase, Z.L. Yang and T. Yamanaka. 2018a. Using mating-type loci to improve taxonomy of the *Tuber indicum* complex, and discovery of a new species, *T. longispinosum*. *PLoS ONE*, 13(3): e0193745.

Kinoshita, A., K. Obase and T. Yamanaka. 2018b. Ectomycorrhizae formed by three Japanese truffle species (*Tuber japonicum, T. longispinosum*, and *T. himalayense*) on indigenous oak and pine species. *Mycorrhiza*, 28: 679–690.

Kinoshita, A., H. Sasaki, T. Orihara, M. Nakajima and K. Nara. 2021. *Tuber iryudaense* and *T. tomentosum*: Two new truffles encased in tomentose mycelium from Japan. *Mycologia*, 113: 653–663.

Kobayashi, H., T. Watahiki, M. Kuramochi, S. Onose and A.Yamada. 2007. Production of pine seedlings with the shiro-like structure of the Matsutake mushroom (*Tricholoma matsutake* (S. Ito et Imai) Sing.) in a large culture bottle. *Mushroom Sci. Biotechnol.*, 15: 151–155. (In Japanese)

Kobayashi, H., M. Terasaki and A. Yamada. 2015. Two-year survival of *Tricholoma matsutake* ectomycorrhizas on *Pinus densiflora* seedlings after outplanting to a pine forest. *Mushroom Sci. Biotechnol.*, 23: 108–113.

Kusuda, M., M. Ueda, M. Nakazawa, K. Miyatake and K. Yamanaka. 2006. Detection of β-glucosidase as saprotrophic ability from an ectomycorrhizal mushroom, *Tricholoma matsutake*. *Mycoscience*, 47: 184–189.

Kusuda, M., M. Ueda, K. Miyatake and T. Terashita. 2008. Characterization of the carbohydrase productions of an ectomycorrhizal fungus, *Tricholoma matsutake*. *Mycoscience*, 49: 291–297.

Li, Q., C. Xiong, X. Li, X. Jin and W. Huang. 2018. Ectomycorrhization of *Tricholoma matsutake* with *Quercus aquifolioides* affects the endophytic microbial community of host plant. *J. Basic Microbiol.*, 58: 238–246.

Lian, C., M. Narimatsu, K. Nara and T. Hogetsu. 2006. *Tricholoma matsutake* in a natural *Pinus densiflora* forest: correspondence between above- and below-ground genets, association with multiple host trees and alteration of existing ectomycorrhizal communities. *New Phytol.*, 171: 825–836.

Mizutani, K. 2005. Fruit body occurrence during 5 years by inoculation of *Lyophyllum shimeji* mycelial culture in *Pinus densiflora* forest. *Bull. Gifu Pref. For. Sci. Res. Inst.*, 34: 1–6. (In Japanese)

Murata, H., A. Ohta, A. Yamada, M. Narimatsu and N. Futamura. 2005. Genetic mosaics in the massive persisting rhizosphere colony "shiro" of the ectomycorrhizal basidiomycete *Tricholoma matsutake*. *Mycorrhiza*, 15: 505–512.

Murata, H., T. Abe, H. Ichida, Y. Hayashi, T. Yamanaka, T. Shimokawa and K. Tahara. 2018. Heavy-ion beam mutagenesis of the ectomycorrhizal agaricomycete *Tricholoma matsutake* that produces the prized mushroom "matsutake" in conifer forests. *Mycorrhiza*, 28: 171–177.

Murata, H., S. Nakano, T. Yamanaka, T. Shimokawa, T. Abe, H. Ichida, Y. Hayashi, K. Tahara and A. Ohta. 2019. Conversion from mutualism to parasitism: A mutant of the ectomycorrhizal agaricomycete *Tricholoma matsutake* that induces stunting, wilting and root degeneration in seedlings of its symbiotic partner *Pinus densiflora in vitro*. *Botany*, 97: 463–474.

Murata, H., T. Yamanaka, T. Shimokawa and A. Ohta. 2020. Morphological changes in a γ-ray irradiation-induced mutant of the ectomycorrhizal agaricomycete *Tricholoma matsutake* during *in vitro* spawning on barley-based substrates. *Bull. FFPRI*, 19: 153–157.

Nakamura, N., J.P. Abe, H. Shibata, A. Kinoshita, K. Obase, J.R.P. Worth, Y. Ota, S. Nakano and T. Yamanaka. 2020. Genotypic diversity of the Asiatic black truffle, *Tuber himalayense*, collected in spontaneous and highly-productive truffle grounds. *Mycol. Prog.*, 19: 1511–1523.

Nakano, S., A. Kinoshta, K. Obase, N. Nakamura, H. Furusawa, K. Noguchi and T. Yamanaka. 2020a. Influence of pH on *in vitro* mycelial growth in three Japanese truffle species: *Tuber japonicum*, *T. himalayense*, and *T. longispinosum*. *Mycoscience*, 61: 58–61.

Nakano, S., T. Yamanaka and H. Kawagishi. 2020b. Effects of 2-azahypoxanthine and imidazole-4-carboxamide on the growth and ectomycorrhizal colonization of *Pinus densiflora* seedlings inoculated with *Tricholoma matsutake*. *Mycoscience*, 61: 259–263.

Nakano, S., A. Kinoshita, K. Obase, N. Nakamura, H. Furusawa, K. Noguchi and T. Yamanaka. 2022a. Physiological characteristics of pure cultures of a white-colored truffle *Tuber japonicum*. *Mycoscience*, 63: 53–57.

Nakano, S., K. Obase, N. Nakamura, A. Kinoshita, K. Kuroda and T. Yamanaka. 2022b. Mitospore formation on pure cultures of *Tuber japonicum* (Tuberaceae, Pezizales) *in vitro*. *Mycorrhiza*, 32: 353–360.

Narimatsu, M., M. Yamaguchi, T. Yamanaka, S. Gisusi, T. Azuma, Y. Tamai, T. Fujita and M. Kawai. 2019. Comparison of mycelial growth of different *Tricholoma matsutake* strains in soil medium at varying temperatures. *Asian J. Biotechnol. Bioresour. Technol.*, 5: 1–8.

Nishino, K., M. Shiro, R. Okura, K. Oizumi, T. Fujita, T. Sasamori, N. Tokitoh, A. Yamada, C. Tanaka, M. Yamaguchi, S. Hiradate and N. Hirai. 2016. The (oxalato) aluminate complex as an antimicrobial substance protecting the "shiro" of *Tricholoma matsutake* from soil micro-organisms. *Biosci. Biotechnol. Biochem.*, 81: 102–111.

Obase, K., S. Yamanaka, A. Kinoshita, Y. Tamai and T. Yamanaka. 2021. Phylogenetic placements and cultural characteristics of *Tuber* species isolated from ectomycorrhizas. *Mycoscience*, 62: 124–131.

Ogawa, M. and M. Hamada. 1975. Primordia formation of *Tricholoma matsutake* (Ito et Imai) Sing. in pure culture. *Trans. Mycol. Soc. Jpn.*, 16: 406–415. (In Japanese)

Ogawa, W., N. Endo, Y. Takeda, M. Kodaira, M. Fukuda and A. Yamada. 2019a. Efficient establishment of pure cultures of the yellow chanterelle, *Cantharellus anzutake*, from ectomycorrhizal root tips, and the morphological characteristics of ectomycorrhizae and cultured mycelium. *Mycoscience*, 60: 45–53.

Ogawa, W., Y. Takeda, N. Endo, S. Yamashita, T. Takayama, M. Fukuda and A. Yamada. 2019b. Repeated fruiting of Japanese golden chanterelle in pot culture with host seedlings. *Mycorrhiza*, 29: 519–530.

Oh, S.Y., J.J. Fong, M.S. Park and Y.W. Lim. 2016. Distinctive feature of microbial communities and bacterial functional profiles in *Tricholoma matsutake* dominant soil. *PLoS ONE*, 11: e0168573.

Ohara, H. and M. Hamada. 1967. Disappearance of bacteria from the zone of active mycorrhizas in *Tricholoma matsutake* (S. Ito et Imai) singer. *Nature*, 213: 528–529.

Ohta, A. 1994a. Some cultural characteristics of mycelia of a mycorrhizal fungus, *Lyophyllum shimeji*. *Mycoscience*, 35: 83–87.

Ohta, A. 1994b. Production of fruit-bodies of a mycorrhizal fungus, *Lyophyllum shimeji*, in pure culture. *Mycoscience*, 35: 147–151.

Ohta, A. 1998. Culture conditions for commercial production of *Lyophyllum shimeji*. *Jpn. J. Mycol.*, 39: 13–20. (In Japanese)

Ohta, A. and N. Fujiwara. 2003. Fruit-body production of an ectomycorrhizal fungus in genus *Boletus* in pure culture. *Mycoscience*, 44: 295–300.

Oikawa, Y., Y. Yoneda, S. Kawai, T. Yamanaka and M. Kawai. 2020. Selection of *Tricholoma bakamatsutake* stains with laccase-like enzyme-secreting and decomposition of phenolic lignin model compounds by these strains. Abstracts presented at the 70th *Annual Meeting Jpn. Wood Res. Soc.*, L17-P2-11. (In Japanese)

Otani, Y. 1976. *Tricholoma bakamatsutake* Hongo collected in New Guinea (In Japanese). *Trans. Mycol. Soc. Jpn.*, 17: 363–365. (In Japanese)

Qu, L., A.M. Quoreshi, K. Iwase, Y. Tamai, R. Funada and T. Koike. 2003. *In vitro* ectomycorrhiza formation on two larch species of seedlings with six different fungal species. *Eurasian J. For. Res.*, 6: 65–73.

Saito, C., W. Ogawa, H. Kobayashi, T. Yamanaka, M. Fukuda and A.Yamada. 2018. *In vitro* ectomycorrhization of *Tricholoma matsutake* strains is differentially affected by soil type. *Mycoscience*, 58: 89–97.

Shibata, H. 1989. Cultivation of *Suillus grevillei* in *Larix leptolepis* forests. *Bull. Yamanashi Pref. For. Res. Exp. Inst.*, 17: 16–23. (In Japanese)

Shimazono, H. 1979. Comparative studies on morphological characteristics of the colonies of *Tricholoma matsutake*, *T. fulvocastaneum* and *T. bakamatsutake* on agar media. *Trans. Mycol. Soc. Jpn.*, 20: 176–184. (In Japanese)

Shimokawa, T., M. Yamaguchi and H. Murata. 2017. Agar plate assays using dye-linked substrates differentiate members of *Tricholoma* sect. *caligata*, ectomycorrhizal symbionts represented by *Tricholoma matsutake*. *Mycoscience*, 58: 432–437.

Shimokawa, T., A. Kinoshita, N. Kusumoto, S. Nakano, N. Nakamura and T. Yamanaka. 2020. Component features, order-active volatiles, and acute oral toxicity of novel white-colored truffle *Tuber japonicum* native to Japan. *Food Sci. Nutr.*, 8: 410–418.

Shimomura, N. 2019. Study of the artificial cultivation of the ectomycorrhizal mushroom, *Rhizopogon roseolus*. *Mushroom Sci. Biotechnol.*, 26: 148–155. (In Japanese)

Terashima, Y., K. Tomiya, M. Takahashi and H. Iwai. 1993. Distribution and characteristics of shiros of *Tricholoma bakamatsutake* in a mixed forest of *Pasania edulis* and *Castanopsis cuspidata* var. *sieboldii*. *Trans. Mycol. Soc. Jpn.*, 34: 229–238. (In Japanese)

Tomikawa, Y. 2006. Effects of inoculation with fruit bodies suspension on cultivation of *Rhizopogon rubescens* in nursery of *Pinus thunbergii*. *Bull. Shimane Mt. Reg. Res. Cent.*, 2: 43–49. (In Japanese)

Vaario, L.M., H. Fritze, P. Spetz, J. Heinonsalo, P. Hanajík and T. Pennanen. 2011. *Tricholoma matsutake* dominates diverse microbial communities in different forest soils. *Appl. Environm. Microbiol.*, 77: 8523–8531.

Vaario, L.M., T. Pennanen, J. Lu, J. Palmén, J. Stenman, J. Leveinen, P. Kilpeläinen and V. Kitunen. 2015. *Tricholoma matsutake* can absorb and accumulate trace elements directly from rock fragments in the shiro. *Mycorrhiza*, 25: 325–334.

Yamada, A., S. Kanekawa and M. Ohmasa. 1999a. Ectomycorrhiza formation of *Tricholoma matsutake* on *Pinus densiflora*. *Mycoscience*, 40: 193–198.

Yamada, A., K. Maeda and M. Ohmasa. 1999b. Ectomycorrhiza formation of *Tricholoma matsutake* isolates on seedlings of *Pinus densiflora in vitro*. *Mycoscience*, 40: 455–463.

Yamada, A., T. Ogura and M. Ohmasa. 2001a. Cultivation of mushrooms of edible ectomycorrhizal fungi associated with *Pinus densiflora* by *in vitro* mycorrhizal synthesis. I. Primordium and basidiocarp formation in open-pot culture. *Mycorrhiza*, 11: 59–66.

Yamada, A., T. Ogura and M. Ohmasa. 2001b. Cultivation of mushrooms of edible ectomycorrhizal fungi associated with *Pinus densiflora* by *in vitro* mycorrhizal synthesis. II. Morphology of mycorrhizas in open pot soil. *Mycorrhiza*, 11: 67–81.

Yamada, A., T. Ogura, Y. Degawa and M. Ohmasa. 2001c. Isolation of *Tricholoma matsutake* and *T. bakamatsutake* cultures from field-collected ectomycorrhizas. *Mycoscience*, 42: 43–50.

Yamada, A., K. Maeda, H. Kobayashi and H. Murata. 2006. Ectomycorrhizal symbiosis *in vitro* between *Tricholoma matsutake* and *Pinus densiflora* seedlings that resembles naturally occurring 'shiro'. *Mycorrhiza*, 16: 111–116.

Yamada, A., H. Kobayashi, T. Ogura and M. Fukuda. 2007. Sustainable fruit-body formation of edible mycorrhizal *Tricholoma* species for 3 years in open pot culture with pine seedling hosts. *Mycoscience*, 48: 104–108.

Yamada, A., H. Kobayashi, H. Murata, E. Kalmiş, F. Kalyoncu and M. Fukuda. 2010. *In vitro* ectomycorrhizal specificity between the Asian red pine *Pinus densiflora* and *Tricholoma matsutake* and allied species from worldwide Pinaceae and Fagaceae forests. *Mycorrhiza*, 20: 333–339.

Yamada, A., N. Endo, H. Murata, A. Ohta and M. Fukuda. 2014. *Tricholoma matsutake* Y1 strain associated with *Pinus densiflora* shows a gradient of *in vitro* ectomycorrhizal specificity with Pinaceae and oak hosts. *Mycoscience*, 55: 27–34.

Yamada, A., H. Furukawa and T. Yamanaka. 2017. Cultivation of edible ectomycorrhizal mushrooms in Japan. *Revista Fitotecnia Mexicana*, 40: 379–389.

Yamada, A., N. Hayakawa, C. Saito, Y. Horimai, H. Misawa, T. Yamanaka and M. Fukuda. 2019. Physiological variation among *Tricholoma matsutake* isolates generated from basidiospores obtained from one basidioma. *Mycoscience*, 60: 102–109.

Yamanaka, K., K. Namba and A. Tajiri. 2000. Fruit body formation of *Boletus reticulatus* in pure culture. *Mycoscience*, 41: 189–191.

Yamanaka, K., T. Aimi, J. Wan, H. Cao and M. Chen. 2011. Species of host trees associated with *Tricholoma matsutake* and close allies in Asia. *Mushroom Sci. Biotechnol.*, 19: 79–87. (In Japanese)

Yamanaka, T., Y. Ota, M. Konno, M. Kawai, A. Ohta, H. Neda, Y. Terashima and A. Yamada. 2014. The host ranges of conifer-associated *Tricholoma matsutake*, Fagaceae-associated *T. bakamatsutake* and *T. fulvocastaneum* are wider *in vitro* than in nature. *Mycologia*, 106: 397–406.

Yamanaka, T., M. Konno, M. Kawai, Y. Ota, N. Nakamura and A. Ohta. 2019. Improved chlamydospore formation in *Tricholoma bakamatsutake* with the addition of amino acids *in vitro*. *Mycoscience*, 60: 319–322.

Yamanaka, T., A. Yamada and H. Furukawa. 2020. Advances in the cultivation of the highly-prized ectomycorrhizal mushroom *Tricholoma matsutake*. *Mycoscience*, 61: 49–57.

2

Mushroom Cultivation in Submerged Culture for Bioactive Compounds Production

Hesham Ali El Enshasy[1,2,3]

1. Introduction

For centuries, mushrooms have been recognized as one of the most important bioresources for food and medicine in different ancient cultures. They have been recognized as part of the treatment plans of different folk medicine practices in the treatment of different diseases (El Enshasy et al. 2013). This is based on their high content of bioactive molecules of functional properties as antioxidants, antimicrobial, anti-diabetic, immunomodulatory, hepatoprotective, anticancer, and many others (El Enshasy 2010, El Enshasy and Hatti-Kaul 2013, Bains et al. 2021, Yadav and Negi 2021). This is based on the presence of a wide range of important chemical substances belonging to different chemical classes as phenolics, terpenoids, proteins, glycoproteins, and polysaccharides (Kalač 2012, Kumar et al. 2021, Campestrini and Salles-Campos 2021).

[1] Institute of Bioproduct Development (IBD), Universiti Teknologi Malaysia (UTM), Johor Bahru, Johor, Malaysia.
[2] Faculty of Chemical and Energy Engineering, Universiti Teknologi Malaysia (UTM), Johor Bahru, Johor, Malaysia.
[3] City of Scientific Research and Technology Applications, New Burg Al Arab, Alexandria, Egypt.

Mushrooms grow naturally in different habitats in almost all parts of the world. However, due to the increased demand for mushrooms in both food and medicine, people start to cultivate mushrooms to make them available in large quantities throughout the year. At first, tree logs were been used as a traditional approach for mushroom cultivation. This is based on the capacity of mushrooms to produce a large number of hydrolytic enzymes which are able to degrade wood lignocellulosic materials to produce consumable carbohydrates for mushroom growth. Further development in cultivation came with the introduction of the greenhouse concept and the growth of mushrooms in a solid substrate composed of different balanced nutrients to support the growth and production of bioactive metabolites. Greenhouse cultivation is the most common type of mushroom large-scale production in many countries. This method is characterized by its ease and cheap approach with higher yield compared to the log method. In addition, the design of medium ingredients in the greenhouse cultivation system not only led to significant improvement in mushroom yield but also make it able to use many agro-industrial wastes as a nutrient source which help in both reducing pollution and improving the overall mushroom production cost. More recently, submerged culture has been introduced as an alternative cultivation system for sustainable mushroom production of mushrooms (Tang et al. 2007, Bakratsas et al. 2021). This is based on the fact that mushrooms as higher fungi and cultivation in submerged culture is possible like many other fungal species. The submerged cultivation systems exhibited many advantages compared to conventional solid-state cultivation in the greenhouse.

2. Solid State Cultivation vs. Submerged Cultivation

Mushroom cultivation using a solid-state cultivation system can be carried out in a greenhouse (under controlled temperature and humidity conditions) or in a small-scale cultivation system (glass vessels) or solid-state bioreactor. However, a greenhouse is the most commonly used for mushroom cultivation, especially when used as a source of food. In the greenhouse cultivation practice, the cultivation medium is usually composed of many lignocellulosic materials as cheap agriculture waste. This is based on the high capacity of mushrooms to produce a wide range of hydrolases enzymes such as lignocellulases and laccases enzymes which are capable of the cells to hydrolyze the complex organic materials to fermentable sugar which is consumed for fungal growth. This makes it possible for the mushroom to be cultivated using many cheap agro-industrial wastes organic substrates such as sawdust, rice husk, corncob, bagasse, coffee husk and many other solid waste plant residues (Hoa et al. 2015, Chai et al. 2021). However, using complex lignocellulosic materials as the main substrate prolong the cultivation time and also exhausts cells which need to produce a wide range of hydrolytic enzymes before the production of biomass and bioactive metabolites.

As shown, the submerged cultivation system is characterized by many advantages over the traditional greenhouse cultivation system. These advantages make it more suitable for the cultivation of rare mushrooms with a high growth rate, ease to extract and purify bioactive mushroom compounds, and full compliance to

Table 1: Comparison between greenhouse and submerged cultivation systems.

	Green house cultivation	**Submerged cultivation**
Advantages	- Easy cultivation system with low investment - Simple Control System (Temperature/humidity) - Scalable process	- Fully controlled production process (controllable/optimizable process). - Short cultivation time (15–25 days cultivation) Scalable Process - Cultivation under complete sterile condition (fully compliable to cGMP) - Less purification step especially if the targeted product produced extracellularly - Easy for process validation - More suitable for cultivation of poisonous mushrooms
Disadvantages/ Limitations	- Many wild types mushroom cannot cultivated in green house - Not suitable for cultivation of poisonous mushrooms. - Different purification steps are required to obtain the final bioactive material in pure form - Not fully sterile cultivation system. Not fulfil the requirements of cGMP for sterile manufacturing	- Large capital investment - Higher production cost - Need more skilled workers

Good Manufacturing Practice (cGMP) which is needed for bioactive compounds production in pharmaceutical industries. In addition, submerged cultivation is a closed cultivation system with complete control and thus can be used for the safe cultivation of poisonous mushrooms for bioactive metabolite production (Garcia et al. 2020, Lee et al. 2021). All these together make submerged cultivation a more attractive system for the production of mushroom bioactive compounds of high medicinal value. In summary, the greenhouse cultivation system is more suitable for mushroom production for the food and nutraceutical market, whereas, submerged cultivation is a more attractive alternative system for bioactive compounds for medical industries.

3. Mushroom Submerged Cultivation

In general, submerged cultivation of mushrooms is mostly needed for the production of high-value bioactive compounds for nutraceutical and medical applications. Medium is usually composed of soluble and easily consumable substrates compared to solid-state cultivation which uses complex substrates which need a mushroom enzymatic system for degradation before utilization. Therefore, this system is not exhausting the mushroom cells by undergoing two steps cultivation system (degradation of the substrate followed by the production of bioactive as in solid-state cultivation system). Thus, the easily consumed substrates can be utilized directly for biomass production and bioactive metabolites biosynthesis. Therefore, the submerged cultivation system is widely used for the production of mushrooms' bioactive compounds of medicinal values such as polysaccharides, terpenoids, lectins, fungal immunomodulatory proteins (FIP), enzymes and many other metabolites (Elisashvili 2012, Dudekula et al. 2020, Bakratsas et al. 2021).

Different types of submerged cultivation vessels have been used for mushroom cultivation on different scales. These include shake flasks, stirred tank bioreactors, air-lift bioreactors and many other specially designed vessels of lower costs. In addition, the cultivation in shake flask has been also carried out using static mode (which favour some metabolites production) or shake mode (which supports more biomass production), or a combination thereof to increase biomass and bioactive metabolites production at the same time (Tang et al. 2015).

3.1 Factors Affecting Mushroom Cultivation in Submerged Cultivation System

Like other submerged cultivation systems of fungal strains, mushroom growth and production of bioactive metabolites are governed by many factors. These factors can be classified under three main groups:

1. Strain-dependent factors: Type of strain, inoculum size and type, inoculum concentration, growth rate, and growth morphology.
2. Medium dependent factor: Carbon and nitrogen sources, C: N ratio, complex organic substrate, phosphate concentration, minerals and trace elements, and precursors.
3. Cultivation condition-dependent factors: Temperature, pH, agitation rate, aeration rate, type and design of cultivation vessel, and shear stress.

All these factors together affect the overall cultivation system. Therefore, it is important to understand the key factors and the weight of individual factors and their effect on biomass growth, desired metabolite production, and by-product formation to optimize the cultivation strategy for the achievement of maximal yield. In addition, these factors also affect growth morphology which is considered one of the key bottlenecks which are important to control to achieve maximal production.

3.2 Mushroom Morphology in Submerged Culture System

Mushrooms exhibited the same associated cultivation challenges related to growth morphology in the submerged culture systems. Like other fungi, mushrooms can grow either in filamentous or in bio-pellet form. In addition, many mushrooms have high adhesion capacity to bioreactor surfaces which causes wall growth and also grow on bioreactor sensors such as (pH, DO, foam, level, and temperature sensors) which lead to false readings of essential bioprocessing parameters. Therefore, for many cultivations, growth in micro-pellet is the preferred morphology in submerged culture. Mushroom growth morphology and bio-pellet production are governed by many factors as those related to other fungi which include: strains-dependent factors, medium-dependent factors, and bioprocess engineering-dependent factors as shown in Figure 1 (El Enshasy 2022). However, for each cultivation process,

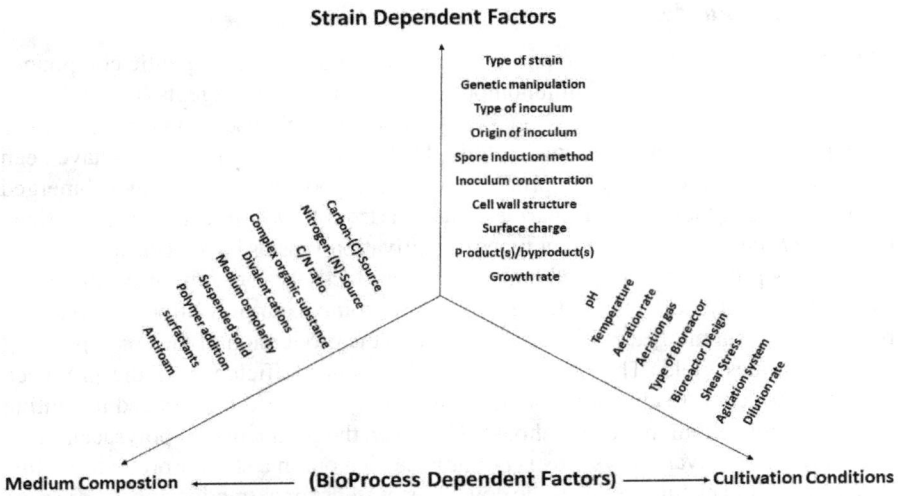

Figure 1: Effect of different strain-dependent factors and bioprocess-dependent factors on the fungal morphology during cultivation in submerged culture (El Enshasy 2022).

it is important to determine the most significant factor for each particular strain to control growth morphology. As reported by many authors, growth in form of a small pellet non-compact pellet structure is more desired as it helps to achieve the maximal productive biomass during the aerobic cultivation process with minimal oxygen transfer limitation in the bio- pellet structure (El Enshasy et al. 2006). In most cases, the type/size/preparation method of inoculum, agitation, and aeration is considered the key factors during scaling up for mushroom cultivation to achieve the desired morphology and maximal productivity as well.

4. Bioactive Metabolites Production in Submerged Culture

For many years, different types of mushrooms have been successfully cultivated in submerged culture for the production of a wide range of bioactive productions of different classes including polysaccharides, lipids, glycoproteins, proteins, lipids, enzymes, terpenoids, phenolics, and many other compounds. However, the production of polysaccharides, ganoderic acid, and cordycepin is the most well-studied metabolites in submerged culture. This is based on their high therapeutic values and increased demand in the medicinal bioactive compound market. During submerged cultivation, a single mushroom can produce more than bioactive compounds such as *Cordyceps* sp. which has the capacity to produce polysaccharides and cordycepin concurrently. *Ganoderma* sp. is also another attractive example that can produce ganoderic acid and polysaccharides concomitantly in the same culture. However, the ratio between the bioactive compounds produced is highly dependent on medium composition and the cultivation parameters. In other cases, a combination between two cultivation modes (static and shake, change from uncontrolled to controlled pH, or from uncontrolled DO to controlled DO) has been used to optimize the cultivation process to increase the production yield.

4.1 *Polysaccharides*

Mushroom polysaccharides are one of the major targeted biotherapeutic compounds based on their bioactivity as immunomodulators and anticancer agents (El-Deeb et al. 2019). Therefore, most submerged culture studies were focused on polysaccharides especially β-glucans in submerged culture. Different types of mushrooms have been reported for their capacity to produce functional polysaccharides in submerged cultures. Among these polysaccharides, Pleuran (from *Pleurotus ostreatus*), Lentinan (from *Lentinus edodes*), Different types of cultivation vessels have been successfully used in this process and showed high system productivity for both intracellular and extracellular polysaccharides. Table 2 shows us some examples of polysaccharides produced in a submerged cultivation system and the production of the corresponding polysaccharides yield. The quality and biotherapeutic efficiency of the produced polysaccharides were almost the same as those polysaccharides produced in fruiting bodies and mycelium in the greenhouse. However, the production of polysaccharides in submerged cultivation systems is not only carried out in a shorter production time with a high yield but also reduces the downstream steps for extraction and purification significantly especially when polysaccharides as an extracellular product. Figure 2 shows a comparative production platform for pleuran production by *P. ostreatus* between submerged cultivation in a large-scale stirred tank bioreactor (Process A) and greenhouse cultivation (Process B) (Modified from El Enshasy and Hatti-Kaul 2013).

Table 2: Production of different bioactive polysaccharides in submerged cultivation systems using different mushroom biofactories.

Mushroom	Polysaccharides type	Maximal product yield	Cultivation vessel	Reference
Agaricus bisporus	IPS	3.43	SF	Argyropoulos et al. (2022)
Agaricus blazei	IPS	0.72 g/L	SF, STR	Liu and Wang (2009)
Auricularia polytricha	EPS	9.42 g/L	SF	Yang et al. (2022)
Cordyceps militaris	EPS	7.30 g/L	SF	Kim et al. (2003)
Ganoderma lucidum	EPS	0.79 g/L	SF	Montoya et al. (2013)
Ganoderma lucidum	EPS	5.00 g/L	SF, STR	Abd Alsaheb et al. (2020)
Lentinula edodes	EPS	1.33 g/L	SF	Liu and Zhang (2019)
Lentinula edodes	EPS	0.94 g/L	SF	Garcia-Cruz et al. (2020)
Phellinus baumii	EPS	150 mg/L	SF, STR	Lung and Deng (2022)
Phellinus gilvus	EPS	5.30 g/L	STR	Hwang et al. (2004)
Pleurotus eryngii	IPS	3.82 g/L	SF	Argyropoulos et al. (2022)
Pleurotus ostreatus	EPS	2.10 g/L	STR	El Enshasy et al. (2010)
Pleurotus ostreatus	EPS	1.98 g/L	STR	Maftoun et al. (2013)
Trametes trogii	EPS	1.86 g/L	STR	Xu et al. (2013)
Tuber sinese	EPS	5.86 g/L	SF	Tang et al. (2008)

Note: EPS: Exopolysaccharides; IPS: Indopolysaccharides; SF: Shake Flask; STR: Stirred Tank reactor.

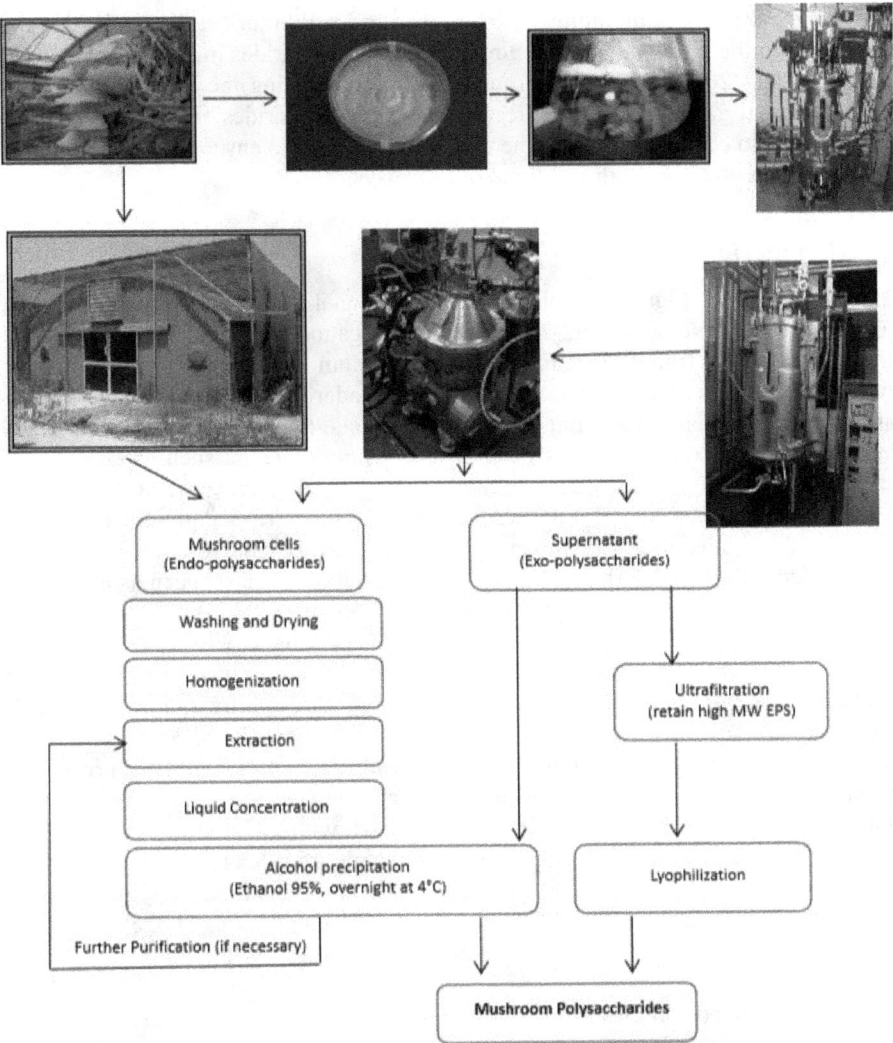

Figure 2: The summarized schematic diagram for bioprocess of mushroom β-glucan polysaccharides production using *Pleurotus ostreatus* for both submerged cultivation in stirred tank bioreactor (Process A) and greenhouse cultivation (Process B) (Modified from El Enshasy and Hatti-Kaul 2013).

In general, the bioactivity of the produced polysaccharides in submerged culture is largely influenced by molecular weight, degree of branching, monosaccharides composition, and degree of glycosylation "in case of glycoprotein" (Bae et al. 2013, Wang et al. 2017, Zhang et al. 2022). Most of the mushroom polysaccharides belong to branched β-glucans. For example, most of the immunomodulator bioactive polysaccharides have the backbone of 1–3, -β-D-glucan with short branched side chains of 1-6, β-linkage. The degree of branching also affects the bioactivity as well which is generally between 20% to 33% for most active polysaccharides. However, it is worth noting, that the immunomodulatory and antitumor activity of lentinan is not linked to the presence of branching in the polysaccharides molecule as reported by Ren et al. (2012). Therefore, in addition to considering the bioprocessing and morphological factors, factors affecting the polysaccharides' molecular activity need to be also considered during the cultivation process to ensure that the produced molecules are effective with full functional properties.

4.2 Ganoderic Acid

This tri-terpenoid high molecular weight functional acid of molecular formula ($C_{30}H_{44}O_7$; CAS No. 81907-62-2). Two forms of ganoderic acids GA-A and GA-B were first isolated from the fruiting bodies, mycelium and spores of *G. lucidum* by Japanese scientists (Kubota and Asaka 1982). Ganoderic acid (GA) has been widely used as part of traditional Chinese medicine in Asia. So far, this compound exists in fruiting bodies of mushrooms belonging to *Ganoderma* sp. such as *G. lucidum* and *G. sinese*. This compound exhibited many proven therapeutic activities as a hepatoprotective, antitumor, antimicrobial, and anti-hyperlipidemic agent (Bryant et al. 2017, Lixin et al. 2019, Guo et al. 2020, Lv et al. 2022). However, in addition to ganoderic acid, more than 430 secondary metabolites have been isolated and identified from different species of *Ganoderma* (Baby et al. 2015).

Production of GA using a greenhouse is a very long process takes 6 months to produce fruiting bodies. This long-term cultivation, makes it sensitive to contamination as it is difficult to control sterility in open cultivation system. With the increased market demand, GA has been produced by *Ganoderma lucidum* in submerged cultivation system in both static and shake cultures. This process is highly governed by medium composition, pH, and oxygen. An early study reported that the maximal production of biomass of 17.3 g/L with 207.9 mg/L was achieved in shake culture at initial pH of 6.5. Lowering the pH from 6.5 to 3.5 to increase the production of the polysaccharide (intracellular- and extracellular-polysaccharides) concomitant with the decrease of GA production (Fang and Zhong 2002a). The same researchers also reported that the best medium composition for biomass and GA production of 16.7 g/L and 212.3 mg/L, respectively, can be achieved in a medium composed of mixed complex nitrogen sources composed of yeast extract and peptone in the concentration of 5 g/L each and glucose as a carbon source of 50 g/L (Fang and Zhong 2002b). Further improvement in GA production was achieved in liquid static culture by oxygen enrichment in the gaseous oxygen level. A maximal biomass of 29.8 g/L and GA production of 1427 mg/L were obtained after 16 days by aeration with an 80% oxygen-enriched gas phase (Zhang et al. 2010). In addition, nitrogen plays a regulatory role in GA production and the production can be enhanced under nitrogen-limiting conditions (Zhao et al. 2011). Other interesting study was reported on the significant influence of light irradiation on GA production in *G. lucidum*. A light shift cultivation strategy in three stages: 2 days of dark cultivation, followed by 6 days with white light irradiation of 0.94 W/m², increased up to 4.70 W/m² for the other four days increasing the production of GA by 548% compared to dark cultivation and reached 466.4 mg/L (Zhang and Tang 2008). On the other hand, using mixed carbon sources of glucose and sucrose enhanced the production of GA production. A fed-batch cultivation strategy was developed starting with a glucose/sucrose mixture in the batch phase, followed by sucrose feeding after 5 days resulting in a significant increase of GA production in a 10-L bioreactor up to 636 mg/L (Wei et al. 2016). It has been also reported that the production of GA by a high-producer isolated strain *G. lucidum* RCKB-2010 can reach up to 2755 mg/L in liquid static culture compared to only 373 mg/L in shaking culture after 25 days (Upadhyay et al. 2014).

On the other hand, other research also reported on the potential enhancement of GA biosynthetic gene expression using different precursors such as phenobarbital (Liang et al. 2010). A recent study reported also the possible control of metabolic flux to affect the distribution in the production of GA-R, GA-S, and GA-T by exogenous addition of oleic acid (Yan et al. 2022).

Recent studies on the production of GA in submerged culture have been carried out using genetic manipulation either using homologous or heterologous approaches. Fei et al. (2019) reported on the overexpression of homologous farnesyl-diphosphate synthase genes to increase GA production in *G. lucidum*. The developed recombinant strain had the capacity to increase the production of all GAs (GA-Me, GA-S, and GA-T) by 2.80-, 2.62-, and 2.27-folds, respectively. To overcome the problems related to fungal growth (growth morphology and low growth rate), and the limitation of the biosynthetic capacity of the mushroom strains, a recombinant *Saccharomyces cerevisiae* has been used for GA production in a submerged culture system. The GA was found to be highly influenced by oxygen supply and the maximal production was only 8.79 mg/L after only 120 h cultivation (Liu and Zhong 2021). However, having GA expressed in a unicellular eukaryotic host with a high growth rate will ease large-scale production even though more studies are needed to increase the production yield.

4.3 Cordycepin

Cordycepin or (9-(3)-Deoxy-β-D-ribofuranosyl adenine) is an adenine-like compound with a molecular formula of $C_{10}H_{13}N_5O_3$ and CAS No. 73-03-0. This potent bioactive compound is produced by the fruiting bodies and in submerged culture of some rare mushrooms belonging to *Cordyceps* sp. such as *C. militaris*, *C. kyushuensis*, and *C. siniensis* (Leung and Wu 2007, Ling et al. 2009). This low molecular weight bioactive molecule exhibited potent biological activities as an anticancer, immunomodulator, anti-inflammatory and hypoglycemic agent (Soltani et al. 2018, Ashraf et al. 2020). Based on the difficulties of growing mushrooms related to *Cordyceps* sp. in the greenhouse, a high level of interest has been paid for the production of bioactive compounds in liquid static culture and submerged culture systems. Much research has been carried out related to the optimization of fermentation medium and cultivation conditions in different scales (Jiapeng et al. 2014, Kang et al. 2014). In one of the early studies, it has been reported that the biosynthesis of cordycepin in submerged culture is highly regulated by carbon and nitrogen sources during the cultivation of *C. militaris* (Mao et al. 2005). Glucose was found to be the most suitable carbon source and the maximal production of 345.5 g/L was achieved when using a medium containing 42 g/L glucose and 15.8 g/L peptone. However, further studies reported also that the production of the surface culture of *C. militaris* can yield up to 640 mg/l in a medium composed of glucose, yeast extract, and peptone (Masuda et al. 2006). Another interesting research studied the effect of amino acids supplementation to a chemically defined medium on cordycepin production. A maximal production of 2.5 g/L was achieved in a medium supplemented with 16 g/L glycine and 1 g/L adenine. The volumetric production of cordycepin in this medium was almost 4 times higher compared to the initial basal medium (Masuda et al. 2007). Another study showed also that medium supplementation with 1 g/L ferrous sulphate can enhance cordycepin production by 70% reaching 596 mg/L (Fan et al. 2012). More recent research also reported that a maximal cordycepin production of 846 mg/L was achieved in submerged culture after 24 days of cultivation at 25°C, and pH of 5.5 in a medium composed of glucose, yeast extract, potassium phosphate monobasic/and dibasic, sodium chloride, and magnesium sulphate when using inoculum size of 8% (Tuli et al. 2014). Another study reported that the optimal medium composition for cordycepin production in submerged culture is composed of (g/L): sucrose, 20.7; peptone, 20.0; $K_2HPO_4 \cdot 3H_2O$, 1.11; $MgSO_4 \cdot 7H_2O$, 0.90, Alanine 12.23; hypoxathine, 5.34 with the addition of VB_1 in the concentration of 10 mg/L. A maximal production of about 2 g/L of cordycepin was achieved in liquid static culture after 35 days without pH control (Kang et al. 2014). Another bit of research was carried out to investigate the effect of the two-step cultivation strategy of the shake stage followed by the static stage. The shake stage promotes more cell growth, while the static stage supports more cordycepin production (Tang et al. 2015). Through this strategy, the production of cordycepin can reach up to 2.62 g/L compared to the maximal production of 1.03 g/L and 0.51 g/L in the case of static culture and shake culture, respectively.

Recent research was focused on using agro-industrial wastes and lignocellulosic biomass as potential substrates for cordycepin production. The potential use of casein hydrolysate as a cheap substrate for cordycepin production in submerged cultivation using a mutant strain of *C. militaris* KYL05 has been studied by Lee et al. (2019). After 6 days of cultivation in the shake flask culture at 150 rpm, 25°C, and pH 6.0, a maximal production of 445 mg/L was achieved (this was almost 3 times improvement in production compared to the initial conventional medium with wild-type strain). Another study by Ha et al. (2021) also reported on the potential use of alkaline-treated pine sawdust for efficient short-time production of cordycepin production in the submerged cultivation system of *C. militaris*. A maximal cordycepin production of 922 mg/L was achieved in after only 72 hours.

5. Conclusion

Submerged cultivation systems not only shorten the cultivation time but also decrease the steps needed for bioactive compound isolation and purification in downstream processes. These all together, decrease the production cost and thus improve the overall economy of the process. In addition, as a closed cultivation system and running under completely sterile conditions, submerged cultivation is suitable for the production of mushroom bioactive compounds according to cGMP. Thus, the produced molecules can be used in the pharmaceutical and biopharmaceutical industries. However, further research is needed to improve the cultivation system and cultivation strategy to improve the yield of production. Scaling up is also one of the other challenges due to the high accumulative behaviour of fungal biomass which usually leads to intense hairy growth and makes it difficult to control culture during cultivation on a large scale. Therefore, controlling growth morphology using different bioprocess engineering approaches is needed to make this process more applied in the mushroom cultivation industry. Further research is also needed to investigate in depth the link between bioprocessing parameters, growth morphology, and polysaccharides (production, glycosylation, helix structure, and molecular weight distribution) to design more efficient production of bioactive molecules. In addition, further research on the design of two modes strategy of cultivation as a combination of static and shaking cultures to develop high cell density culture with maximal desired metabolites production is needed to be further investigated to design the most efficient strategy for the production of particular metabolites.

References

Abd Alsaheb, R., K.Z. Zjeh, R. Abd Malek, J.K. Abdullah, A. El Baz, N. El Deeb, D. Dailin, S.Z. Hanapi, D. Sukmawati and H. El Enshasy. 2020. Bioprocess optimization for exopolysaccharides production by *Ganoderma lucidum* in semi-industrial scale. *Rec. Pat. Food Nutr. Agric.*, 11: 211–218.

Argyropoulos, D., C. Psallida, P. Sitareniou, E. Flemetakis and P. Diamantopulou. 2022. Biochemical evaluation of *Agaricus* and *Pleurotus* strains in batch cultures for production optimization of valuable metabolites. *Microorganisms*, 10: 964.

Ashraf, S.A., A.E.O. Elkhalifa, A.J. Siddiqui, M. Patel, A.M. Awadelkareem, M. Snoussi, M.S. Ashraf, M. Adnan and S. Hadi. 2020. Cordycepin for health and wellbeing: A potent bioactive metabolite of an entomopathogenic medicinal fungus *Cordyceps* with its nutraceutical and therapeutic potential. *Molecules*, 25: 2735.

Baby, S., A.J. Johnson and B. Govindan. 2015. Secondary metabolites from *Ganoderma*. *Phytochem.*, 114: 66–101.

Bae, I.Y., H.W. Kim, H.J. Yoo, E.S. Kim, S. Lee, D.Y. Park and H.G. Lee. 2013. Correlation of branching structure of mushroom β-glucan with its physiological activities. *Food Res. Int.*, 51: 195–200.

Bains, A., P. Chawla, S. Kaur, A. Najda, M. Fogarasi and S. Fogarasi. 2021. Bioactives from mushroom: Health attributes and food industry applications. *Materials*, 14: 7640.

Bakratsas, G., A. Polydera, P. Katapodis and H. Stamatis. 2021. Recent trends in submerged cultivation of mushrooms and their application as a source of nutraceuticals and food additives. *Future Foods*, 4: 100086.

Bryant, J.M., M. Bouchard and A. Haque. 2017. Anticancer activity of Ganoderic acid DM: Current status and future perspective. *J. Clin. Cell. Immunol.*, 8: 535.

Campestrini, L.H. and C. Salles-Campos. 2021. Aspects of mushroom cultivation to obtain polysaccharides in submerged cultivation. *Afr. J. Biotechnol.*, 20: 100–107.

Chai, W.Y., U.G. Krishnan, V. Sabaratnam and J.H.L. Tan. 2021. Assessment of coffee waste in formulation of substrate for oyster mushrooms *Pleurotus ostreatus* and *Pleurotus floridanus*. *Future Foods.*, 4: 100075.

Dudekula, U.T., K. Doriya and S.K. Devarai. 2020. A critical review on submerged production of mushroom and their bioactive metabolites. *3 Biotech.*, 10: 337.

El-Enshasy, H., J. Kleine and U. Rinas. 2006. Agitation effects on morphology and protein productive fractions of filamentous and pelleted growth forms of recombinant *Aspergillus nige*. *Process Biochem.*, 41(10): 2103–2112.

El-Enshasy, H., A. Daba, M. El-Demellawy, A. Ibrahim, S. El Sayed and I. El-Badry. 2010. Bioprocess development for large scale production of anticancer exo-polysaccharide by *Pleurotus ostreatus* in submerged culture. *J. Appl. Sci.*, 10: 2523–2529.

El-Enshasy, H.A. and R. Hatti-Kaul. 2013. Mushroom immunomodulators: Unique molecules with unlimited applications. *Trends Biotechnol.*, 31: 668–677.

El-Enshasy, H.A., E.A. Elsayed, R. Aziz and M.A. Wadaan. 2013. Mushrooms and truffles: Historical biofactories for complementary medicine in Africa and in the middle East. *Evid. Based Compl. Altern. Med.*, 2013: Article ID 620451.

El-Enshasy, H.A. 2022. Fungal morphology: A challenge in bioprocess engineering industries for product development. *Curr. Opin. Chem. Eng.*, 35: 100729.

El-Enshasy, H.E. 2010. Immunomodulators. pp. 165–194. *In*: Hofrichter, M. (ed.). Volume X, Industrial applications. 2nd. Edition, Springer Verlag, Heidelberg.

El-Deeb, N.M., H.I. El-Adawi, A. Abd El-wahab, A.M. Haddad, H.A. El Enshasy, Y-W. He and K.R. Davis. 2019. Modulation of NKG2D, KIR2DL and cytokine production by *Pleurotus ostreatus* glucan enhances natural killer cell cytotoxicity toward cancer cells. *Front. Cell Develop. Biol.*, 7: 165.

Elisashvili, V. 2012. Submerged cultivation of medicinal mushrooms: Bioprocesses and products (review). *Int. J. Med. Mushrooms.*, 14: 211–239.

Fan, D.D., W. Wang and J.J. Zhong. 2012. Enhancement of cordycepin production in submerged cultures of *Cordyceps militaris* by addition of ferrous sulfate. *Biochem. Eng. J.*, 60: 30–35.

Fang, Q.-H. and J.-J. Zhong. 2002a. Effect of initial pH on production of ganoderic acid and polysaccharide by submerged fermentation of *Ganoderma lucidum*. *Process Biochem.*, 37: 769–774.

Fang, Q.-H. and J.-J. Zhong. 2002b. Submerged fermentation of higher fungus *Ganoderma lucidum* for production of valuable bioactive metabolites-ganoderic acid and polysaccharide. *Biochem. Eng. J.*, 10: 61–65.

Fei, Y., N. Li, D.-H. Zhang and J.-W. Xu. 2019. Increased production of ganoderic acids by overexpression of homologous farnesyl diphosphate synthase and kinetic modelling of ganoderic acid production in *Ganoderma lucidum*. *Microb. Cell Fact.*, 18: 115.

Garcia, B.L., J.R. Undan, R.M.R. Dulay, S.P. Kalaw and R.G. Reyes. 2020. Molecular identification and optimization of cultural conditions for mycelial biomass production of wild strain of *Chlorophyllum molybdites* (G. Mey) Massee from the Philippines. *J. Appl. Biol. Biotechnol.*, 8: 1–6.

Garcia-Cruz, F., E. Durá-Párano, M.A. Garín-Aguilar, G. Valencia del Toro and I. Chairez. 2020. Parametric chracterization of the initial pH effect on the polysaccharides production by *Lentinula edodes* in submerged culture. *Food Bioprod. Proc.*, 119: 170–178.

Guo, W.-L., J.-B. Guo, B.-Y. Liu, J.-Q. Lu, M. Chen, B. Liu, W.-D. Bai, P.-F. Rao, L. Ni and X.-C. Lv. 2020. Ganoderic acid A from *Ganoderma lucidum* ameliorates lipid metabolism and alters gut microbiota composition in hyperlipidemic mice fed a high-fat diet. *Food Funct.*, 11: 6818–6833.

Ha, S., J. Jung, J. Park, C. Yu, J. Park and J. Yang. 2021. Effects of pine (*Pinus densiflora*) sawdust on cordycepin yield from medicinal fungus *Cordyceps militaris* in submerged culture. *BioResour.*, 16: 6643–6660.

Hoa, H.T., C.L. Wang and C.-H. Wang. 2015. The effects of different substrates on the growth, yield, and nutritional composition of two oyster mushrooms (*Pleurotus ostreatus* and *Pleurotus cystidiosus*) *Mycobiol.*, 43: 423–434.

Hwang, H.J., S.W. Kim, C.P. Xu, J.W. Choi and J.W. Yun. 2004. Morphological and rheological properties of the three different species of basidiomycetes *Phellinus* in submerged cultures. *J. Appl. Microbiol.*, 96: 1296–1305.

Jiapeng, T., L. Yiting and Z. Li. 2014. Optimization of fermentation conditions and purification of cordycepin from *Cordyceps militaris*. *Prep. Biochem. Biotechnol.*, 44: 90–106.

Kalač, P. 2012. A review of chemical composition and nutritional value of wild-growing and cultivated mushrooms. *J. Sci. Food. Agric.*, 93: 209–218.

Kang, C., T.C. Wen, J.C. Kang, Z.B. Meng, G.R. Li and K.D. Hyde. 2014. Optimization of large-scale culture conditions for the production of cordycepin with *Cordyceps militaris* by liquid static culture. *Sci. World J.*, 2014: 510627.

Kim, S.-W., H.-J. Hwang, C.-P. Xu, J.-M. Sung, J.-W. Choi and J.W. Yun. 2003. Optimization of submerged culture process for the production of mycelial biomass and exopolysaccharides by *Cordyceps militaris* C738. *J. Appl. Microbiol.*, 94: 120–126.

Kubota, T. and Y. Asaka. 1982. Structures of ganoderic acid A and B, two new Ianostane type bitter interpenes from *Ganoderma lucidum* (FR.) Karst. *Helv. Chim. Acta*, 65: 611–619.

Kumar, K., R. Mehra, R.P.F. Guiné, M.J. Lima, N. Kumar, R. Kaushik, N. Ahmed, A.N. Yadav and H. Kumar. 2021. Edible Mushrooms: A comprehensive review on bioactive compounds with health benefits and processing aspects. *Foods*, 10(12): 2996.

Lee, D.W., S.Y. Ha, J.Y. Jung and J.-K. Yang. 2021. Optimization of submerged culture conditions for roridin E production from the poisonous mushroom *Podostroma cornudamae*. *J. Mushrooms*, 19: 81–87.

Lee, S.K., J.H. Lee, H.R. Kim, Y. Chun, J.H. Lee, H.Y. Yoo, C. Park and S.W. Kim. 2019. Improved Cordycepin production by *Cordyceps militaris* KYL05 using casein hydrolysate in submerged conditions. *Biomolecules*, 9: 461.

Leung, P.H. and J.Y. Wu. 2007. Effects of ammonium feeding on the production of bioactive meta bolites (cordycepin and exopolysaccharides) in mycelial culture of a *Cordyceps sinensis* fungus. *J. Appl. Microbiol.*, 103: 1942–1949.

Liang, C.-X., Y.-B. Li, Y.-W. Xu, J.-L. Wang, X.-L. Miao, Y.-J. Tang, T. Gu and J.-J. Zhong. 2010. Enhanced biosynthetic gene expression and production of ganoderic acids in static liquid culture of *Ganoderma lucidum* under phenobarbital induction. *Appl. Microbiol. Biotechnol.*, 86: 1367–1374.

Ling, J.Y., G.Y. Zhang, J.Q. Lin, Z.L. Cui and C.K. Zhang. 2009. Supercritical fluid extraction of cordycepin and denosine from *Cordyceps kyushuensis* and purification by high-speed counter-current chromatography. *Sep. Purif. Technol.*, 66: 625–629.

Liu, G.-Q. and X.-L. Wang. 2009. Selection of a culture medium for reducing costs and enhancing biomass and intracellular polysaccharide production by *Agaricus blazei* AB2003. *Food Technol. Biotechnol.*, 47: 210–214.

Liu, S.-R. and W.-R. Zhang. 2019. Optimization of submerged culture conditions involving a developed fine powder solid seed for exopolysaccharide production by the medicinal mushroom *Ganoderma lucidum*. *Food Sci. Biotechnol.*, 28: 1135–1145.

Liu, T.-T. and J.-J. Zhong. 2021. Impact of oxygen supply on production of a novel ganoderic acid in *Saccharomyces cerevisiae* fermentation. *Process Biochem.*, 106: 176–183.

Lixin, X., Y. Lijun and H. Songping. 2019. Ganoderic acid A against cyclophosphamide-induced hepatic toxicity in mice. *Biochem. Mol. Toxicol.*, 33: e22271.

Lv, X.C., Q. Wu, Y.J. Cao, Y.C. Lin, W.L. Guo, P.F. Rao, Y.Y. Zhang, Y.T. Chen, L.Z. Ai and L. Ni. 2022. Ganoderic acid A from *Ganoderma lucidum* protects against alcoholic liver injury through ameliorating the lipid metabolism and modulating the intestinal microbial composition. *Food Funct.*, 13: 5820–5837.

Maftoun, P., R. Malek, M. Abdel-Sadek, R. Aziz and H. El-Enshasy. 2013. Bioprocess for semi-industrial production of immunomodulator polysaccharide Pleuran by *Pleurotus ostreatus* in submerged culture. *J. Sci. Ind. Res.*, 72: 655–662.

Mao, X.B., T. Eksriwong, S. Chauvatcharin and J.J. Zong. 2005. Optimization of carbon source and carbon/nitrogen ratio for cordycepin production by submerged cultivation of medicinal mushroom *Cordyceps militaris*. *Process Biochem.*, 40: 1667–1672.

Masuda, M., E. Urabe, H. Honda, A. Sakurai and M. Sakakibara. 2007. Enhanced production of cordycepin by surface culture using the medicinal mushroom *Cordyceps militaris*. *Enz. Microb. Technol.*, 40: 1199–1205.

Masuda, M., E. Urabe, A. Sakurai and M. Sakakibara. 2006. Production of cordyceps by surface culture using medicinal mushroom *Cordyceps militaris*. *Enz. Microb. Technol.*, 39: 641–646.

Montoya, S., O.J. Sanchez and L. Levin. 2013. Polysaccharide production by submerged and solid-state cultures from several medicinal higher Basidiomycetes. *Int. J. Med. Mushrooms.*, 15: 71–79.

Ren, L., C. Perera and Y. Hemar. 2012. Antitumor activity of mushroom polysaccharides: A review. *Food Funct.*, 3: 1118–1130.

Soltani, M., R. Abd Malek, N.A. Elmarzugi, M.F. Mahomoodally, D. Uy, O.M. Leng and H.A. El-Enshasy. 2018. Cordycepin: A biotherapeutic molecule from medicinal mushroom. pp. 319–349. *In*: Singh, H.P. and Lallawmsanga, Passari, A.K. (eds.). Biology of Macrofungi. Springer Nature Switzerland.

Tang, J., Z. Qian and L. Zhu. 2015. Two-step shake-static fermentation to enhance cordycepin production by *Cordyceps militaris. Chem. Eng. Trans.*, 46: 19–24.

Tang, Y.-J., L.W. Zhu and D.-S. Li. 2007. Submerged culture of mushrooms in bioreactors-challenges, current state-of-the-art, and future prospects. *Food Technol. Biotechnol.*, 45: 221–229.

Tang, Y.-J., D.-S. Zhu, R.-S. Liu, H.-M. Li, D.-S. Li and Z.-Y. Mi. 2008. Quantitative response of cell growth and *Tuber* polysaccharides biosynthesis by medicinal mushroom Chinese truffle *Tuber sinense* to metal ion in culture medium. *Bioresour. Technol.*, 99: 7606–7615.

Tuli, H.S., A.K. Sharma and S.S. Sandhu. 2014. Optimization of fermentation conditions for cordycepin production using *Cordyceps militaris* 393b. *J. Biol. Chem. Sci.*, 1: 35–47.

Upadhyay, M., B. Shrivastava, A. Jain, M. Kidwai, S. Kumar, J. Gomes, D.G. Goswami, A.K. Panda and R.C. Kuhad. 2014. Production of ganoderic acid by *Ganoderma lucidum* RCKB-2010 and its therapeutic potential. *Ann. Microbiol.*, 64: 839–846.

Wang, Q., F. Wang, Z. Xu and Z. Ding. 2017. Bioactive mushroom polysaccharides: A review on monosaccharide composition, biosynthesis and regulation. *Molecules*, 22: 955.

Wei, Z.H., L. Liu, X.F. Guo, Y.J. Li, B.C. Hou, Q.L. Fan, K.X. Wang, K.X., Y. Luo and J.J. Zhong. 2016. Sucrose fed-batch strategy enhanced biomass, polysaccharide, and ganoderic acids production in fermentation of *Ganoderma lucidum* 5.26. *Bioproc. Biosyst. Eng.*, 39: 37–44.

Xu, C., L. Geng and W. Zhang. 2013. Production of extracellular polysaccharides by the medicinal mushroom *Trametes trogii* (higher Basiodiomycetes) in stirred-tank and airlift bioreactors. *Int. J. Med. Mushrooms*, 15: 183–189.

Yadav, D. and P.S. Negi. 2021. Bioactive components of mushrooms: Processing effects and health benefits. *Food Res. Int.*, 148: 110599.

Yan, M.-Q., X.-W. Su, Y.-F. Liu, C.-H. Tang, Q.-J. Tang, S. Zhou, Y. Tan, L.-P. Liu, J.-S. Zhang and J. Feng. 2022. Effect of oleic acid addition methods on the metabolic flux distribution of ganoderic acids R, S and T's biosynthesis. *J. Fungi*, 8: 615.

Yang, Z., Y. Hu, J. Wu, J. Liu, F. Zhang, H. Ao, Y. Zhu, L. He, W. Zhang and X. Zeng. 2022. High-efficiency production of *Auricularia polytricha* polysaccharides through yellow slurry water fermentation and its structure and antioxidant properties. *Front. Microbiol.*, 13: 811275.

Zhang, S., L. Lei, Y. Zhou, F.-Y. Ye and G.-H. Zhao. 2022. Roles of mushroom polysaccharides in chronic disease management. *J. Integr. Agric.*, 21: 1839–1866.

Zhang, W. and Y.-J.Tang. 2008. A novel three-stage light irradiation strategy in the submerged fermentation of medicinal mushroom *Ganoderma lucidum* for the efficient production of ganoderic acid and *Ganoderma* polysaccharides. *Biocehnol. Prog.*, 24: 1249–1261.

Zhang, W.-X., Y.-J. Tang and J.-J. Zhong. 2010. Impact of oxygen level in gaseous phase on gene transcription and ganoderic acid biosynthesis in liquid static cultures of *Ganoderma lucidum. Bioproc. Biosyst. Eng.*, 33: 683–690.

Zhao, W., J.-W. Xu and J.-J. Zhong. 2011. Enhanced production of ganoderic acids in static liquid culture of Ganoderma lucidum under nitrogen-limiting conditions. *Bioresour. Technol.*, 102: 8185–8190.

3

Spent Mushroom Substrate Perspectives and Applications

Deepak Kumar Rahi,[1,]* *Sonu Rahi*[2]
and *Devender Kumar Sharma*[1]

1. Introduction

The biomass that is left after the mushroom harvesting time is known as spent mushroom substrate (SMS). It was calculated that the production of one kilogram of mushrooms generates about five kilograms of SMS. The estimated yearly global production of spent mushroom compost is 10 to 50 million metric tonnes (Lau et al. 2003, Philippoussis et al. 2004). SMS are bulky, therefore making disposal more difficult. Unregulated SMS dumping contributes to an environmental nuisance as well as air, water, and land pollution. Agricultural land is one of the natural locations for the big amounts of SMS (Rinker 2017), which has several advantages over chemical fertilizers because it distributes nutrients gradually and won't burn crops. Composted SMS is also added to soils to reduce acidity and make it easier to cultivate even difficult soils (Ahalawat et al. 2007). Due to its typical high nutritional content, delayed mineralization, and high cation exchange capacity, SMS can be a useful source of plant nutrients.

SMS are generally employed as fertilizers due to their abundance in calcium, nitrogen, ash, and protein (Lou et al. 2017). As a result of their abundance in nitrogen, SMSs have come to be considered for reuse in soil conditioning. Additionally, it is

[1] Department of Microbiology, Panjab University, Chandigarh, Punjab, India.
[2] Department of Botany, Government Post-Graduate Girls College, Awadhesh Pratap Singh University, Rewa, Madhya Pradesh, India.
* Corresponding author: deepakraahi10@gmail.com

plentiful in organic components and essential nutrients for plant growth, making it a great material for mulching and enhancing soil. SMS is always applied to agricultural land to increase the organic matter and nutritional contents because it is high in phosphorus (Zhu et al. 2012). It is well known that a deficiency of certain micronutrients can lead to serious diseases in plants, animals, and even human beings. Hidden hunger is a micronutrient insufficiency defined as an inadequate balance of certain micronutrients (Murgia et al. 2012). SMS is a promising biofertilizer to increase crop output since it has a huge quantity of micronutrient availability, is less expensive, is biodegradable, and has a greater bioavailability of nutrients. Therefore, SMS is regarded as a valuable source of organic matter and is rich in macro and micro-elements for plants, which contribute to boosting the biological activity of the soil (Debosz et al. 2002, Vandenkoornhuyse et al. 2002). It is also well known that a variety of fungal communities may be found living on the roots of most plants, both intracellular and intercellular. In addition, ectomycorrhizal and ectotrophic interactions between fungi and plants are also typical. These fungi are known as arbuscular mycorrhizal fungi (AMF). Plant growth-promoting fungi are another category of fungi that are frequently present in soil and aid in encouraging root colonization. Spent mushroom substrate obtained by cultivating the mushroom species such as *Ganoderma, Pleurotus Trametes, Schyzophyllum, Lentinus* and *Agaricus* has a bioavailability of nutrients and can be used as biofertilizer for the growth of various cash crops such as paddy, maize, wheat, brinjal, tomato, Capsicum, Okra, Tomato and others. The SMS can be used as such or can be blended with other manure to increase the mineral content. *Pleurotus* spent substrate frosted with pig manure could increase its N, P and K value (Meng et al. 2018). Different microbial community in the SMS also contributes to the increase of bioactive compounds such as polysaccharides, vitamins and macro- and micro-elements (Fe, Ca, Zn, and Mg) (Zhu et al. 2012). Cellulose, hemicellulose, lignin and remnant of edible fungi in the SMS serve as by-products of the mushroom production industry. The substrate used

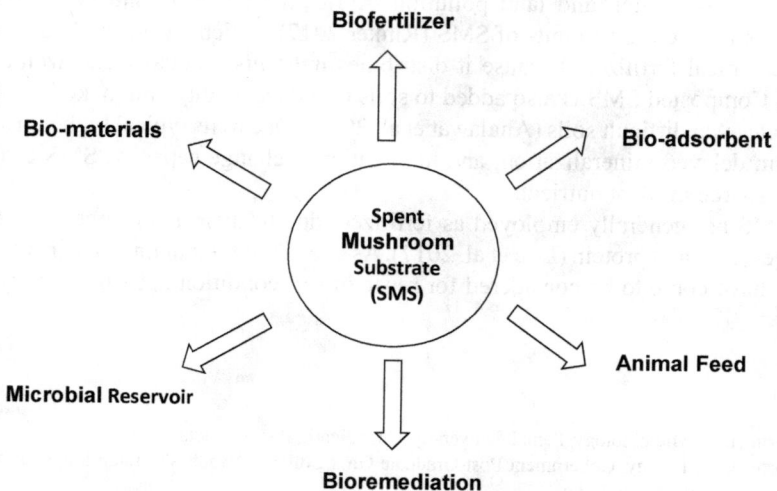

Figure 1. The multifaceted role of spent mushroom substrates.

as a growth medium for mushrooms is mainly composed of corn cobs, cotton seed hulls, wheat straw, grass, bagasse, and hay.

The SMS could be regarded as a significant low-cost adsorbent for wastewater treatment due to the abundance of surface carbonyl, hydroxyl, amide, carboxyl, and phosphate groups. Printing, dyeing, cosmetics, and other businesses all employ dyes extensively (Chequer et al. 2009). Due to the rapid growth of the dye industry and widespread use, roughly 10–15% of the dyes were discharged into the environment, which led to the issue of dye wastewater pollution. SMS of white rot fungi such as *Pleurotus*, *Trametes*, and *Ganoderma* (Wu et al. 2018) can be used to remove heavy metals (Hg, Fe, Cd, Ni, As, and Pb) and dye from textile industrial wastewater.

A significant issue with the environment is heavy metal pollution. Without proper treatment, biodiversity and public health are seriously threatened by the release of these metals into the environment. These metals can be found in industrial effluent frequently, especially in the mining, textile, metal refining, automotive, and other industries (Amarasinghe et al. 2007, Hydari et al. 2012). The majority of these operations produce enormous amounts of wastewater that are highly concentrated in heavy metals. Industrial wastewater treatment has commonly used techniques such as chemical precipitation, electrochemical, adsorption, membrane filtration, photocatalysis, ion exchange, phytoremediation, chemical oxidation, and chemical reduction (Barakat 2011, Fu et al. 2011). However, for the past two decades, it has been discovered that biosorption techniques are preferable to other approaches for removing heavy metals because they are less expensive, easier to use, and less sensitive to harmful compounds (Vijayaraghavan et al. 2008). The passive uptake of toxins by dead or inert biological materials or materials originating from biological sources is known as biosorption. The SMS, which also has a tremendous biosorption potential, can be evolved into more efficient and cost-effective techniques for the removal of metals and dyes from industrial effluent.

2. Cultivable Mushrooms and Substrate

Many species of mushrooms are cultivated worldwide (Table 1). The amount of SMS produced is 3–5 folds higher than the mushroom produced. For instance, in button mushrooms, it is 5-folds times, oysters 3-folds and 3–8 folds in paddy straw mushrooms. According to the data released by ICAR, DMR and Solan (India), estimated world mushroom production during 2018–19 was 43 million tons with 26% *Lentinula edodes* (shiitake), 21% *Auricularia* spp., 16% *Pleurotus ostreatus*, 11% *Agaricus bisporus*, 7% *Flammulina velutipes*, 5% *Pleurotus eryngii*, 1% *Volvariella volvacea* and 13% other mushrooms. Royse (2014) estimates *Agaricus bisporus*, *A. subrufrescens* (together 30%), *Pleurotus* sp. (27%), *Lentinula edodes* (17%), *Auricularia* sp. (6%), and *Flammulina* (5%) accounting for 85% of global production. Various lignocellulosic biowaste such as paddy straw, wheat straw, sawdust, leaf litter and sugarcane bagasse is used for the cultivation of mushrooms. The residue that remained after harvesting these mushrooms known as spent mushroom substrate can be used as a biofertilizer in agriculture. Additionally, *Ganoderma*, *Trametes*, and *Schyzophyllum* spent substrate have strong biofertilizer potential as well.

Table 1. Types of substrates used for cultivation of important mushroom species.

White Rot Mushroom	Substrate type	Reference
Lyophyllum decastes	Soybean pulp, sorghum, Japanese cedar, wheat bran and rice bran	Parada et al. (2012), Arase et al. (2013)
Agaricus subrufrescens	Coast-cross hay, sugarcane bagasse, wheat bran, superphosphate fertilizer, limestone, gypsum and ammonium sulfate	Marques et al. (2014)
Pleurotus ostreatus	Maize stalk, maize husk, maize cob, paddy straw, sawdust and rice bran	Omokaro and Ogechi (2013), Adjapong et al. (2015)
Agaricus bisporus	Paddy straw, wheat bran and poultry manure	Roy et al. (2015), Muchena et al. (2021)
Ganoderma Lucidum	Wheat bran, sugarcane bagassae and corn cob rice straw	Gonzalez et al. (2015)
Lentinula edodes	Oak sawdust substrate, rice bran and barley straw	Bastida et al. (2016), Kang et al. (2017)
Pleurotus eryngii	Sugarcane bagasse, sweet potatoes, sawdust, wheat bran and rice straw	Lou et al. (2017)
Flamulina velutipes	Sawdust, cottonseed hull, corncob, sugarcane bagasse, wheat bran, corn powder and gypsum	Zhang et al. (2018)
Hypsizigus marmoreus	Corncob, cottonseed hull, sugarcane bagasse, wheat bran, sawdust and gypsum	Zhang et al. (2018)

Agro-biomass production in India is expected to be around 750 MMT and if left untreated, this biomass can have a variety of negative effects on the environment. Mushroom cultivation on these bio-masses and the conversion of mushroom spent substrate into value-added products like biofertilizers, biosorbent agents, and biomaterials are two alternatives to their sustainable recycling. Overall, lignocellulosic biomass is made up of roughly 40% cellulose, 25% hemicellulose, and 18% lignin, and can be degraded into monomeric units by enzymatic machinery of fungi which play a significant role in the growth of mushrooms, plant growth as biofertilizers, and are more likely to be used as animal feed, and as biosorbent materials for the removal of heavy metals.

In 2015, the global market for mushrooms was approximately $35 billion. The market is expected to grow by 9.2 percent from 2016 to 2025. As a result, the market is anticipated to grow to around $60 billion by 2025. Canada, China, France, Ireland, Italy, Netherlands, Poland, Spain, the United Kingdom and the United States are among the top nations that produce mushrooms. India is ranked eleventh on the list, while China is at the top. According to Meng et al. (2018), China is one of the world's top producers of edible fungus, producing 41.88 million tons annually, whereas India's production as of 2018 was 0.13 million tons (Raman et al. 2018). In India, Himachal Pradesh, Haryana and Punjab are the largest producers of mushrooms while North-Eastern states showed comparatively less production. Excluding China, the rest of the world produces approximately 58 million tons of mushrooms.

Countries like Italy, the USA, Poland, and the Netherlands have a lesser contribution to mushroom production and produce only 3.36, 2.26, 1.4, and 1.16 million tonnes of mushrooms, respectively. According to reports, about five tons of SMS are produced per ton of fresh edible mushrooms (Paredes et al. 2009, Aziera et al. 2015). Thus, the spent mushroom substrate after cultivation of the mushroom will be 3–5 times more than the total yield of mushrooms. According to Agrotex Global & Food and Agriculture Organisation (FAO), the worldwide share of mushroom production by different countries is given in Figure 1.

3. Global Generation of Spent Mushroom Substrate

According to ICAR data, India produces 0.5 million tonnes of spent mushroom substrate. China produced 185.7 million tonnes of SMS (Aziera et al. 2015), whereas Ireland produces about 0.254 million tonnes annually (Barry et al. 2012) and the Netherlands and Spain produce more than 800,000 tonnes of SMS annually (Oei and Albert 2012). After Asia and North America, Europe is the third-largest producer of discarded mushroom substrates globally. Recycling SMS into value-added products is urgently needed. Countries in Europe are spending a lot of money to dispose of this enormous amount of biowaste. Additionally, the price of SMS disposal currently ranges from 10 to 50 euros per tonne in Europe, putting a huge financial strain on the mushroom business of up to 150 million euros annually.

Table 2. Top spent mushroom substrate-producing countries in the world.

Countries	SMS (Million Ton)	Reference
China	185	Aziera et al. (2015)
USA	1.6	Beckers et al. (2019)
Netherland	0.8	Beckers et al. (2019)
Spain	0.8	Beckers et al. (2019)
India	0.5	Andrew et al. (2021)
Ireland	0.25	Beckers et al. (2019)
South Korea	0.1	Soylu et al. (2016)

4. Mechanism and Enzyme Involved in the Production of SMS

Mushrooms such as *Pleurotus*, *Ganoderma*, *Agaricus*, *Trametes*, and *Schyzophyllum* have a greater tendency to degrade lignocellulolytic biowaste and convert polysaccharides into simpler compounds for their growth and after their harvest, the residue of the substrate containing enzymes, proteins, sugars, and nutrients are used as spent mushroom substrate. The lignocellulosic biomass is mainly composed of lignin (5%–35%), cellulose (9%–80%), and hemicellulose (10%–50%), and their biodegradability is highly influenced by its composition. Various enzymes such as laccase, xylanase, mannanse, pectinase and cellulase are produced by fungi to degrade biomass and convert them into a usable form for their growth and development. The

Table 3. Enzymes are produced by important species of higher fungi.

	Enzyme produced	References
Pleurotus ostreatus	Laccase, MnP, VP, xylanase and cellulase	Cohen et al. (2001), Khalil et al. (2011), Vishwakarma (2012)
Pleurotus chrysosporium	MnP, xylanase and cellulase	Couto et al. (1999)
Phlebia radiata	MnP, laccase and LiP	Perez (1990), Arora and Gill (2001)
Ganoderma lucidum	Laccase, VP and MnP	D'souza et al. (1999)
Lentinula edodes	MnP, laccase and cellulase	Boer et al. (2004)
Coriolus versicolor	Laccase	Levin et al. (2004)
Daedalea quercina	Laccase	Baldrian (2004)
Trametes versicolor	LiP	Christan et al. (2005)
Funalia trogii	Laccase	Mazmanci and Unyayar (2005)
Irpex lacteus	MnP and laccase	Unyayar (2005)
Phenerocheate chrysosporium	MnP and cellulase	Karimi et al. (2006)
Cerrena unicolor	Laccase	Michniewicz et al. (2008)
Scyzophyllum cummune	MnP and laccase	Asgher (2013)
Pleurotus eryngii	MnP, laccase and VP	Akpinar et al. (2014)

MnP, Manganese peroxidase; VP, Versatile peroxidase; LiP, Lignin peroxidase.

enzymes produced by these fungi are capable of degrading lignocellulolytic biomass which in turn are crucial for the production of the spent mushroom substrate also (Table 3).

5. National and International Status of Lignocellulosic Biomass Generation

India is one of the largest producers of lignocellulosic waste (Table 4). Every year, India produces 750 million tonnes of lignocellulosic wastes according to information provided by the Ministry of New and Renewable Energy while globally it is approximately 5.5 billion tons. If it is left untreated or mismanaged, poses a

Table 4. Generation of different types of lignocellulosic biomass annually in India (million metric tons).

Lignocellulosic waste	Annual production	References
Corn stalk	13.59	Purohit et al. 2009
Rice husk	24	Swetha et al. 2014
Wheat straw and sorghum straw	135.59	Chandel et al. 2017
Paddy straw & forest waste	114 & 200	Millati et al. 2019
Corn waste	26	Bhuvaneshwari et al. 2019
Sugarcane bagasse	90	Dutta et al. 2020

constant threat to our ecosystem. Lignocellulosic biomass has greater importance in the production of spent mushrooms. So recycling that much of the biowaste released every year will protect our environment and will be managed and dumped in a healthier manner.

Developing countries are leading producers of agricultural waste and they have a non-availability of robust technology to convert this huge amount of biowaste into value-added products. A little chunk of biowaste as SMS is used as biofertilizers, while the remaining lignocellulosic biowaste is either burnt or dumped on land which causes a constant threat to our environment. Recycling of lignocellulosic waste into value-added products by innovating various techniques is the need of the hour. Country-wise production of lignocellulosic waste is summarized below in Table 5.

Table 5. Countries produce the majority of lignocellulosic waste per year.

Substrate	Production (Million tons)	Country	Reference
Rice straw	149 114 39 36 30	China India Indonesia Bangladesh Vietnam	Millati et al. 2019
Rice Husk	39 30 10 9 8	China India Indonesia Bangladesh Vietnam	Millati et al. 2019
Corn Cob	81 49 20 13 9	USA China Brazil EU Argentina	Millati et al. 2019
Oil Palm Waste	37 19 2 1 0.9	Indonesia Malaysia Thailand Columbia Nigeria	Millati et al. 2019
Wheat straw	128 110 100 40 25	EU China India USA Canada	Millati et al. 2019
Sugarcane bagassae	94 93 55 38 29	Brazil India EU Thailand China	Birru 2016, Millati et al. 2019

6. Composition of Spent Mushroom Substrate

The nutritive value of mushroom spent substrate is extremely enriched in elements like carbon, nitrogen, phosphorus, potassium, sodium, magnesium, proteins, and carbohydrates, among others. SMS is a powerful biosorbent material due to the abundance of free functional groups it contains, such as hydroxyl, amide, and formyl. The Nutritional value of a spent mushroom is 1.5–2.16% N, 2.16–0.69% P, and 2.2–3% K (Corral-Bobadilla et al. 2019, Andrews et al. 2021). Additionally, SMS contains C, Ca, Mg and Na in amounts of around 54.2, 10, 0.83 and 0.24%, respectively. The type of biomass utilized in the mushroom cultivation process defines the composition of N, P, and K values. Recent, FTIR and SEM studies have demonstrated that lignocellulosic fibers have been broken down by white rot fungus and carry several free functional groups that have a higher capacity to entrap metals and minerals.

7. Applications of SMS

Due versatile properties of spent mushroom substrates, it finds applications in a number of fields discussed below.

7.1 *Biofertilizer*

As discussed above, SMS is highly enriched with a number of nutritional components. It has a higher concentration of N-P-K value which is crucial for the growth of plants in agriculture. SMS has tremendous potential in soil enrichment by supplying essential minerals to the soil. It is an ideal supportive nutrient media for vegetables like cucumbers, tomatoes, broccoli, tulips, cauliflower, peppers, spinach and others. The SMS can be used as a biofertilizer in field crops such as corn, wheat, barley, cucumber, onion, and potatoes, as well as in greenhouse crops like tomato and capsicum, and in nurseries as well. It is inexpensive and eco-friendly for agricultural uses as it acts as an alternative means of waste disposal. According to Paredes et al. (2016), using SMS-derived compost as mineral fertilizer increased lettuce yield and increased the soil's overall content of organic carbon, nitrogen, and phosphorus. Similarly, Roy et al. (2015) used SMS compost in capsicum plants and reported a remarkable yield. SMS can curtail the use of chemical fertilizers and other conventional used fertilizers to a certain extent.

7.2 *Biocontrol Agents*

Excessive use of chemical fertilizers and pesticides in agricultural activities as bio-control agents results in expensive and slowly progressing results. For integrated agricultural disease management systems, bio-control agents can replace pesticides. As a result, there is a need for innovative, low-cost bioactive compounds with high bioactive ingredient concentrations as bio-control agents. It does demonstrate that using bioactive chemicals selectively as bio-control agents is a successful method for achieving that goal. It is well recognized that SMS contains bioactive substances with pathogen-fighting abilities. It may be used to treat contaminated and infertile soils as well as produce bio-control agents against plant diseases. There have been numerous investigations conducted to identify SMS rich in antimicrobial properties against bacterial and fungal pathogens including control of plant pathogens. In certain research, the substrate for mushrooms and the species of mushrooms used to produce antimicrobial compounds against clinical and foodborne diseases, including fungi, are correlated. The majority of SMS is proving to be efficient bio-control agents, either when used directly as a soil conditioner or when combined with extracts to combat soil-borne and plant infections.

7.3 Bioremediation

SMS has highly entrapment or biosorbing tendency due to their free functional group and ionic charges. Exploring this property of SMS for the removal of heavy metals which are the real cause of air, water, and soil pollution can be considered an inexpensive and qualitative approach. It can be used for the treatment of effluents of textile and other industries which contain Cu, Pb, Fe, As, Co and dyes which are harmful to our environment. It is also a very important technology for the bioremediation of pesticides, fungicides and weedicides. Xu et al. (2013) reported the application of spent *Agaricus bisporus* to remove heavy metals like lead and cadmium from aqueous solutions. Likewise, Dai et al. (2012) have used spent mushrooms of *Tricholoma lobayense* for the removal of lead, and spent mushrooms from *Lentinus edodes* have reportedly been capable of removing cadmium from wastewater (Chen et al. 2008). Recently, Menaga et al. (2021) reported the removal of ferrous ions using *Pleurotus* mushroom substrate. Utilising this inexpensive and eco-friendly approach to remove the heavy metals from effluents can curtail the havoc created by these metals in environmental pollution.

7.4 Biomaterial Production

The SMS has superior mechanical strength and may be converted into a variety of useful construction materials, including insulated board, mycobricks, mycotiles, and mycofoam (Deshmukh et al. 2022). Recently, Tacer-Caba et al. (2020) made mycofoam from a composite of fungal mycelium that possesses a strong compressive strength. Similarly, mushroom spent substrate was used to make insulated brick, which has exceptional mechanical strength and fire resistance properties (Velasco et al. 2014). For interior home design or building, mycobricks and insulated boards can be employed, and mycofoam can be used in place of polystyrene foam for packing and shipping in-animate objects. It is indeed relatively new to transform SMS into biomaterials; now, only one or two businesses worldwide produce biomaterials like particle board from the used mushroom substrate, however, the area requires more study.

7.5 Cattle Feed

The spent mushroom substrate contains nutrients that are necessary for ruminant animals, such as protein, sugar, enzymes, feed supplements, and minerals. As they are rich in vital nutritional components and offer therapeutic effects which protect animals from various diseases. SMS can be demonstrated as a robust animal diet or as a feed additive. Various genera of mushrooms have been used as a spent mushroom substrate for animal feed after harvesting fruiting bodies. Numerous studies have looked into using *Pleurotus* spp. the waste substrate as a feedstock, for cattle feed (Kim et al. 2011). Another benefit of using SMS as cattle feed is that it has a lot of heavy ingredients like protein, carbs, and a high C/N ratio, allowing animals to release a lot of calories and fulfil their hunger with a smaller amount of feed. This benefits the farmer because administering less of these substances can lower animal hunger levels and boost output. Another advantage of using it as cattle feed is that it is inexpensive and less labour-intensive than other types of animal feed. Less nutritive Agricultural waste like corn stalks, pigeon pea stalks, soybean straw, and cotton stalks can be converted into highly proteinaceous feed with white rot fungi such as *Pleurotus sajor-caju* (Mane et al. 2007).

7.6 Source of Different Micro Flora

In addition to fungi, the substrate from decomposed mushrooms serves as a habitat for mesophilic or thermophilic bacteria, which can be beneficial to researchers as they produce industrial important enzymes. It was reported that Gram-positive bacteria predominate in SMS. There are bacteria that are ligninolytic and produce enzymes like laccase, xylanase, cellulase, and amylase, including *Actinobacteria, Arthrobacter, Bacillus, Brevibacterium, Carnobacterium, Exiguobacterium, Firmicutes, Microbacterium, Paenibacillus* and *Staphylococcus* (Song et al. 2001, Ntougias et al. 2004, Sharma et al. 2013). *Actinomycetes* and *Streptomyces*, two thermophilic bacteria, have also been reported in several mushroom composts.

8. Future Prospects

The SMS are the typical by-product of the mushroom industry, which produces a range of nutrients and bioactive substances, including extracellular enzymes, secondary metabolites, amino acids, carbohydrates, and organic matter. In addition, it possesses a lot of free-end functional groups, making it a powerful biosorbent material that helps in the removal of heavy metals from industrial effluents. As a result, there have been a lot of discussions nowadays concerning the potential of employing SMS to produce products with added value. It is susceptible to being transformed into biofertilizer, pollutant removal, biocontrol agents, and the development of biomaterials. The discovery of bioactive substances in the SMS as a result of the microbial community's biological activity served as an effective biocontrol agent against soil and plant infections, where biocontrol agents provide a significant alternative to synthetic chemicals. Besides, the application of SMS has positive effects on the physicochemical characteristics of soil, including improvements in soil structure, availability of nutrients, and potential for bioremediation of pollutants. The application of SMS would be noteworthy to develop new biomaterials for safe and eco-friendly practices. In near future, the value-added products derived from SMS such as biofertilizers, animal feed, biomaterials, bio-adsorbents can have robust demand among people and can be commercialized by many industries. SMS derived value-added products can be proved as a replacement for conventionally used products such as chemical fertilizer, electric filter chambers for heavy water removal in effluents, expensive cattle feed and construction materials.

References

Adjapong, A.O., K.D. Ansah, F. Angfaarabung and H.O. Sintim. 2015. Maize residue as a viable substrate for farm scale cultivation of oyster mushroom (*Pleurotus ostreatus*). *Adv. Agric.*, 2015: 1–6.

Ahalawat, O.P. and M.P. Sagar. 2007. Management of spent mushroom substrate. National Research Centre for Mushroom, Indian Council of Agricultural Research, New Delhi.

Amarasinghe, B.M.W.P.K. and R.A. Williams. 2007. Tea waste as a low cost adsorbent for the removal of Cu and Pb from wastewater. *Chem. Eng. J.*, 132(1-3): 299–309.

Andrews, A., S. Singh, A.C. Ishani and Y.V. Kumar. 2021. Utilization of spent mushroom substrate: A review. *Pharma Innovation*, 10(5): 1017–1021.

Arase, S., Y. Kondo, R. Parada, H. Otani, M. Ueno and J. Kiiharai, J. 2013. Suppression of rice blast disease by autoclaved water extract from the spent mushroom substrate of *Lyophyllum decastes*. *Mushroom Sci. Biotechnol.*, 21(2): 79–83. https://doi.org/10.24465/msb.21.2_79.

Arora, D.S. and P.K. Gill. 2001. Effects of various media and supplements on laccase production by some white rot fungi. *Biores. Technol.*, 77(1): 89–91.

Asgher, M., M. Irshad and I.H.M. Nasir. 2013. Purification and characterization of novel manganese peroxidase from *Schizophyllum commune* IBL-06. *Int. J. Agric. Biol.*, 15(4): 749–754.

Akpinar, M. and R.O. Urek. 2014. Extracellular ligninolytic enzymes production by *Pleurotus eryngii* on agro industrial wastes. *Prep. Biochem. Biotechnol.*, 44(8): 772–781.

Aziera, N., A. Rasib, Z. Zakaria, M.F. Tompang and H. Othman. 2015. Characterization of biochemical composition for different types of spent mushroom substrate in Malaysia. *Malay. J. Anal.*, 19(1): 41–45.

Baldrian, P. 2004. Increase of laccase activity during interspecific interactions of white rot fungi. *FEMS Microbiol. Ecol.*, 50(3): 245–253.

Barakat, M.A. 2011. New trends in removing heavy metals from industrial wastewater. *Arabian J. Chem.*, 4(4): 361–377.

Barry, J., O. Doyle, J. Grant and H. Grogan. 2012. Supplementary of spent mushroom substrate (SMS) to improve the structure and productivity as a casing material. pp. 735–742. *In:* 18th Congress of the International Society for Mushroom Science. Beijing, China.

Bastida, A.Z., D.O. Ramírez, S.S. Simental, N.R. Perez and M.A. Martínez. 2016. Comparison of antibacterial activity of the spent substrate of *Pleurotus ostreatus* and *Lentinula edodes*. *J. Agric. Sci.*, 8(4): 43–49.

Beckers, S.J., I.A. Dallo, I. Del Campo, C. Rosenauer, K. Klein and F.R. Wurm. 2019. From compost to colloids-valorization of spent mushroom substrate. *ACS Sust. Chem. Eng.*, 7(7): 6991–6998.

Bhuvaneshwari, S., H. Hettiarachchi and J.N. Meegoda. 2019. Crop residue burning in India: Policy challenges and potential solutions. *Int. J. Environ. Res. Public Health*, 16(5): 832. doi: 10.3390/ijerph16050832.

Birru, E. 2016. Sugar cane industry overview and energy efficiency considerations. Literature Survey Document, Division of Heat and Power Technology, WE-100 44 Stockholm, p. 61.

Boer, C.G., L. Obici, C.G.M. de Souza and R.M. Peralta. 2004. Decolorization of synthetic dyes by solid state cultures of *Lentinula* (*Lentinus*) *edodes* producing manganese peroxidase as the main ligninolytic enzyme. *Biores. Technol.*, 94(2): 107–112.

Chandel, S. and S. Kaushal. 2017. Enhancing the shelf life of *Trichoderma* species by adding antioxidants producing crops to various substrates. *J. Crop Prot.*, 6(3): 307–314.

Chen, G., G. Zeng, L.Tang, C. Du, X. Jiang, G. Huang, H. Liu and G.Shen. 2008. Cadmium removal from simulated wastewater to biomass byproduct of *Lentinus edodes*. *Biores. Techno.*, 99(15): 7034–7040.

Chequer, F.M.D., J.P.F. Angeli, E.R.A. Ferraz, M.S. Tsuboy, J.C. Marcarini, M.S. Mantovani and D.P. de Oliveira. 2009. The azo dyes Disperse Red 1 and Disperse Orange 1 increase the micronuclei frequencies in human lymphocytes and in HepG2 cells. *Mut. Res. Gen. Toxicol. Environ. Mutagen.*, 676(1-2): 83–86.

Christian, V., R. Shrivastava, D. Shukla, H. Modi and B.R.M. Vyas. 2005. Mediator role of veratryl alcohol in the lignin peroxidase-catalyzed oxidative decolorization of Remazol Brilliant Blue R. *Enz. Microb. Technol.*, 36(2-3): 327–332.

Cohen, R. and Y. Hadar. 2001. The roles of fungi in agricultural waste conversion. pp. 305–334. *In*: Gadd, G.M. (ed.). Fungi in Bioremediation. Cambridge University Press, Cambridge.

Corral-Bobadilla, M., A. González-Marcos, E.P. Vergara-González and F. Alba-Elías. 2019. Bioremediation of wastewater to remove heavy metals using the spent mushroom substrate of *Agaricus bisporus*. *Water*, 11(3): 454. https://doi.org/10.3390/w11030454.

Couto, S.R., M.A. Longo, C. Cameselle and A. Sanromán. 1999. Ligninolytic enzymes from corncob cultures of *Phanerochaete chrysosporium* under semi-solid-stat conditions. *Acta Biotechnol.*, 19(1): 17–25.

D'souza, T.M., C.S. Merritt and C.A. Reddy. 1999. Lignin-modifying enzymes of the white rot basidiomycete *Ganoderma lucidum*. *Appl. Environ. Microbiol.*, 65(12): 5307–5313.

Dai, J., F. Cen, J. Ji, W. Zhang and H. Xu. 2012. Biosorption of lead (II) in aqueous solution by spent mushroom *Tricholoma lobayense*. *Water Environ. Res.*, 84(4): 291–298.

Debosz, K., S.O. Petersen, L.K. Kure and P. Ambus. 2002. Evaluating effects of sewage sludge and household compost on soil physical, chemical and microbiological properties. *Appl. Soil Ecol.*, 19(3): 237–248.

Deshmukh, S.K., M.V.K.R. Deshpande and R.S. Kandikere (eds.). 2022. Fungal Biopolymers and Biocomposites: Prospects and Avenues. Springer, pp. 421.

Dutta, A., M. Iisa, M. Talmadge, C. Mukarakate, M. Griffin, E. Tan, N. Wilson, M.M. Yung, M.R. Nimlos, J. Schaidle, H. Wang, M. Thorson, D. Hartley, J. Klinger and H. Cai. 2022. *Ex situ* catalytic fast pyrolysis of lignocellulosic biomass to hydrocarbon fuels: 2019 state of technology and future research. Golden, CO: National Renewable Energy Laboratory. NREL/TP-5100-76269. https://www.nrel.gov/docs/fy20osti/76269.pdf.

Food and Agriculture Organization of the U.N.O. 2018. Rice Market Monitor. http://www.fao.org/economic/ RMM (released date: 27 April 2018).

Fu, F. and Q. Wang. 2011. Removal of heavy metal ions from wastewaters: A review. *J. Environ. Management*, 92(3): 407–418.

González-Marcos, A., F. Alba-Elías, F.J. Martínez-de-Pisón, J. Alfonso-Cendón and M. Castejón-Limas. 2015. Composting of spent mushroom substrate and winery sludge. *Compos. Sci. Util.*, 23(1): 58–65. https://doi.org/10.1080/1065657X.2014.975868.

Hydari, S., H. Sharififard, M. Nabavinia and M. Reza Parvizi. 2012. A comparative investigation on removal performances of commercial activated carbon, chitosan biosorbent and chitosan/activated carbon composite for cadmium. *Chem. Eng. J.*, 193: 276–282.

Kang, D.S., K.J. Min, A.M. Kwak, S.Y. Lee and H.W. Kang. 2017. Defense response and suppression of *phytophthora* blight disease of pepper by water extract from spent mushroom substrate of *Lentinula edodes*. *Plant Pathol. J.*, 33(3): 264275. doi.org/10.5423/PPJ.OA.02.2017.0030.

Karimi, A., F. Vahabzadeh and B. Bonakdarpour. 2006. Use of *Phanerochaete chrysosporium* immobilized on Kissiris for synthetic dye decolourization: Involvement of manganese peroxidase. *World J. Microbiol. Biotechnol.*, 22(12): 1251–1257.

Khalil, M.I., M. Hoque, M.A. Basunia, N. Alam and M.A. Khan. 2011. Production of cellulase by *Pleurotus ostreatus* and *Pleurotus sajor-caju* in solid state fermentation of lignocellulosic biomass. *Turk. J. Agric. For.*, 35(4): 333–341.

Kim, Y.I., W.M. Cho, S.K. Hong, Y.K. Oh and W.S. Kwak. 2011. Yield, nutrient characteristics, ruminal solubility and degradability of spent mushroom (*Agaricus bisporus*) substrates for ruminants. *Asian-Aust. J. Anim. Sci.*, 24(11): 1560–1568.

Lau, K.L., Y.Y. Tsang and S.W. Chiu. 2003. Use of spent mushroom compost bioremediate PAH-contaminated samples. *Chemosphere*, 52: 1539–1546.

Levin, L., L. Papinutti and F. Forchiassin. 2004. Evaluation of Argentinean white rot fungi for their ability to produce lignin-modifying enzymes and decolorize industrial dyes. *Biores. Technol.*, 94(2): 169–176.

Lou, Z., Y. Sun, S. Bian, S. Ali Baig, B. Hu and X. Xu. 2017a. Nutrient conservation during spent mushroom compost application using spent mushroom substrate derived biochar. *Chemosphere*, 169: 23–31.

Lou, Z., Y. Sun, X. Zhou, S.A. Baig, B. Hu and X. Xu. 2017b. Composition variability of spent mushroom substrates during continuous cultivation, composting process and their effects on mineral nitrogen transformation in soil. *Geoderma*, 307: 30–37.

Mane, V.P., S.S. Patil, A.A. Syed and M.M.V. Baig. 2007. Bioconversion of low quality ligno-cellulosic agricultural waste into edible protein by *Pleurotus sajor-caju* (Fr.) Singer. *J. Zhejiang Univ. Sci.* B, 8(10): 745–751.

Marques, E.L.S., E.T. Martos, R.J. Souza, R. Silva, D.C. Zied and E. Souza Dias. 2014. Spent mushroom compost as a substrate for the production of lettuce seedlings. *J. Agric. Sci.*, 6(7): 138–143. https://doi.org/10.5539/jas.v6n7p138.

Mazmanci, M.A. and A. Ünyayar. 2005. Decolourisation of reactive black 5 by Funalia trogii immobilised on *Luffa cylindrica* sponge. *Process Biochem.*, 40(1): 337–342.

Meng, X., B. Liu, C. Xi, X. Luo, X. Yuan, X. Wang, W. Zhu, H. Wang and Z. Cui. 2018. Effect of pig manure on the chemical composition and microbial diversity during co-composting with spent mushroom substrate and rice husks. *Biores. Technol.*, 251: 22–30.

Michniewicz, A., S. Ledakowicz, R. Ullrich and M. Hofrichter. 2008. Kinetics of the enzymatic decolorization of textile dyes by laccase from *Cerrena unicolor*. *Dyes and Pigments*, 77(2): 295–302.

Millati, R., R.B. Cahyono, T. Ariyanto, I.N. Azzahrani, R.U. Putri and M.J. Taherzadeh. 2019. Agricultural, industrial, municipal, and forest wastes: An overview. *Sustain. Res. Rec. Zero Waste Approc.*, 2019: 1–22.

Murgia, I., P. Arosio, D. Tarantino and C. Soave. 2012. Bio fortification for combating 'hidden hunger' for iron. *Trends Plant Sci.*, 17(1): 47–55.

Muchena, F.B., C. Pisa, M. Mutetwa, C. Govera and W. Ngezimana. 2021. Effect of spent button mushroom substrate on yield and quality of baby spinach (*Spinacia oleracea*). *Int. J. Agron.*, 2021: https://doi.org/10.1155/2021/6671647.

Menaga, D., S. Rajakumar and P.M. Ayyasamy. 2021. Spent mushroom substrate: A crucial biosorbent for the removal of ferrous iron from groundwater. *SN Appl. Sci.*, 3(1): 1–17.

Ntougias, S., G.I. Zervakis, N. Kavroulakis, C. Ehaliotis and K.K. Papadopoulou. 2004. Bacterial diversity in spent mushroom compost assessed by amplified rDNA restriction analysis and sequencing of cultivated isolates. *Appl. Microbiol.*, 27: 746–754.

Oei, P. and G. Albert. 2012. Recycling casing soil. 18th Congress of the International Society for Mushroom Science. Beijing, China, pp. 757–765.

Omokaro, O. and A. Ogechi. 2013. Cultivation of mushroom (*Pleurotus ostreatus*) and the microorganisms associated with the substrate used. *J. Sci. Technol.*, 8(4): 49–59.

Purohit, P. 2009. Economic potential of biomass gasification projects under clean development mechanism in India. *J. Clean. Prod.*, 17(2): 181–193.

Parada, R.Y., S. Murakami, N. Shimomura and H. Otani. 2012. Suppression of fungal and bacterial diseases of cucumber plants by using the spent mushroom substrate of *Lyophyllum decastes* and *Pleurotus eryngii*. *J. Plant Phytopathol.*, 160(7-8): 390–396.

Paredes, C., E. Medina, M.A. Bustamante and R. Moral. 2016. Effects of spent mushroom substrates and inorganic fertilizer on the characteristics of a calcareous clayey-loam soil and lettuce production. *Soil Use Manag.*, 32(4): 487–494.

Paredes, C., E. Medina, R. Moral, M.D. Perez-Murcia, J. Moreno-Caselles, M.A. Bustamante and J.A. Cecilia. 2009. Characterization of the different organic matter fractions of spent mushroom substrate. *Commun. Soil Sci. Plant Anal.*, 40(1e6): 150e161. https://doi.org/10.1080/00103620802625575.

Perez, J. and T.W. Jeffries. 1990. Mineralization of 14C-ring-labeled synthetic lignin correlates with the production of lignin peroxidase, not of manganese peroxidase or laccase. *Appl. Environ. Microbiol.*, 56(6): 1806–1812.

Philippoussis, A., G.I. Zervakis, P. Diamantpoulou, K. Papadopoulou and C. Ehaliotis. 2004. Use of spent mushroom compost as a substrate for plant growth and against plant infections caused by *Phytophthora. Mushr. Sci.*, 16: 579–584.

Raman, J., S.K. Lee, J.H. Im, M.J. Oh, Y.L. Oh and K.Y. Jang. 2018. Current prospects of mushroom production and industrial growth in India. *Journal of Mushroom*, 16(4): 239–49.

Rinker, D.L. 2017. Spent mushroom substrate uses. *Edible and Medicinal Mushrooms: Technology and Applications*, 1: 427–454.

Roy, S., S. Barman, U. Chakraborty and B. Chakraborty. 2015. Evaluation of spent Mushroom substrate as biofertilizer for growth improvement of *Capsicum annuum* L. *J. App. Biol. Biotech.*, 3(3): 022–027.

Royse, D.J. 2014. November. A global perspective on the high five: *Agaricus, Pleurotus, Lentinula, Auricularia & Flammulina.* In Proceedings of the 8th Int. Conference on Mushroom Biology and Mushroom Products (ICMBMP8), (1): 1–6.

Ruttimann, C., E. Schwember, L. Salas, D. Cullen and R. Vicuna. 1992. Ligninolytic enzymes of the white rot basidiomycetes *Phlebia brevispora* and *Ceriporiopsis subvermispora. Biotechnol. Appl. Biochem.* (USA) 16(1): 64–76.

Sharma, A., A.V. Singh and B.N. Johri. 2013. Functional and genetic characterization of culturable bacteria associated with late phase of mushroom composting assessed by amplified rDNA restriction analysis. *Int J. Curr. Microbiol. App. Sci.*, 2(6): 162–175.

Shweta, K. and H. Jha. 2014. Rice husk extracted lignin–TEOS biocomposites: Effects of acetylation and silane surface treatments for application in nickel removal. *Biotechnol. Reports*, 7: 95–106.

Song, J., H.Y. Weon, S.H. Yoon, D. Park, S.S.J. Go and J.W. Suh. 2001. Phylogenetic diversity of thermophilic Actinomycetes and Thermoactinomyces spp. isolated from mushroom composts in Korea based on 16S rRNA gene sequence analysis. *FEMS Microbiol.*, 202: 97–102.

Soylu, M.K., K. Boztok and D. Esiyok. 2014. Mycelial growth performance of the Pleurotus eryngii species complex strains on different temperatures. In XXIX Int. Horticultural Cong. on Horticulture: Sustaining Lives, Livelihoods and Landscapes (IHC2014), 1123: 207–214.

Tacer-Caba, Z., J.J. Varis, P. Lankinen and K.S. Mikkonen. 2020. Comparison of novel fungal mycelia strains and sustainable growth substrates to produce humidity-resistant biocomposites. *Mater. Des.*, 192: 108728.

Vandenkoornhuyse, P., R. Husband, T.J. Daniell, I.J. Watson, J.M. Duck, A.H. Fitter and J.P.W. Young. 2002. Arbuscular mycorrhizal community composition associated with two plant species in a grassland ecosystem. *Mol. Ecol.*, 11(8): 1555–1564.

Vishwakarma, S.K., M.P. Singh, A.K. Srivastava and V.K. Pandey. 2012. Azo dye (direct blue 14) decolorization by immobilized extracellular enzymes of Pleurotus species. *Cell. Mol. Biol.*, 58(1): 21–25.

Velasco, P.M., M.P.M. Ortiz, M.A.M. Giro, M.C.J. Castelló and L.M. Velasco. 2014. Development of better insulation bricks by adding mushroom compost wastes. *Energy Build.*, 80: 17–22.

Vijayaraghavan, K. and Y.S. Yun. 2008. Bacterial biosorbents and biosorption. *Biotechnol. Adv.*, 26(3): 266–291.

Wu, J., T. Zhang, C. Chen, L. Feng, X. Su, L. Zhou, Y. Chen, A. Xia and X. Wang. 2018. Spent substrate of *Ganoderma lucidum* as a new bio-adsorbent for adsorption of three typical dyes. *Biores. Technol.*, 266: 134–138.

Xu, H., Y. Chen, H. Huang, Y. Liu and Z. Yang. 2013. Removal of lead (II) and cadmium (II) from aqueous solutions using spent Agaricus bisporus. *Can. J. Chem. Eng.*, 91(3): 421–431.

Zhang, B., L. Yan, Q. Li, J. Zou, H. Tan, W. Tan, W. Peng, X. Li and X. Zhang. 2018. Dynamic succession of substrate-associated bacterial composition and function during *Ganoderma lucidum* growth. *Peer J.*, 6: e4975. https://doi.org/10.7717/peerj.4975.

Zhu, H.-J., J.-H. Liu, L.-F. Sun, Z.-F. Hu and J.-J. Qiao. 2013. Combined alkali and acid pre-treatment of spent mushroom substrate for reducing sugar and biofertilizer production. *Biores. Technol.*, 13(6): 257–266.

Bioactive Compounds

4

Bioactive Compounds and Medicinal Value of the Rare Mushroom *Agaricus blazei*

Maizatulakmal Yahayu,[1] Solleh Ramli,[1] Zaitul Iffa Abd Rasid,[1] Daniel Joe Dailin,[1,2] Dalia Sukmawati,[3] Neo Moloi[4] and Hesham Ali El Enshasy[1,2,5]

1. Introduction

For many centuries, edible mushrooms have been appreciated as a valuable source of food and medication purposes. Extensive research has been carried out up to the molecular level to reveal the potential uses of edible mushrooms as a therapeutic agent in the treatment of various diseases. A recent study has described the anti-angiogenic effect of mushroom extract and purified compounds for cancer treatment in advanced stages (Jana and Acharya 2022). One of those beneficial mushrooms

[1] Institute of Bioproduct Development (IBD), Universiti Teknologi Malaysia (UTM), Johor Bahru, Johor, Malaysia.
[2] Faculty of Chemical and Energy Engineering, Universiti Teknologi Malaysia (UTM), Johor Bahru, Johor, Malaysia.
[3] Faculty of Mathematics and Natural Science, Universitas Negeri Jakarta, Jakarta Timur 13220, Indonesia.
[4] Sawubona Mycelium Co., Centurion, Gauteng, South Africa.
[5] City of Scientific Research and Technology Applications, New Burg Al Arab, Alexandria, Egypt.

which is known as the Brazillian mushroom, *Agaricus blazei* Murill (AB) has gained more attention from scientists worldwide. It is widely cultivated in Japan and becomes the most important edible mushroom and is often prescribed in that country for medical purposes (Kumar and Kumari 2021). This basidiomycete fungus is belonging to the family of Agaricaceae, which is native to southern Brazil and was introduced and cultivated in Japan during the 1960s. In Japan, it is commonly named "Himematsutake", considering 60.6% of the medicinal mushroom used (Matsushita et al. 2018). Therefore, as a rich source of many functional molecules, this mushroom is considered one of the main ingredients in the nutraceutical, cosmeceutical, and pharmaceutical industries (Taofiq et al. 2019a). This chapter will provide up-to-date comprehensive information about the bioactive ingredients of AB and their potential uses in biotherapeutic industries.

2. *Agaricus blazei*

Agaricus is a large mushroom genus with a distribution range extended to all continents except Antarctica (Callac and Chen 2018). This species is widely cultivated in the boreal region, grassland, forest, manures and any damp places with lignicolous debris or decaying organic matter (Medel-Ortiz et al. 2022). Up to September 2018, the number of species described under this genus exceeded 500 and many presumed new species have not yet been named and categorized in many regions (Callac and Chen 2018).

Agaricus blazei Murill (AB) is an *Agaricus* mushroom species characterized by a brownish-gold cap of 7–25 cm diameter and chocolate brown basidiospores (5 × 4 μm) which is close to *A. subrufescens*. It is generally described as a mushroom of small to large fruit bodies, convex, fleshy, short-stem and hard, with white, yellow or brown pileus; free pallid or pinkish young lamellae, chocolate-brown matured lamellae and smooth basidiospores (Heinemann 1978). AB was first discovered in the highland areas of Sao Paolo, Brazil and raised great interest in Asian and North American markets as one of the sources of protein-rich food and contributed to the discoveries of effective medicinal products (Chang and Wasser 2018). It was found by a Japanese farmer and researcher in 1965 and further investigated before being identified as AB by a Belgian botanist namely Heinemann in 1967 (Firenzuoli et al. 2008). The mushroom has a variety of names including ABM (for *Agaricus blazei* Murrill), *cogumelo do sol* (mushroom of the sun), *cogumelo de Deus* (mushroom of God), *cogumelo de vida* (mushroom of life), *himematsutake*, royal sun agaricus, *Mandelpilz*, and almond mushroom (da Eira et al. 2002).

2.1 *Chemical Constituents*

Generally, mushrooms compositions will include water (60–90%), carbohydrates (1–55%), protein (2-40%), fiber (3–32%), fat (2–8%) and ash (8–10%) (Firenzuoli et al. 2008). Nowadays, much research has been conducted using submerged fermentation to isolate the bioactive metabolites in fruit bodies, culture filtrate and pure culture mycelia. The first attempt to fractionate the antitumor fraction in fruiting bodies of AB was carried out by Kawagishi et al., in 1989. The author has identified polysaccharides with prominent antitumor properties in the fraction which comprised a glycol-protein complex, composed of 50.2% carbohydrates and 43.4% protein. The fraction was also detected to contain simple (1-6)-β-D-glucopyranosyl chains (Firenzuoli et al. 2008, Kawagishi et al. 1989). In addition, whole-mushroom extracts are mainly consisting of several other compounds such as ergosterol and linoleic acid, palmitic acid, pro-vitamin D2 and amino acid (Shimizu et al. 2016, Cho et al. 2008). A large amount of phenolic and flavonoid constituents also has been detected in AB including catechin, syringic acid, gallic acid, protocatechuic acid, quercetin and myricetin through microwave-assisted extraction technique (Zhang et al. 2012). These metabolites are reported to exhibit several chemopreventive and chemotherapeutic properties such as antioxidant, anti-proliferative and anti-carcinogenic effects (Tungmunnithum et al. 2018), hypoglycemic and anti-diabetic effects (Chen et al. 2019) (Table 1). The mode of action comprises inhibition of cell division, antioxidant activity, carcinogens deactivation, cell differentiation and cell-death induction, apoptosis induction, angiogenesis inhibition and reversal of multidrug resistance (Costea et al. 2020, Eguchi et al. 2017, Kimura et al. 2004).

2.1.1 The α and β Glucans

Based on the previous literature findings, AB mainly contains β-(1,3)-D-glucan, β-(1,4)-D-glucan, β-(1,6)-D-glucan and various kinds of active polysaccharides molecules with anti-tumour activity mainly isolated from mushroom fruiting bodies. The metabolites include FII-a-β (β-glucan), FIII2-β (β-glucan-protein), FA-1a-β (hetero-β-glucan), FA-2b-β (RNA-protein complex) and FV-1 (insoluble β-glucan) (Bertollo et al. 2022). β-glucans seem to be the most important compounds which are responsible for therapeutic actions on cancer and primarily initiate modulation of the immune system, although studies are showing another mode of action involved (Ohno et al. 2001). In general, AB contains both α and β-glucans in the fruiting bodies and the contents are different with the maturity stages of AB. Therefore, the harvesting time of AB and its conservation is important to obtain the best extract with the highest yield of glucans content.

Polysaccharides are carbohydrate polymers built-up of long chains of monosaccharide units, binding together through glycosidic linkages. Previous studies have confirmed that insoluble (13/1-6)-β-glucan, has superior biological functions compared to those soluble ones such as (1-3/1-4)-β-glucans equivalents (Ooi and Liu 2000). The differences between β-glucan association and chemical structure are substantial due to the mode of action, solubility as well as overall biological activity. These polysaccharides are of different chemical compositions, with most belonging to the group of β-glucans; these have β-(1-3) linkages in the main chain of the glucan and additional β-(1-6) branch points which are needed for anti-tumour behaviour. High molecular weight glucans appear to be more effective than those with low molecular weight (Yuminamochi et al. 2007, Smiderle et al. 2011).

β-glucans extracted from AB are well-known leukocyte activators that have shown their potential in stimulating immune cells such as macrophages, NK cells, dendritic cells and granulocytes (polymorponuclear leukocytes and mononuclear

Table 1: Metabolite constituents in *A. blazei* Murill.

Metabolites	Chemical structure	References
Ergosterol		Misgiati et al. 2021, Monteiro et al. 2020, Rózsa et al. 2019, Taofiq et al. 2019b, Gąsecka et al. 2018, Cho et al. 2008
5-dihydroergosterol		Kim et al. 2020
Cerevisterol		Kim et al. 2020
Cerebroside B		Kim et al. 2020
Cerebroside D		Kim et al. 2020
Adenosine		Kim et al. 2020
Benzoic acid		Kim et al. 2020

Table 1 contd. ...

...Table 1 contd.

Metabolites	Chemical structure	References
Linoleic acid		Soares et al. 2020, Sande et al. 2019, Cho et al. 2008
Palmitic acid		Weber et al. 2022, Sande et al. 2019, Corrêa et al. 2018, Cho et al. 2008
Oleic acid		Weber et al. 2022, Li et al. 2020, Corrêa et al. 2018
Stearic acid		Castañeda-Ramírez et al. 2020, Corrêa et al. 2018
Catechin		Soares et al. 2020, Zhang et al. 2012
Syringic acid		Bertollo et al. 2022, Gobi et al. 2019, Zhang et al. 2012
Gallic acid		Bertollo et al. 2022, Balik et al. 2020, Wei et al. 2020, Gobi et al. 2019, Zhang et al. 2012

Table 1 contd. ...

...Table 1 contd.

Metabolites	Chemical structure	References
Protocatechuic acid		Bertollo et al. 2022, Balik et al. 2020, Zhang et al. 2012
Quercetin		Bertollo et al. 2022, Zhang et al. 2012
Myricetin		Bertollo et al. 2022, Roychoudhury 2020, Zhang et al. 2012
Pyrogallol		Roychoudhury 2020, Gobi et al. 2019, Gąsecka et al. 2018
Brefeldin A		Dong et al. 2013

cells) (Gozaga et al. 2000, Albeituni and Yan 2013, Roudi et al. 2017). Proteoglycans are also believed to contribute in AB's effect on the immune system (Oliveira Lima et al. 2011). Generally, the unique properties of AB in comparison with other mushrooms are due to the high amount of β-glucans content in AB despite its anti-tumour properties (Ohno et al. 2001). For example, champignon (*Agaricus bisporus*) has much less β-glucans content but a high amount of mannan sugar content. Instead of that, AB includes β-glucans, ergosterol (pro-vitamin D2) derivatives, glucomannan, mannogalactoglucan, proteoglucans and riboglucans (Ohno et al. 2001, Takaku et al. 2001). Scientifically, it has been confirmed that particular arrangements of β-glucans can play a significant role in human health (Weitberg 2008).

In addition, a neutral α-glucan (denoted as ABM40-1) with a carbohydrate content of 96% was purified from AB extract using ethanol precipitation and gravity column chromatography (Zhang et al. 2018). In comparison with the β-glucan content in the fruiting bodies of AB, α-glucan showed a lower percentage than β-glucan which is almost the same with other mushroom species with the range from

0.41–5.61% of dry weight. According to previous studies, α-glucan is typically low in normally cultivated mushrooms, which is less than 10% (McCleary and Draga 2016). However, little is known about the bioactivity of α-glucan in mushroom cell walls, while their effects are studied extensively by researchers (Wunjuntuk et al. 2022).

3. Therapeutic Values of *A. blazei*

3.1 Anticancer and Angiogenic Activity

A number of insightful reviews on the therapeutic activity of mushrooms have been done especially to evaluate their potential as an anticancer agent. The implementation of several strategies to prevent and fight cancer are including of the exploration of natural products (National Cancer Institute 2020). Several modern approaches to cancer prevention and treatment by using mushroom extracts were conducted including DNA vaccine therapy with mushroom immunomodulatory adjuvants, pro-drugs development by using mushroom lectin to recognize glycoconjugates on the cancer cell surface and development of nano-vectors (Ivanova et al. 2014, Santos 2018, Kothari et al. 2018). A commonly cultivated mushroom, AB species has also proven anti-tumor and anti-metastatic activities for a few types of cancers and the use of natural product compounds may improve the health condition and increase the immunity of patients (Ahn et al. 2004, Wang et al. 2013, Lin et al. 2019). The β-glucans are suggested to be the main contributor to these properties, with the main biological target on modulation of the immune system (Bertollo et al. 2022).

The inhibitory mechanism of AB on the growth of prostate cancer *in vitro* and *in vivo* has been studied by (Yu et al. 2009). In this study, the broth fraction of AB was found to inhibit cell proliferation in different types of prostate cancer cell lines. The AB broth induced lactate dehydrogenase leakage in three cancer cell lines, although the caspase 3 and DNA fragmentation activities were enriched the most in androgen-independent PC3 cells. Oral consumption of AB broth (with higher content of β-glucan) suppressed the tumour growth significantly without inducing adverse effects in severe combined immune-deficient mice with PC3 tumor xenograft. The xenograft from AB mice displayed a decrease in proliferating cell nuclear antigen-positive cells and reduced the tumor micro-vessel density and these results proved that AB directly inhibit the growth of prostate cancer cell via an apoptotic pathway and suppress the prostate tumor growth through anti-proliferative and anti-angiogenic mechanisms (Yu et al. 2009, Sovrani et al. 2017). An extensive study by another researcher was also carried out for melanoma and sarcoma using a mouse model. The results showed that low molecular weight (LMW) polysaccharides in AB are involved in reducing lung metastasis and the mechanisms for reduced tumor growth were known as the anti-angiogenetic effect (Niu et al. 2009a, Niu et al. 2009b).

Conventionally, AB has been consumed for cancer prevention and treatment because of its anti-tumorigenic effects (Lee et al. 2003). The substantial effects on the immune systems and anticancer activity of AB have been proved through a series of laboratory examinations by scientists around the world (Kim et al. 2011,

Matsushita et al. 2018, Hetland et al. 2020, Sun et al. 2020, Misgiati et al. 2021, Yasuma et al. 2021). Matsushita and co-workers have studied the bioactivity of hot water extract of AB by employing human pancreatic cancer cell lines, MIAPaCa-2, PCI-35 and PK-8, as well as the immortalized human pancreatic duct-epithelial cell line, namely HPDE. Based on the observation, the extract successfully inhibited the cell proliferation of cultured pancreatic cancer cell lines through the induction of G0/G1 cell cycle arrest and caspase-dependent apoptosis, however, the effect was lesser on HDPE cell lines (Matsushita et al. 2018). The research on the ability of AB in inhibiting human leukaemia in vitro still remains unclear. Thus, a comprehensive study on the anti-leukemic effect of AB was conducted in 2018 to evaluate its potential. The acidic RNA protein complex, namely FA-2-b-β was extracted from wild edible AB and tested against primary CML bone marrow, CML K562 cells at different concentration and time. The results have demonstrated that FA-2-b-β possessed high anti-proliferative potency and strong pro-apoptotic effects (Sun et al. 2020).

In 2009, a group of researchers revealed the anti-tumor effects of hydro-alcoholic extract of AB against human acute promyelocytic leukemic cell line NB-4 using an MTT assay. Amongst all, ethanol-water extract (70% v/v), 80°C (JAB80E70) possessed the highest suppression of NB-4 growth with a value of 82.6% and the lowest IC_{50} value of 82.2 µg/ml (Kim et al. 2009). Another research was also conducted to evaluate the preventive effect of *Agaricus brasiliensis* KA21 (AGA) in hepatic oxidative stress. the result showed that AGA is capable of preventing non-alcoholic steatohepatitis (NASH) development in the liver (Nakamura et al. 2019). A recent study has been carried out to evaluate the anticancer activity of a compound isolated from AB n-hexane extract against MCF-7 cell lines. From the findings, the isolated compound, namely ergosterol exhibited strong cytotoxicity activity against MCF-7 cell lines with IC_{50} value of 43.10 µg/ml (Misgiati et al. 2021).

The orally controlled β-glucans were treated via macrophages through the receptor surface like Dectin-1 with or without TLR-2/6 and complement receptor (CR)-3, provoking a response by the immune system. The most important activity is the phagocytosis of antibody-tagged tumor cells. NK cell activation is another treatment using AB in order to treat the patients. A group of research studies have reported the use of active metabolites such as proteoglycans, β-glucans and ergosterol to be in charge of the induction of tumor regression in mice (Ohno et al. 2001, Kim et al. 2005, Shimizu et al. 2016). Moreover, daily β-glucans supplemented mice, which were isolated from AB, showed a decline in the spontaneous metastasis level (Kobayashi et al. 2005). Boosting the immune system through the activation of white blood cells involving "immune directors" improve the activity during the infection (Hsu et al. 2008, Martins et al. 2008). Therefore, natural killer cells including macrophages have been improved efficiently to destroy the tumor cells or microbes after body system enhancement through innate immunity and adaptive immunity via dendritic cell activation, and accordingly specialized lymphocyte assignation (Førland et al. 2010). In 2011, Fernandes and co-workers have revealed the submission of AB to vacuum, spray and freeze-drying for evaluation of *in vitro* to maintain the antitumor effect. The AB drying extract showed that total sugars and protein were reduced to 33% of the inhibition rate over Whrlic tumor cell *in vitro* (Fernandes et al. 2011).

3.2 Antimicrobial Activity

Infectious diseases are often associated with antimicrobial resistance and often represent a crucial public health problem. Currently, antibiotic development plays an important role in the treatment and prevention of infectious diseases triggered by several pathogens (Lima et al. 2016). However, the increase in antibiotic resistance has resulted in an increase in morbidity, mortality, prolonged hospital stays and increased hospital service costs (Thabit et al. 2015, Dadgostar 2019, Morrison and Zembower 2020). Data showed that in most countries, the antimicrobial resistance problem has risen and is not only found in hospital-acquired but also involved community-acquired factors (Fistarol et al. 2015, Abella et al. 2015). The development of new antimicrobial agents has become crucial and prioritized due to the outspread of multi-drug resistance microorganisms. This antimicrobial resistance property is facilitated by various factors, closely related to both humans and bacteria. For example, antibiotic resistance is associated with bacteria progression and cannot be interrupted by humans, where the bacteria mutation becomes one of the factors (Johnson et al. 2009, Lima et al. 2016, Arenz and Wilson 2016).

As reported by many researchers, the antimicrobial properties of AB are correlated with the presence of different substances including, glucans (Carneiro et al. 2013, Uyanoglu et al. 2014, Lima et al. 2016). Some studies conducted on animals and in-vitro have indicated that pro-inflammatory mediators can help in inducing phagocytosis against different pathogenic bacteria such as *Mycobacterium tuberculosis* and *Staphylococcus aureus* (Kisich et al. 2002). Current literature findings also suggested that the basidiomycete AB comprises several functional properties, either acting individually or synergistically it is crucial to identify which active compounds play a role in the antimicrobial activity. A recent study has reported the findings of five compounds isolated from the mycelia of AB with their inhibition against plant pathogens such as *Peptobacterium carotovorum, Burkholderia glumae* and *Clavibacter michiganensis* (Wang et al. 2022).

In 2014, a direct antimicrobial effect of AB hot water extract has been evaluated against various bacteria. Both MIC and MBC results of these extracts turned out equal to or better for the inactivation of *Pseudomonas aeruginosa* than the standard drugs, namely ampicillin and streptomycin (Soković et al. 2014). A significant reduction in virulence factors (pyocyanin production and mortality) and biofilm formation of *P. aeruginosa* was also observed. It was also found that the extract possessed anti-quorum sensing activity against the tested microbe, where the inhibition offered a new strategy for the treatment of bacterial infections. Meanwhile, the ethanolic extract of AB displayed greater antimicrobial activity against the *Listeria monocytogenes* growth profile compared to other mushroom species (Soković et al. 2014, Mazzutti et al. 2012). Another continuous study in 2013 has also compared the antimicrobial effect of β-glucans, pectin and commercial AB mushroom extract namely AndoSan™ and proposed that these polysaccharides are having anti-infective properties in various models of mice against microorganisms including bacteria (Tangen et al. 2017). The antimicrobial mode of action presumed for AndoSan™ is that the increment of serum levels of pro-inflammatory cytokines MIP-2, which is equal to interleukin-8 (IL-8) in humans, and tumor necrosis factor (TNF-α) in mice that consumed AB extract (Bernardshaw et al. 2005, Tangen et al. 2017). Some other researchers have proposed that the antimicrobial action of AB was promoted by β-glucans which stimulate the synthesis and secretion of cytokines by macrophages. A significant number of β-glucans in AB extract has become an added value to stimulate the innate immune system and exert an antimicrobial effect (Chan et al. 2009, Al-Faqeeh et al. 2020).

In addition, other studies have reported that there was no antimicrobial effect of both methanolic and water extracts of AB using *in vitro* method. The reason behind these negative results is probably due to the fact that the antimicrobial properties of AB bioactive require engagement with the immune system, especially the innate immune system which cannot be displayed in vitro testing (Bernardshaw et al. 2005, Kim et al. 2022). In other studies, a group of scientists conducted a review to explain the antimicrobial properties of extracts acquired from several other mushroom species and highlight the active compounds isolated, including both high and low molecular weight. The research revealed that the low molecular weight secondary metabolites possessed antimicrobial activity such as terpenes, sesquiterpenes, benzoic acid, anthraquinones and quinolone as well as a primary metabolite, namely oxalic acid (Alves et al. 2012, Lima et al. 2016). Another metabolite in AB extract is linoleic acid, which is formerly known to exhibit the bactericidal activity of this mushroom species (Mazzutti et al. 2012). This essential polyunsaturated fatty acid is mostly found in plant oils due to the inability of the human body to synthesize it (Sanhueza et al. 2002).

3.3 Anti-hepatitis Activity

The consumption of AB is also beneficial in the treatment of various diseases including hepatitis B and C (Grinde et al. 2006, Hsu et al. 2008). Based on epidemiological data and statistical reports, excluding drug and alcohol overdose, the main cause of liver damage is the infection caused by the hepatitis virus, particularly B (HBV) and C (HCV) forms. The liver is a blood-rich tissue having great numbers of NK and phagocytic cells, which can be triggered by β-glucans, over and above the macrophages Kuppfer cells-relatives, in addition to endothelial cells supporting the sinusoids that can be moved by AB (Mowat 2003).

3.4 Gastroprotective Activity

Food nutrients normally will absorb into the bloodstream in the digestive system of the human body. Inflammation in an intestinal system may cause a disorder, which is known as Inflammatory Bowel Disease (IBD). It is reported that this disorder is due to a problem in the autoimmune system as the aetiology of these types of inflammatory diseases is still unknown. In Norway, some clinical trials were carried out on patients who suffered a lot from IBD. The results showed that Ulcerative Colitis (UC) and Crohn's disease (CD) related patients suffer from pro-inflammatory cytokines with a down-regulated level like serum TNF as a signal for the local consequence in the colon wall itself. This is a potential effectiveness sign of AB to hinder pro-inflammatory cytokines formation as well as decrease calprotectin (a marker for IBD) in patients with CD and UC (Takaku et al. 2001).

3.5 Anti-diabetic Activity

Over 6.8% or 4 million global death in 2010 is contributed to diabetic disease (Zafar and Malik 2014). Glucose level reduction in the blood system of diabetic rat models by AB has shown positive effects (Oh et al. 2010). Furthermore, clinical proof of the combination of AB with an anti-diabetic drug reported the enhancement of patients with insulin resistance in type II diabetes (Hsu et al. 2007). The authors have claimed that the growth in the pretended concentration of adiponectin might be the mechanism behind the AB effect. This is also supported by another group due to the oxidative stress of AB suppression and the production of the pro-inflammatory cytokine, which later resulted in pancreatic beta-cells mass enhancement (Niwa et al. 2011, Chen et al. 2019).

4. Submerged Fermentation Technique for *A. blazei* Cultivation

The expensive price of AB fruiting bodies has led to the utilization of relatively cheap and sustainable sources of AB mycelium on the market shelves (Lin and Yang 2006). In terms of productivity improvement and cost reduction strategy, extensive research had studied the utilization of the submerged fermentation process of AB in lab and pilot scales. In view of the importance of the medicinal properties of this mushroom, an attempt was made to study the exopolysaccharide production in submerged culture.

Submerged fermentation has been defined as the fermentation or growth of mushrooms in the presence of excess water, complemented with nutrients and oxygen supply for aerobic conditions. In submerged fermentation, the mushroom or fungal will grow as a pellet or filamentous propagation form. Compared with solid-state fermentation, submerge fermentation offers more biomass yield in limited space, less time and less probability of contamination. The production of exopolysaccharides was also high in submerged fermentation systems (Lin and Yang 2006). Liquid culture using mushrooms provides greater advantages that encourage more production of mycelia in a closed space and requires short incubation time with less probability of contamination (Choi et al. 2007). The product yield is highly dependent on the fungal morphology during the fermentation stage where the morphology of mycelia normally depends on the fermentation conditions. In this biotechnology approach, the manipulation of chemical and mechanical variables in culture conditions is helpful in controlling the morphology of the organisms tested.

Its widely known that polysaccharides are in high content in the fruiting bodies of macrofungi. Due to its long history around the world, mushroom cultivation became limited in recent biotechnological applications due to its capability to accumulate an extensive number of heavy metals like lead, cadmium, nickel, copper, arsenic, chromium and mercury content in polluted soil. Therefore, using a submerged culture system in the production of bioactive constituents in a controlled environment has become crucial.

4.1 Media Optimization

4.1.1 Effect of Carbon, Nitrogen, Different Carbon to Nitrogen Ratios and Phosphate Sources on the Growth and Production of Exopolysaccharide

One of the most important factors which help in becoming an energy source for the growth of cell and carbohydrate, protein, lipid and nucleic acid synthesis in fungi is the carbon source (Choi et al. 2007). In many studies, carbon source in the medium was added in form of different sugars such as glucose, sucrose, fructose, maltose, or complex carbohydrates such as starch, rice bran extract and corn flour extract. Other researchers also manipulated the combination of two types of carbon sources such as glucose with corn flour extract, glucose with rice bran extract, sucrose with corn flour extract and sucrose with rice bran extract (Lin and Yang 2006). The preparation of corn flour and rice brand extract by using a mass of 100 g of dried and milled corn flour or rice bran powder was collected in a solution containing 10 L of distilled water and 50 mL of H_2SO_4 and then autoclaved for 20 min. The separation process of liquid fraction was conducted where the pH was adjusted to 6.5 using sodium hydroxide and undergoes a solvent removal process to yield the extract. From previous reports on AB cultivation, glucose was regarded as the best carbon source.

It was known that both carbon and nitrogen sources are the main factors in determining mycelial growth and production of exopolysaccharides in different types of mushrooms. To achieve the optimum content of both sources, a few nitrogen sources including malt extract, tryptophan, peptone, polypeptone, yeast extract and inorganic nitrogen sources such as $Ca(NO_3)_2$, $NaNO_3$, $(NH_4)_2SO_4$, NH_4NO_3, $NH_4H_2PO_4$, $(NH_4)_2HPO_4$, and KNO_3 were utilized. Amongst all, the yeast and malt extracts gave the best results on the growth of mycelial and exopolysaccharides production (Choi et al. 2007). Inorganic nitrogen sources were found to give lower mycelial biomass as well as EPS production. Yeast extract was discovered to be responsible for the highest EPS production (Hamedi et al. 2007). This has revealed that complex nitrogen sources are better than inorganic ones in the development of fungal growth and production of mycelial. Likewise, another literature has reported that organic nitrogen sources are better for EPS and biomass production in different concentrations of yeast extract. The maximal production of EPS was observed when using 4 g/L of yeast extract where a high concentration of biomass resulted from a lower concentration of yeast extract. The results were in agreement with the increasing concentration of yeast extract led to the decrease of EPS concentration as reported by other authors (Hamedi et al. 2007).

In addition, another important factor in media optimization was the effect of carbon to nitrogen ratio in the development of AB growth as well as the secondary metabolites compositions. Therefore, it's important to determine the best ratio of carbon to nitrogen for optimum growth and exopolysaccharide production. The exopolysaccharides yield did not change significantly when the carbon-to-nitrogen ratio was between 5 to 10. In the meantime, a negative trend was observed when the carbon-to-nitrogen ratio was increased above 10 and this result has shown that the optimum condition for carbon to nitrogen ratio should be more than 10 to achieve high polysaccharides production.

In the submerged culture fermentation of *A. brasiliensis*, medium supplementation with KH_2PO_4 significantly affected the growth and polysaccharides production. The optimum conditions were found at 24 days of culture, where 5.41 g/L was obtained with an initial KH_2PO_4 concentration of 2.0 g/L. The value of biomass concentrations was about 10% higher than the culture with the absence of KH_2PO_4. The maximal production of polysaccharides of about 3 g/L was achieved after 24 days with an initial KH_2PO_4 concentration of 3.0 g/L. This exopolysaccharide concentration was over 20% higher than that obtained in the culture without the added KH_2PO_4. Instead of that, former research has revealed that the addition of inorganic phosphate has contributed to the enhancement of exopolysaccharide yield in submerged culture of other macrofungi as well (Kim et al. 2003, Hsieh et al. 2006). However, less attention has been paid to the effect of this supplementation on biological activity of the microorganism. In terms of the stimulation of TNF-α release by macrophages, of the exopolysaccharide samples produced with different initial KH_2PO_4 concentrations. The highest biological activity (1440 pg of TNF-α/mL/5 × 104 cells) was obtained with the exopolysaccharide produced in the culture supplemented with 3.0 g/L of KH_2PO_4. The content of β-(1-3)-glucan in the ABEP obtained from different KH_2PO_4 levels was also closely correlated with its biological activity (R_2 = 0.96). On the other hand, there was little correlation (R_2 = 0.42) between the biological activity of the exopolysaccharide samples obtained in the KH_2PO_4 supplementation experiment.

5. Batch Cultivation

5.1 Effect of pH, Temperature, Aeration and Agitation on the Growth and Production of Exopolysaccharide

Other than that, another variable set in batch cultivation was the temperature and initial pH of the biomass and EPS production. Based on the results, the optimum condition for pH measurement was achieved at pH7, and the optimum temperature was 25°C for the growth and production of mycelial biomass. Other than that, the temperature of 20°C was found to be optimized for EPS production (Hamedi et al. 2007).

The Dissolved Oxygen (DO) levels at 50 and 150 rpm were found to be significant in reducing 100% saturation to 10% saturation at the beginning to the end of the fermentation process. In comparison to that, the DO level at 300 rpm was preserved at level 35% saturation on day 9 and slowly increase to around 50% at the end of fermentation (Kim et al. 2006). It is important to realize that the DO level at 50 and 150 rpm did not result in a lower mycelia growth and shorter log-growth phase. On contrary, lower agitation speed has resulted in higher growth of fungi due to the prevention of mycelium fragmentation mediated by a severe shearing effect. The crucial effect of the aeration rate on the mycelia biomass and exopolysaccharide production is at a high level of mycelia biomass was accumulated at the lowest aeration rate (0.5 vvm), whereas the maximum exopolysaccharide production was obtained at 1.0 vvm (Kim et al. 2005).

The mycelia growth and biomass production can be increased by increasing the agitation speed. However, higher sugar concentration usually promotes mushroom polysaccharides production. The optimum biomass production was dependent on the degree of homogenization process and heat transfer with the presence of oxygen (Mantzouridou et al. 2002). Low agitation at 50 rpm was optimum to help in increasing the biomass and polysaccharides production by *Phellinus linteus* (Hwang et al. 2003), while agitation activity at 100rpm was found to be optimal for culture conditions of *Antrodia cinnamomea* which help in increasing the mycelium growth. Generally, a high agitation speed is highly significant in the mycelium growth of submerging culture process, where it is affected by increasing the oxygen uptake rate in nutrients. However, agitation is one of the main factors governing mycelial morphology, bio-pellet structure, and metabolite production (Kim et al. 2006, El Enshasy 2022).

6. Conclusion

The previous extensive researches and scientific literature reviews have strongly contributed the evidence for therapeutic uses of medicinal fungi, *Agaricus blazei* species. This popular Sun Mushroom have numerous bioactive secondary metabolites that actively participate in the tumoricidal, anti-carcinogenic, anti-hepatitis, anti-diabetic, gastroprotective, anti-mutagenic and antimicrobial properties of this mushroom. However, there are many uncertainties, limitations and details studies needed in an attempt to further evaluate its biological properties in the human organism in health and diseases. More clinical trials and using reliable statistical methods and standardized preparations up to the pilot scale stage are desirable to establish the efficacy of AB as a future medicinal natural product drug.

References

Abella, J., A. Fahy, R. Duran and C. Cagnon. 2015. Integron diversity in bacterial communities of freshwater sediments at different contamination levels. *FEMS Microbiol. Ecol.*, 91(12): fiv140.

Ahn, W.S., D.J. Kim, G.T. Chae, J.M. Lee, S.M. Bae, J.I. Sin, Y.W. Kim, S.E. Namkoong and I.P. Lee. 2004. Natural killer cell activity and quality of life were improved by consumption of a mushroom extract, *Agaricus blazei* Murill Kyowa, in gynecological cancer patients undergoing chemotherapy. *Int. J. Gynecol. Cancer*, 14(4): 589–594.

Albeituni, S.H. and J. Yan. 2013. The effects of β-glucans on dendritic cells and implications for cancer therapy. *Anticancer Agents Med. Chem.*, 13(5): 689–698.

Al-Faqeeh, L.A.S., R. Naser, S.R. Kagne and S.W. Khan. 2020. Review on anticancer and antimicrobial activities of mushrooms. *World J. Pharm. Pharm. Sci.*, 10(2): 1922–1936.

Arenz, S. and D.N. Wilson. 2016. Blast from the past: Reassessing forgotten translation inhibitors, antibiotic selectivity, and resistance mechanisms to aid drug development. *Mol. Cell*, 61(1): 3–14.

Balik, M., K.J. Sułkowska-Ziaja, M. Ziaja and B. Muszyńska. 2020. Phenolic acids-occurrence and significance in the world of higher fungi. *Med. Int. Rev.*, 29(115): 72–81.

Bernardshaw, S., E. Johnson and G. Hetland. 2005. An extract of the mushroom *Agaricus blazei* Murill administered orally protects against systemic *Streptococcus pneumoniae* infection in mice. *Scand. J. Immunol.*, 62(4): 393–398.

Bertollo, A.G., M.E.D. Mingoti, M.E. Plissari, G. Betti, W.A.R. Junior, A.R. Luzardo and Z.M. Ignácio. 2022. *Agaricus blazei* Murrill Mushroom: A review on the prevention and treatment of cancer. *Pharmacol. Res. Mod. Chinese Med.*, 2: 100032.

Callac, P. and J. Chen. 2018. Tropical species of *Agaricus*. Updates on tropical mushrooms. Basic and Applied Research. El Colegio de la Frontera Sur, 227 p., 2018, 978-607-8429-60-8. (hal-02785368).

Carneiro, A.A., I.C. Ferreira, M. Dueñas, L. Barros, R. Da Silva, E. Gomes and C. Santos-Buelga. 2013. Chemical composition and antioxidant activity of dried powder formulations of *Agaricus blazei* and *Lentinus edodes*. *Food Chem.*, 138(4): 2168–2173.

Castañeda-Ramírez, G.S., J.F.D.J. Torres-Acosta, J.E. Sánchez, P. Mendoza-de-Gives, M. González-Cortázar, A. Zamilpa, L.K.T. Al-Ani, C. Sandoval-Castro, F.E. de Freitas Soares and L. Aguilar-Marcelino. 2020. The possible biotechnological use of edible mushroom bioproducts for controlling plant and animal parasitic nematodes. *BioMed Res. Int.*, 2020. Article ID 6078917.

Chan, G.C.F., W.K. Chan and D.M.Y. Sze. 2009. The effects of β-glucan on human immune and cancer cells. *J. Hematol. Oncol.*, 2(1): 1–11.

Chang, S.T. and S.P. Wasser. 2018. Current and future research trends in agricultural and biomedical applications of medicinal mushrooms and mushroom products. *Int. J. Med. Mushrooms*, 20(12): 1121–1133.

Chen, L., C. Gnanaraj, P. Arulselvan, H. El-Seedi and H. Teng. 2019. A review on advanced microencapsulation technology to enhance bioavailability of phenolic compounds: Based on its activity in the treatment of Type 2 Diabetes. *Trends Food Sci. Technol.*, 85: 149–162.

Cho, S.M., K.Y. Jang, H.J. Park and J.S. Park. 2008. Analysis of the chemical constituents of *Agaricus brasiliensis*. *Mycobiology*, 36(1): 50–54.

Choi, D.B., J.M. Maeng, J.L. Ding and W.S. Cha. 2007. Exopolysaccharide production and mycelial growth in an air-lift bioreactor using *Fomitopsis pinicola*. *J. Microbiol. Biotechnol.*, 17(8): 1369–1378.

Corrêa, R.C., L. Barros, Â. Fernandes, M. Sokovic, A. Bracht, R.M. Peralta and I.C. Ferreira. 2018. A natural food ingredient based on ergosterol: Optimization of the extraction from *Agaricus blazei*, evaluation of bioactive properties and incorporation in yogurts. *Food Funct.*, 9(3): 1465–1474.

Costea, T., O.C. Vlad, L.C. Miclea, C. Ganea, J. Szöllősi and M.M. Mocanu. 2020. Alleviation of multidrug resistance by flavonoid and non-flavonoid compounds in breast, lung, colorectal and prostate cancer. *Int. J. Mol. Sci.*, 21(2): 401.

Dadgostar, P. 2019. Antimicrobial resistance: Implications and costs. *Infect. Drug Resist.*, 12: 3903.

da Eira, A.F., M.Y. Didukh, P.E. Stamets, S.P. Wasser and M.A.L. de Amazonas. 2002. Is a widely cultivated culinary-medicinal royal sun *Agaricus* (the himematsutake mushroom) indeed *Agaricus blazei* Murrill?. *Int. J. Med. Mushrooms*, 4(4): 267–290.

Dong, S., Y. Furutani, S. Kimura, Y. Zhu, K. Kawabata, M. Furutani, T. Nishikawa, T. Tanaka, T. Masaki, R. Matsuoka and R. Kiyama. 2013. Brefeldin A is an estrogenic, Erk1/2-activating component in the extract of *Agaricus blazei* mycelia. *J. Agric. Food. Chem.*, 61(1): 128–136.

Eguchi, N., K. Fujino, K. Thanasut, M. Taharaguchi, M. Motoi, A. Motoi, K. Oonaka and S. Taharaguchi. 2017. *In vitro* anti-influenza virus activity of *Agaricus brasiliensis* KA21. *Biocontrol Sci.*, 22(3): 171–174.

El Enshasy, H. 2022. Fungal morphology: A challenge in bioprocess engineering industries for product development. *Curr. Opin. Chem. Eng.*, 35: 100729.

Fernandes, M.B.A., S. Habu, M.A. De Lima, V. Thomaz-Soccol and C.R. Soccol. 2011. Influence of drying methods over *in vitro* antitumoral effects of exopolysaccharides produced by *Agaricus blazei* LPB 03 on submerged fermentation. *Bioproc. Biosyst. Eng.*, 34(3): 253–261.

Firenzuoli, F., L. Gori and G. Lombardo. 2008. The medicinal mushroom *Agaricus blazei* Murrill: Review of literature and pharmaco-toxicological problems. *Evid. Based Complement. Altern. Med.*, 5(1): 3–15.

Fistarol, G.O., F.H. Coutinho, A.P.B. Moreira, T. Venas, A. Cánovas, S.E. de Paula Jr, R. Coutinho, R.L. de Moura, J.L. Valentin, D.R. Tenenbaum, R. Paranhos, R. do Valle, A.B. de, A.C.P. Vicente, G.M.A. Filho, R.C. Pereira, R. Kruger, C.E. Rezende, C.C. Thompson, P.S. Salomon and F.L. Thompson. 2015. Environmental and sanitary conditions of Guanabara Bay, Rio de Janeiro. *Front. Microbiol.*, 6: 1232.

Førland, D.T., E. Johnson, A.M.A. Tryggestad, T. Lyberg and G. Hetland. 2010. An extract based on the medicinal mushroom *Agaricus blazei* Murill stimulates monocyte-derived dendritic cells to cytokine and chemokine production *in vitro*. *Cytokine*, 49(3): 245–250.

Gąsecka, M., Z. Magdziak, M. Siwulski and M. Mleczek. 2018. Profile of phenolic and organic acids, antioxidant properties and ergosterol content in cultivated and wild growing species of *Agaricus*. *Eur. Food Res. Technol.*, 244(2): 259–268.

Gobi, V.V., S.R. Sankar, W.M.S. Johnson, K. Prabu and M. Ramkumar. 2019. Antiapoptotic role of *Agaricus blazei* extract in rodent model of Parkinson's disease. *Front. Biosci.* (Elite Ed), 11(1): 12–19.

Grinde, B., G. Hetland and E. Johnson. 2006. Effects on gene expression and viral load of a medicinal extract from *Agaricus blazei* in patients with chronic hepatitis C infection. *Int. Immunopharmacol.*, 6(8): 1311–1314.

Hamedi, A., H. Vahid and F. Ghanati. 2007. Optimization of the medium composition for production of mycelial biomass and exo-polysaccharide by *Agaricus blazei* Murill DPPh 131 using response surface methodology. *Biotechnology*, 6: 456–464.

Heinemann, P. 1978. Essai d'une clé de détermination des genres *Agaricus* et Micropsalliota. *Sydowia*, 30: 6–37.

Hetland, G., J.M. Tangen, F. Mahmood, M.R. Mirlashari, L.S.H. Nissen-Meyer, I. Nentwich, S.P. Therkelsen, G.E. Tjonnfjord and E. Johnson. 2020. Antitumor, anti-inflammatory and antiallergic effects of *Agaricus blazei* mushroom extract and the related medicinal basidiomycetes mushrooms, *Hericium erinaceus* and *Grifola frondosa*: A review of preclinical and clinical studies. *Nutrients*, 12(5): 1339.

Hsieh, C., M.H. Tseng and C.J. Liu. 2006. Production of polysaccharides from *Ganoderma lucidum* (CCRC 36041) under limitations of nutrients. *Enz. Microb. Technol.*, 38(1-2): 109–117.

Hsu, C.H., K.C. Hwang, Y.H. Chiang and P. Chou. 2008. The mushroom *Agaricus blazei* Murill extract normalizes liver function in patients with chronic hepatitis B. *J. Altern. Complement. Med.*, 14(3): 299–301.

Hwang, H.J., S.W. Kim, C.P. Xu, J.W. Choi and J.W. Yun. 2003. Production and molecular characteristics of four groups of exopolysaccharides from submerged culture of *Phellinus gilvus*. *J. Appl. Microbiol.*, 94(4): 708–719.

Ivanova, T.S., T.A. Krupodorova, V.Y. Barshteyn, A.B. Artamonova and V.A. Shlyakhovenko. 2014. Anticancer substances of mushroom origin. *Exp. Oncol.*, 36: 58–66.

Jana, P. and K. Acharya. 2022. Mushroom: A new resource for anti-angiogenic therapeutics. *Food Rev. Int.*, 38(1): 88–109.

Johnson, E., D.T. Førland, L. Saetre, S.V. Bernardshaw, T. Lyberg and G. Hetland. 2009. Effect of an extract based on the medicinal mushroom *Agaricus blazei* Murill on release of cytokines, chemokines and leukocyte growth factors in human blood *ex vivo* and *in vivo*. *Scand. J. Immunol.*, 69(3): 242–250.

Kawagishi, H., R. Inagaki, T. Kanao, T. Mizuno, K. Shimura, H. Ito, T. Hagiwara and T. Nakamura. 1989. Fractionation and antitumor activity of the water-in-soluble residue of fruiting bodies. *Carbohydr. Res.*, 186(2): 267–273.

Kim, C.F., J.J. Jiang, K.N. Leung, K.P. Fung and C. Bik-San Lau. 2009. Inhibitory effects of *Agaricus blazei* extracts on human myeloid leukemia cells. *J. Ethnopharmacol.*, 122(2): 320–326.

Kim, G.Y., M.Y. Lee, H.J. Lee, D.O. Moon, C.M. Lee, C.Y. Jin, Y.H. Choi, Y.K. Jeong, K.T. Chung, J.Y. Lee, I.H. Choi and Y.M. Park. 2005. Effect of water-soluble proteoglycan isolated from *Agaricus blazei* on the maturation of murine bone marrow-derived dendritic cells. *Int. Immunopharmacol.*, 5(10): 1523–1532.

Kim, J.E., S.Y. Lee, Y. Chang and M.H. Jin. 2020. Effect of glucosylceramides and sterols isolated from *agaricus blazei* extract on improvement of skin cell. *J. Soc. Cosmet. Sci. Korea*, 46(2): 105–117.

Kim, J.H., C.C. Tam, K.L. Chan, N. Mahoney, L.W. Cheng, M. Friedman and K.M. Land. 2022. Antimicrobial efficacy of edible mushroom extracts: Assessment of fungal resistance. *Appl. Sci.*, 12(9): 4591.

Kim, M.O., D.O. Moon, J.M. Jung, W.S. Lee, Y.H. Choi and G.Y. Kim. 2011. *Agaricus blazei* extract induces apoptosis through ROS-dependent JNK activation involving the mitochondrial pathway and suppression of constitutive NF-κB in THP-1 cells. *Evid-Based Complement. Altern. Med.*, 2011: 838172.

Kimura, Y., T. Kido, T. Takaku, M. Sumiyoshi and K. Baba. 2004. Isolation of an anti-angiogenic substance from *Agaricus blazei* Murill: Its antitumor and antimetastatic actions. *Cancer Sci.*, 95(9): 758–764.

Kisich, K.O., M. Higgins, G. Diamond and L. Heifets. 2002. Tumor necrosis factor alpha stimulates killing of *Mycobacterium tuberculosis* by human neutrophils. *Infect. Immunol.*, 70(8): 4591–4599.

Kobayashi, H., R. Yoshida, Y. Kanada, Y. Fukuda, T. Yagyu, K. Inagaki, T. Kondo, N. Kurita, M. Suzuki, N. Kanayama and T. Terao. 2005. Suppressing effects of daily oral supplementation of beta-glucan extracted from *Agaricus blazei* Murill on spontaneous and peritoneal disseminated metastasis in mouse model. *J. Cancer Res. Clin. Oncol.*, 131(8): 527–538.

Kothari, D., S. Patel and S.K. Kim. 2018. Anticancer and other therapeutic relevance of mushroom polysaccharides: A holistic appraisal. *Biomed. Pharmacot.*, 105: 377–394.

Kumar, S. and R. Kumari. 2021. Traditional uses, phytochemistry and biological activities of *Agaricus blazei* Murill: A Comprehensive Review. *Am. J. Pharmacol. Pharmacother.*, 8(3): 14.

Lee, Y.L., H.J. Kim, M.S. Lee, J.M. Kim, J.S. Han, E.K. Hong, M.S. Kwin and M.J. Lee. 2003. Oral administration of *Agaricus blazei* (H1 strain) inhibited tumor growth in a sarcoma 180 inoculation model. *Exp. Anim.*, 52(5): 371–375.

Li, Y., Y. Sheng, X. Lu, X. Guo, G. Xu, X. Han, L. An and P. Du. 2020. Isolation and purification of acidic polysaccharides from *Agaricus blazei* Murill and evaluation of their lipid-lowering mechanism. *Int. J. Biol. Macromol.*, 157: 276–287.

Lima, C.U., E.F. Gris and M.G. Karnikowski. 2016. Antimicrobial properties of the mushroom *Agaricus blazei*-integrative review. *Rev. Bras. Farmacogn.*, 26: 780–786.

Lin, M.H., K.M. Lee, C.Y. Hsu, S.Y. Peng, C.N. Lin, C.C. Chen, C.K. Fan and P.C. Cheng. 2019. Immunopathological effects of *Agaricus blazei* Murill polysaccharides against *Schistosoma mansoni* infection by Th1 and NK1 cells differentiation. *Intern. Immunopharmacol.*, 73: 502–514.

Lin, J.H. and S.S. Yang. 2006. Mycelium and polysaccharide production of *Agaricus blazei* Murill by submerged fermentation. *J. Microbiol. Immunol. Infect.*, 39(2): 98–108.

Mantzouridou, F., T. Roukas and P. Kotzekidou. 2002. Effect of the aeration rate and agitation speed on β-carotene production and morphology of *Blakeslea trispora* in a stirred tank reactor: Mathematical modeling. *Biochem. Eng. J.*, 10(2): 123–135.

Martins, P.R., M.C. Gameiro, L. Castoldi, G.G. Romagnoli, F.C. Lopes, A.V.F.D.S. Pinto, W. Loyola and R. Kaneno. 2008. Polysaccharide-rich fraction of *Agaricus brasiliensis* enhances the candidacidal activity of murine macrophages. *Mem. Inst. Oswaldo Cruz*, 103: 244–250.

Matsushita, Y., Y. Furutani, R. Matsuoka and T. Furukawa. 2018. Hot water extract of *Agaricus blazei* Murill specifically inhibits growth and induces apoptosis in human pancreatic cancer cells. *BMC Complement. Altern. Med.*, 18(1): 1–11.

Mazzutti, S., S.R. Ferreira, C.A. Riehl, A. Smania Jr, F.A. Smania and J. Martínez. 2012. Supercritical fluid extraction of *Agaricus brasiliensis*: Antioxidant and antimicrobial activities. *J. Supercrit. Fluids*, 70: 48–56.

McCleary, B.V. and A. Draga. 2016. Measurement of β-Glucan in mushrooms and mycelial products. *J. AOAC Int.*, 99(2): 364–373.

Medel-Ortiz, R., R. Garibay-Orijel, A. Argüelles-Moyao, G. Mata, R.W. Kerrigan, A.E. Bessette, J. Geml, C. Angelini, L.A. Parra. and J. Chen. 2022. *Agaricus macrochlamys*, a New Species from the (Sub) tropical Cloud Forests of North America and the Caribbean, and *Agaricus fiardii*, a New Synonym of *Agaricus subrufescens*. *J. Fungi*, 8(7): 664.

Misgiati, M., A. Widyawaruyanti, S.J. Raharjo and S. Sukardiman. 2021. Ergosterol isolated from *Agaricus blazei* Murill n-hexane extracts as potential anticancer MCF-7 activity. *Pharmacogn. Mag.*, 13(2): 418–426.

Monteiro, H., B. Moura, M. Iten, T.M. Mata and A.A. Martins. 2020. Life cycle energy and carbon emissions of ergosterol from mushroom residues. *Energy Rep.*, 6: 333–339.

Morrison, L. and T.R. Zembower. 2020. Antimicrobial resistance. Gastrointest. *Endosc. Clin.*, 30(4): 619–635.

Mowat, A.M. 2003. Anatomical basis of tolerance and immunity to intestinal antigens. *Nat. Rev. Immunol.*, 3(4): 331–341.

Nakamura, A., Q. Zhu, Y. Yokoyama, N. Kitamura, S. Uchida, K. Kumadaki, K. Tsubota and M. Watanabe. 2019. *Agaricus brasiliensis* KA21 may prevent diet-induced nash through its antioxidant, anti-inflammatory, and anti-fibrotic activities in the liver. *Foods*, 8(11): 546.

National Cancer Institute. 2020. Metastatic Cancer: When Cancer Spreads- National Cancer Institute. *Natl. Cancer Inst.* https://www.cancer.gov/types/metastatic-cancer.

Niu, Y.C., J.C. Liu, X.M. Zhao and X.X. Wu. 2009a. A low molecular weight polysaccharide isolated from *Agaricus blazei* suppresses tumor growth and angiogenesis *in vivo. Oncol. Rep.*, 21(1): 145–152.

Niu, Y.C., J.C. Liu, X.M. Zhao and J. Cao. 2009b. A low molecular weight polysaccharide isolated from *Agaricus blazei* Murill (LMPAB) exhibits its anti-metastatic effect by down-regulating metalloproteinase-9 and up-regulating Nm23-H1. *Am. J. Chinese Med.*, 37(05): 909–921.

Niwa, A., Tajiri, T. and H. Higashino. 2011. *Ipomoea batatas* and *Agarics blazei* ameliorate diabetic disorders with therapeutic antioxidant potential in streptozotocin-induced diabetic rats. *J. Clin. Biochem. Nutr.*, 1101050064-1101050064.

Oh, T.W., Y.A. Kim, W.J. Jang, J.I. Byeon, C.H. Ryu, J.O. Kim and Y.L. Ha. 2010. Semipurified fractions from the submerged-culture broth of *Agaricus blazei* Murill reduce blood glucose levels in streptozotocin-induced diabetic rats. *J. Agric. Food Chem.*, 58(7): 4113–4119.

Ohno, N., M. Furukawa, N.N. Miura, Y. Adachi, M. Motoi and T. Yadomae. 2001. Antitumor β-glucan from the cultured fruit body of *Agaricus blazei. Biol. Pharm. Bull.*, 24(7): 820–828.

Oliveira Lima, C.U.J., C.O. de Almeida Cordova, O. de Tolêdo Nóbrega, S.S. Funghetto and M.G. de Oliveira Karnikowski. 2011. Does the *Agaricus blazei* Murill mushroom have properties that affect the immune system? An integrative review. *J. Med. Food*, 14(1-2): 2–8.

Ooi, V.E. and F. Liu. 2000. Immunomodulation and anticancer activity of polysaccharide-protein complexes. *Curr. Med. Chem.*, 7(7): 715–729.

Roudi, R., S.R. Mohammadi, M. Roudbary and M. Mohsenzadegan. 2017. Lung cancer and β-glucans: Review of potential therapeutic applications. *Invest New Drugs*, 35(4): 509–517.

Roychoudhury, A. 2020. Mushrooms as medicinal and therapeutic agents. Indian J. Pharmaceut. *Biol. Res.*, 8(4): 1–6.

Rózsa, S., D.N. Măniuţiu, G. Poşta, T.M. Gocan, I. Andreica, I. Bogdan, M. Rózsa and V. Lazăr. 2019. Influence of the culture substrate on the *Agaricus blazei* Murrill mushrooms vitamins content. *Plants*, 8(9): 316.

Sande, D., G.P. de Oliveira, M.A.F. e Moura, B. de Almeida Martins, M.T.N.S. Lima and J.A. Takahashi. 2019. Edible mushrooms as a ubiquitous source of essential fatty acids. *Food Res. Int.*, 125: 108524.

Sanhueza, J., S. Nieto and A. Valenzuela. 2002. Ácido linoleico conjugado: un ácido graso con isomería trans potencialmente beneficioso. *Revista chilena de nutrición*, 29(2): 98–105.

Santos, M.O. 2018. Estimativa 2018: Incidência de câncer no Brasil. *Rev. Bras. Cancerol.*, 64(1): 119–120.

Shimizu, T., J. Kawai, K. Ouchi, H. Kikuchi, Y. Osima and R. Hidemi. 2016. Agarol, an ergosterol derivative from *Agaricus blazei*, induces caspase-independent apoptosis in human cancer cells. *Intern. J. Oncol.*, 48(4): 1670–1678.

Smiderle, F.R., A.C. Ruthes, J. van Arkel, W. Chanput, M. Iacomini, H.J. Wichers and Van L.J. Griensven. 2011. Polysaccharides from *Agaricus bisporus* and *Agaricus brasiliensis* show similarities in their structures and their immunomodulatory effects on human monocytic THP-1 cells. *BMC Complement. and Altern. Med.*, 11(1): 1–11.

Soares, A.A., F.A.P. Ramos, P.M. Favetta, I.C. Dorneles, L.K. Otutumi, R. de Melo Germano and R.M. Peralta. 2020. Comparison of total phenolic content and antioxidant activity of different extracts of *Agaricus blazei* Murril. *Braz. J. Dev.*, 6(3): 13561–13573.

Soković, M., A. Ćirić, J. Glamočlija, M. Nikolić and L.J. Van Griensven. 2014. *Agaricus blazei* hot water extract shows anti quorum sensing activity in the nosocomial human pathogen *Pseudomonas aeruginosa*. *Molecules*, 19(4): 4189–4199.

Sovrani, V., J. da Rosa, M. de Paula Drewinski, F.G. Colodi, T.T. Tominaga, H.S. Dalla Santa and R. Rebeca. 2017. *In vitro* and *in vivo* antitumoral activity of exobiopolymers from the royal sun culinary-medicinal mushroom *Agaricus brasiliensis* (agaricomycetes). *Int. J. Med. Mushrooms*, 19(9): 767–775.

Sun, Y., M. Cheng, L. Dong, K. Yang, Z. Ma, S.Yu, P. Yan, K. Bai, X. Zhu and Q. Zhang. 2020. *Agaricus blazei* extract (FA2bβ) induces apoptosis in chronic myeloid leukemia cells. *Oncol. Lett.*, 20(5): 270. doi: 10.3892/ol.2020.12133.

Takaku, T., Y. Kimura and H. Okuda. 2001. Isolation of an antitumor compound from *Agaricus blazei* Murill and its mechanism of action. *J. Nutr.*, 131(5): 1409–1413.

Tangen, J.M., T. Holien, M.R. Mirlashari, K. Misund and G. Hetland. 2017. Cytotoxic effect on human myeloma cells and leukemic cells by the *Agaricus blazei* Murill based mushroom extract, Andosan™. *BioMed. Res. Int.*, 2017: 2059825.

Taofiq, O., F. Rodrigues, L. Barros, R.M. Peraita, M.F. Barreiro, I.C.F.R. Ferreira and M.B.P.P. Oliverira. 2019a. *Agaricus blazei* Murrill from Brazil: An ingredient for nutraceutical and cosmeceutical applications. *Food Func.*, 10: 565–572.

Taofiq, O., R.C. Corrêa, L. Barros, M.A. Prieto, A. Bracht, R.M. Peralta, A.M. González-Paramás, M.F. Barreiro and I.C. Ferreira. 2019b. A comparative study between conventional and non-conventional extraction techniques for the recovery of ergosterol from *Agaricus blazei* Murrill. *Food Res. Int.*, 125: 108541.

Thabit, A.K., J.L. Crandon and D.P. Nicolau. 2015. Antimicrobial resistance: Impact on clinical and economic outcomes and the need for new antimicrobials. *Expert Opin. Pharmacother.*, 16(2): 159–177.

Tungmunnithum, D., A. Thongboonyou, A. Pholboon and A. Yangsabai. 2018. Flavonoids and other phenolic compounds from medicinal plants for pharmaceutical and medical aspects: An overview. *Medicines*, 5(3): 93.

Uyanoglu, M., M. Canbek, L.J. van Griensven, M. Yamac, H. Senturk, K. Kartkaya, A. Oglakcı, O. Turgak, and G. Kanbak. 2014. Effects of polysaccharide from fruiting bodies of *Agaricus bisporus, Agaricus brasiliensis*, and *Phellinus linteus* on alcoholic liver injury. *Int. J. Food Sci. Nutr.*, 65(4): 482–488.

Wang, J., J. Wu, R. Ogura, H. Kobori, J.H. Choi, H. Hirai, Y. Takikawa and H. Kawagishi. 2022. Anti-phytopathogenic-bacterial fatty acids from the mycelia of the edible mushroom *Agaricus blazei*. *Biosci. Biotechnol. Biochem.*, 86(10): 1327–1332.

Wang, H., Z. Fu and C. Han. 2013. The medicinal values of culinary-medicinal royal sun mushroom (*Agaricus blazei* Murrill). *Evid.-based Complement. Altern. Med.*, 2013: 1–6.

Weber, S.S., A.C.S. de Souza, D.C.L. Soares, C.C. Lima, A.C.R. de Moraes, S.V. Gkionis, T. Arenhart, L.G.G. Rodrigues, S.R.S. Ferreira, D.B. Silva and E.B. Parisotto. 2022. Chemical profile, antimicrobial potential, and antiaggregant activity of supercritical fluid extract from *Agaricus bisporus*. *Chem. Pap.*, 76(10): 6205–6214.

Wei, Q., Y. Zhan, B. Chen, B. Xie, T. Fang, S. Ravishankar and Y. Jiang. 2020. Assessment of antioxidant and antidiabetic properties of *Agaricus blazei* Murill extracts. *Food Sci. Nutr.*, 8(1): 332–339.

Weitberg, A.B. 2008. A phase I/II trial of beta-(1,3)/(1,6) D-glucan in the treatment of patients with advanced malignancies receiving chemotherapy. *J. Exp. Clin. Cancer Res.*, 27(1): 1–4.

Wunjuntuk, K., M. Ahmad, T. Techakriengkrai, R. Chunhom, E. Jaraspermsuk, A. Chaisri, R. Kiwwongngam, S. Wuttimongkolkul and S. Charoenkiatkul. 2022. Proximate composition, dietary fibre, beta-glucan content, and inhibition of key enzymes linked to diabetes and obesity in cultivated and wild mushrooms. *J. Food Compost. Anal.*, 105: 104226.

Yasuma, T., M. Toda, H. Kobori, N. Tada, C.N. D'Alessandro-Gabazza and E.C. Gabazza. 2021. Subcritical water extracts from *Agaricus blazei* Murrill's Mycelium inhibit the expression of immune checkpoint molecules and Axl receptor. *J. Fungi*, 7(8): 590.

Yu, C.H., S.F. Kan, C.H. Shu, T.J. Lu, L. Sun-Hwang and P.S. Wang. 2009. Inhibitory mechanisms of *Agaricus blazei* Murill on the growth of prostate cancer *in vitro* and *in vivo*. *J. Nutr. Biochem.*, 20(10): 753–764.

Yuminamochi, E., T. Koike, K. Takeda, I. Horiuchi and K. Okumura. 2007. Interleukin-12-and interferon-γ-mediated natural killer cell activation by *Agaricus blazei* Murill. *Immunology*, 121(2): 197–206.

Zafar, S. and B. Malik. 2014. Trends on the use of E-learning in continuing medical education: A review. *J. Islamic Int. Med. Coll.*, 58–64.

Zhang, A., J. Deng, X. Liu, P. He, L. He, F. Zhang, R.J. Linhardt and P. Sun. 2018. Structure and conformation of α-glucan extracted from *Agaricus blazei* Murill by high-speed shearing homogenization. *Int. J. Biol. Macrom.*, 113: 558–564.

Zhang, Z., G. Lv, H. Pan and L. Fan. 2012. Optimisation of the microwave-assisted extraction process for six phenolic compounds in *Agaricus blazei* Murrill. *Int. J. Food Sci. Tech.*, 47(1): 24–31.

5

Antimicrobial and Antioxidant Properties of Cultivated Macrofungi Extraction and Characterization

Raquel Peña[1] and Jaime Carrasco[2,3,]*

1. Introduction

It is estimated that the Fungi Kingdom consists of 1.5 million species, of which only 70.000 have been described, and 650 have pharmacological properties (Blackwell 2011). Within this broad group, the basidiomycetes stand out, with promising biological activities, due to a wide range of metabolites that act as immunomodulatory, antitumor, antiviral or antimicrobial substances (Varghese et al. 2019). This is due to its enormous number of molecules with biological activities both in the mycelium and in the fruiting body (Suarez Arango and Nieto 2013). Mushrooms, therefore, contain a wide range of bioactive compounds with different activities and properties. It is of the utmost importance to find efficient methods for the extraction and purification of each of their metabolites, in order to develop cost-effective and efficient techniques for the isolation of the bioactive compounds. Eventually, the production of the extracts and purified compounds from extracts, and later industrial scaling-up can be the basics for designing functional foods and as a source of physiologically beneficial and non-invasive medicines. In this

[1] Centro Tecnológico de Investigación del Champiñón de La Rioja (CTICH), 26560 Autol, Spain.
[2] Centro de Investigación Agroforestal de Albaladejito (IRIAF-JCCM), Carretera Toledo-Cuenca km 174, 16194 Cuenca, Spain.
[3] Department of Biology, University of Oxford, Oxford OX1 3RB, United Kingdom.
* Corresponding author: carraco.jaime@gmail.com

regard, non-edible mushrooms are also considered a source of medicinally beneficial compounds. This chapter reviews and discusses the state-of-the-art regarding the antimicrobial and antioxidant properties of macrofungi that can be used to treat medical conditions, used as bio-based products for agriculture or employed as novel ingredients for functional foods.

2. Antimicrobial Properties in Cultivated Mushrooms

Many fungal species have been known since ancient times to have medicinal properties. Among the medicinal macrofungi (mushrooms) historically described are for instance *Fomes fomentarius* which was used by ancient Greeks and Calvatia genus was used by the first North Americans (Stamets and Zwickey 2014). In recent years, multiple researchers have been generating relevant knowledge about their barely described mechanisms driving the noted healthy properties.

Depending on the species, the activity of interest can appear specifically in different parts of the toadstool, the cap, the stipe or specifically in the outer tissue ("skin") of the cap.

Due to the rich environmental niche in which mushroom grow, they rely on the production of multiple secondary metabolites with antimicrobial properties that promote, and favour microbial selection and facilitate growth and development. As a source of novel antibacterial and antifungal compounds, researchers are now examining the antimicrobial activities of an increasing number of mushroom species.

2.1 *Extraction of Mushroom Antimicrobials*

It can be divided into two main groups, conventional methods and new methods.

2.1.1 *Conventional Methods*

In this group water extraction and hydroalcoholic extraction are included.

On the one hand, *water extraction* is a cheap method, but it requires high temperatures (50–80°C) along the treatment, so perhaps thermolabile compounds can be lost.

On the other hand, the *hydroalcoholic extraction* technique requires lower temperatures (25–60°C) but a high concentration of solvents.

Risks associated with these types of extraction include the loss of bioactive material by temperature degradation or either the breakdown into another molecule that reacts with the solvents.

The soxhlet method is also considered another type of conventional extraction. The method involves inserting thimbles (cellulose) with ground-up materials in the extraction chamber directly above the collection flask and beneath a reflux condenser. The heating flask's already-added solvent is then heated to create steam, which condenses under the influence of cold water and returns to the thimbles carrying the sample. Continuous reflux is maintained until the aqueous extract is eventually recovered from the heating flask once more. The advantages of this method consist of using a lower amount of solvents and higher speed due to its continuous approach. However, heat affects the extraction.

2.1.2 New Methods

This extraction set methods include enzyme-assisted extraction, pulsed electric fields and ultrasound-assisted extraction.

Enzyme-assisted extraction initially targets breaking the cell wall. It is performed by the action of hydrolytic enzymes with chitinase or glucanase activity. Since the fungal cell wall is made up of chitin (b-1,4-N-acetylglucosamine) and glucans (b-1,3 and b-1,6) the cited enzymes are required. The purity of the extracts and the speed of the process offer significant advantages (Roselló-Soto et al. 2016).

Pulsed electric fields consist of a treatment system used to generate more clear extracts with higher colloid stability than the conventional methods. It can be used combined with pressure extraction for a great yield of fresh proteins and polysaccharides. The method springs from the transmembrane potential in nature. The cytoplasm always has a negative electric potential opposite to the positive potential of the intracellular space.

The electric pulse that has to be applied depends on the geometrical and electrical properties of the cell. The rate of permeation or cellular disintegration Z is data that serves to know the damage in the tissue. It is calculated using the following equation (σ electrical conductivity at the time, σi electrical conductivity in the intact tissue, σd electrical conductivity in the damaged tissue) giving a value equal to 0 for intact tissue and 1 for completely damaged tissue (Roselló-Soto et al. 2016).

$$Z = \frac{\sigma - \sigma i}{\sigma d - \sigma i}$$

Ultrasound-assisted extraction has the advantage that it is cheaper and simple. The process consists of shock waves from cavitation bubbles, which facilitates mass transfer through the cycles of expansion and contraction in the biomass of the material experienced. It favours the kinetics and the extraction yield in addition to facilitating larger mass transfer (Roselló-Soto et al. 2016).

The surface microstructure of *Flammulina velutipes* polysaccharides has been observed to change with ultrasound-assisted extraction, the triple helix structure was disrupted, and the proportion of low molecular weight polysaccharide components rose (Chen et al. 2019).

If combined with other extraction solvents ultrasound-assisted extraction allows extracting faster and at a lower temperature, which can be useful for tempolabile compounds. Parameters such as power and absorbed energy must be controlled in order to monitor the process. In comparison to traditional extraction methods, the polysaccharides extracted using ultrasound technology displayed considerably better antioxidant activity (Chen et al. 2019).

Microwave-assisted extraction. This extraction is dependent on two parameters specific to each material: dielectric constant and dielectric loss. The process works by heating the material through electromagnetic waves between wavelengths from far infrared light to radio frequencies. Its spectrum is located 300 MHz up to 300 GHz.

The heating promotes evaporation and creates significant pressure which fractures the cells releasing the intracellular content. While the conventional heating method heats to the medium, the microwave heats the middle up the exterior. The mushroom's fruiting body undergoes morphological change as a result of microwave-assisted extraction, and the surface microstructure is more expansive, porous, and loose. Also, you can modify the temperature of the extraction, in this way the rate of β-glucan extraction increases with extraction temperature, and extraction time decreases in the case of *Ganoderma lucidum* and *Pleurotus ostreatus*. *Morchella conica*'s polysaccharide is extracted using the innovative technique of ultrasonic microwave synergistic extraction (UMSE), which combines ultrasonic and microwave energy for greater efficiency and less time (Leong et al. 2021).

Subcritical Water Extraction (SWE) or hot pressurised water extraction. Considered an eco-friendly option to conduct extractions. It consists of high temperatures of boiling water around 100–374°C (374°C is the subcritical temperature). In this method, the variability of the dielectric constant is related to the temperature.

A temperature of 200°C is the most suitable for extracting water-soluble polysaccharides. At this point batch systems can extract up to three times more soluble material than semi-continuous systems. At higher temperature, the result of the extraction shows more content of the antioxidant activity. A temperature of extraction of 250°C showed higher antioxidant activity (Roselló-Soto et al. 2016).

Though the investigation is focused on phenolic and polysaccharides, it is interesting to investigate the polarity of compounds. For example, in *Lentinula edodes*, 90% of polysaccharides were obtained by SWE with 10.1 MPa, while only 29% were obtained with boiling water with 0.1 MPa (Morales et al., 2019). This improvement is most likely due to the partial destruction of the hydrogen bonding.

Supercritical Fluid Extraction (SFE) is performed over supercritical fluids. At a certain temperature and pressure above its critical point for a particular organic matrix, separate liquid and gas phases do not exist. Although a supercritical fluid has a viscosity and density that are similar to those of a liquid and a gas, its diffusivity is in between the two, favouring the extraction of intracellular chemicals. In spite of the apparent improvement in the extraction of intracellular chemicals, it has been found that it does not retain the most desirable properties of the mushroom. However, it can be a good way to obtain hydrophilic antioxidants in addition to co-solvent (ethanol-CO_2) but because of the polarity, reduce the fatty acids.

Deep eutectic solvent extraction technology (DESs) that is based on solvents has minimal or no negative environmental consequences. They have kindred properties to ionic solvents, but in addition, they are cheaper and easy to produce.

The low pressures and extraction temperatures of the deep eutectic solvents have some advantages over the technology of subcritical and supercritical fluid extraction. Since carbohydrates are very soluble in these solvents, polymers are recovered at high rates. DES, such as choline chloride:acetic acid (CC-AA), choline chloride:formic acid (CC-FA), and alanine/lactic acid, can be prepared from natural compounds or purchased using straightforward heating and mixing techniques. The cationic and anionic groups used are: quaternary ammonium (hydrogen bond acceptor), amines, amides, carboxylic acids and polyols (hydrogen donor) (Smith et al. 2014).

Low vapor pressure, chemical and thermal stability, non-flammability, high dissolvability, low melting point, biodegradability, non-toxicity, polarity, recyclability, and low cost are further characteristics of the DES technique.

3. Antibacterial Properties

Prokaryotes are divided into two groups: bacteria (gram+ bacteria and gram- bacteria) and archaeas. The most important pathogenic bacteria for animals and humans are gram+ and gram-bacteria. Some secondary metabolites obtained from mushrooms exhibit antibacterial properties and reduce the growth or kill the existing bacteria.

3.1 Against Gram+ Bacteria

Different mushrooms show antibacterial activity against gram+ bacteria. For instance, *Agaricus bisporus* is effective against *Bacillus subtilis*, *Bacillus cereus*, *Micrococcus luteus*, *Micrococcus flavus*, *Staphylococcus aureus* and *Staphylococcus epidermidis* (Rezaeian et al. 2016). *Agaricus bitorquis* and *Agaricus essettei* also have activity vs a broad range of gram+ bacteria. *Agaricus silvicola* against *Bacillus cereus*, *Bacillus subtilis* and *Staphylococcus aureus*. *Agaricus cf. nigrecentulus* and *Tyromyces duracinus* only vs. *Staphylococcus saprophyticus*. *Sarcina lutea* is susceptible to *Armillaria mellea*. Concerning the Boletus genus, the wild mushroom *Boletus edulis* is active against *Staphylococcus aureus* and *Bacillus cereus*. Also, the wild *Cantharellus cibarius* is active against *Bacillus subtilis* and *Staphylococcus aureus* (Alves et al. 2012).

There are so many active metabolites produced by mushrooms, and the screening of the one responsible for the antibacterial activity is challenging. Meanwhile, there are already a good number of molecules that have been described.

Among the molecules from mushrooms described to pose antibacterial activity against pathogenic gram+ bacteria are:

Four molecules belonging to the terpenes group (confluentin, grifolin, neogrifolin and Ganomycin) (Alves et al. 2012). One steroid, 3,11-dioxolanosta-8,24(Z)-diene-26-oic acid active against *Bacillus cereus* and *Enterococcus faecalis*. Four sesquiterpenes known as enokipodins A, B, C, and D. Enokipodins A and C were the only ones to have activity against *Staphylococcus aureus* among the four enokipodins (A, B, C and D) that had activity against *Bacillus subtilis* (Alves et al. 2012). Oxalic acid, which is an organic acid, was found to be active against *Bacillus cereus*, *Staphylococcus aureus*, and *Streptococcus faecalis* when it was isolated from the *Lentinus edodes* mycelium (Taofiq et al. 2016). Coloratin A, 6-Methylxanthopurpurin-3-O-methyl ether, (1S,3S)-austrocortilutein, (1S,3R)-austro-cortilutein, (1S,3S)-austrocortirubin, torosachrysone and a ribonuclease from *Pleurotus sajor-caju* are active against *Staphylococcus aureus*. The peptide plectasin from *Pseudoplectania nigrella* possesses some efficacy against the following bacteria: *Bacillus cereus*, *Bacillus thuringiensis*, *Corynebacterium diphtheriae*, *Corynebacterium jeikeium*,

Enterococcus faecalis, Enterococcium (VREF) and, *Staphylococcus aureus* (MRSA) (Alves et al. 2012).

3.2 Against Gram-Bacteria

Methanolic extract from *Agaricus bisporus* is active against *S. typhimurium* and *B. cereus*, similar to the activity of *P. ostreatus* and *Amarilla mellea* in ethanolic extracts (Fogarasi et al. 2020). Methanolic extract from *Clitocybe alexandri* has shown activity against *Enterobacter aerogenes* and *Escherichia coli. Lactifluus piperatus* has fairly activity against *E. coli.* and *P. aeruginosa. Boletus edulis* also has activity again *P. aeruginosa. Escherichia coli* is inhibited by *Leucogaricus cinerus, Marasmius cf. bellus* and *Marasmius* sp. (Kumar et al. 2015). In addition, species from the family *Tricholomataceae* show significant interest due to the range of the secondary metabolites produced. *Hydnum repandum* through methanolic extract is active against *Pseudomonas aeruginosa.* Glucans from the *Pleurotus florida* blue variant in the form of nanoparticles result in effective against multiple antibiotic-resistant (MAR) bacteria, specifically *Klebsiella pneumoniae* (Friedman 2016). *Lentinus squarrosulus* has a heteropolysaccharide which put together with nanoparticles is also active against MAR *E. coli* (Fiedman 2016). Finally, *Pleurotus sajor-caju* inhibits *Escherichia coli, Enterococcus aerogenes, Pseudomonas aeruginosa,* and *Klebsiella pneumoniae.*

Among the compounds identified with antibacterial properties are terpenes (from *Ganoderma pfeifferi*), organic acids (from *Lentinus edodes*), benzoic acids (from *Xylaria intracolarata*), ribonucleases (*Pleurotus sajor-caju*) and a fraction B from *Pycnoporus sanguineus.* However, the mechanism of action of these compounds is not described or is unknown.

Additionally, *P. ostreatus* phenolic and tannin components may have antibacterial activity via a variety of modes of action characterized by cell membrane lysis, protein synthesis inhibition, proteolytic enzymes, and microbial adhesins. It was noted that this mushroom has an activity that covers a broad spectrum of bacteria. It inhibits both gram- and gram+ bacteria (Younis et al. 2015)

Only the bacteria with the lowest MIC, *B. cereus* and *Cryptococcus neoformans,* were responsive to *Sarcodon imbricatus.* It is important to note that the *S. imbricatus* stipe extract was ineffective against the tested microorganisms whereas the cap extract was selective for *B. cereus.* These studies indicate that both stipe and pileus may have different antibacterial properties (Elisashvili et al. 2022).

Bacterial species were resistant to *Phellinus torulosus, Fomes fomentarius, Trametes versicolor, Pisolithus albus,* and *Fomitopsis pinicola* extracts on a wide spectrum (*E. coli, P. aeruginosa, A. hydrophila, B. subtilis, E. faecalis, S. typhimurium* and *L. monocytogenes*) (Elisashvili et al. 2022).

4. Antifungal Properties

Among the antimicrobials of mushroom origin, the one that arises from interspecies competitiveness is the antifungal property. Within those antifungals, there are many compounds that inhibit some other fungi such as those listed below.

Fusarium oxysporum and *Mycosphaerella arachidicola* are inhibited by eryngin (from *P. eryngii*), while *F. oxysporum* is inhibited by ganodermin, hypsin, and *P. sajor-caju* ribonuclease with IC_{50} values of 12.4, 14.2, and 9.5 µM, respectively (Wong et al. 2010).

Physalospora piricola, *Botrytis cinerea*, and *M. arachidicola* were shown to have their microscopic growth inhibited by the antifungal lentin protease that was isolated from *L. edodes* (Alves et al. 2013).

Fresh *Pleurotus sajor-caju's* ribonuclease has been demonstrated to be inhibitory of *F. oxysporum* and *M. arachidicola*. Another antifungal protein from *G. lucidum*, ganodermin, showed inhibitory efficacy against *B. cinerea*, *F. oxysporum*, and *P. piricola* mycelial growth (Alves et al. 2013).

Hypsin (from *Hypsizigus mamoreus*), Ganodermin (from *Ganoderma lucidum*), *P. sajor-caju* RNase, and Lyophyllum antifungal protein (from *Lyophyllum* shimeji) all inhibit *P. pyricola* with IC_{50} values of 2.5, 18.1, 72, and 70 µM, respectively (Sanodiya et al. 2009).

B. cinerea growth is inhibited by hypsin and ganodermin, while *M. arachidicola* is inhibited by *P. sajor-caju* ribonuclease with an IC_{50} of 0.06, 15.2, 2.7, and 72 M, respectively (Wong et al. 2010).

Pleurostrin, trichogin, and alveolarin prevent the mycelial development of the fungi *F. oxysporum*, *M. arachidicola* and *Physalospora pyricola*. Eryngin is an inhibitor of *M. arachidicola* and *F. oxysporum*. Hypsin, ganodermin, and *P. sajor-caju* RNAsa all stop *F. oxysporum* from growing (Wong et al. 2010).

M. arachidicola and *P. pyricola* are inhibited by the Lyophyllum antifungal protein, while *Coprinus comatus*, *Colletotrichum gossypii*, and *Rhizoctonia solani* are unaffected (Wong et al. 2010).

Various *Fusarium* species have been shown to be sensitive to extracts from cultivated species of *P. ostreatus*, wild *Pleurotus*, *Ganoderma*, *Cordyceps militaris*, wild *Fomitopsis pinicola* and *Lactarius vellereus* mushrooms (Shen et al. 2017).

Due to their ability to inhibit harmful fungi like *Trichoderma* spp., *Aspergillus* spp., and *Penicillium* spp. during growth, mushrooms with antifungal qualities also have the added benefit of minimizing the risks to the safety of food connected with the use of fungicides in cultivated mushrooms. Eleven mushroom taxa, including *Agaricus*, *Agrocybe*, *Coprinus*, *Cordyceps*, *Dictyophora*, *Hygrophorus*, *Lactarius*, *Leucoagaricus*, *Ganoderma*, *Pleurotus*, and *Phellinus*, have been found having antifungal capabilities against *Aspergillus* spp. or *Penicillium* spp. Only the *Agrocybe*, *Lactarius*, and *Phellinus* genera of mushrooms exhibit antifungal action against species of *Aspergillus* and *Penicillium* (Girma et al. 2018).

The cupareno-type sesquiterpenoids, such as the enokipodinas C and D of *Flammulina velutipes*, also inhibit the microscopic growth of *Cladosporium herbarum* (Passari et al. 2020).

5. Antiviral Properties

Viruses are much more complex organisms with larger diversity. The main classification is based on the type of genetic material they harbour.

Firstly, in the group of enveloped DNA viruses, antiviral activity against the Herpes simplex virus has been reported. Depending on the type of mushroom, various mushroom component substances, including polysaccharide, sulfated polysaccharide, proteoglycan, peptide RC28, and triterpenoid (ganoderona A, lucialdehyde B, and ganodermadiol), have been shown to have effects prior to, concurrent with, in combination with, and after treatment. All steps of viral replication, including entry, decapitation, replication, assembly, and release of the Herpes simplex virus, may be affected by them. The viral glycoprotein, replication, and transcription of the E and L genes were all impacted by sulfated polysaccharide from *Agaricus brasiliensis*. In mice, peptide RC28's antiherpetic action from *Rozipes caperata* was just as effective as ganciclovir (Seo et al. 2021).

Secondly, inside the group of enveloped RNA viruses, including influenza virus, human immunodeficiency virus and hepatitis C virus. Bioactive compounds from mushrooms such as *P. pulmonarius, C. militaris, L. edodes, P. baumii, P. ignarius, G. pfeifferi*, and *P. linteus* have antiviral effects against influenza viruses. Compounds described include acidic polysaccharide from *C. militaris*, polysaccharide fraction from *P. pulmonarius*; peptidomannan from *L. edodes*; polyphenols from *P. baumii*; pyrone (3-hydroxy-2-methyl-4-pyrone), and sesquiterpenoids from *P. ignarius*; polyphenols, hispidin, inoscavin A, davallia lactone, and phelligridin D; Triterpenes (applanoxidic acid G) and triterpenoids (ganodermadiol and lucidadiol) from *G. pfeifferi*, as well as other chemical compounds from *P. linteus*, such as inotilone and 4-(3,4-dihydroxyphenyl)- 3-buten-2-one. Sesquiterpenoids, inotilone, polyphenols, and 4-(3,4-dihydroxyphenyl)-3-buten-2-one that target neuraminidase (NA) (Seo et al. 2021).

It has been determined that the amino acid (Asn 170) of neuraminidase (NA) interacts with the sesquiterpenoid's hydroxyl group. Flu virus cytopathic impact was decreased by polyphenols, sesquiterpenoid, pyrone, ganodermadiol, lucidadiol, and applanoxidic acid G (Song et al. 2014).

Studies on the human immunodeficiency virus with the mushrooms *P. abalonus, Coriolus versicolor, A. bisporus, P. citrinopileatus, L. edodes, P. ostreatus, R. paludosa*, and *Tricholoma giganteum* have proved that all produced anti-HIV mushroom chemicals. Targets for HIV reverse transcriptase include polysaccharide, polysaccharopeptide, lectin, lentin, a ubiquitin-like protein, peptides, and laccase. Moreover, compounds originating from mushrooms may have antiviral properties similar to those of antiviral medications. Among them, *Inonstiu obliqueis* water-soluble lignin-blocked HIV protease and *Trametes versicolor* (L. fr) protein-bound polysaccharides exhibit antiviral effects against HIV and cytomegalovirus (CMV) (Abugri et al. 2019).

P. ostreatus and *A. bisporus* laccase and tyrosinase enzymes, respectively, show antiviral activity against the Hepatitis C virus (HCV). In PBMCs and HepG2 cells, *P. ostreatus* laccase can prevent viral entry and reproduction. HepG2 cells and PBMCs. In Huh-5-2 cells with replicons, tyrosinase from *A. bisporus* prevented viral replication. Tyrosinase also focused on viral nonstructural proteins that are crucial for viral replication, including NS3, NS4A, and NS5A (Seo et al. 2021).

The group of non-enveloped RNA viruses to which noroviruses and enteroviruses 71 belong.

The norovirus substitutes murine norovirus (MNV) and feline calicivirus (FCV) were both susceptible to *Inonotus obliquus*' antiviral effects. *I. obliquus* and its polysaccharide decreased norovirus surrogates by pre-, co-, post-, and simultaneous treatment into cells, therefore they are likely to obstruct various steps in the replication process (Seo et al., 2021).

To continue we have to talk about enterovirus 71 (EV 71). According to reports, triterpenoids and heteropolysaccharides have antiviral effects against EV 71. By regulating apoptosis, the heteropolysaccharide from *G. frondosa* may prevent the expression of the capsid protein and lessen viral infections. Triterpenoids from *G. lucidum* decreased the viral RNA level in human rhabdomyosarcoma by pre- and co-treatment. Triterpenoids' potential antiviral action involved preventing viral entry to prevent viral penetration into cells (Zhao et al. 2017).

The last group of viruses are Poliovirus and Coxsackievirus to which antiviral properties among mushrooms have been reported. PV-1 was resistant to the antiviral effects of polysaccharides produced by *A. brasiliensis* and *L. edodes*. PV-1 and polysaccharides were treated together at the same time to lower the viral titer in human laryngeal epithelial cell carcinoma (HEp-2) cells (Seo et al. 2021).

When coxsackievirus B3 (CVB3) and *Phellinus pini* polysaccharide were administered to HeLa cells at the same time, CVB3 titer was reduced. Therefore, the first stage of viral replication may be influenced by polysaccharides from *A. brasiliensis*, *L. edodes*, and *P. pini* (Seo et al. 2021).

Lentinan, an antiviral compound found in *Lentinula edodes* mycelia, inhibits the propagation of the infectious hematopoietic necrosis virus (IHNV). Additionally, injection of LNT-1 resulted in a notable reduction in the expression of cytokines that promote inflammation. Since COVID-19 patients have high levels of inflammatory cytokines, LNT-1's impact on SARS-COV-2 should be taken into consideration (Shahzad et al. 2020).

Cat viruses such as feline calicivirus, feline herpesvirus 1, feline influenza, feline infectious peritonitis virus, and feline panleukopenia virus are affected by *Inonotus obliquus* polysaccharides (Tian et al. 2017). All five viral subtypes showed inhibition of RNA viruses and DNA viruses.

6. Diversity of Antimicrobial Metabolites

Numerous primary and secondary metabolites, including phenolic compounds, polyketides, terpenoids, steroids, nonprotein amino oxide, antibacterial or antifungal proteins, and volatile fatty acids, can accumulate as intracellular and extracellular products during the growth of mushrooms. Secondary metabolites (acids, terpenoids, polyphenols, polysaccharides, sesquiterpenes, alkaloids, lactones, sterols, metals, chelating agents, nucleotide analogs, and vitamins), glycoproteins, and polysaccharides, primarily glucans, make up the majority of the bioactive compounds found in mushrooms (Figure 1).

Terpenoids such as ganoderic acids, ganoderals, ganoderols, ganodermanontriol, lanostane, lucidone, carotenoids and ganodermanondiol. Polysaccharides like lentinan, protein-bound PSK and β-(1-6)-D-glucan (We et al. 2016).

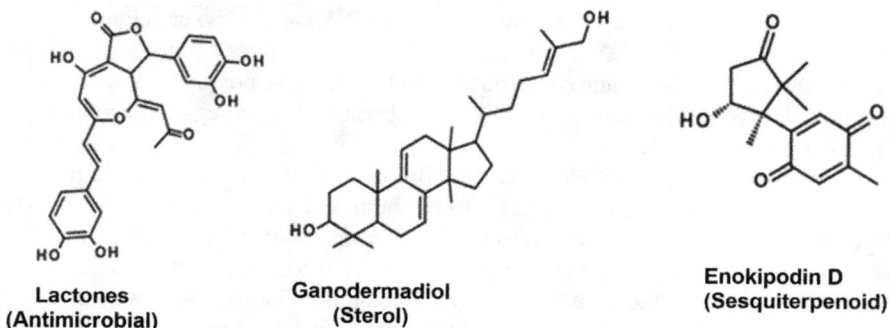

**Lactones
(Antimicrobial)**

**Ganodermadiol
(Sterol)**

**Enokipodin D
(Sesquiterpenoid)**

Figure 1: Molecules from fungal extracts of different families of compounds with antimicrobial properties (molecules have been drawn by Raquel Peña based on Alves et al. 2012).

Such metabolite creation, especially secondary metabolite production, is linked to nutrition supplies and mushroom life conditions, all of which contribute to the enormous diversity of metabolites. Despite the fact that they may exhibit comparable antimicrobial effects, various mushroom species typically have distinctive metabolite profiles (We et al. 2016).

7. Application of Antimicrobials

Antimicrobials have applications in many industries and sectors including sectors such as agriculture, the food industry and the health sector.

In agriculture, mushroom antimicrobials can be used to control crop diseases and pests caused by nematodes or insects and control food decay during postharvest.

For example, for the control of plant diseases, eight extracts from the mushrooms *Ganoderma resinaceum*, *Laetiporus sulphureus*, *Dictyopanus pusillus*, and *Bjerkandera adusta* showed antibiotic efficacy against bacteria and filamentous fungi, *Xanthomonas vesicatoria*, *Aspergillus oryzae*, *Penicillium expansum*, *Botrytis cinerea*, and *Rhizopus stolonifer*. In addition, it is demonstrated that the MIC values for *Laethiporus sulphureus* against *X. vesicatoria* are low (Petrović et al. 2014, Barneche et al. 2016).

Mushroom such as *Lepista nuda* has an impact on the microbial structure. When the ability to produce active metabolites by *L. nuda* was investigated, results demonstrated that wild populations consistently possess the ability to produce antibiotics (Sidovora et al. 2000). The majority of bacteria that were isolated exclusively from soils outside of their colonies were blocked from growing, however, bacteria that emerged in the vicinity of this species' active growing mycelia were entirely resistant. In a variety of soil micromycetes species, it resulted in total or partial inhibition of spore germination, but this activity was fungistatic. This species' antibiotics play a significant role in controlling the micro- and mycobiota's structural composition in forest soils and litter.

In order to limit foodborne germs, the use of synthetic preservatives has for many years become a standard technique in the food processing industry. However, a widespread dislike for manufactured chemicals in food has recently emerged, but not for naturally occurring compounds. Researchers' efforts aim to design safe antimicrobials as alternatives, particularly from traditional food ingredients themselves, like edible mushrooms.

Water extracts of mushrooms such as *Ganoderma*, *Tremella*, and *Agaricus* species, which mostly contain antimicrobial polysaccharides, have been transformed into polysaccharide powders as food supplements using a rotational concentration technique and spray-drying (Zhang et al. 2021).

To replace synthetic antibiotics and preservatives, prepared mushroom antimicrobial agents can be used directly in food. Such properties of mushroom antimicrobials have been proven for their impact on antibiotic treatment-induced modulation of the gut microbiota and antibiotic treatment-induced extension of product shelf-life.

Mushroom extracts are also used as feed supplements for farm animals. It decreased the quantity of undesired or harmful bacteria like *Bacteroides* spp. and *E. coli* while promoting the growth of helpful bacteria like bifidobacteria and lactobacilli. Advantages in replacing synthetic antibiotics in cattle have been proved and lowering the hazards associated with various foods of animal origin in terms of food safety when compared to the use of antibiotics alone (Guo et al. 2004).

Foods can be preserved by directly adding materials or extracts from the different origins that have antimicrobial compounds to increase their shelf life by inhibiting deterioration-causing microorganisms such as bacteria. The multifunctional characteristics of fungi frequently lead to their antibacterial effects. Nutraceutical and pharmacological applications can be frequently exploited from fungi that have antibacterial properties.

Ultimately, as Venturella et al. have recently reviewed (Venturella et al. 2021), in the helth sector mushroom derivatives can be used as medicinal foods that can help to prevent conditions including cancer, hypertension, diabetes, and high cholesterol because dietary fiber, specifically chitin, and polysaccharides such as β-glucans are present. Some mushroom species may be able to lower high blood sugar levels because they possess anticancer, antiviral, antithrombotic, and immunomodulating properties. Furthermore, they have antimicrobial effects that can be exploited as new drugs to fight pathogenic fungi, bacteria or viruses and in turn combat the current multi-antibiotic resistance problem. For animals and humans, many of these compounds are safe just because they conform to edible mushrooms.

8. Antioxidant Properties of Cultivated Mushrooms

The study of the antioxidant activity of mushrooms requires to development of the optimal methodology for the preparation of extracts from mycelium and fruiting bodies. Mushrooms can be therefore used not exclusively as food but as a rich source of suitable antioxidant bioactive compounds in the food industry (to obtain functional foods), the pharmaceutical industry (bioactive extracts with medicinal activity and nutraceuticals), the cosmetic industry (fibers and proteins), chemical (biofungicides for agriculture), among other applications.

The pipeline for the production of rich antioxidant fractions from mushrooms requires: (a) the identification of the most suitable macro-fungi as potential carriers of antioxidant bioactive compounds; (b) the development and optimization of methodologies capable of evaluating the antioxidant capacity; (c) identification and quantification of the main antioxidants.

8.1 Antioxidant Sources and their Health Benefits

The pathophysiology of many diseases involves reactive oxygen and nitrogen species. DNA oxidative damage can start the carcinogenesis process. The majority of preparations and chemicals obtained from mushrooms are employed as new classes of dietary supplements (DS) or nutraceuticals, which fit extremely well within the idea of functional foods, rather than as medications. In addition to providing traditional medicines for the prevention or treatment of human cancer. For instance, dietary chemotherapeutic drugs can act as powerful agents to boost the therapeutic effect of chemotherapy and radiotherapy.

Fruit bodies, mycelium, and fungal broth were found to contain antioxidant substances that were later identified as phenolics, flavonoids, glycosides, polysaccharides, tocopherols, ergothioneine, carotenoids, and ascorbic acid (Figure 2) (Kozarsky et al. 2015).

The two main categories of fungal antioxidants are primary (chain-breaking and free radical scavenging) and secondary or preventive antioxidants, which are produced as a result of metal deactivation, inhibition or decomposition of lipid hydroperoxides, regeneration of primary antioxidants, singlet oxygen, and free radical scavenging (Flieger et al. 2021).

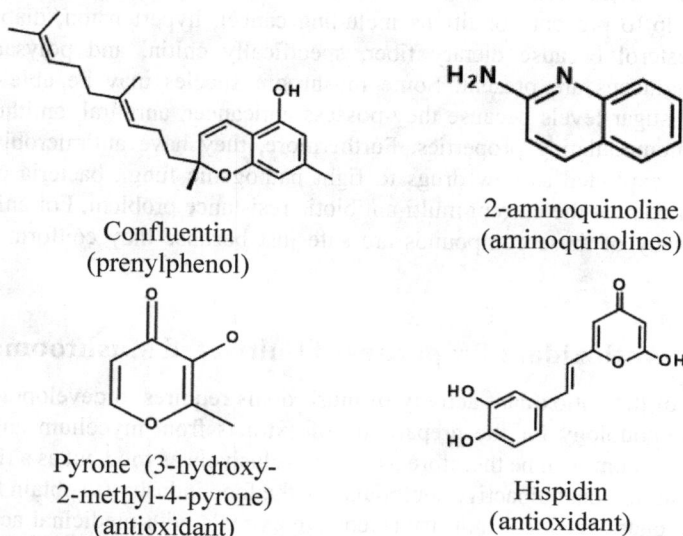

Confluentin
(prenylphenol)

2-aminoquinoline
(aminoquinolines)

Pyrone (3-hydroxy-
2-methyl-4-pyrone)
(antioxidant)

Hispidin
(antioxidant)

Figure 2: Molecules from fungal extracts of different families of compounds with antioxidant properties (molecules have been drawn by Raquel Peña based on Alves et al. 2012).

Certain elements found in a fungus that have antioxidant action operate as inducers and/or biological signals, causing alterations in gene expression and the activation of ROS-deactivating enzymes.

Phellinus rimosus

The family Hymenochetaceae contains the vast and widely dispersed genus *Phellinus*. For the formation of basidiocarps, environmental conditions like temperature, humidity, light, and host trees are crucial. *Phellinus* (*P. senex*), *P. rimosus, P. badius P. fastuosus, P. adamantinus, P. caryophylli* and *P. durrissimus* are the species that are most common (Ajith and Janardhanan 2007).

In a dose-dependent way, *P. rimosus* were successful in eliminating O_2 produced by photo illumination of riboflavin, OH produced by the Fenton reaction, and nitric radical liberated from an aqueous solution of sodium nitroprusside. In rat whole liver homogenate, the extracts reduced ferrous ion-induced lipid peroxidation in a dose-dependent manner (Ajith and Janardhanan 2007).

Ganoderma lucidum

The oldest known usage of therapeutic characteristics for *Ganoderma lucidum* and related species dates back at least four thousand years. *In vitro* antioxidant activity has been discovered in extracts of *G. lucidum* fruiting bodies and mycelia, which are native to southern India (Venturella et al. 2021).

Ethyl acetate, methanol, and the aqueous extract of *G. lucidum* were found to produce O_2- and -OH radicals, according to the results of the antioxidant assays. However, the aqueous extract was ineffective in preventing lipid peroxidation brought on by ferrous ions. Instead, the extract demonstrated strong reducing power and radical scavenging abilities (Ajith and Janardhanan 2007).

Pleurotus Species

Significant antioxidant, anti-inflammatory, and anticancer activity are present in *Pleurotus* species (Venturella et al. 2021). Methanol extract from *Pleurotus florida* fruiting bodies has the ability to prevent lipid peroxidation and scavenge OH-radicals. Additionally, the extract demonstrated strong reducing power and radical scavenging abilities (Ajith and Janardhanan 2007).

Other Poisonous Mushrooms

A significant antioxidant potential has been demonstrated for *Agaricus xanthodermus* (Sevindik 2020). Researchers have used many techniques to determine that *Amanita muscaria* possesses antioxidant capability; DPPH has an activity EC_{50} value of 2.87 mg/mL and in another assay a value of 40.05 µL/mL (Sevindik 2020).

The methanol extract of *Aamanita pantherina* has a DPPH EC_{50} scavenging activity that ranges from 3.89 mg/mL to 42.04 mg/mg (Sevindik 2020). Alpha amanitis and phalloidin extracted from *Aamnita phalloides* have excellent antioxidant potentials, according to studies. According to reports, the ethanol extract of *A. porphyria* has an IC_{50} value of 5.07 mg/mg in the DPPH test (Sevindik 2020).

Extracts from *Cortiniarus sanguineus* in ethanol, *Chlorophyllum molybdites* and *Gymnopilus junonius* in methanol are seen to have antioxidant properties. Other poisonable mushrooms such as *Gyromitra esculenta* (methanol extract), *Hypholoma fasciculare* (methanol extract), *Omphalotus olearius* (cyclohexane, chloroform, ethanol, and water extracts), *Paxillus involutus* (ethanol extract), *Sarcosphaera coronaria* (ethanol extract), *Rubroboletus satanas* (water extract), and *Russula emetica* (water extract) also have antioxidant properties demonstrated by DPPH and/ or ABTS tests (Sevindik 2020).

8.2 *Extraction of Antioxidant Fraction*

By extracting mushrooms with cyclohexane, dichloromethane, methanol, and water, antioxidant fractions can be obtained. In terms of antioxidant activity and extraction effectiveness, ethanol and acetone (or ethyl acetate) are highly efficient solvents.

Different fractions were extracted with a sequence of increasingly polar solvents: n-hexane, chloroform, ethyl acetate, acetone, ethanol, and clean water obtained from five wild edible macrofungi (Sezgin et al. 2020). Briefly, using a magnetic stirrer, the powdered fungus samples were stirred for two hours at room temperature (22°C) while being diluted with a 10-fold volume of n-hexane (g/ml). After that, they underwent a 20-minute centrifugation. A collection of supernatants was made. Ten times each, the pellet was dissolved in chloroform, ethyl acetate, acetone (80%), ethanol (80%), and pure water (g/ml). The supernatants were then collected. Under reduced pressure and 37°C, the fractions produced from the various solvents were evaporated. Of these compounds, water or polar solvents are best for extraction (Sezgin et al. 2020).

Polysaccharides

β-glucans are principally responsible for mushrooms' antioxidant effects (Leong et al. 2021).

The nature of fungus and the physical characteristics of the polysaccharides influence the extraction process and the technology used to purify the polysaccharide fractions in many different ways. Typically, ethanol precipitation is needed to separate these polysaccharides from the hot water extract (Aguiló-Aguayo 2017, Chen et al. 2019, Leong et al. 2021).

Phenolic Compounds

They consist of hydroxybenzoic acids, flavonoids, phenolic acids, tannins, lignans, stilbenes, hydroxycinnamic acids, and oxidized polyphenols (Sanchez 2017).

Hydroxybenzoic acids are a part of intricate structures like those made of lignins and hydrolyzable tannins. P-hydroxybenzoic acid, protocatechuic acid, gallic acid, gentisic acid, homogentisic acid, vanillic acid, 5-sulphosalicylic acid, syringic acid, veratric acid, and vanillin are the most prominent (common) benzoic acid derivatives found in mushrooms (Sanchez 2017).

Hydroxycinnamic acids are bonded to the cellulose, lignin, and proteins that make up the cell wall as well as to organic acids like tartaric or quinic acids (Sanchez 2017). Caffeic, ferulic, sinapic, p-coumaric and o-coumaric are the majority of cinnamic acid derivatives found in fungi, including three forms of caffeoylquinic acid: 3, 4 and 5 (Sanchez 2017). Flavonoids are a big group of molecules with antioxidant properties. The best solvents to extract them were ethanol and methanol, although in some cases distilled water might be the best option (Sanchez 2017).

Pressurized liquid extraction (PLE), subcritical water extraction (SWE), supercritical fluid extraction (SFE), microwave-assisted extraction (MAE), solid phase extraction (SPE), ultrasonic-assisted extraction (UAE), high hydrostatic pressure extraction (HHPE), solid-supported matrix dispersion (MSD), and ultrasonic-assisted extraction (UAE) are non-traditional extraction techniques (CCC) for the extraction of phenolic compounds (Alara et al. 2021). Conventional techniques for the extraction of phenolic compounds are maceration, percolation, digestion, and soxhlet methods (Alara et al. 2021).

However, it is generally accepted that solvents of higher polarity often perform better in terms of extracting polyphenols due to the high solubility of polyphenols in such solvents. The literature demonstrated that there is no generally acceptable solvent as the best solvent for polyphenol extraction. One of the causes is the diversity of solubility of phenolic compounds, they will dissolve better in one solvent or another depending on the molecular structure (Smolskalte et al. 2015).

Tocopherol

It is incorporated into the cell membrane. In the ester moiety formed between acetic acid and α-tocopherol is tocopherol acetate, which has an inhibitory effect on tyrosinase (Vamanu 2014).

Tocopherols were more prevalent throughout the fluidized bed ethanol extraction process. Its process consists in using a fluidized bed extractor, the lyophilized mycelium is dissolved in 70% ethanol to create an extract, and two extraction cycles are created. In a rotary evaporator with a vacuum controller, the alcohol extracts are concentrated. To produce the solid fraction, the concentrated solution is freeze-dried. The dried fractions can be dissolved in 80% ethanol to the desired concentration (e.g., 0.2–1.0 mg/mL) (Vamanu 2014).

Sterols

Within the sterol group of fungal origin are vitamin D_2 and ergosterol. To obtain ergosterol and vitamin D_2 supercritical fluid extraction (SFE) is used. UV application produce vitamin D_2 due to the photolysis of ergosterol into ergocalcipherol (vit. D_2). High temperatures (55–75°C) might cause the conversion of ergosta-7,22-dienol and fungisterol into ergosterol, which in turn had an impact on extraction yields and ergosterol concentrations (Morales et al. 2017). Ergosterol-enriched extracts must be extracted, then dissolved in methanol or ethanol, and then exposed to radiation to partially convert ergosterol to vitamin D_2.

Because of its mild critical temperature (31.3°C) and pressure, supercritical carbon dioxide is the most often employed supercritical fluid (72.9 atm.). Furthermore, since carbon dioxide is a gas at room temperature, it is simple to separate or remove it in order to create an extract without the need for solvents. Supercritical carbon dioxide has a low polarity and may not be as efficient in extracting polar molecules, which is one of its disadvantages. Modifiers or cosolvents like methanol, ethanol, or water can be added to supercritical carbon dioxide to increase its solvating power in order to get over this restriction.

Carotenoids

The main components of this group are β-carotene and lycopene. For these components, supercritical fluid extraction is used from fungal matrix, like in the case of sterols. SFE (supercritical fluid extraction) avoids excessive processing temperatures by performing the selective isolation of carotenoids in a single phase (Camara et al. 2013). This makes it advantageous for the extraction of carotenoids that are susceptible to heat. The SFE for the carotenoids is better technique than the conventional soxhlet extraction.

Lipid Fraction

Although lipids are not a very important fraction in mushrooms, Folch's method is the best way to extract them. Folch's extraction method is the most used for biological samples. It uses a biphasic solvent system with a volumetric ratio of 8:4:3 for chloroform, methanol, and water.

The material is first homogenized using a solution with two times as much chloroform as methanol. Small molecules are produced as a result, along with a residue that cannot be extracted.

The method's effectiveness is based on the ability to separate the methanol, water, and chloroform mixture into two phases. The top phase was mostly hydrophilic and made up of methanol and water. While the chloroform that makes up the majority of the bottom phase will keep the lipids in place (Eggers et al. 2016).

Both polar and nonpolar solvents can cause lipids to form micelles and aggregates. The direct application of Nernst's law, which states that "the ratio of concentrations of a solubilized compound in a two-phase system is constant at a given temperature" is constrained by this. This combination does, however, produce very little aggregate formation. Then, to wash the lipids, they are subjected to successive washes with water.

Polar lipids and neutral lipid fractions are separated by solvent fractionation with acetone.

8.3 Assays to Check Antioxidant Properties (Munteanu and Apetrei 2021)

Different biological samples can have their antioxidant potential determined using a variety of techniques. However, there is no quick way to assess antioxidant activity and no single technique that works for every biological matrix.

Efficient methodologies for the quantification of antioxidant activity include analytical methods to determine antioxidant levels depending on the types of compounds of interest. Antioxidants can be analysed either as a functional group, a group of antioxidants or individual antioxidants. The antioxidant capacity has been quantified as a functional group by means of different tests: ABTS (2, 2'-azino-bis -3-ethylbenzothiazolin-6-sulfonic acid), DPPH (2, 2- Diphenyl-1-picrylhydrazyl), FRAP (FRAP Assay Kit, Ferric Reducing Antioxidant Power). As a group of antioxidants, the main tests include the determination of the content of total phenols by the Folin-Ciocalteu method. While liquid chromatography techniques with ultraviolet detection are used to determine individual antioxidants.

Total Phenols Assay

The analysis is carried out according to the Folin-Ciocalteu colorimetric reaction (Singleton and Rossi 1965). To determine the content of total phenols in mushroom extracts, a calibration curve is made from a stock solution of gallic acid (GA). Subsequently, 10 μL of extract, 150 μL of distilled water and 20 μL of FC reagent are used. The mixture is homogenized and pre-incubated for five minutes. Next, 30 μL of a sodium carbonate solution (140 g/L) is added, and it is incubated for 1 h at room temperature, protected from light. Finally, the absorbance of the sample is measured at 725 nm in a UV-Vis spectrophotometer.

DPPH Test (Kedare et al. 2011)

The DPPH method quantifies the discoloration of the 1,1-diphenyl-2-picrylhydrazyl radical, this is a stable radical with violet coloration, which absorbs at 517 nm. The addition of antioxidants to the DPPH· radical reduces it as a function of concentration.

In a 96-well plate, 100 μL of the different extract concentrations (0.2, 0.4, 0.6, 0.8, 1, 1.2, 1.4, 1.6, 2, 3 and 5 mg/mL) and 150 μL of DPPH·solution is added at 76 μM. Subsequently, OD is measured at 517 nm in a plate reader with distilled water as a reference. The reference antioxidant used can be vitamin C in a concentration range of 0.05 to 1 mg/mL.

The sequestering activity of the extract is expressed by the percentage of inhibition of the absorbance of DPPH·, according to the equation: Inh DPPH• (%) = [1 – (ODsample/ODref)] * 100

FRAP Assay

The FRAP assay measures the reduction of a complex consisting of a chromogen, typically TPTZ (2, 4, 6-tripyridyl-s-triazine), and colorless ferric iron (Fe^{3+}) to a deep blue ferrous (Fe^{2+}) complex. greenish in the presence of antioxidants. TPTZ (2,4,6 tripyridyl-s-triazine), which at acidic pH (pH = 3.6) binds to the ferrous ion forming a blue compound that has a maximum absorbance at 593 nm for 30 minutes. This color intensity is directly related to the total reducing capacity of the antioxidant and can be quantitatively measured by spectrophotometry.

In a 96-well plate, 10 μL of the extract and 240 μL of the FRAP reagent, previously incubated at 37°C, are added. Incubate for 10 min and read OD at 593 nm on a plate reader. The final absorbance of the samples is compared with the standard curve of Trolox (0–5000 μmol/L) dissolved with methanol.

ABTS Test

The addition of the antioxidants in the sample reduces the ABTS·+ radical. In this way, the degree of discoloration as a percentage of inhibition of the ABTS cation radical is determined as a function of the concentration as well as the corresponding value using Trolox as a standard, under the same conditions. A chemical method is chosen for the generation of the radical: using potassium persulfate at 70 mM as an oxidant. The radical is generated from its precursor, 2,2'-azinobis(3-ethylbenzothiazolin)-6-sulfonic acid (ABTS) at a concentration of 2 mM. The ABTS·+ radical generated is a bluish-green, stable cationic radical with an absorption spectrum in the UV-vis (734 nm).

The mushroom extracts can be evaluated at concentrations of 0.4, 1, 2, 3 and 5 mg/mL in order to calculate the inhibition line and calculate the EC_{50} of TROLOX and the EC_{50} extract. To later calculate the TEAC (Trolox equivalent antioxidant capacity) value (EC_{50} of TROLOX/EC_{50} extract) expressing the results as μmol Trolox/g extract.

In a 96-well plate, 2.5 μL of the different concentrations of extract and 250 μL of diluted solution of the ABTS·+ radical can be added. OD is subsequently measured at 734 nm after 1 hour of incubation in a plate reader.

9. Concluding Remarks

The presence of antimicrobials and antioxidants in macrofungi is confirmed by many pieces of research evidence. Fruiting bodies or mycelium can differ in their bioactivity exhibiting more or less antibiotic or antioxidant capacity respectively.

Bioactive extracts with antibiotic or antioxidant activity can be prepared from mushrooms by means of different methodologies that require optimisation to achieve the best yield and reduce the cost of extraction. From this extract, through a previous process of analytical fractionation, molecules of interest can be isolated. Eventually, the structural determination of the compounds of interest can derive into novel drugs for medical therapy, nutraceuticals and active ingredients for the food industry or bio-pesticides to cope with crop diseases.

Acknowledgements

This work has received funding from the H2020 program under the grant agreement 101000651 (BIOSCHAMP project). Dr. Jaime Carrasco is the grant holder of the contract Ramon y Cajal 2021-032796-I funded by the Ministerio de Ciencia e Innovacion (Spain).

References

Abugri, D.A., J.A. Ayariga, B.J. Tiimob, C.G. Yedjou, F. Mrema and W.H. Witola. 2019. Medicinal mushrooms as novel sources for new antiparasitic drug development medicinal mushrooms (pp. 251–273). Springer Singapore. https://doi.org/ 10.1007/978-981-13-6382-5_9.

Aguiló-Aguayo, I., J. Walton, I. Viñas and B.K. Tiwari. 2017. Ultrasound assisted extraction of polysaccharides from mushroom by-products. *Lwt*, 77: 92–99.

Ajith, T.A. and K.K. Janardhanan. 2007. Indian medicinal mushrooms as a source of antioxidant and antitumor agents. *J. Clin. Biochem. Nutr.*, 40(3): 157–162.

Alara, O.R., N.H. Abdurahman and C.I. Ukaegbu. 2021. Extraction of phenolic compounds: A review. *Curr. Res. Food Sci.*, 4: 200–214.

Alves, M., I. Ferreira, J. Dias, V. Teixeira, A. Martins and M. Pintado. 2012. A review on antimicrobial activity of mushroom (basidiomycetes) extracts and isolated compounds. *Planta Medica*, 78(16): 1707–1718.

Alves, M.J., I. Ferreira, J. Dias, V. Teixeira, A. Martins and M. Pintado. 2013. A review on antifungal activity of mushroom (basidiomycetes) extracts and isolated compounds. *Current Topics in Medicinal Chemistry*, 13(21): 2648–2659. doi: 10400.14/12863/5/0002R.

Arango, C.S. and I.J. Nieto. 2013. Cultivo biotecnológico de macrohongos comestibles: una alternativa en la obtención de nutracéuticos. *Rev. Iberoam. Micol.*, 30(1): 1–8. doi: 10.1016/j.riam.2012.03.011.

Barneche, S., G. Jorcin, G. Cecchetto, M.P.i. Cerdeiras, A. Vázquez and S. Alborés. 2016. Screening for antimicrobial activity of wood rotting higher basidiomycetes mushrooms from uruguay against phytopathogens. *Int. J. Med. Mushrooms*, 18(3): 261–267. doi: 10.1615/IntJMedMushrooms.v18. i3.90.

Blackwell, M. 2011. The fungi: 1, 2, 3... 5.1 million species? *Am. J. Bot.*, 98(3): 426–438. doi: 10.3732/ajb.1000298.

Cámara, M., M. de Cortes Sánchez-Mata, V. Fernández-Ruiz, R.M. Cámara, S. Manzoor and J.O. Caceres. 2013. Chapter 11—Lycopene: A review of chemical and biological activity related to beneficial health effects. *Stud. Nat. Prod. Chem.*, 40: 383–426. https://doi.org/10.1016/B978-0-444-59603-1.00011-4.

Chen, X., D. Fang, R. Zhao, J. Gao, B.M. Kimatu, Q. Hu, G. Chen and L. Zhao. 2019. Effects of ultrasound-assisted extraction on antioxidant activity and bidirectional immunomodulatory activity of *Flammulina velutipes* polysaccharide. *Int. J. Biol. Macromol.*, 140: 505–514. doi: 10.1016/j. ijbiomac.2019.08.163.

Elisashvili, V., M.D. Asatiani, T. Khardziani and M. Rai. 2022. Natural antimicrobials from basidiomycota mushrooms. Promising antimicrobials from natural products (pp. 323–353). Springer International Publishing. DOI:10.1007/978-3-030-83504-0_13.

Flieger, J., W. Flieger, J. Baj and R. Maciejewski. 2021. Antioxidants: Classification, natural sources, activity/capacity measurements, and usefulness for the synthesis of nanoparticles. *Materials*, 14(15): 4135. doi: materials-14-04135-v2.pdf.

Fogarasi, M., Z.M. Diaconeasa, C.R. Pop, S. Fogarasi, C.A. Semeniuc, A.C. Fărcaş, D. Ţibulcă, C. Sălăgean, M. Tofană and S.A. Socaci. 2020. Elemental composition, antioxidant and antibacterial properties of some wild edible mushrooms from Romania. *Agronomy*, 10(12): 1972. https://doi. org/10.3390/agronomy10121972.

Friedman, M. 2016. Mushroom polysaccharides: Chemistry and antiobesity, antidiabetes, anticancer, and antibiotic properties in cells, rodents, and humans. *Foods*, 5(4): 80. https://doi.org/10.3390/foods5040080.

Girma, W. and T. Tasisa. 2018. Application of mushroom as food and medicine. *Adv. Biotech. & Micro.*, 11(4): 555817. DOI: 10.19080/ AIBM.2018.11.555817.

Guo, F.C., B.A. Williams, R.P. Kwakkel, H.S. Li, X.P. Li, J.Y. Luo, W.K. Li and M.W.A. Verstegen. 2004. Effects of mushroom and herb polysaccharides, as alternatives for an antibiotic, on the cecal microbial ecosystem in broiler chickens. *Poult. Sci.*, 83(2): 175–182. doi: 10.1093/ps/83.2.175.

Kedare, S.B. and R.P. Singh. 2011. Genesis and development of DPPH method of antioxidant assay. *J. Food Sci. Technol.*, 48(4): 412–422. doi: 10.1007/s13197-011-0251-1.

Kozarski, M., A. Klaus, D. Jakovljevic, N. Todorovic, J. Vunduk, P. Petrović, M. Niksic, M.M. Vrvic and L. Van Griensven. 2015. Antioxidants of edible mushrooms. *Molecules*, 20(10): 19489–525. doi: 10.3390/molecules201019489.

Kumar, K. 2015. Role of edible mushroom as functional foods—A review. *South Asian Journal of Food Technology and Environment*, 1(3&4): 211–218.

Leong, Y.K., F. Yang and J. Chang. 2021. Extraction of polysaccharides from edible mushrooms: Emerging technologies and recent advances. *Carbohydr. Polym.*, 251: 117006. doi: 10.1016/j. carbpol.2020.117006.

Morales, D., A. Gil-Ramirez, F.R. Smiderle, A.J.A. Piris, Ruiz-Rodriguez and C. Soler-Rivas. 2017. Vitamin D-enriched extracts obtained from shiitake mushrooms (*Lentinula edodes*) by supercritical fluid extraction and UV-irradiation. *Innov. Food Sci. Emerg. Technol.*, 41: 330–336. https://doi. org/10.1016/j.ifset.2017.04.008.

Morales, D., F.R. Smiderle, M. Villalva, H. Abreu, C. Rico, S. Santoyo, M. Iacomini and C. Soler-Rivas. 2019. Testing the effect of combining innovative extraction technologies on the biological activities of obtained β-glucan-enriched fractions from *Lentinula edodes. J. Func. Foods*, 60: 103446. https:// doi.org/10.1016/j.jff.2019.103446.

Munteanu, I.G. and C. Apetrei. 2021. Analytical methods used in determining antioxidant activity: A review. *Int. J. Mol. Sci.*, 22(7): 3380. https://doi.org/10.3390/ijms22073380.

Passari, A.K. and S. Sánchez. 2020. An introduction to mushroom. https://doi.org/10.5772/intechopen.86908.

Petrović, J., M. Papandreou, J. Glamočlija, A. Ćirić, C. Baskakis, C. Proestos, F. Lamari, P. Zoumpoulakis and M. Soković. 2014. Different extraction methodologies and their influence on the bioactivity of the wild edible mushroom *Laetiporus sulphureus* (Bull.) Murrill. *Food Funct.*, 5(11): 2948–2960. https://doi.org/10.1039/C4FO00727A.

Rezaeian, S. and H.R. Pourianfar. 2016. Antimicrobial properties of the button mushroom, Agaricus bisporus: A mini-review. *Int. J. Adv. Res.*, 4(1): 426–429.

Roselló-Soto, E., O. Parniakov, Q. Deng, A. Patras, M. Koubaa, N. Grimi, N. Boussetta, B.K. Tiwari, E. Vorobiev, N. Lebovka and F.J. Barba. 2016. Application of non-conventional extraction methods: toward a sustainable and green production of valuable compounds from Mushrooms. *Food Eng. Rev.*, 8(2): 214–234. https://doi.org/10.1007/s12393-015-9131-1.

Sánchez, C. 2017. Reactive oxygen species and antioxidant properties from mushrooms. *Synth. Syst. Biotechnol.*, 2(1): 13–22. https://doi.org/10.1016/j.synbio.2016.12.001.

Sanodiya, B., G. Thakur, R. Baghel, G. Prasad and P. Bisen. 2009. Ganoderma lucidum: A potent pharmacological macrofungus. *Curr. Pharm. Biotechnol.*, 10(8): 717–742. doi: 10.2174/138920109789978757.

Seo, D.J. and C. Choi. 2021. Antiviral bioactive compounds of mushrooms and their antiviral mechanisms: A review. *Viruses*, 13(2): 350. doi:10.3390/v13020350.

Sevindik, M. 2020. Poisonous mushroom (nonedible) as an antioxidant source. *Plant Antioxidants and Health*, 1–25.

Sezgin, S., A. Dalar and Y. Uzun. 2020. Determination of antioxidant activities and chemical composition of sequential fractions of five edible mushrooms from Turkey. *J. Food Sci. Technol.*, 57(5): 1866–1876. doi:10.1007/s13197-019-04221-7.

Shahzad, F., D. Anderson and M. Najafzadeh. 2020. The antiviral, anti-inflammatory effects of natural medicinal herbs and Mushrooms and SARS-CoV-2 infection. *Nutrients*, 12(9): 2573. doi:10.3390/nu12092573.

Shen, H., S. Shao, J. Chen and T. Zhou. 2017. Antimicrobials from Mushrooms for assuring food safety. *Compr. Rev. Food Sci. Food Saf.*, 16(2): 316–329. doi:10.1111/1541-4337.12255.

Sidorova, I.I. and L.L. Velikanov. 2000. Bioactive substances of agaricoid basidiomycetes and their possible role in regulation of myco- and microbiota structure in soils of forest ecosystems. II. Antibiotic activity in cultures of litter saprotrophic mushroom *Lepista nuda. Mikol Fitopatol.*, 34(4): 10–16. https://eurekamag.com/research/003/659/003659328.php.

Singleton, V.L. and J.A. Rossi. 1965. Colorimetry of total phenolics with phosphomolybdic-phosphotungstic acid reagents. *Am. J. Enol. Vitic.*, 16(3): 144–158.

Smith, E.L., A.P. Abbott and K.S. Ryder. 2014. Deep eutectic solvents (DESs) and their applications. *Chemical Reviews*, 114(21): 11060–11082. doi: 10.1021/cr300162p.

Smolskaitė, L., P.R. Venskutonis and T. Talou. 2015. Comprehensive evaluation of antioxidant and antimicrobial properties of different mushroom species. *LWT - Food Sci. Technol.*, 60(1): 462–471. https://doi.org/10.1016/j.lwt.2014.08.007.

Song, A.R., X.L. Sun, C. Kong, C. Zhao, D. Qin, F. Huang and S. Yang. 2014. Discovery of a new sesquiterpenoid from Phellinus ignarius with antiviral activity against influenza virus. *Archives of Virology*, 159: 753–760. doi: 10.1007/s00705-013-1857-6.

Stamets, P. and H. Zwickey. 2014. Medicinal mushrooms: Ancient remedies meet modern science. *Integr. Med. (Encinitas)*, 13(1): 46–47.

Taofiq, O., S. Heleno, R. Calhelha, M. Alves, L. Barros, M. Barreiro, A. González-Paramás and I. Ferreira. 2016. Development of mushroom-based cosmeceutical formulations with anti-inflammatory, anti-tyrosinase, antioxidant, and antibacterial properties. *Molecules*, 21(10): 1372. doi:10.3390/molecules21101372.

Tian, J., X. Hu, D. Liu, H. Wu and L. Qu. 2017. Identification of *Inonotus obliquus* polysaccharide with broad-spectrum antiviral activity against multi-feline viruses. *Int. J. Biol. Macromol.*, 95: 160–167. doi: 10.1016/j.ijbiomac.2016.11.054.

Vamanu, E. 2014. Antioxidant properties of mushroom mycelia obtained by batch cultivation and tocopherol content affected by extraction procedures. *Biomed. Res. Int.*, 2014, Article ID 974804 | https://doi.org/10.1155/2014/974804.

Varghese, R., Y.B. Dalvi, P.Y. Lamrood, B.P. Shinde and C.K.K. Nair. 2019. Historical and current perspectives on therapeutic potential of higher basidiomycetes: an overview. *3 Biotech*, 9(10): 1–15. doi: 10.1007/s13205-019-1886-2.

Venturella, G., V. Ferraro, F. Cirlincione and M.L. Gargano. 2021. Medicinal mushrooms: Bioactive compounds, use, and clinical trials. *Int. J. Mol. Sci.*, 22(2): 634. doi:10.3390/ijms22020634.

Wong, J.H., T.B. Ng, R.C.F. Cheung, X.J. Ye, H.X. Wang, S.K. Lam, P. Lin, Y.S. Chan, E.F. Fang, P.H.K. Ngai, L.X. Xia, X.Y. Ye, Y. Jiang and F. Liu. 2010. Proteins with antifungal properties and other medicinal applications from plants and mushrooms. *Appl. Microbiol. Biotechnol.*, 87(4): 1221–1235. doi:10.1007/s00253-010-2690-4.

Wu, Y., M. Choi, J. Li, H. Yang and H. Shin. 2016. Mushroom cosmetics: The present and future. *Cosmetics*, 3(3): 22. doi: 10.3390/cosmetics3030022.

Younis, A.M., F.S. Wu and H.H. El Shikh. 2015. Antimicrobial activity of extracts of the oyster culinary medicinal mushroom *Pleurotus ostreatus* (higher basidiomycetes) and identification of a new antimicrobial compound. *Int. J. Med. Mushrooms*, 17(6): doi: 10.1615/IntJMedMushrooms.v17. i6.80.

Zhang, Y., D. Wang, Y. Chen, T. Liu, S. Zhang, H. Fan, H. Liu and Y. Li. 2021. Healthy function and high valued utilization of edible fungi. *Food Sci. Hum. Wellness*, 10(4): 408–420. doi: 10.1016/j. fshw.2021.04.003.

Zhao, C., L. Gao, C. Wang, B. Liu, Y. Jin and Z. Xing. 2016. Structural characterization and antiviral activity of a novel heteropolysaccharide isolated from *Grifola frondosa* against enterovirus 71. *Carbohydr. Polym.*, 144: 382–389. doi: 10.1016/j.carbpol.2015.12.005.

6

Bioactive Compounds from Macrofungi and their Potential Applications

*Hoda M El-Gharabawy** and *Mamdouh S Serag*

1. Introduction

Macrofungi are presented in two phyla of the Fungi kingdom; Ascomycota and Basidiomycota (Ali et al. 2017). They are fleshy fungi with distinct above-ground or hypogeal fruiting bodies, including mushrooms or "toadstools", puffballs, hoofed mushrooms, coral mushrooms, and other large fungal species (Govorushko et al. 2019). Macrofungi are highly multicellular eukaryotic microorganisms that live as saprophytes degrading lignocellulose and pectin compounds, in a symbiosis of ectomycorrhizal associations with plants or parasites. The term "mushroom" refers to the fruiting body formed by piling up piles of mycelium under different types of habitats (Okan et al. 2014).

Macrofungi are ubiquitous in nature and have been consumed by mankind for their medicinal and nutraceutical properties. Mushrooms were used as a tea and dietary supplements in Eastern culture for their distinctive aroma and texture. Wild mushrooms are potential nutritional and health attributes and can be compared to various meat, fish, egg and dairy products (Nagy et al. 2017, Dudekula et al. 2020). They have been called the "meat of the forest" or the "meat of the poor" (Dimitrijevic et al. 2018). The fruiting bodies of dried macrofungi contain carbohydrates (65%) and protein (35%) with trace amounts of fats (6%) (Rathore et al. 2017). Furthermore,

Botany and Microbiology Department, Faculty of Science, Damietta University, New Damietta, 34517, Egypt.
* Corresponding author: hoda_mohamed@du.edu.eg/mserag@du.edu.eg

they contain various bioactive constituents such as biologically active proteins and polysaccharides, dietary fiber, unsaturated fatty acids (palmitic acid, oleic acid and linoleic acid), phenolic compounds, minerals and vitamins (El-Gharabawy et al. 2012, Lu et al. 2020).

Nearly 14,000 species of mushrooms are known worldwide, of which approximately 2,000 species are considered edible (Garofalo et al. 2017). Approximately 200 species of mushrooms have been cultivated for medicinal purposes and for human consumption (Mleczek et al. 2018).

Wild mushrooms have a high economic value and are considered an important profitable activity for both nutritional and medicinal purposes, such as mantis and truffle mushrooms. Lingzi (*Ganoderma lucidum*) extracts have a wide range of products from cosmetics to dietary supplements, while shiitake (*Lentinula edodes*) has been used as a medicinal source (Liao et al. 2015). The consumption of medicinal mushrooms as a healthy food and as a source of bioactive compounds should increase in the future. According to recent studies, their market size could grow by $13.88 billion between 2018 and 2022 (Niego et al. 2021). The main purpose of this chapter is to highlight the bioactive compounds of macrofungi and their potential applications.

2. Medicinal Value of Macrofungi

2.1 Folk Medicine

The world's love for mushrooms goes back thousands of years. Mushrooms are a rising star in traditional medicine and therapies, their health and nutritional value have long been understood and continue to be the subject of exciting research (Buller 1914, Power et al. 2015). Ancient Egyptians believed that mushrooms were the plants of immortality or "the gift of the god Osiris". The pharaohs were particularly fond of mushrooms and stated that they were a frequent part of their diet. Mushrooms were reserved only for royalty and there were rules and regulations for their cultivation and production (Niksic et al. 2015). The ancient Greeks also recognized its medicinal properties, while the ancient Romans diagnosed the taste of early boletus (*Boletus edulis*), truffles (*Tuber* spp.), Caesar mushrooms (*Amanita ceasarea*), mushrooms (*Agaricus* spp.) and cow balls (*Lycoperdon* spp.) (Buller 1914, Kotowski 2019).

In Asian countries (Korea, Japan, and China), about 300 mushroom species are used as medicines (Garibay-Orijel et al. 2007). 270 species of mushrooms are considered therapeutic in traditional Chinese medicine. Some traditionally used edible mushrooms, such as *Lentinula edodes*, *Grifola frondosa*, *Hericium erinaceus*, *Flammulina velutipes*, *Pleurotus ostreatus* and *Tremella mesenterica* are also sources of pure bioactive compounds that are not suitable for edible medicinal purposes such as *G. lucidum* and *Schizophyllum*. Macrofungi have been used in folk medicine throughout history in ancient Chinese, Egyptian, Roman and Greek civilizations. In traditional Chinese medicine, *Hericium erinaceus* is used to fight *Helicobacter pylori* and treat stomach ulcers (Liu et al. 2016).

In Asia, Russia, the United States, Canada, Mexico and Venezuela, mushrooms are known to have a long history in the treatment of various diseases (Grienke et al. 2014). Morel is used in traditional medicine as an antiseptic for the healing of wounds (Mahmood et al. 2011), as an immunostimulator, as a general tonic (Sher et al. 2015) and for the treatment of colds and coughs and as gastrointestinal symptoms (Sola. et al. 2012). *Podaxis pistillaris* (desert shepherd man) desert flower globe (Figure 1). This fungus has a global distribution and grows in a symbiotic association with termites in semi-arid regions (Buys et al. 2018, El-Fallal et al. 2019). Their fruit bodies have been used in traditional folk medicine to treat skin diseases, sunburn and wounds. In addition, it has been used as face paint and to darken white hair (Muhsin et al. 2012). Recently, they have been used medicinally for inflammation and antimicrobial activity (Feleke and Doshi 2018).

Figure 1: *Podaxis pistillaris* was collected from Sinai in Egypt by Prof Dr. A. Khedr; (A) mushroom growing in groups in the sandy desert; (B) external and internal view of mushroom in the laboratory.

2.2 Modern Medicine

The second most important property of macrofungi after food is the medicinal value of their fruiting bodies and cultured mycelium. Polysaccharide proteins extracted from the culture of macrofungi showed antihypertensive and relaxing effects, immunomodulatory and antitumor effects (Valverde et al. 2015), while some lectins obtained from their fruiting bodies showed immunomodulatory and antitumor effects (Zhao et al. 2020). Edible mushrooms such as *Flammulina*, *Lyophyllum*, *Agaricus*, *Boletus*, *Letinula* and *Pleurotus* have been reported to have crucial biological activities including antioxidant, antiviral, antimicrobial, anticancer, anti-inflammatory, hypocholesterolemic, antihyperglycemic (El-Khateeb and Daba 2022). Oyster mushrooms (*Pleurotus* spp.) contain a high ratio of potassium to sodium, making mushrooms an ideal food for those suffering from blood pressure and heart disease. Various species of *Pleurotus* have been shown to have several therapeutic properties, acting as antitumor agents (Patel et al. 2012), as well as their ability to lower cholesterol and reduce colon carcinoma, immunomodulatory, antigenotoxic, anti-inflammatory, antigenic effects, anticancer and antihypertensive (Anusiya et al. 2021). These species are considered to improve the secretion of insulin, an effect on the heart muscles and are useful for diabetes, ulcers and lung diseases. They have hypoglycemic and antithrombotic effects and lower blood lipid level concentrations (Patel et al. 2012). In addition, methanol extracts have a high content of antioxidants (such as phenols and flavonoids, carotenoids, and anthocyanins), vitamins (ascorbic acid, vitamin D_3) and antimicrobial (El-Gharabawy 2012). *Ganoderma* is highly valued in modern medicine for the treatment of many diseases. It has been used as a popular remedy to treat a series of diseases, including hepatitis, nephritis, arthritis, bronchitis, asthma, atherosclerosis, hypertension, diabetes and gastric ulcers (Batra et al. 2018). Among the *Ganoderma* species, *G. applanatum* and *G. lucidum* have been reported in modern medicine and pharmacology for their medicinal properties and bioactive compounds (Mohammadi et al. 2020). Total phenols and flavonoid content, betulinic acid and also antioxidant activity were determined. *G. lucidum* had a higher content of total polysaccharides and proteins with low contents of terpenoids (oleanolic acid and ursolic acid). Some examples of medicinal macrofungi are presented in Table 1.

Table 1: Medicinal values of some examples of mushrooms.

Mushroom	Bioactive compound	Medicinal properties	Reference
Agaricus bisporus	Lectins	Enhance insulin secretion	Tirta et al. 2020
Ganoderma lucidum	Ganoderic acid. Beta-glucan	Augments immune system Liver protection Antibiotic properties Inhibits cholesterol synthesis	Moradali and Mostafavi 2006 El Sheikha 2022
Lentinula edodes	Eritadenine Lentinan	Lower cholesterol Anti-cancer agent	Bisen et al. 2010
Morchella spp.	Aqueous extract	Relieves stomach pain Cures pneumonia Reducing respiratory problems Acne treatment	Lakhanpal 2010 Thakur et al. 2021
Pleurotus sajor-caju	Lovastatin	Lower cholesterol	Ying et al. 2018

3. Bioactive Potentials of Macrofungi

Medicinal mushrooms, including wood-decaying fungi such as polypores, can be used as an excellent natural source of bioactive components. These compounds can have a great effect on maintaining human health, strengthening the immune system and preventing dangerous diseases. They have various physiological and pharmacological properties such as antioxidant, antimicrobial, antidiabetic, antitumor, immunomodulatory, cardiovascular and hepatoprotective (Garofalo et al. 2017, Fogarasi et al., 2020).

Bioactive compounds of mushrooms are grouped into lectins, polysaccharides, phenolic and polyphenolic compounds, terpenoids, ergosterols and volatile organic substances such as ceramides (Dudekula et al. 2020). Bioactive potentials can be strictly stress-specific, while their chemical composition largely depends on the environment, in particular the dominant climatic and edaphic variables. Mushrooms produce antibacterial and antifungal compounds to survive in the natural environment. Therefore, antifungal compounds with lower or higher potential activity could be isolated from several fungal species beneficial for humans (El-Gharabawy et al. 2021, El-Fallal et al. 2022). The difference in activity between *Agaricus* species may be due to the number of bioactive compounds secreted by these fungi that affect their growth. These compounds may contain phenols and flavonoids, which may have antifungal activity (Ramos et al. 2019). A rapid increase in the number of cultivated mushrooms for food or medicinal purposes. For example, common sable (*Grifola frondosa*), *Coriolus versicolor*, *Lentinula edodes*, *Schizophyllum commune*, *Hericium erinaceus* and *Ganoderma lucidum* (Kumar et al. 2021). Bioactive compounds of *Ganoderma* possess antioxidant, antiviral, antimicrobial, anticancer, anti-allergic, anti-hypertension, anti-ageing, anti-inflammatory, hypoglycemic and hepatoprotective activities. They are used in the treatment of many diseases such as liver disease, nephritis, bronchitis, asthma, hypertension, hyperlipemia, arteriosclerosis, arthritis, insomnia, neurasthenia, peptic ulcers and diabetes (Cör Andrejč et al. 2022). Many bioactive compounds of some *Ganoderma* species protect the body against free radicals and have other health benefits, such as inhibition of cancer cell growth or antiviral activity (Bhat et al. 2019). *Lepista* is a variegated, purple, excellent-tasting edible mushroom (El-Fallal et al. 2017). Polysaccharides extracted from *L. sordida* exhibited antioxidant, antitumor potential, and antiproliferative activity for human laryngeal carcinoma Hep-2 cells (Miao et al. 2013). In this chapter, we will focus on the following biological activities of some macrofungi.

3.1 *Antioxidant Activity*

An antioxidant is a molecule that can prevent the oxidation of other molecules, especially reactive oxygen species (ROS), stop oxidation chain reactions by scavenging free radicals and prevent other oxidation reactions, preventing many diseases (Unsal et al. 2020). Most living cells in the body produce antioxidants and repair systems that can protect the body from oxidative damage; however, these agents are not sufficient to completely prevent or repair the damage. Therefore, more antioxidants are needed in food, which helps protect cells from oxidative damage and maintain the health of the body (Kurutas 2016). Synthetic antioxidants, phenolic compounds such as butyl hydroxytoluene (BHT), butylated hydroxylanisome (BHA) and gallic acid, etc., are often used as food additives, but they have had some adverse effects on health (Shahidi and Ambigaipalan 2015). Mushrooms are a natural source of antioxidants that protect the endogenous system. Such properties increase the growing interest in the use of mushrooms in various nutritional products (Kozarski et al. 2015). The content of these antioxidants can be phenolic compounds (α-tocopherols, flavonoids and phenolic acids), nitrogen compounds (alkaloids, chlorophyll derivatives, amines and amino acids) or carotenoids and ascorbic acid (El-Gharabawy 2012, Fallal et al. 2022). Extracts of *G. lucidum* and *G. tsugae*, were found to have high antioxidants and exhibited superoxide and hydroxyl radical scavenging activity. Hot water extract of *G. lucidum* exhibited antioxidative effects on mouse liver and kidney lipid peroxidation (Lin and Deng 2019).

3.1.1 Total Phenolics

Phenolic compounds are aromatic compounds having one or more aromatic rings containing one or more hydroxyl groups. They encompass numerous flavonoids, and phenolic acids including hydroxybenzoic and hydroxycinnamic acids, lignans, tannins, and oxidized polyphenols (D'Archivio et al. 2010). Polyphenolic compounds play a role in stabilizing lipid oxidation, inhibiting cancer and acting as anti-aging agents. Mushrooms are rich in phenols, which are highly efficient scavengers of peroxy radicals. Many studies attribute the antioxidant properties to the presence of phenols and flavonoids (Shaffique et al. 2021). The presence of phenolic compounds in *Agaricus bisporus* and *A. bitorquis* supports their antioxidant activity (Ahmed 2014).

3.1.2 Total Flavonoids

Flavonoids are involved in plant free radical scavenging activity by retaining hydroxyl groups and are known to have significant antioxidant effects on human health. Dietary flavonoids are usually glycosylated and can be classified as anthocyanidins, flavanols (catechins), flavones, flavanones and flavonols (Pukalski and Latowski 2022).

Both mushrooms and the culture of macrofungi can be excellent sources of flavonoids. Extract of scleroderma warts has high antioxidant activity in reducing DPPH free radicals and significant contents of total flavonoids and phenols (de Menezes et al. 2022). The mycelium of *Pleurotus floridanus* produced high levels of phenols and flavonoids via a solid-state fermentation system in response to various environmental factors (El-Fallal et al. 2022). Methanol extract of the fungus *Pleurotus columbinus* contained valuable levels of phenols and flavonoids, levels of which were higher in the UV-C-induced mutant (R20) (El-Gharabawy 2012).

3.2 Antimicrobial Potential

Pathogenic infections are often caused by multi-drug-resistant microorganisms, leading to disease resistance and difficulty in drug therapy. As a result, pharmacoeconomic healthcare costs are rising rapidly and are a public health concern in many countries (Friedman 2015, Xin et al. 2018). This situation has increased the search for new antimicrobial agents from various synthetic or natural sources. Macrofungi are a rich source of natural antimicrobial compounds that fight bacteria (antibacterial) and fungi (antifungal). Proteins produced by these fungi exhibit multiple biological activities, including antiproliferative, immunomodulatory, antiviral, antibacterial, and antifungal activities. Polysaccharides or triterpenes as fungal-derived chemical groups act as bioactive compounds (Das et al. 2020).

Basidiomycetes have several active secondary metabolites, including phenolic compounds, polyketides, terpenoids, and steroids (Campi et al. 2019). Bioactive compounds have been isolated from basidiomycetes that inhibit the growth of a wide range of pathogenic fungi. Furthermore, these fungi are able to inhibit the development of bacteria and other fungi from their microhabitat and exhibit antibacterial activity (Winnie et al. 2019). Two new lanostane triterpenes, termed ganoderic acids AW_1 and AW_2, have been isolated from *Ganoderma* sp. It also showed good malaria activity against chloroquine-sensitive strains of Plasmodium falciparum (Wahba et al. 2019).

3.2.1 Antibacterial Properties

Many pathogenic bacterial species have acquired antibiotic resistance mechanisms that provide limited therapeutic options. *Schizophyllum commune* is a producer of antibiotics against *Pseudomonas aeruginosa*, *Staphylococcus aureus*, *Escherichia coli* and *Klebsiella pneumonia*. *Armillaria mellea* showed activity against *Staphylococcus aureus*, *Bacillus cereus* and *Bacillus subtilis*, while *Dendropollyporus umbellatus* was active against *Staphylococcus aureus* and *Escherichia coli* (Wasser 2010). Cultures extracted from the genus *Agaricus* showed significant antibacterial activity against *Staphylococcus aureus* (Ahmed 2014). *Flammulina velutipes* mycelia proved antibacterial activity against the Gram-positive bacteria of *Bacillus subtilis* and *Staphylococcus aureus* (Nicolcioiu et al. 2017). A known example of the diterpenoid antibiotic; pleuromutilin (94) was isolated and obtained from *Clitopilus passeckerianus* (Novak and Shlaes 2010). Oyster mushroom extract has shown good antibacterial activity against several Gram-positive and Gram-negative bacteria (El-Gharabawy 2012).

3.2.2 Antifungal Properties

Antifungal activity has been studied for bioactive compounds such as unsaturated fatty acids from mycelia and cultured liquid extracts of *Pleurotus* spp. Secondary metabolites of *P. eryngii* were active against *Candida albicans*, *C. glabrata*, and *Epidermophyton* species (Akyuz and Ayse 2018). Mycelium and liquid filtrate from *Pleurotus* spp. showed various inhibitory activities against *Trichoderma harzianum*, *Pythium* sp. and *Verticillium* sp. and other plant pathogens (El-Gharabawy 2012, Kadhim et al. 2020). Presence of tannins, saponins and flavonoids in *P. ostreatus* var. Florida may be responsible for the positive antifungal activity against *Trichoderma* species. Antifungal agents such as chitinases and proteases from *P. ostreatus*, *P. florida* and *P. sajor-caju* culture filtrates successfully combated soil fungi (Hassan et al. 2011). An antifungal peptide that inhibits the hyphal growth of pathogenic fungi has been isolated from the fruiting body of the fungus *P. eryngii* (Patel et al. 2012). Production of p-anisaldehyde by *P. ostreatus* has been described as a defense mechanism against other organisms by providing antibacterial and antifungal activity (Parameswari and Chinnaswamy 2011).

Mycelial plugs of *Ganoderma mbrekobenum* (Figure 2) showed high inhibitory activity against the growth of *Monascus ruber*, *Aspergillus ochraceus* and *Penicillium* sp. Antifungal activities of aqueous and organic extracts of the *Ganoderma* mushroom were investigated, and the methanol-chloroform extract was shown to have the maximum activity, making it a good antifungal against mycotoxins producing fungi (El-Fallal et al. 2021).

Figure 2: *Ganoderma mbrekobenum* (collected and confirmed identification by Dr. H.M. El-Gharabawy) from El-Senaniah orchards, Damietta, Egypt, growing widely on two different hosts; (A) date palm tree, (B) lemon tree.

3.2.3 *Antiviral Properties*

Diseases caused by viruses cannot be treated with conventional antibiotics, so targeted drugs are needed. Antiviral properties have been investigated not only for whole mushroom extracts but also for isolated compounds derived from secondary metabolites. These antiviral compounds can act directly by inhibiting viral enzymes, viral nucleic acid synthesis, or viral adsorption and uptake into mammalian cells (Diaz 2015, Le Daré et al. 2021). Several triterpenes from *G. lucidum*, such as bioactive compounds with antiviral properties, ganoderiol F, ganodermanonetriol, and ganoderic acid B, all act as antiviral agents against human immunodeficiency virus (HIV) type 1 (Friedman 2015). Antiviral activity against influenza virus types A and B has been identified in mycelium extracts from *Kuehneromyces mutabilis* (Vlasenko et al. 2021) and two isolated phenolic compounds from *Inonotus hispidus*, as well as ergosterol peroxide, derived from many mushrooms (Li et al. 2022).

3.4 *Anticancer Potential*

A glycosylated form of ergosterol peroxide obtained from a methanol extract of cordyceps inhibited the growth of cell lines such as the tumor cell lines WM-1341, HL-60, K562, Jurkat and RPMI-8226 (Zhang et al. 2022). An anticancer sterol from *Sarcodon aspratus* inhibits the proliferation of HT29 cancer cells (Ivanova et al. 2014). Anticancer studies have shown that sterols induce the expression of the cyclin-dependent kinase inhibitor 1A, causing cell cycle arrest and apoptosis in HT29 cells (Yoshikawa et al. 2010). A crude protein-bound polysaccharide substance extracted from the Egyptian mushroom; *Volvariella speciosa*, proved to have a significant antitumor effect in tumor-bearing mice (Hereher et al. 2018). Ergosterol peroxide, a common fungal steroid extracted from the Egyptian mushroom, *Ganoderma resinaceum*, showed a preferred inhibition of MCF-7 over MDA-MB-231 breast cancer (El-Sherif et al. 2020). Antitumor activity of the phytosterol α-spinasterol isolated from this mushroom was assessed on human breast cancer (MCF-7, MDA-

MB-231), and ovarian cancer cell line (SKOV-3). The compound exhibited a higher antitumor activity on MCF-7 cells relative to SKOV-3 cells, while its lowest antitumor activity was against MDA-MB-231 cells. An increase in p53 and Bax expression was also reported in treated cells, demonstrating potent inhibitory activity against breast and ovarian cancer (Sedky et al. 2018). Phenols can also exhibit anticancer activity as prophylactic agents with antioxidant activity that can induce direct cytotoxicity against cancer cells. Some examples of bioactive compounds are identified in medicinal mushrooms and their anticancer therapeutic potential (Table 2).

Table 2: Bioactive compounds recorded in medicinal mushrooms and their therapeutic potential against cancer.

Species	Bioactive compound	Mechanisms of action	Cancer types affected	Experimental models	References
Fomitopsis pinicola	Methanol extract	Cytotoxicity	Hepatocarcinoma, cervical cancer	Cell lines	(Xin et al. 2018)
	Ergosterol	ROS-mediated apoptosis	Colorectal cancer	Cell lines	(Wang et al. 2014)
	Ethanol extract	Tumor growth arrest	Sarcoma	Mouse xenograft tumors	Friedman 2015)
Hericium erinaceus	Ethanol extract	Tumor growth arrest	Gastric, liver, colon cancer	Cell lines, mouse xenograft tumors	(Wang et al. 2014)
	Aqueous extract	Metalloproteinase mediated migration inhibition, suppression of ERK and JNK kinase activation	Colon carcinoma	Mouse xenograft tumors	(Saitsu et al. 2019)
		Apoptosis via downregulation of antiapoptotic proteins	Leukemia	Cell lines	(Zhang et al. 2019)
	Erinacine A (ditherpenoid)	ROS-mediated cell cycle arrest cancer antiinvasive	cancer, colorectal	Cell lines, mouse xenograft tumors	(Zhang et al. 2019)
	Cerebroside E	Angiogenesis blocker	HUVEC cell line	(Lu et al. 2016)
	HEP3 protein	Immuno-stimulation via gut microbiota	adenocarcinoma	Mouse xenograft tumors	(Elkhateeb et al. 2019)
	Ethanol extract	Gastric ulcer cytoprotective (carcinogenic condition)		Rat model	(Phan et al. 2015)
Inonotus obliquus	Aqueous extracts	Colon cancer, cytotoxic/cytostatic liver cancer		Cell lines	(Lee et al. 2015)

3.5 Vitamins

Vitamins are essential nutrients for the human body and play important roles in various functions such as metabolism, immunity and digestion. Mushrooms are an excellent source of vitamins, such as thiamine (vitamin B_1), riboflavin (vitamin B_2), pyridoxine (vitamin B_6), pantothenic acid (vitamin B_5), nicotinic acid/niacin and nicotinamide (vitamin B_3), folic acid (vitamin B_9) and cobalamin (vitamin B_{12}). It also contains other vitamins such as biotin (vitamin B_8), tocopherols (Vitamin E), and ergosterol, a precursor of vitamin D2 (Niego et al. 2021). A deficiency of vitamin D represents a risk of a large spectrum of chronic diseases including cancer, heart diseases, diabetes, obesity, arthritis, hypertension and psoriasis (Malik et al. 2022). Vitamin D also helps in maintaining a healthy immune system (El-Sharkawy et al. 2020). Vitamin D2 level is likely to remain above 10 µg/100 g of fresh weight, which is higher than the level in most foods and similar to the daily requirement for vitamin D recommended (Cardwell et al. 2018). Vitamin D2 concentrations were increased depending on the time of post-harvest exposure to UV-C irradiation (El-Fallal et al. 2013). The effects of storage and cooking on vitamin D2 content, and the bioavailability of vitamin D2 from mushrooms (Cardwell et al. 2018).

4. Examples of Medicinal Macrofungi

Some examples of macrofungi that have been studied for their bioactive compounds with medicinal properties are discussed:

4.1 Ganoderma spp. (Medicinal Mushroom)

Ganoderma lucidum is widely classified as a medicinal mushroom in traditional medicine and is widely used to treat chronic infections (Nithya et al. 2013). The main active compounds isolated from *G. lucidum* are polysaccharides, triterpenes, protein-bound polysaccharides, phenols, alkaloids, lectins, and amino acids (Das et al. 2020).

4.1.1 Antibacteria Potential of Ganoderma

Two new hydroquinones named ganomycin A (95) and ganomycin B (96), have been isolated from *G. lucidum*. Both compounds showed antibacterial activity against several Gram-positive and Gram-negative bacteria (Kamble et al. 2011). *G. applanatum* contains the sterols 5α-ergost-7-en-3β-ol, 5α-ergost-7, 22-dien-3βol, 5,8-epidioxy-5α-ergost-6, 22-diene-3β, and a novel *Bacillus cereus* and *Staphylococcus aureus* These sterols also inhibited Gram-negative bacteria such as *Escherichia coli* and *Pseudomonas aeruginosa* (Smania et al. 1999). A bioactive extract of *G. lucidum* was tested against an isolate of *Xantomonas campestris* pv. *vesicatoria*, which causes bacterial spot disease in pepper; the broth inhibited phytopathogenic isolates by 100% and reduced leaf infections (Robles-Hernández et al. 2017).

4.1.2 Polysaccharides from Ganoderma

Polysaccharides derived from *Ganoderma* represent a structurally diverse class of biopolymers with a wide range of physic-chemical properties. Polysaccharides have the greatest potential for structural variability, offering the highest capacity to carry biological information compared to proteins and nucleic acids. The other is matrix-like β-glucans, α-glucans, and glycoproteins (Ferreira et al. 2015). The main polysaccharides isolated from Reishi species are glucans, β-1-3 and β-1-6 D-glucans. Other immunomodulatory polysaccharides have been identified, especially glycopeptides and proteoglycans. Studies have shown that the most active immunomodulatory polysaccharides are water-soluble β-1-3-D and β-1-6-D glucans, which can be precipitated by ethanol (Boh et al. 2007). Other antitumor polysaccharides of *G. lucidum* are a group of polysaccharides known as heteropolysaccharides, glycoproteins (protein-bound polysaccharides), or ganoderan A, B, and C (Lawal et al. 2017). *Ganoderma* is best known for its immune-stimulating effects in aiding cancer treatment and for its anti-HIV activity (Wasser 2011). The action of the host's immune system by polysaccharides of *Ganoderma* to show anticancer and antitumor effects was via increasing interleukin, interferon and antibody production, also the stimulation of cytotoxic T-lymphocytes (Friedman 2016). Intracellular and extracellular polysaccharides of *G. lucidum* inhibited the growth of cancer in the human body (Ferreira et al. 2015).

4.1.3 Triterpenes from Ganoderma

Triterpenes are a class of naturally occurring bioactive compounds whose carbon backbone consists of isoprene C-5 units (Boh et al. 2007). The fruiting body of *G. lucidum* contains bitter triterpenoids. About 30 compounds, major components include lucidenic acid, leucidic acid, ganodelenic acid, lucidic acid, ganorcidic acid, aplanoxidic acid, lucidone, ganoderal, and ganoderol. Among all other medicinal macrofungi, *G. lucidum* is the only known source of a group of triterpenes known as ganoderic acids, which have molecular structures similar to steroid hormones (Boh et al. 2007). The mechanism of action of triterpenes is different from that of polysaccharides. Polysaccharides and triterpenes have been shown to have direct cytotoxicity against tumor cells (Li et al. 2020). Triterpenes have potential anticancer agents due to the biological activity they have shown against the actively growing prostate (Qu et al. 2017).

4.2 True Morels (*Morchella* spp.)

Morel (*Morchella* sp.) has economic and scientific importance because of its functional and medicinal properties. An important discovered edible morel (*M. galilaea*) grows mainly in the Mediterranean region, including Egypt (El-Gharabawy et al. 2019) (Figure 3), Turkey and Europe (Taşkın et al. 2015). *M. esculenta* is a delicious edible mushroom with health based properties. Many biologically active compounds have been studied such as poly- and monosaccharides, polyphenols, and proteins (Wen et al. 2019). These compounds possess biological activities, including antioxidant, anti-inflammatory, immunomodulatory, hypoglycemic, anti-atherosclerotic, and anti-tumor effects (Li et al. 2021).

Figure 3: *Morchella galilaea* recorded in El-Senaniah orchards at Damietta, Egypt, 2013. (A) Young immature pale coloured mushroom, (B) Mature dark brown coloured mushroom (Photos were taken by Dr. H. M. El-Gharabawy).

4.2.1 *Antioxidant Properties of Morchella*

Morchella mushrooms are an excellent source of natural antioxidants and phenolic compounds. Their antioxidant activity has been studied. *M. esculenta* is a promising source of antioxidant, immunomodulatory, anticancer, and anti-inflammatory applications. Furthermore, morel's anti-inflammatory activity against HT-29 colon cancer cells and significant inhibition of activation of the pro-inflammatory cytokine NF-kB has been reported (Kim et al. 2011). Radical scavenging activity and antiradical activity of methanol extracts of *M. conica* and *M. anguisticeps* have been reported (Vieira et al. 2016). Free radical scavenger, anti-inflammatory and anti-arthritic anti-edema properties of *M. elata* bioactive extract (Ramya et al. 2021).

4.2.2 *Antitumor and Anti-inflammatory Properties of Morchella*

Morchella has potent anti-inflammatory and cytotoxic effects against human cancer cells (Lee et al. 2018, Zhao et al. 2018). Several bioactive compounds with antioxidant, anti-inflammatory, immunomodulatory, and antitumor activity have been identified by *M. esculenta* (Wu et al. 2021). Ethanolic extract of *M. esculenta* mycelium was protective against cisplatin- and gentamicin-induced nephrotoxicity. Methanol extract of *M. esculenta* showed antimutagenic and antimitotic potential (Stojkovic et al. 2013). The submerged fermented endopolysaccharides of *M. esculenta* have been reported to induce significant antihyperlipidemic and antiatherosclerotic activity and the potential for hyperlipidemia (Zhao et al. 2022).

4.3 *Desert Truffles*

Truffles are the fruiting bodies of underground ascomycetes, including many genera of ectomycorrhizal fungi that grow in close association with tree roots (Barea et al. 2011). The desert truffle 'Terfass' grows in mostly semiarid and arid countries in sand under herbaceous angiosperms, under *Cistus* and *Helianthemum* (Khabar et al. 2014). These truffles grow naturally in semi-arid and arid regions of hot climates in most of the Mediterranean Basin, particularly in North Africa and Middle Eastern countries, after rains in September and October. Two genera of desert truffles; *Terfezia* sp. and *Tirmania* sp. are common in the Middle East (Fortas et al. 2022).

Desert truffles have a high unique flavor and high nutritional value and have been used as functional foods in traditional and modern medicine (Volpato et al. 2012). In addition, due to their content of phenols, carotenoids, anthocyanins, ascorbic acid, flavonoids, tannins, glycosides, ergosterol, polysaccharides, etc., they are immunomodulatory, possessing hepatoprotective, antidepressant, antibacterial, antifungal, antiviral, antioxidant, antiradical and anti-inflammatory properties. Moreover, desert truffles are important in pharmacological applications, especially in the treatment of eye infections and cancer (Veeraraghavan et al. 2021).

4.3.1 *Antioxidant and Anti-inflammatory Activities of Truffles*

Desert truffles contain powerful antioxidant metabolites such as phenolics, flavonoids, tocopherol, carotenoids, ascorbic acid, anthocyanins and phytosterols (Patel 2012). These compounds are important natural products that have diverse biological potentials involving antioxidative, anti-inflammatory, anti-mutagenic, and anticancer properties (Gil-Ramírez et al. 2016). Ergosterol, the biological precursor of vitamin D_2, is also one of the abundant bioactive compounds in diverse truffle species (Tripathi et al. 2017).

Aqueous extract from *Tirmania melanosporum* showed a significant hypoglycemic effect on the streptozotocin-induced hyperglycemic rat and reduced levels of glucose. This effect of truffle extract was significantly correlated with Nrf_2 and NF-kB pathways, and the regulation of enzymatic and non-enzymatic antioxidants including superoxide dismutase, catalase, vitamin E, and vitamin C (Zhang et al. 2018). Aqueous extract from *T. claveryi* proved a hepatoprotective effect on liver damage caused by oxidative stress. *T. claveryi* and *T. nivea* can be used as a source of natural therapeutic agents in the treatment of eye infections which are caused by some resistant bacteria such as *Pseudomonas aeruginosa* and *Staphylococcus aureus* (Gouzi et al. 2011).

4.3.2 *Antimicrobial Activity of Truffles*

The antimicrobial activity of desert truffles has been well-reported for a few decades. Silver nanoparticles synthesized by *Tirmania* sp. revealed the antibacterial activity of Gram-negative and Gram-positive bacteria (Owaid et al. 2018). Aqueous extract of *T. claveryi* proved antimicrobial properties against *Escherichia coli*, *Staphylococcus epidermidis*, and *Staphylococcus aureus*, and consequently, their usage was suggested to treat eye infections caused by those three bacterial species (Casarica et al. 2016). Another desert truffle, *T. nivea* from Tunisia arid zone also exhibited remarkable inhibitory activity against three species of Gram-positive and four species of Gram-negative bacteria. including *Salmonella typhimurium*, *Escherichia coli*, *Pseudomonas aeruginosa*, *Enterococcus faecalis*, *Staphylococcus aureus*, *Staphylococcus epidermis*, and *Bacillus subtilis* (Hamza et al. 2016). The growth of *Bacillus subtilis* and *Staphylococcus aureus* was inhibited by ethyl acetate extract of *T. pinoyi* confirming its antimicrobial activity (Dib-Bellahouel and Fortas 2011).

Truffles (*Terfezia* sp. and *Tirmania* sp.) exhibited a high inhibitory activity in extracted tannins and glycosides against the growth of pathogenic bacteria such as

P. aeruginosa, K. pneumonia, S. aureus, E. coli, Proteus sp., *Enterobacter* sp. and *Aeromonas* sp. and fungi such as *Fusarium* sp., *A. nigar*, and *A. terrus* (Camin et al. 2010). *Terfezia* sp. and *Tirmania* sp. showed anticancer activity, immune-modulating activity, antiviral activity, hepato protective activity, and antidepressant activity (Vahdani et al. 2017). Furthuremore, desert truffles are important as medicinal drugs, especially in the treatment of eye infections (Bradai et al. 2015).

4.3.3 Antitumor Activity of Truffles

Many studies have reported the anticancer properties of desert truffles. Methanol extracts of *T. aestivum* and *T. magnatum* possessed significant in vitro cytotoxic effects against different cancer cell lines (HeLa, MCF-7, and HT-29), while their water extracts showed a prominent activity against breast cancer (MCF-7) (Beara et al. 2014). Silver nanoparticles synthesized from aqueous extract of *T. claveryi* exhibited significant *in vitro* cytotoxic effects against the MCF-7 assay (Khadri et al. 2017).

Truffle phytosterols have a role in the inhibition of cancer, angiogenesis, and stimulation of apoptosis. Squalene involved in natural triterpene potentially inhibits cancer cell proliferation without disturbing the biochemical pathways (Ramprasath and Awad 2015). The oleic acid content in truffles also contributes to anticancer activities. Oleic acid suppresses the over-expression of HER_2, an oncogene correlated to the invasion and metastasis of cancer cells (Carrillo et al. 2012).

Truffles polysaccharides as β-glucan polymers possessing anticancer activity (Friedman et al. 2016). These polysaccharides could stimulate lymphocyte and macrophage division and synthesis of cytokines such as interferons, interleukins, and immunoglobulins directed against cancer antigens (Li et al. 2018). *Terfezia* extracts were able to suppress the proliferation of the NCI-H460, HeLa, HepG2, and MCF-7 cell lines (GI50 values ranged from 19 to 78, 33 to 301, 83 to 321, and 102 to 321 g/mL, respectively) (Tejedor-Calvo et al. 2021).

4.3.4 Antiviral Activity of Truffles

Terfezia boudieri extracts had a significant antiviral effect on both human simplex virus-2 and vesicular stomatitis virus (Mekawey 2015). The bioactivity of desert truffels' 44 compounds were detected against the COVID-19 virus using molecular docking (Al-Mazaideh et al. 2021). Ten bioactive substances, including catechin hydrate, caffeic acid, trans-cinnamic acid, luteolin, quercetin, naringenin, and hesperetin, were found to have drug-like effects similar to those of the HIV inhibitors; lopinavir and indinavir. Moreover, catechin hydrate, naringenin, and hesperetin could bind to the P-glycoprotein substrate.

4.3.5 Other Potential Applications of Truffles

Truffles have many other medicinal properties, including antidepressants, cholesterol-lowering agents, and immunostimulants (Ramsden et al. 2018). Truffles are rich in amino acids, including L-tyrosine, a precursor of catecholamines, neurotransmitters that are involved in neural circuits (Kıvrak 2015). Heteroglycan polysaccharides extracted from *T. rufum* showed immunostimulatory effects (Pattanayak et al.

2017). Truffles have been used as an effective sexual enhancer because they contain androstenol, a steroidal pheromone that significantly increases levels of male sex hormones (luteinizing hormone and testosterone) (Veeraraghavan et al. 2021). *T. boudieri* demonstrated a substantial function in treating male sexual impotence and infertility by boosting serum levels of testosterone, testis weight, sperm count, and sperm motility in rats (Zabihi et al. 2017). Anti-angiogenic activities of *Terfezia claveryi* extract were demonstrated (Dahham et al. 2018). It prevented the synthesis of glucose-bound haemoglobin, reduced the action of the amylase enzyme, and slowed down the hydrolysis of starch to glucose (AlAhmed and Khalil 2019). Two rice pests (*Sitophilus oryzae* and *Rhyzopertha dominica*) were successfully eradicated by the Algerian desert trufle *T. claveryi* extracts (Neggaz et al. 2020). The insecticidal action against *S. oryzea* had the highest potency, with an LD50 value of 162.11 g/mL. After 24 hours of exposure, this extract eliminated *S. oryzea* at concentrations of 250 and 300 g/mL. In mice treated with the extracts, only *Tirmania pinoyi* (TP) and *T. nivea* (TN) exhibited anticonvulsant, sedative, central, and peripheral analgesic effects (Aboutabl et al. 2022).

5. Poisonous Macrofungi

Many studies of mushroom poisoning have been made all over the world. Some of the 70–80 species of poisonous mushrooms are actually lethal if ingested. Clinical efficacy depends on the type of fungus and toxin (White et al. 2015). These dangerous mycotoxins may yield a vast array of beneficial natural compounds for people all over the world (Wu et al. 2019). Consuming toxic macrofungi can result in a variety of health problems, including rhabdomyolysis symptoms and erythromelalgia syndrome, as well as hepatotoxicity, nephrotoxicity, gastroenteritis, diarrhoea, and neurotoxicity (Vişneci et al. 2019). In rare cases, it can even be fatal. According to clinical signs, there are six different categories of deadly mushrooms (Xiaoxiao 2022) as detailed below.

5.1 Cytotoxic Macrofungi

Cytotoxicity of some poisonous macrofungi targets mainly the liver and kidney organs, presented in two categories; hepatotoxicity or nephrotoxicity. *Amanita* and *Cortinarius* spp. produce amanitin, aminohexadienoic acid, and orellanine toxins that harm the liver and kidneys (He et al. 2022).

5.1.1 Hepatotoxic Macrofungi

Some toxic macrofungi can cause complete liver failure (coagulopathy, bleeding, and hepatic encephalopathy). These mushrooms can cause pre-hepatotoxic symptoms in small doses, which first appear as severe gastrointestinal symptoms such as nausea, vomiting, diarrhoea, and abdominal pain, followed by secondary dehydration six hours later. Liver damage may happen within 2–3 days, and hepatic failure-related death may happen after 7 days (Puschner and Wegenast 2012). It is possible for liver function enzymes to rise abruptly and erratically before soon returning to normal

(Azeem et al. 2020). These species include *Amanita virosa, A. phalloides, Lepiota castanea, L. boudieri* and *Galerina marginata. A. phalloides* contains several toxic cyclopeptides, including the octopeptide amatoxin, which causes hepatocyte damage and hepatocyte loss. Phalloidin, a toxic heptapeptide found in the genus *Amanita* causes disruption of intracellular actin filaments, leading to cell damage or death (Vo et al. 2017).

5.1.2 Nephrotoxic Macrofungi

Those mushrooms seriously harm human kidneys. High concentrations of aminohexadienoic acid are produced by *Amanita smithiana* and *A. pseudoporphyria* (White et al. 2019). After 30 minutes or 12 hours of intake, these toxins might cause pre-renal symptoms, including diarrhoea, vomiting, stomach cramps, headaches, weakness, and tiredness. Orellanine poison that is produced by *Cortinarius* species (Dinis-Oliveira et al. 2016) causes significant renal damage, within two days of consumption, even for a very small quantity of this mushroom. Within 12 hours of ingesting, nausea, abdominal discomfort, exhaustion, headaches, chills, parasthesia, drowsiness, sweating, rash, dyspnea, and lumbar pain take place. Acute renal failure sets in within 3–10 days. These species include *Craterellus tubaeformis, Cortinarius rubellus*, and *C. orellanus* (Diaz 2021).

5.2 Myotoxic Macrofungi

Rhabdomyolysis poisoning is the primary symptom of certain mushrooms (Wu et al. 2019). Strong rhabdomyolysis symptoms are present in the mycotoxic mushrooms *Russula* and *Tricholoma* (He et al. 2022). The mechanism of rhabdomyolysis may differ depending on the species of mushroom and toxin. Rapid onset with fast mycotoxicity and delayed onset are the two primary subtypes of myotoxicity.

5.2.1 Rapid Onset

The poisonous chemical is purportedly identified as cycloprop-2-ene carboxylic acid, a highly strained carboxylic acid. In this species, cyclopropylacetyl-(R)-carnitine has also been identified (Cho and Han 2016). In most cases, symptoms appear within 2 hours of consumption and usually start with nausea, vomiting, and diarrhoea. Rhabdomyolysis, myalgias, hypertension, renal failure, hyperkalemia, and cardiovascular collapse may infrequently occur after more severe GIT side effects (He et al. 2022).

5.2.2 Delayed Onset

This form of poisoning is caused by *Tricholoma* species, including *T. equestre, T. terreum*, and possibly *T. auratum* (He et al. 2022). Slow-onset fatigue appears 1–3 days after consuming mushrooms for several consecutive meals and resolves in 15 days or less. Progressive weakness, stiffness in the legs, myoglobinuria (black urine), sporadic facial erythema, mild nausea, and perspiration will happen after that. Despite the possibility for a significant increase in plasma CK, the liver, kidneys, serum potassium, and coagulation continue to operate normally. It is possible to experience dyspnea, fever, acute myocarditis, heart failure, hyperkalemia, and a catastrophic cardiac collapse (White et al. 2019).

5.3 Acute Gastroenteritis Macrofungi

A large number of macrofungi can produce gastrointestinal irritation poisoning with a wide range of effects on the gastrointestinal tract, and dehydration problems due to fluid loss in serious cases (Wu et al. 2019). Mainly secondary to one of various "backyard fungi" such as *Chlorophyllum molybdites* (green spore-bearing parasol). It is a common virulent fungus, commonly found growing in lawns and parks (see Figure 4), and has been described as a gastrointestinal irritant for over a century (Sushila et al. 2012). A chemical survey identified nine steroid derivatives, two pyrrolidine alkaloids, lepiotin A and B, and α-methylglycerate. A toxic protein, molybdophyllidin, has also been isolated from this fungus (Lincoff 2011). Other species are reported to be gastrointestinal irritants such as *Agaricus arvensis*, *A. subrufescens*, *Lactarius atrobrunneus*, *Russula japonica*, *Scleroderma* sp., *Sutorius flavidus*, *Tricholoma terreum*, *Tylopilus neofelleus* (Li et al. 2020).

Figure 4: *Chlorophyllum molybdites* recorded by Prof. Dr. M. Serag, and identification was confirmed by Dr. H.M. El-Gharabawy during a field survey in New Damietta Botanical Garden, Egypt, Oct. 2022; (A) A group of fruiting mushrooms (B) Mushroom dorsal view, (C) Mushroom venteral view, (D) Characteristic green spore print.

5.4 Hallucinations Macrofungi

The main cause is produced by the bioactive compounds of psilocybin and psilocin from mushroom species such as *Psilocybe*, *Conocybe*, *Gymnopilus*, and *Panaeolus*. These chemically bioactive compounds act as agonists or partial agonists at 5-hydroxytryptamine (5-HT) subtype receptors (Dehariya et al. 2020). The mushroom can be ingested as fresh mushroom caps or dried. The main culprits are the bioactive compounds psilocybin and psilocin from fungal species such as *Psilocybe*, *Conocybe*, *Gymnopilus* and *Panaeolus*. These active biological compounds act as agonists or partial agonists at the 5-hydroxytryptamine (5-HT) subtype receptor. As an alternative to mushrooms, fresh mushroom caps or dried can be ingested and can produce sensory changes and euphoria 30 minutes to 2 hours after ingestion (Xin et al. 2018).

5.5 Disulfiram-like Reaction Macrofungi

Effects similar to those of disulfiram are produced by species possessing coprine, such as the inky cap mushroom *Coprinus atramentarius* (Xu 2022). Aldehyde dehydrogenase is inhibited by this toxin, which results in headaches, nausea, vomiting, flushing, tachycardia, and in rare instances, hypotension (Verma et al. 2019).

Unfavourable reaction is quite noticeable and only happens when alcohol is consumed a few hours to days after eating coprine-containing mushrooms. Due to coprine's delayed conversion to its poisonous secondary metabolites when consumed with alcohol, co-ingesting the toxin can have less hazardous consequences (Wong et al. 2020).

5.6 Neurotoxic Macrofungi

Muscarine, ibotenic acid, psilocybin are common toxins produced by some mushrooms and cause neurological effects when ingested. Symptoms include profuse sweating, coma, hallucinations, excitement, depression, convulsions, stirring the central and automatic nervous system (Dehariya et al. 2020). *Inocybe* or *Clitocybe* mushrooms contain significant amounts of muscarine poison, while *Amanita muscaria* and *A. pantherina* produce ibotenic acid and *Psilocybe cubensis*, *P. caerulescens*, *P. mexicana* and *P. caerulipes* are major psilocybin-producing mushrooms (Lowe et al. 2021).

5.7 Medicinal Potential of Poisonous Macrofungi

The majority of toxic mushrooms produce bioactive substances that are poisonous by nature, antibiotics, antiviral, hallucinogenic, or have the potential to glow in the dark (Elkhateeb et al. 2019). Mushroom toxicity can safeguard the basidiocarp's ability to function by producing toxins that render the mushroom unfit for human eating. There may be potential medical uses for the poisonous cyclic peptides that mushrooms create that can enter the human bloodstream and target certain cells (de Mattos-Shipley et al. 2016, Fokunang et al. 2022).

The majority of hallucinogenic mushrooms are well-known to be used extensively in traditional folk medicine across the globe. *Psilocybe zapotecorum*, a hallucinogenic mushroom, is one example of a psychoactive mushroom that has been sacramentally utilized in ceremonies to promote physical and mental healing as well as visionary abilities (Xin et al. 2018). *Lecocoprinus birnbaumii*, often known as the flowerpot parasol mushroom, grows naturally on decomposing organic matter in Africa, North America, and Europe (El-Fallal et al. 2019). Agaritine, a chemical found in *Agaricus* species is an aromatic hydrazine derivative and is cancer-causing. Although this substance has the potential to be antiviral and has an antitumor, it may be hazardous to both animals and humans (Vignesh and Ghosh 2021). The most important bioactive component that *Cordyceps* produces is called cordycepin, or 3'-deoxyadenosine (Ashraf et al. 2020). This compound has numerous nutraceutical and medicinal benefits, including those for anti-aging, anti-arthritis, anti-cancer, antidiabetic, anti-fungal, anti-hyperlipidemia, anti-inflammatory, anti-osteoporotic, antioxidant, antiviral, hepato-protective, hypo-sexuality, immunomodulatory, weight-regulating, and many more pharmaceutical applications in cardiovascular diseases (Qin et al. 2019, Tan et al. 2021).

6. Conclusion and Future Trends

Macrofungi are widely known in both traditional and modern medicine. An interest in mushrooms as a source of bioactive compounds with therapeutic potential— including drugs with antiviral, antimicrobial anticancer, immunopotentiating, hypocholesterolemic, and hepatoprotective properties. The most prevalent bioactive compounds, including vitamins, antimicrobials, and chemicals against cancer—are covered. The use of three macrofungi as medical examples, including *Ganoderma*, true morels, and desert truffles, are discussed. Furthermore, it is important to consider the medical potential of poisonous and medicinal macrofungi. The best way to exploit the bioactive chemicals produced by macrofungi for upcoming biotechnological applications is recommended.

References

Ahmed, E.M.S. 2014. Molecular and physiologicalstudies on growth and fruiting of some wild and cultivated *Agaricus* species. MSc. Thesis, Faculty of Science, Damietta University, Egypt.

Akyuz, M. and E.R.E.N. Ayşe. 2018. Bazı makrofungus misellerin antimikrobiyal aktivitelerinin belirlenmesi. *Mantar Dergisi*, 9(2): 196–205.

AlAhmed, A. and H.E. Khalil. 2019. Antidiabetic activity of *Terfezia claveryi*; an *in vitro* and *in vivo* study. *Biomed. Pharmacol. J.*, 12: 603–608.

Allı, H., S.S. Candar and I. Akata. 2017. Macrofungal diversity of Yalova Province. *J. Fungus*, 8(2): 76–84. 10.15318/fungus.2017.36.

Al-Mazaideh, G.M., F.K. Al-Swailmi and M.U.R. Parrey. 2021. Molecular docking study reveals naringenin and hesperetin from desert truffles as promising potential inhibitors for coronavirus (COVID-19). *Ann. Clin. Anal. Med.*, 12(9): 980–985.

Anusiya, G., Prabhu U. Gowthama, N.V. Yamini, N. Sivarajasekar, K. Rambabu, G. Bharath and F. Banat. 2021. A review of the therapeutic and biological effects of edible and wild mushrooms. *Bioengineered*, 12(2): 11239–11268. 10.1080/21655979.2021.2001183.

Ashraf, S.A., A.E.O. Elkhalifa, A.J. Siddiqui, M. Patel, A.M. Awadelkareem, M. Snoussi, M.S. Ashraf, M. Adnan and S. Hadi. 2020. Cordycepin for health and wellbeing: A potent bioactive metabolite of an entomopathogenic medicinal fungus *Cordyceps* with its nutraceutical and therapeutic potential. *Molecules*, 25: 2735. 10.3390/molecules25122735.

Azeem, U., K.R. Hakeem and M. Ali. 2020. Toxigenic fungi. In Fungi for Human Health. Springer, Cham., 41–48.

Balandaykin, M.E. and I.V. Zmitrovich. 2015. Review on Chaga medicinal mushroom, *Inonotus obliquus* (Higher Basidiomycetes): Realm of medicinal applications and approaches on estimating its resource potential. *Inter. J. Med. Mushrooms*, 17(2): 95–104. 10.1615/IntJMedMushrooms.v17.i2.10.

Batra, P., A.K. Sharma and R. Khajuria. 2013. Probing Lingzhi or Reishi medicinal mushroom *Ganoderma lucidum* (higher Basidiomycetes): A bitter mushroom with amazing health benefits. *Inter. J. Med. Mushrooms*, 15(2): 127–143. 10.1615/IntJMedMushr.v15.i2.20.

Beara, I.N., M.M. Lesjak, D.D, Četojević-Simin, Z.S. Marjanović, J.D. Ristić and Z.O. Mrkonjić. 2014. Phenolic profile, antioxidant, anti-inflammatory and cytotoxic activities of black (*Tuber aestivum* Vittad.) and white (*Tuber magnatum* Pico) truffles. *Food Chem.*, 165: 460–466. 10.1016/j.foodchem.2014.05.116.

Bernas, E., G. Jaworska and Z. Lisiewska. 2006. Edible mushrooms as a source of valuable nutritive constituents. *Acta Scientiarum Polonorum, Technol. Alimentaria*, 5: 5–20.

Bhat, Z.A., A.H. Wani, M.Y. Bhat and A.R. Malik. 2019. Major bioactive triterpenoids from *Ganoderma* species and their therapeutic activity: A review. *Asian J. Pharm. Clin. Res.*, 12(4): 22–30.

Bisen, P.S., R.K. Baghel, B.S. Sanodiya, G.S. Thakur and G.B.K.S. Prasad. 2010. *Lentinus edodes*: a macrofungus with pharmacological activities. *Current Med. Chemist.*, 17(22): 2419–2430. 10.2174/092986710791698495.

Boh, B., M. Berovic, J. Zhang and L. Zhi-Bin. 2007. *Ganoderma lucidum* and its pharmaceutically active compounds. *Biotechnol. Ann. Rev.*, 13: 265–301. 10.1016/S1387-2656(07)13010-6.

Bradai, L., S. Neffar, K. Amrani, S. Bissati and H. Chenchouni. 2015. Ethnomycological survey of traditional usage and indigenous knowledge on desert truffles among the native Sahara Desert people of Algeria. *J. Ethnopharmacol.*, 13(162): 31–38. 10.1016/j.jep.2014.12.031.

Brkljača, R. and S. Urban. 2015. Rapid dereplication and identification of the bioactive constituents from the fungus, *Leucocoprinus birnbaumii*. *Nat. Prod. Comm.*, 10: 95–98. 10.1177/1934578X1501000124.

Buller, A.H.R. 1914. The fungus lore of the Greeks and Romans. *Trans. Br. Mycol. Soc.*, 5: 21–66.

Buys, M., B. Conlon, H.D.F. Licht, D. Aanen, M. Poulsen and Z. De Beer. 2018. Searching for *Podaxis* onthe trails of early explorers in southern Africa. *South Afr. J. Bot.*, 115: 317. 10.1016/j.sajb.2018.02.150.

Camin, F., R. Larcher, G. Nicolini, L. Bontempo, D. Bertoldi, M. Perini, C. Schlicht, A. Schellenberg, F. Thomas, K. Heinrich and S. Voerkelius. 2010. Isotopic and elemental data for tracing the origin of European olive oils. *J. Agric. Food Chem.*, 58: 570–577. 10.1021/jf902814s.

Campi, M., C. Mancuello, F. Ferreira, Y. Maubet, E. Cristaldo and D. Benítez. 2019. Preliminary evaluation of phenolic compounds, antioxidant activity and bioactive compounds in some species of basidiomycetes fungi from Paraguay. *Steviana*, 11(1): 26–41. 10.56152/StevianaFacenV11N1A3_2019.

Cardwell, G., J.F. Bornman, A.P. James and L.J. Black. 2018. A review of mushrooms as a potential source of dietary vitamin D. *Nutrients*, 10(10): 1498.

Carrillo, C., M. del Cavia and S.R. Alonso-Torre. 2012. Antitumor effect of oleic acid; mechanisms of action. A review. *Nutr. Hosp.*, 27: 1860–1865. 10.3305/nh.2012.27.6.6010.

Casarica, A., M. Moscovici, M. Daas, I. Nicu, M. Panteli and I. Rasit. 2016. A purified extract from brown truffles of the species *Terfezia claveryi* chatin and its antimicrobial activity. *Farmacia*, 64: 298–301.

Chang, S.T. and P.G. Miles. 2004. Mushrooms: Cultivation, nutritional value, medicinal effect, and environmental impact, 264–265. 10.1201/9780203492086.

Cho, J.T. and J.H. Han. 2016. A case of mushroom poisoning with russula subnigricans: development of rhabdomyolysis, acute kidney injury, cardiogenic shock, and death. *J. Korean Med. Sci.*, 31: 1164. 10.3346/jkms.2016.31.7.1164.

Cör Andrejč, D., Ž. Knez and M. Knez Marevci. 2022. Antioxidant, antibacterial, antitumor, antifungal, antiviral, anti-inflammatory, and nevro-protective activity of *Ganoderma lucidum*: An overview. *Front. Pharm.*, 2757.

Dahham, S.S., S.S. Al-Rawia, A.H. Ibrahim, A.S.A. Majid and Abdul A.M.S. Majid. 2018. Antioxidant, anticancer, apoptosis properties and chemical composition of Black Truffle *Terfezia claveryi*. *Saudi J. Biol. Sci.*, 25: 1524–1534. 10.1016/j.sjbs.2016.01.031.

D'Archivio, M., C. Filesi, R. Vari, B. Scazzocchio and R. Masella. 2010. Bioavailability of polyphenols: Status and controversies. *Int. J. Mol. Sci.*, 11(4): 1321–1342.

Das, B., B. De, R. Chetree and S.C. Mandal. 2020. Medicinal aspect of mushrooms: A view point. pp. 509–532. *In*: Herbal Medicine in India. Springer, Singapore. 10.1007/978-981-13-7248-3_31.

de Mattos-Shipley, K.M., K.L. Ford, F. Alberti, A.M. Banks, A.M. Bailey and G.D. Foster. 2016. The good, the bad and the tasty: The many roles of mushrooms. *Stud. Mycol.*, 85(1): 125–157. 10.1016/j.simyco.2016.11.002.

de Menezes Filho, A.C.P., M.V.A Ventura, I. Alves, A.S. Taques, H.R.F Batista-Ventura, C.F. de Souza Castro, M.B. Teixeira and F.A.L. Soares. 2022. Phytochemical prospection, total flavonoids and total phenolic and antioxidant activity of the mushroom extract *Scleroderma verrucosum* (Bull.) Pers. *Brazil. J. Sci.*, 1(1): 1–4. 10.14295/bjs.v1i1.2.

Dehariya, P., A. Kushwaha and A. Dehariya. 2020. Effect of poisonous mushrooms on humankind. *In*: An Overview of Toxicants, Vishwagayan Prakashan Delhi, India. ISBN: 978-93-83837-97-7.

Diaz, J.H. 2021. Nephrotoxic mushroom poisoning: Global epidemiology, clinical manifestations, and management. *Wilderness Environ. Med.*, 32(4): 537–544. 10.1016/j.wem.2021.09.002.

Dib-Bellahouel, S. and Z. Fortas. 2011. Antibacterial activity of various fractions of ethyl acetate extract from the desert truffle, *Tirmania pinoyi*, preliminarily analysed by gas chromatography–mass spectrometry (GC–MS). *Afr. J. Biotechnol.*, 10: 9694–9699. 10.5897/AJB10.2687.

Dimitrijevic, M.V., V.D. Mitic, O.P. Jovanovic, V.P. Stankov Jovanovic, J.S. Nikolic, G.M. Petrovic and G.S. Stojanovic. 2018. Comparative study of fatty acids profile in eleven wild mushrooms of boletacea and russulaceae families. *Chem. Biodiv.*, 15(1): e1700434. 10.1002/cbdv.201700434.

Dinis-Oliveira, R.J., M. Soares, C. Rocha-Pereira and F. Carvalho. 2016. Human and experimental toxicology of orellanine. *Human Exper. Toxicol.*, 35: 1016–1029. 10.1177/0960327115613845.

Dudekula, U.T., K. Doriya and S.K. Devarai. 2020. A critical review on submerged production of mushroom and their bioactive metabolites. *Biotech.*, 10: 337. 10.1007/s13205-020-02333-y.

El-Fallal, A.A., A.K.A. El-Sayed and H.M. El-Gharabawy. 2013. Induction of low sporulating-UV mutant of oyster mushroom with high content of vitamin D2. Eg. J. Bot. (3rd. International Conf.), 17–18.

El-Fallal, A.A., A.K.A. Elsayed and H.M. El-Gharabawy. 2017. First record of *Lepista sordida* (Schumach) singer in Eastern North Africa. *Eg. J. Bot.*, 57 (7th. International Conf.), 111–118. 10.21608/ejbo.2017.980.1087.

El-Fallal, A., A.K. El-Sayed and H.M. El-Gharabawy. 2019. *Podaxis pistillaris* (L.) Fr. and *Leucocoprinus birnbaumii* (Corda) Singer; new addition to macrofungi of Egypt. *Eg. J. Bot.*, 59(2): 413–423. 10.21608/ejbo.2019.5990.1255.

El-Fallal, A.A., A.K. El-Sayed, M.F. El-Fawal and H.M. El-Gharabawy. 2021. Biocontrol of mycotoxigenic fungi in feedstuff using spices and *Ganoderma* mushroom. *Catrina: The International J. Environ. Sci.*, 24(1): 65–73. 10.21608/cat.2022.107327.1112.

El-Fallal, A.A., T.M. El-Katony, M.M. Nour El-Dein, N.G. Ibrahim and H.M. El-Gharabawy. 2022. Optimizing the antioxidant activity of solid state fermentation systems with *Pleurotus floridanus* and *Paecilomyces variotii* on rice straw. *Eg. J. Bot.*, 62(2): 523–535.

El-Gharabawy, H.M. 2012. Study on Enhanced UV Radiation Effect on Some Physiological, Biochemical and Molecular Activities of Oyster Mushroom. M.Sc. Thesis, Botany and Microbiology Department, Faculty of Science, Damietta University, Egypt.

El-Gharabawy, H.M., A.A. El-Fallal and K.A.A. El-Sayed. 2019. Description of a yellow morel from Egypt using morphological and molecular tools. *Nova Hedwigia*, 109(1): 95–110. 10.1127/nova_hedwigia/2019/0535.

El-Gharabawy, H.M., C.A. Leal-Dutra and G.W. Griffith. 2021. *Crystallicutis* gen. nov. (Irpicaceae, Basidiomycota), including *C. damiettensis* sp. nov., found on *Phoenix dactylifera* (date palm) trunks in the Nile Delta of Egypt. *Fungal Boil.*, 125(6): 447–458. 10.1016/j.funbio.2021.01.004.

El-Sharkawy A. and A. Malki. 2020. Vitamin D signaling in inflammation and cancer: Molecular mechanisms and therapeutic implications. *Molecules*, 25: 3219. 10.3390/molecules25143219.

El Sheikha, A.F. 2022. Nutritional profile and health benefits of *Ganoderma lucidum* "Lingzhi, Reishi, or Mannentake" as functional foods: Current scenario and future perspectives. *Foods*, 11(7): 1030. 10.3390/foods11071030.

El-Sherif, N.F., S.A. Ahmed, A.K. Ibrahim, E.S. Habib, A.A. El-Fallal, A.K. El-Sayed and A.E. Wahba. 2020. Ergosterol peroxide from the Egyptian red lingzhi or reishi mushroom, *Ganoderma resinaceum* (Agaricomycetes), showed preferred inhibition of MCF-7 over MDA-MB-231 breast cancer cell lines. *Inter. J. Med. Mushrooms*, 22(4): 389–396. 10.1615/IntJMedMushrooms.2020034223.

Elkhateeb, W. and G.M. Daba. 2022. Bioactive potential of some fascinating edible mushrooms *Flammulina, Lyophyllum, Agaricus, Boletus, Letinula, Pleurotus* as a treasure of multipurpose therapeutic natural product. *J. Pharm. Res.*, 6(1): 1–10. 10.23880/oajpr-16000263.

Elkhateeb, W.A., G.M. Daba, P.W. Thomas and T.-C. Wen. 2019. Medicinal mushrooms as a new source of natural therapeutic bioactive compounds. *Eg. Pharm. J.*, 18: 88–101. 10.4103/epj.epj_17_19.

Feleke, H.T. and A. Doshi. 2018. Antimicrobialactivity and bioactive compounds of Indian wild Mushrooms. *Indian J. Natur. Products and Res.*, 8(3): 254–262.

Ferreira, I.C., S.A. Heleno, F.S. Reis, D. Stojkovic, M.J.R. Queiroz, M.H. Vasconcelos and M. Sokovic. 2015. Chemical features of *Ganoderma* polysaccharides with antioxidant, antitumor and antimicrobial activities. *Phytochem.*, 114: 38–55. 10.1016/j.phytochem.2014.10.011.

Fogarasi, M., Z.M. Diaconeasa, C.R. Pop, S. Fogarasi, C.A. Semeniuc, A.C. Fărcaș, D. Tibulcă, C.-D. Sălăgean, M. Tofană and S.A. Socaci. 2020. Elemental composition, antioxidant and antibacterial properties of some wild edible mushrooms from romania. *Agronomy*, 10(12): 1972. https://doi.org/10.3390/agronomy10121972.

Fokunang, E.T., M.G. Annih, L.E. Abongwa, M.E. Bih, T.M. Vanessa, D.J. Fomnboh and C. Fokunang. 2022. Medicinal mushroom of potential pharmaceutical toxic importance: Contribution in phytotherapy. *In*: Shiomi, N. and Savitskaya, A. (eds.). Current Topics in Functional Food. IntechOpen. 10.5772/intechopen.103845.

Fortas, Z., D.B. Soulef and C. Aibeche. 2022. Characterization of a rare habitat of Terfezia boudieri Chatin in the coastal dunes of northwestern Algeria. *South Asian J. Exper. Biol.*, 12(5): 651–660.

Friedman, M.J. 2015. Chemistry, nutrition, and health-promoting properties of *Hericium erinaceus* (Lion's Mane) Mushroom fruiting bodies and mycelia and their bioactive compounds. *Agricultural Food Chem.*, 63(32): 7108–7123. 10.1021/acs.jafc.5b02914.

Friedman, M. 2016. Mushroom polysaccharides: Chemistry and antiobesity, antidiabetes, anticancer, and antibiotic properties in cells, rodents, and humans. *Foods*, 5(80): 1–40. 10.3390/foods5040080.

Garibay-Orijel, R., J. Caballero, A. Estrada-Torres and J. Cifuentes. 2007. Understanding cultural significance, the edible mushrooms case. *J. Ethnobiol. Ethnomed.*, 3: 1–18. 10.1186/1746-4269-3-4.

Garofalo, C., A. Osimani, V. Milanović, M. Taccari, F. Cardinali, L. Aquilanti, P. Riolo, S. Ruschioni, N. Isidoro and F. Clementi. 2017. The microbiota of marketed processed edible insects as revealed by high-throughput sequencing. *Food Microbiol.*, 62: 15–22. 10.1016/j.fm.2016.09.012.

Gil-Ramírez, A., C. Pavo-Caballero, E. Baeza, N. Baenas, C. Garcia-Viguera, F.R. Marín and C. Soler-Rivas. 2016. Mushrooms do not contain flavonoids. *J. Funct. Foods*, 25: 1–13. 10.1016/j.jff.2016.05.005.

Gouzi, H., L. Belyagoubi, K.N. Abdelali and A. Khelifi. 2011. *In vitro* antibacterial activities of aqueous extracts from Algerian desert truffles (*Terfezia* and *Tirmania*, Ascomycetes) against *Pseudomonas aeruginosa* and *Staphylococcus aureus*. *Int. J. Med. Mushrooms*, 13(6): 553–558. 10.1615/intjmedmushr.v13.i6.70.

Govorushko, S., R. Rezaee, J. Dumanov and A. Tsatsakis. 2019. Poisoning associated with the use of mushrooms: A review of the global pattern and main characteristics. *Food. Chem. Toxicol.*, 128: 267–279. 10.1016/j.fct.2019.04.016.

Grienke, U., M. Zöll, U. Peintner and J.M. Rollinger. 2014. European medicinal polypores—A modern view on traditional uses. *J. Ethnopharmacol.*, 154(3): 564–583. 10.1016/j.jep.2014.04.030.

Hamza, A., H. Jdir and N. Zouari. 2016. Nutritional, antioxidant and antibacterial properties of *Tirmania nivea*, a wild edible desert truffle from Tunisia arid zone. *Med. Aromat. Pl.*, 5: 258. 10.4172/2167-0412.1000258.

Hassan, A.A., A.E. Al-Kurtany, E.M. Jbara and K.F. Saeed. 2011. Evaluation of an edible mushroom *Pleurotus* sp. efficiency against to plant pathogens: Nematodes and soil fungi. pp. 431–47. *In* Proceedings of the 5th Scientific Conference of College of Agriculture, Tikrit University, Iraq.

He, M.-Q., M.-Q. Wang, Z.-H. Chen, W.-Q. Deng, T.-H. Li, A. Vizzini, R. Jeewon, K.D. Hyde and R.-L. Zhao. 2022. Potential benefits and harms: A review of poisonous mushrooms in the world. *Fungal Biol. Rev.*, 10.1016/j.fbr.2022.06.002

Hereher, F., A. ElFallal, E. Toson and M. Abou-Dobara. 2018. Antitumor effect of a crude protein-bound polysaccharide substance extracted from *Volvariella speciosa*. *Bioactive Carbohydrates and Dietary Fibre*, 16: 75–81. 10.1016/j.bcdf.2018.07.001.

Ivanova, T.S., T.A. Krupodorova, V.Y. Barshteyn, A.B. Artamonova and V.A. Shlyakhovenko. 2014. Anticancer substances of mushroom origin. *Exp. Oncol.*, 36(2): 58–66. dspace.nbuv.gov.ua/handle/123456789/145333.

Kamble, R., S. Venkata and A.M. Gupte. 2011. Antimicrobial activity of *Ganoderma lucidum* mycelia. *J. Pure Appl. Microbiol.*, 5(2): 983–986.

Khabar, L. 2014. Mediterranean basin: North Africa. pp. 143–158. *In*: Kagan-Zur, V., Roth-Bejerano, N., Sitrit, Y. and Morte, A. (eds.). Desert Truffles, Soil Biology, vol 38, Springer, Berlin, Heidelberg, 10.1007/978-3-642-40096-4_10.

Khadri, H., Y.H. Aldebasi and K. Riazunnisa. 2016. Truffle mediated (*Terfezia claveryi*) synthesis of silver nanoparticles and its potential cytotoxicity in human breast cancer cells (MCF-7). *Afr. J. Biotechnol.*, 16: 1278–84. 10.5897/AJB2017.16031.

Kim, S.P., M.Y. Kang, J.H. Kim, S.H. Nam and M. Friedman. 2011. Composition and mechanism of antitumor effects of *Hericium erinaceus* mushroom extracts in tumor-bearing mice. *J. Agric. Food Chem.* 59(18): 9861–9869. 10.1021/jf201944n.

Kirchmair, M., P. Carrilho, R. Pfab, B. Haberl, J. Felgueiras, F. Carvalho, J. Cardoso, I. Melo, J. Vinhas and S. Neuhauser. 2012. Amanita poisonings resulting in acute, reversible renal failure: New cases, new toxic *Amanita* mushrooms. *Nephrology, Dialysis, Transplantation*, 27(4): 1380–1386. 10.1093/ndt/gfr511.

Kıvrak, İ. 2015. Analytical methods applied to assess chemical composition, nutritional value and in vitro bioactivities of *Terfezia olbiensis* and *Terfezia claveryi* from Turkey. *Food Anal. Methods*, 8: 1279–12793. 10.1007/s12161-014-0009-2.

Kotowski, M.A. 2019. History of mushroom consumption and its impact on traditional view on mycobiota—An example from Poland. *Microbial Biosyst.*, 4(3): 1–13. 10.21608/mb.2019.61290.

Kozarski, M., A. Klaus, D. Jakovljevic, N. Todorovic, J. Vunduk, P. Petrović, M. Niksic, M. Vrvic and L.V. Griensven. 2015. Antioxidants of edible mushrooms. *Molecules*, 20(10): 19489–19525. 10.3390/molecules201019489.

Kumar, K., R. Mehra, R.P.F. Guiné, M.J. Lima, N. Kumar, R. Kaushik, N. Ahmed, A.N. Yadav and H. Kumar. 2021. Edible Mushrooms: A comprehensive review on bioactive compounds with health benefits and processing aspects. *Foods*, 10(12): 2996. 10.3390/foods10122996.

Kurutas, E.B. 2016. The importance of antioxidants which play the role in cellular response against oxidative/nitrosative stress: Current state. *Nutr. J.*, 15(1): 71. 10.1186/s12937-016-0186-5.

Lakhanpal, T.N. 2010. Biology of Indian Morels, IK International Pvt. Ltd., New Delhi.

Lawal, T.O., S.M. Wicks and G.B. Mahady. 2017. *Ganoderma lucidum* (Lingzhi): The impact of chemistry on biological activity in cancer. *Curr. Bioactive Compounds*, 13(1): 28–40. 10.2174/15734072126 66160614074801.

Le Daré, B., P.-J. Ferron and T. Gicquel. 2021. Toxic effects of amanitins: Repurposing toxicities toward new therapeutics. *Toxins*, 13(6): 417. 10.3390/toxins13060417.

Lee, H.S., E.J. Kim and S.H. Kim. 2015. Ethanol extract of *Innotus obliquus* (Chaga mushroom) induces G1 cell cycle arrest in HT-29 human colon cancer cells. *Nutr. Res. Pract.*, 9(2): 111–116. 10.4162/nrp.2015.9.2.111.

Lee, S.R., H.S. Roh, S. Lee, H.B. Park, T.S. Jan, Y-J. Ko, K-H. Baek and K.H. Kim. 2018. Bioactivity-guided isolation and chemical characterization of antiproliferative constituents from morel mushroom (*Morchella esculenta*) in human lung adenocarcinoma cells. *J. Fun. Foods*, 40(12): 249–260. 10.1016/j.jff.2017.11.012.

Li, F., H. Lei and H. Xu. 2021. Influences of subcritical water extraction on the characterization and biological properties of polysaccharides from *Morchella sextelata*. *J. Food Process. Preserv.*, 46: e16024. 10.1111/jfpp.16024.

Li, H., H. Zhang, Y. Zhang, K. Zhang, J. Zhou, Y. Yin, S. Jiang, P. Ma, Q. He, Y. Zhang and K. Wen. 2020. Mushroom poisoning outbreaks—China, 2019. *China CDC Weekly*, 2(2): 19–24.

Li, L.F., H.-B. Liu, Q.W. Zhang, Z.P. Li, T.L. Wong, H.Y. Fung, J.X. Zhang, S.P. Bai, A.P. Lu and Q.B. Han. 2018. Comprehensive comparison of polysaccharides from *Ganoderma lucidum* and *G. sinense*: Chemical, antitumor, immunomodulating and gut-microbiota modulatory properties. *Sci. Rep.*, 8(1): 6172. 10.1038/s41598-018-22885-7.

Li, P., L. Liu, S. Huang, Y. Zhang, J. Xu and Z. Zhang. 2020. Anti-cancer effects of a neutral triterpene fraction from *Ganoderma lucidum* and its active constituents on SW620 human colorectal cancer cells. *Anticancer Agents Med. Chem.*, 20(2): 237–244. 10.2174/1871520619666191015102442.

Li, Z., H. Bao, C. Han and M. Song. 2022. The regular pattern of metabolite changes in mushroom *Inonotus hispidus* in different growth periods and exploration of their indicator compounds. *Sci. Rep.*, 12(1): 1–15. 10.1038/s41598-022-18631-9.

Liao, B., X. Chen, J. Han, Y. Dan, L. Wang, W. Jiao, J. Song and S. Chen. 2015. Identification of commercial *Ganoderma* (Lingzhi) species by ITS2 sequences. *Chin. Med.*, 10(22): 1–9. 10.1186/s13020-015-0056-7.

Lin, Z. and A. Deng. 2019. Antioxidative and free radical scavenging activity of *Ganoderma* (Lingzhi). *Ganoderma and Health*, 1182: 271–297. 10.1007/978-981-32-9421-9_12.

Lincoff, G. 2011. The Complete Mushroom Hunter: An Illustrated Guide to Finding, Harvesting, and Enjoying Wild Mushrooms. Quarry Books.

Lone, F.A., S. Lone, M.A. Aziz and F.A. Malla. 2012. Ethnobotanical studies in the tribal areas of district Kupwara, Kashmir, India. *Int. J. Pharma. Bio. Sci.*, 3(4): 399–411.

Lowe, H., N. Toyang, B. Steele, H. Valentine, J. Grant, A. Ali, W. Ngwa and L. Gordon. 2021. The therapeutic potential of psilocybin. *Molecules*, 26(10): 2948. 10.3390/molecules26102948.

Lu, H., H. Lou, J. Hu, Z. Liu and Q. Chen. 2020. Macrofungi: A review of cultivation strategies, bioactivity, and application of mushrooms. *Comprehensive Reviews in Food Sci. Food Saf.*, 19(5): 2333–2356. 10.1111/1541-4337.12602.

Mahmood, A., R.N. Malik, Z.K. Shinwari and A. Mahmood. 2011. Ethnobotanical survey of plants from Neelum, AzadJammu and Kashmir, Pakistan. *Pakistan J. Bot.*, 43: 105–110.

Malik, M.A., Y. Jan, L.A. Al-Keridis, A. Haq, J. Ahmad, M. Adnan, N. Alshammari, S.A. Ashraf and B.P. Panda. 2022. Effect of Vitamin-D-Enriched edible Mushrooms on Vitamin D status, bone health and expression of CYP2R1, CYP27B1 and VDR gene in Wistar rats. *J. Fungi*, 8(8): 864. 10.3390/jof8080864.

Mekawey, A.A.I. 2015. *Terfezia boudieri* as sources of antitumor and antiviral agent. *World J. Pharm. Pharm. Sci.*, 4: 294–315.

Miao, S., X. Mao, R. Pei, S. Miao, C. Xiang, Y. Lv, X. Yang, J. Sun, S. Jia and Y. Liu. 2013. Antitumor activity of polysaccharides from *Lepista sordida* against laryngocarcinoma *in vitro* and *in vivo*. *Inter. J. Biol. Macromol.*, 60: 235–240. 10.1016/j.ijbiomac.2013.05.033.

Mleczek, M., P. Rzymski, A. Budka, M. Siwulski, A. Jasińska, P. Kalač, B. Poniedziałek, M. Gąsecka and P. Niedzielski. 2018. Elemental characteristics of mushroom species cultivated in China and Poland. *J. Food Compos. Anal.*, 66: 168–178.

Mohammadifar, S., S. Fallahi Gharaghoz, M.R. Asef Shayan and A. Vaziri. 2020. Comparison between antioxidant activity and bioactive compounds of *Ganoderma applanatum* (Pers.) Pat. and *Ganoderma lucidum* (Curt.) P. Karst from Iran. *Iranian J. Plant Physiol.*, 11(1): 3417–3424.

Moradali, M.F., H. Mostafavi, G.A. Hejaroude, A.S. Tehrani, M. Abbasi and S. Ghods. 2006. Investigation of potential anti-bacterial properties of methanol extracts from fungus *Ganoderma applanatum*. *Chemotherapy*, 52(5): 241–244. 10.1159/000094866.

Muhsin, T.M., A.F. Abass and E.K. Al-Habeeb. 2012. *Podaxis pistillaris* (Gasteromycetes) from the desert of southern Iraq, an addition to the known mycota of Iraq. *J. Basrah Res.*, 38(3): 29–35.

Nagy, M., S. Socaci, M. Tofana, E.S. Biris-Dorhoi, D.T. IbulcĂ, G. PetruT and C.L. Salanta. 2017. Chemical composition and bioactive compounds of some wild edible mushrooms. *Bull. Univ. Agric. Sci. Vet. Med. Cluj-Napoc. Food Sci. Technol.*, 74(1): 1–8. 10.15835/buasvmcn-fst:12629.

Neggaz, S., M. Chenni, F.E.H. Zitouni-Haouar and X. Fernandez. 2020. Mycochemical composition and insecticidal bioactivity of Algerian desert truffles extract against two stored-product insects: *Sitophilus oryzae* (L.) (Coleoptera: Curculionidae) and *Rhyzopertha dominica* (F.) (Coleoptera: Bostrychidae). *3 Biotech.*, 10(11): 1–9.

Nicolcioiu, M.B., G. Popa and F. Matei. 2017. Antimicrobial activity of ethanolic extracts made of mushroom mycelia developed in submerged culture. *Scientific Bulletin Series F. Biotechnol.*, 21: 159–164.

Niego, A.G., S. Rapior, N. Thongklang, O. Raspé, W. Jaidee, S. Lumyong and K.D. Hyde. 2021. Macrofungi as a nutraceutical source: Promising bioactive compounds and market value. *J. Fungi (Basel)*, 7(5): 397. 10.3390/jof7050397.

Niksic, M., A. Klaus and D. Argyropoulos. 2015. Safety of foods based on mushrooms. *Reg. Saf. Trad. Ethnic Foods*, 421.

Nithya, M., V. Ambikapathy and A. Panneerselvam. 2013. Studies on antimicrobial potential of different strains of *Ganoderma lucidum* (Curt.: Fr.) P. Karst. *Inter. J. Pharm. Sci. Rev. Res.*, 21(2): 317–320.

Novak, R. and D.M. Shlaes. 2010. The pleuromutilin antibiotics: A new class for human use. *Current Opinion in Investigational Drugs (London, England: 2000)*, 11(2): 182–191.

Okan, O.T., S. Yildiz, A. Yilmaz, J. Barutciyan and I. Deniz. 2014. Wild edible mushrooms having an important potential in east black sea region. *Int. Cauc. For Symp.*, 2014 (October): 673–680.

Ooi, V.E.C. 2008. Antitumor and immunomodulatory activities of mushroom polysaccharides. *Mushrooms Funct. Food*, 147–198. 10.1002/9780470367285.ch5.

Owaid, M.N., R.F. Muslim and H.A. Hamad. 2018. Mycosynthesis of silver nanoparticles using *Terminia* sp. Desert truffle, Pezizaceae, and their antibacterial activity. *Jordan J. Biol. Sci.*, 11(4): 401–405.

Parameswari, V. and P. Chinnaswamy. 2011. An *in vitro* study of the inhibitory effect of *Pleurotus florida* a higher fungi on human pathogens. *J. Pharm. Res.*, 4 (6): 1948–1949. 10.1039/c7fo00895c.

Patel, Y., R. Naraian and V.K. Singh. 2012. Medicinal properties of *Pleurotus* species (oyster mushroom): A review. *W. J. Fungal Pl. Biol.*, 3(1): 1–12.

Pattanayak, M., S. Samanta, P. Maity, D.K. Manna, I.K. Sen, A.K. Nandi, B.C. Panda, S. Chattopadhyay, S. Roy, A.K. Sahoo, N. Gupta, S.S. Islam and S.S. Islam. 2017. Polysaccharide of an edible truffle *Tuber rufum*: Structural studies and effects on human lymphocytes. *Inter. J. Biol. Macromolecules*, 95: 1037–1048. 10.1016/j.ijbiomac.2016.10.092.

Phan, C.W., P. David, M. Naidu, K.H. Wong and V. Sabaratnam. 2015. Therapeutic potential of culinary-medicinal mushrooms for the management of neurodegenerative diseases: Diversity, metabolite, and mechanism. *Crit. Rev. Biotechnol.*, 35(3): 355–368. 10.3109/07388551.2014.887649.

Pogoń, K., A. Gabor, G. Jaworska and E. Bernaś. 2017. Effect of traditional canning in acetic brine on the antioxidants and vitamins in *Boletus edulis* and *Suillus luteus* mushrooms. *J. Food Process. Preserv. 2017*, 41: e12826. 10.1111/jfpp.12826.

Poucheret, P., F. Fons and S. Rapior. 2006. Biological and pharma-cological activity of higher fungi: 20-Year retrospective analysis. *Mycologie*, 27: 311–333.

Power, R.C., D.C. Salazar-García, L.G. Straus, M.R.G. Morales and A.G. Henry. 2015. Microremains from El Mirón Cave human dental calculus suggest a mixed plant–animalsubsistence economy during the Magdalenian in Northern Iberia. *J. Archaeol. Sci.*, 60: 39–46. 10.1016/j.jas.2015.04.003.

Pukalski, J. and D. Latowski. 2022. Secrets of flavonoid synthesis in mushroom cells. *Cells*, 11(19): 3052. 10.3390/cells11193052.

Puschner, B. and C. Wegenast. 2012. Mushroom poisoning cases in dogs and cats: Diagnosis and treatment of hepatotoxic, neurotoxic, gastroenterotoxic, nephrotoxic, and muscarinic mushrooms. *Veterinary Clinics: Small Animal Practice*, 42(2): 375–387. 0.1016/j.cvsm.2011.12.002.

Puttaraju, N.G., S.U. Venkateshaiah, S.M. Dharmesh, S.M. Urs and R. Somasundaram. 2006. Antioxidant activity of indigenous edible mushrooms. *J. Agricul. Food Chem.*, 54: 9764–9772. 10.1021/jf0615707.

Qin, P., X. Li, H. Yang, Z.-Y. Wang and D. Lu. 2019. Therapeutic potential and biological applications of cordycepin and metabolic mechanisms in cordycepin-producing fungi. *Molecules*, 24: 2231. 10.3390/molecules24122231.

Qu, L., S. Li, Y. Zhuo, J. Chen, X. Qin and G. Guo. 2017. Anticancer effect of triterpenes from *Ganoderma lucidum* in human prostate cancer cells. *Oncol. Lett.*, 14(6): 7467–7472. 10.3892/ol.2017.7153.

Ramprasath, V.R. and A.B. Awad. 2015. Role of phytosterols in cancer prevention and treatment. *J. AOAC Int.*, 98(3): 735–738. 10.5740/jaoacint.

Ramos, M., N. Burgos, A. Barnard, G. Evans, J. Preece, M. Graz, A.C. Ruthes, A. Jiménez-Quero, A. Martínez-Abad, F. Vilaplana and L.P. Ngoc. 2019. Agaricus bisporus and its by-products as a source of valuable extracts and bioactive compounds. *Food Chem.*, 292: 176–187. 10.1016/j.foodchem.2019.04.035.

Ramsden, C.E., D. Zamora, B. Leelarthaepin, S.F. Majchrzak-Hong, K.R. Faurot, C.M. Suchindran, A. Ringel, J.M. Davis and J.R. Hibbeln. 2013. Use of dietary linoleic acid for secondary prevention of coronary heart disease and death: Evaluation of recovered data from the Sydney Diet Heart Study and updated meta-analysis. *BMJ*, 346: 1–18. 10.1136/bmj.e8707.

Ramya, H., K.S. Ravikumar, Z. Fathimathu, K.K. Janardhanan, T.A. Ajith, M.A. Shah, R. Farooq and Z.A. Reshi. 2022. Morel mushroom, *Morchella* from Kashmir Himalaya: A potential source of therapeutically useful bioactives that possess free radical scavenging, anti-inflammatory, and arthritic edema-inhibiting activities. *Drug Chem. Toxicol.*, 45(5): 2014–2023. 10.1080/01480545.2021.1894750.

Rathore, H., S. Prasad and S. Sharma. 2017. Mushroom nutraceuticals for improved nutrition and better human health: A review. *Pharma Nutr.*, 5(2): 35–46. 10.1016/j.phanu.2017.02.001.

Robles-Hernández, L., D. Ojeda-Barrios, A. González-Franco, J. Hernández-Huerta, N. Salas-Salazar and O.A. Hernández-Rodríguez. 2017. Susceptibilidad de aislados de *Xanthomonas campestris* pv. vesicatoria a Streptomyces y extractos bioactivos de *Ganoderma*. *Acta Universitaria*, 27(6): 30–39 10.15174/au.2017.1417.

Saitsu, Y., A. Nishide, K. Kikushima, K. Shimizu and K. Ohnuki. 2019. Improvement of cognitive functions by oral intake of *Hericium erinaceus*. *Biomed. Res.*, 40(4): 125–131. doi: 10.2220/biomedres.40.125.

Sedky, N.K., Z.H. El Gammal, A.E. Wahba, E. Mosad, Z.Y. Waly, A.A. El-Fallal, R.K. Arafa and N. El-Badri. 2018. The molecular basis of cytotoxicity of α-spinasterol from *Ganoderma resinaceum*: Induction of apoptosis and overexpression of p53 in breast and ovarian cancer cell lines. *J. Cell Biochem.*, 119(5): 3892–3902. doi: 10.1002/jcb.26515.

Shaffique, S., S.M. Kang, A.Y. Kim, M. Imran, M. Aaqil Khan and I.J. Lee. 2021. Current knowledge of medicinal mushrooms related to anti-oxidant properties. *Sustainability*, 13(14): 7948. 10.3390/su13147948.

Shahidi, F. and P. Ambigaipalan. 2015. Phenolics and polyphenolics in foods, beverages and spices: Antioxidant activity and health effects–A review. *J. Functional Foods*, 18: 820–897. 10.1016/j.jff.2015.06.018.

Sher, H., A. Aldosari and R.W. Bussmann. 2015. Morels of Palas Valley, Pakistan: A potential source for generating income and improving livelihoods of mountain communities. *Economic Bot.*, 69(4): 345–359. 10.1007/s12231-015-9326-7.

Shiao, M.S. 2003. Natural products of the medicinal fungus *Ganoderma lucidum*: Occurrence, biological activities, and pharmacological functions. *The Chemical Record*, 3(3): 172–180. 10.1002/tcr.10058.

Shikongo, L.T. 2012. Analysis of the mycochemicals components of the indigenous Namibian *Ganoderma* Mushrooms. M.Sc. Thesis, Faculty of Science, Department of Biological Sciences, University of Namibia, Namibia.

Smania, A., F. Delle Mondache, E.F. Smania and R.S. Cuneo. 1999. Antibacterial activity of steroidal compounds from *Ganoderma applanatum* (Pers.) Pat. (Aphyllophoromycetideae) fruit body. *Int. J. Med. Mushr.*, 1: 325–330.

Stojković, D.S., S. Davidović, J. Živković, J. Glamočlija, A. Ćirić, M. Stevanović and M. Soković. 2013. Comparative evaluation of antimutagenic and antimitotic effects of *Morchella esculenta* extracts and protocatechuic acid. *Frontiers in Life Science*, 7(3-4): 218–223. 10.1080/21553769.2014.901925.

Sushila, R., R. Dharmender, D. Rathee, V. Kumar and P. Rathee. 2012. Mushrooms as therapeutic agents. *Brazilian J. Pharmacognosy*, 22(2): 459–474.

Tan, L., X. Song, Y. Ren, M. Wang, C. Guo, D. Guo, Y. Gu, Y. Li, Z. Cao and Y. Deng. 2021. Anti-inflammatory effects of cordycepin: A review. *Phyther. Res.*, 35: 1284–1297. 10.1002/ptr.6890.

Taşkin, H., H.H. Doğan and S. Büyükalaca. 2015. *Morchella galilaea*, an autumn species from Turkey. *Mycotaxon*, 130(1): 215–221.

Tejedor-Calvo, E., K. Amara, F.S. Reis, L. Barros, A. Martins, R.C. Calhelha, M.E. Venturini, D. Blanco, D. Redondo, P. Marco and I.C. Ferreira. 2021. Chemical composition and evaluation of antioxidant, antimicrobial and antiproliferative activities of Tuber and Terfezia truffles. *Food Res. Inter.*, 140: 110071.

Thakur, M., I. Sharma and A. Tripathi. 2021. Ethnomedicinal aspects of morels with special reference to *Morchella esculenta* (Guchhi) in Himachal Pradesh (India): A Review. *Curr. Res. Environ. & App. Mycol.*, 11(1): 284–293. 10.5943/cream/11/1/21.

Tirta Ismaya, W., R.R. Tjandrawinata and H. Rachmawati. 2020. Lectins from the edible mushroom *Agaricus bisporus* and their therapeutic potentials. *Molecules*, 25(10): 2368. 10.3390/molecules25102368.

Tripathi, N.N., P. Singh and P. Vishwakarma. 2017. Biodiversity of macrofungi with special reference to edible forms: A review. *J. Indian. Bot. Soc.*, 96: 144–87.

Turkoglu, A., M.E. Duru, N. Mercan, I. Kivrak and K. Gezer. 2007. Antioxidant and antimicrobial activities of *Laetiporus sulphureus* (Bull.) *Murill. Food Chem.*, 101: 267–273. 10.1016/j.foodchem.2006.01.025.

Unsal, V., T. Dalkıran, M. Çiçek and E. Kölükçü. 2020. The role of natural antioxidants against reactive oxygen species produced by cadmium toxicity: A review. *Adv. Pharm. Bull.*, 10(2): 184–202. 10.34172/apb.2020.023.

Vahdani, M., S. Rastegar, M. Rahimzadeh, M. Ahmadi and A. Karmostaji. 2017. Physicochemical characteristics, phenolic profile, mineral and carbohydrate contents of two truffle species. *J. Agric Sci. Technol.*, 19(5): 1091–101.

Valverde, M.E., T. Hernández-Pérez and O. Paredes-López. 2015. Edible mushrooms: Improving human health and promoting quality life. *Int. J. Microbiol.*, 2015: 376387. 10.1155/2015/376387.

Veeraraghavan, V.P., S. Hussain, J. Papayya Balakrishna, L. Dhawale, M. Kullappan, J. Mallavarapu Ambrose and S. Krishna Mohan. 2021. A comprehensive and critical review on ethnopharmacological importance of desert truffles: *Terfezia claveryi*, *Terfezia boudieri*, and *Tirmania nivea*. *Food Rev. Int.*, 1–20. 10.1080/87559129.2021.1889581.

Verma, N., A. Bhalla and S. Singh. 2019. Mushroom poisoning. *Principles and Practice of Critical Care Toxicology*, p. 367.

Vieira, V., Â. Fernandes, L. Barros, J. Glamočlija, A. Ćirić, D. Stojković, A. Martins, M. Soković and I.C. Ferreira. 2016. Wild *Morchella conica* Pers. from different origins: A comparative study of nutritional and bioactive properties. *J. Sci. Food. Agric.*, 96: 90–98. 10.1002/jsfa.7063.

Vignesh, K. and M. Ghosh. 2021. Effect of anti-cancerous compounds on cancer cells extracted from edible mushrooms: A review. *Preprints*, 2021090341. 10.20944/preprints202109.0341.v1.

Vlasenko, V.A., T.N. Ilyicheva, I.V. Zmitrovich, D. Turmunkh, B. Dondov, T.V. Teplyakova, O. Enkhtuya, K. Nyamsuren, J. Samiya, U. Altangerel and S.V. Asbaganov. 2002. First data on antiviral activity of aqueous extracts from medicinal mushrooms from the Altai mountains in Russia against Influenza Virus Type A. *Inter. J. Med. Mushrooms*, 23(12).

Vo, K.T., M.E. Montgomery, S.T. Mitchell, P.H. Scheerlinck, D.K. Colby, K.H. Meier, S. Kim-Katz, I.B. Anderson, S.R. Offerman, K.R. Olson and C.G. Smollin. 2017. *Amanita phalloides* mushroom poisonings—Northern California. *Morbidity and Mortality Weekly Report*, 66(21): 549–553. 10.15585/mmwr.mm6621a1.

Volpato, G.P., P. Kourková and V. Zelený. 2012. Healing war wounds and perfuming exile: Theuse of vegetal, animal, and mineral products for perfumes, cosmetics, and skin healing among Sahrawi refugees of Western Sahara. *J. Ethnobiol. Ethnomed.*, 8: 49. 10.1186/1746-4269-8-49.

Wahba, A.E., A.K.A. El-Sayed, A.A. El-Falal and E.M. Soliman. 2019. New antimalarial lanostane triterpenes from a new isolate of Egyptian *Ganoderma* species. *Med. Chem. Res.*, 28(12): 2246–2251. doi.org/10.1007/s00044-019-02450-1.

Wang, X.M., J. Zhang, L.H. Wu, Y.L. Zhao, T. Li and J.Q. Li. 2014. A mini-review of chemical composition and nutritional value of edible wild-grown mushroom from China. *Food Chem.*, 151: 279–285. 10.1016/j.foodchem.2013.11.062.

Wasser, S.P. 2010. Medicinal mushroom science: History, current status, future trends, and unsolved problems. *Inter. J. Med. Mushrooms*, 12(1): 1–16. 10.1615/IntJMedMushr.v12.i1.10.

Wasser, S.P. 2011. Current findings, future trends, and unsolved problems in studies of medicinal mushrooms. *Appl. Microbiol. Biotechnol.*, 89(5): 1323–1332. 10.1007/s00253-010-3067-4.

Wen, Y., D. Peng, C. Li, X. Hu, S. Bi, L. Song, B. Peng J. Zhu, Y. Chen and R. Yu. 2019. A new polysaccharide isolated from *Morchella importuna* fruiting bodies and its immunoregulatory mechanism. *Int. J. Biol. Macromol.*, 137: 8–19. 10.1016/j.ijbiomac.2019.06.171.

White, J., S. Weinstein, L. De Haro, R. Bedry, A. Schaper, B. Rumack and T. Zilker. 2015. Mushroom poisoning: A proposed new clinical classification. *Toxicon.*, 53(4): 341–342. 10.1016/j.toxicon.2018.11.007.

White, J., S.A. Weinstein, L. De Haro, R. Bédry, A. Schaper, B.H. Rumack and T. Zilker. 2019. Mushroom poisoning: A proposed new clinical classification. *Toxicon.*, 157: 53–65. 10.1016/j.toxicon.2018.11.00.

Winnie, C.S., A.I. Stephen and C.M. Josphat. 2019. Antimicrobial activity of Basidiomycetes fungi isolated from a Kenyan tropical forest. *African J. Biotechnol.*, 18(5): 112–123.

Wong, J.H., T.B. Ng, H.H.L. Chan, Q. Liu, G.C.W. Man, C.Z. Zhang, S. Guan, C.C.W. Ng, E.F. Fang, H. Wang and F. Liu. 2020. Mushroom extracts and compounds with suppressive action on breast cancer: Evidence from studies using cultured cancer cells, tumor-bearing animals, and clinical trials. *Appl. Microbiol. Biotechnol.*, 104(11): 4675–4703. 10.1007/s00253-020-10476-4.

Wu, F., L.W. Zhou, Z.L. Yang, T. Bau, T.H. Li and Y.C. Dai. 2019. Resource diversity of Chinese macrofungi: Edible, medicinal and poisonous species. *Fungal Diversity*, 98(1): 1–76.

Wu, H., J. Chen, J. Li, Y. Liu, H.J. Park and L. Yang. 2021. Recent advances on bioactive ingredients of *Morchella esculenta*. *App. Biochem. Biotechnol.*, 193(12): 4197–4213. 10.1007/s12010-021-03670-1.

Xiaoxiao, Y.U. 2022. The Progress of the Quantum Hall Effect, Highlights in Science, Engineering and Technology: Vol. 5 (2022): International Conference on Earth Science, Energy Technology and Engineering Physics (ESETEP 2022).

Xin, M., X. Ji, L.K. DeLa Cruz, S. Thareja and B. Wang. 2018. Strategies to target the hedgehog signaling pathway for cancer therapy. *Medical Res. Rev.*, 38: 870–913. 10.1002/med.21482.

Xu, J. 2022. Assessing global fungal threats to humans. *Life*, 1(3): 223–240.

Yaltirak, T., B. Aslim, S. Ozturk and H. Alli. 2009. Antimicrobial and antioxidant activities of *Russula delica* Fr. *Food Chem. Toxicol.*, 47(8): 2052–2056. 10.1016/j.fct.2009.05.029.

Yaziji, M. and R. Saoud. 2008. A study of anti-fungal effectiveness of different extracts of *Lactarius* sp. against some pathogenic fungi. Tishreen University J. Res. *Scientific Studies-Biological Sci. Series*, 30(2): 91–104.

Ying, J., L.D. Du and G.H. Du. 2018. Lovastatin. pp. 93–99. *In* Natural Small Molecule Drugs from Plants. Springer, Singapore. doi.org/10.1007/978-981-10-8022-7_15.

Yoon S., S. Park and Y. Park. 2018. The anticancer properties of cordycepin and their underlying mechanisms. *Int. J. Mol. Sci.*, 19: 3027. 10.3390/ijms19103027.

Yoshikawa, N., N.H. Tsuno, Y. Okaji, K. Kawai, Y. Shuno, H. Nagawa, N. Oshima and K. Takahashi. 2010. Isoprenoid geranylgeranylacetone inhibits human colon cancer cells through induction of apoptosis and cell cycle arrest. *Anticancer Drugs*, 21(9): 850–860. 10.1097/CAD.0b013e32833e53cf.

Zhang, B.B., Y.Y. Guan, P.F. Hu, L. Chen, G.R. Xu, L. Liu and P.C. Cheung. 2019. Production of bioactive metabolites by submerged fermentation of the medicinal mushroom *Antrodia cinnamomea*: Recent advances and future development. *Crit. Rev. Biotechnol.*, 39(4): 541–554. 10.1080/07388551.2019.1577798.

Zhang, T., M. Jayachandran, K. Ganesan and B. Xu. 2018. Black truffle aqueous extract attenuates oxidative stress and inflammation in STZ-induced Hyperglycemic rats via Nrf2 and NF-κB pathways. *Front Pharmacol.*, 9: 1257. 10.3389/fphar.2018.01257.

Zhang, Y., G. Zhang and J. Ling. 2022. Medicinal fungi with antiviral effect. *Molecules*, 27(14): 4457. 10.3390/molecules27144457.

Zhao, S., Q. Gao, C. Rong, S. Wang, Z. Zhao, Y. Liu and J. Xu. 2020. Immunomodulatory Effects of edible and medicinal Mushrooms and their bioactive immunoregulatory products. *J. Fungi*, 6(4): 269. 10.3390/jof6040269.

Zhao, X., E. Hengchao, H. Dong, Y. Zhang, J. Qiu, Y. Qian and C. Zhou. 2022. Combination of untargeted metabolomics approach and molecular networking analysis to identify unique natural components in wild *Morchella* sp. by UPLC-Q-TOF-MS. *Food Chem.*, 366(2022): 130642. 10.1016/j.foodchem.2021.130642.

7

On the Secondary Metabolites and Biological Activities of Bird's Nest Fungi

Sunil K Deshmukh,[1,]* Manish K Gupta[2] and Hemraj Chhipa[3]

1. Introduction

The genus *Cyathus* Haller (*Nidulariaceae*) is an agaricoid clade of the Basidiomycota (Matheny et al. 2006), and is characterized by a three-layered bell or vase-shaped basidiomata with lenticular structures (peridioles) attached to the inner wall by the funicular cord (Brodie 1975). The genus currently has 61 spp. (Accioly et al. 2018, He et al. 2019, Góis et al. 2020) with a cosmopolitan distribution in temperate as well as tropical countries, but rarely found in polar regions (Brodie 1975).

Cyathus stercoreus (Schw.) de Toni Small bird's nest fungus, occurring on the ground in dung heaps and among dead branches. Used by the Baiga for soothing eye disorders (pain, redness, conjunctivitis). The peridioles are ground up with water, filtered through cotton, and used as eye drops, two drops twice a day. *Cyathus limbatus* Tul. A second bird's nest fungus was found especially on decaying branches of *Dendrocalamus strictus*. Used by the Bharia in the same way that the Baiga use *C. stercoreus* (Ayachi et al. 1993).

[1] R & D Division, Greenvention Biotech Pvt. Ltd., Uruli-Kanchan, Pune 412202, Maharashtra, India.
[2] SGT College of Pharmacy, SGT University, Gurugram 122505, Haryana, India.
[3] Agriculture University, Kota Rajasthan-324001, India.
* Corresponding author: sunil.deshmukh1958@gmail.com

A number of cyathane diterpenoids with an unusual 5/6/7-tricyclic skeleton, including their xylosides, were isolated from a diverse variety of higher basidiomycetes of the genera *Cyathus*. These diterpenes were demonstrated to display a wide range of biological properties, including antibacterial (Anke et al. 1977, Liu and Zhang 2004), antifungal (Hecht et al. 1978, Nitthithanasilp et al. 2018), anti-inflammatory (Han et al. 2012, Xu et al. 2013) anticancer (He et al. 2016) properties. Several cyathane diterpenoids were found to stimulate the synthesis of NGF in human nerve cells (Bai et al. 2015, Wei et al. 2018, Kou et al. 2019). The present review highlights the bioactive metabolites reported from the genus *Cyathus* and their biological activities.

2. Bioactive Metabolites from Genus *Cyathus*

2.1 *Cyathus helenae*

The early work on bioactive metabolites was started at the University of Alberta Canada by Professor Harold Johnston Brodie a Canadian mycologist, and his co-workers Dr. William A. Ayer and Prof. B.N. Johri.

Cyathus helenae is responsible for the production of the complex of antibiotics collectively called cyathin, which include cyathin A3 (**1**), cyathin A4 (**2**), cyathin B3 (**3**), cyathin B4 (**4**), and cyathin C5 (**5**), allocyathin A4 (**6**), and 2,4,5-trihydroxybenzaldehyde (**7**) (Figure 1). All species of tested *Micrococcuscus* proved to be sensitive along with *Staphylococcus aureus* penicillin-sensitive and penicillin-resistant strains. However, *Escherichia coli*, *Proteus mirabilis* and *Serratia* sp. showed meagre sensitivity. Pathogenic *Corynebacterium diphtheriae*, *Clostridium welchii*, *Diplococcus pneumoniae*, *Haemophilus parainfluenzae*, *Neisseria meningitides*, *Pneumococcus multocida* and *Streptococcus pyogenes* were sensitive. Crown gall disease causing *Agrobacterium tumefaciens* was extremely sensitive of all the organisms screened. Cyathin was most active against *Aspergillus*, *Penicillium*, *Fusarium*, *Trichoderma*, *Chaetomium*, and *Gliocladium*. Among the six dermatophytes tested, all of them showed a certain degree of sensitivity to cyathin. The most important examples include *Microsporum canis*, *Trichophyton rubrum*, and *T. terrestre* (Allbutt et al. 1971).

Simultaneously the brownish antibiotic complex named cyathin, which includes cyathin A3 (**1**), cyathin A4 (**2**), allocyathin A4 (**6**), cyathin B3 (**3**), cyathin B4 (**4**), cyathin C5 (**5**), and chromocyathin (=2,4,5-trihydroxybenzaldehyde) (**7**) (Figure 1) were purified from *Cyathus helenae* by preparative thin-layer chromatography as well as column chromatography (Johri et al. 1971). These compounds showed antimicrobial activity in spite of differences in the action spectrum as well as biologically effective doses (Johri et al. 1971). Letter cyathin A3 (**1**), cyathin A4 (**2**), cyathin B3 (**3**), cyathin B4 (**4**), cyathin C5 (**5**), allocyathin A4 (**6**), allocyathin B3/B (**8**), cyathin C3 (**9**), neoallocyathin A4 (**10**), glochidone (**11**) (Figure 1) were purified from *Cyathus helenae* (Ayer and Taube 1972, Ayer and Carstens 1973, Ayer et al. 1978).

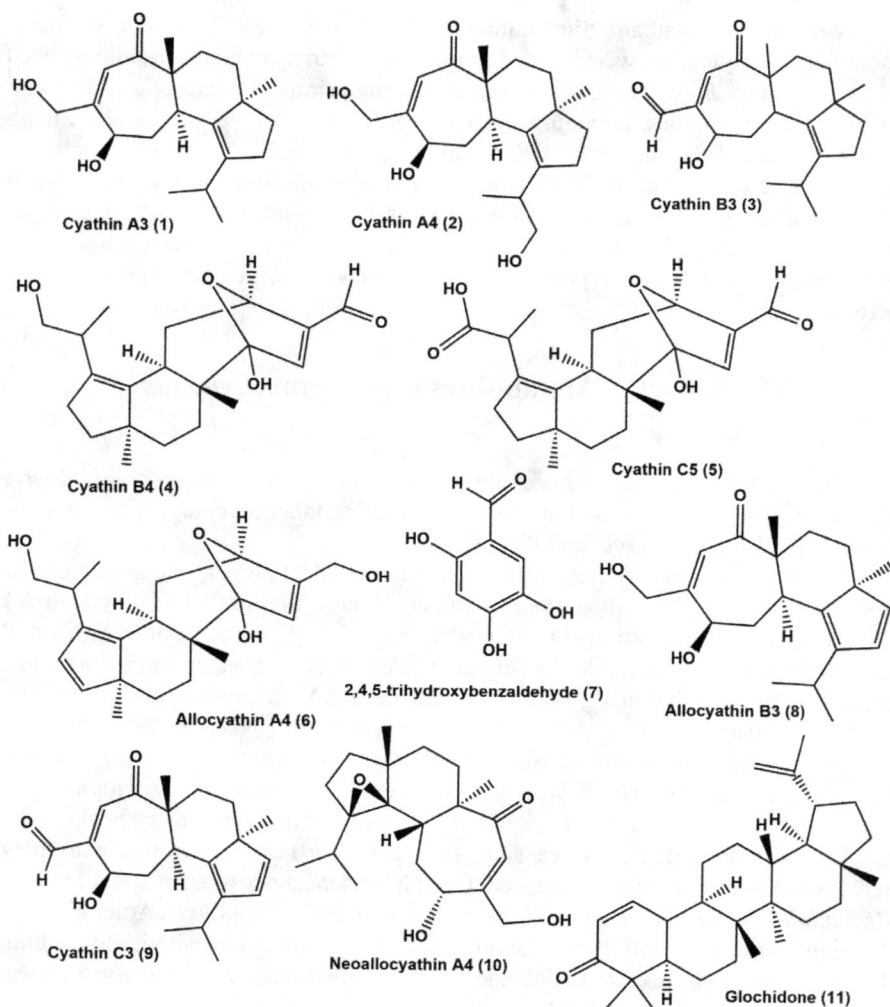

Figure 1: Structural details of metabolites isolated from *Cyathus helenae* (1–11).

2.2 *Cyathus africanus*

Cyathus africanus is another extensively explored genus for bioactive metabolites. Four new diterpenoids which have been designated cyafrin A4 (**12**), cyafrin B4 (**13**), allocyafrin B4 (**14**), and cyafrin A5 (**15**) (Figure 2), along with known diterpenoid metabolites cyathin A3, (**1**) and allocyathin B3, (**8**) were isolated from *C. africanus* is grown in liquid culture. Cyafrin A4, (**12**), allocyafrin B4, (**14**), and cyafrin A5, (**15**) have been correlated with cyathin A3 (**1**), by chemical transformations (Ayer et al. 1978).

Five unique cyathane diterpenes, cyathins D–H (**16–20**), and three known diterpenes, neosarcodonin O (**21**), cyathatriol (**22**), and 11-O-acetylcyathatriol (**23**)

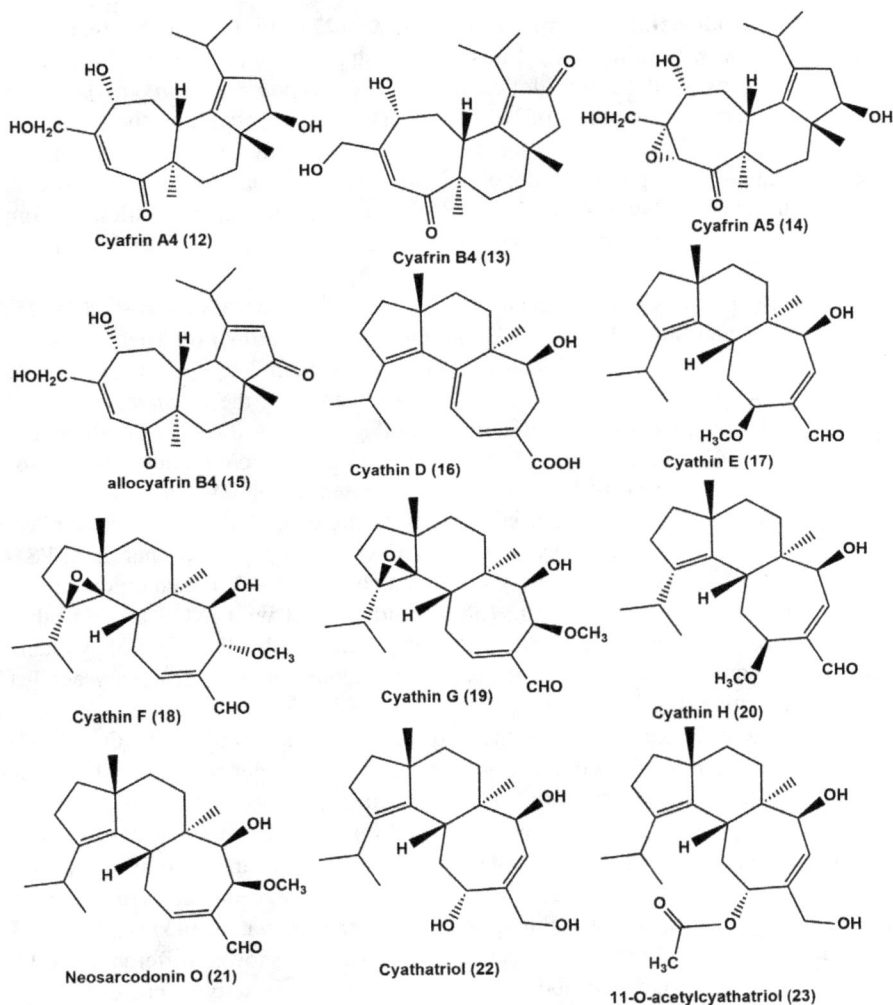

Figure 2: Structural details of metabolites isolated from *C. africanus* (12–23).

(Figure 2), were isolated from the *C. africanus* solid culture. Compounds (**18, 20, 21,** and **23**) possess potent inhibition of nitric oxide production in lipopolysaccharide-activated macrophages having IC_{50} values of 2.57, 1.45, 12.0, and 10.73 μM, respectively. Compounds (**21,** and **23**) showed robust cytotoxicity against Hela as well as K562 cell lines with an IC_{50} less than 10 μM (Han et al. 2013).

Three new cyathane diterpenoids cyathin W (**24**), cyathin V (**25**) and cyathin T (**26**) (Figure 3), were obtained from the *C. africanus* solid culture. Compounds (**24** and **26**), exhibited moderate inhibition of nitric oxide production in lipopolysaccharide-activated macrophages having IC_{50} values of 80.07 and 88.87 μM, respectively. Compound (**24**) exhibited weak cytotoxicity against K562 cell line having an IC_{50} of 12.1 μM (Han et al. 2015).

A new cyathane-like diterpene, cyathin Q (**27**) (Figure 3) purified from *Cyathus africanus*. Cyathin Q (**27**) possesses a strong anticancer potential against HCT116 cells as well as Bax-deficient HCT116 (*in vitro* and *in vivo*). It has induced hallmarks of apoptosis of HCT116 cells, caspase activation, the release of cytochrome c, poly (ADP-ribose) polymerase (PARP) cleavage, and mitochondrial inner transmembrane potential depolarization. It is accompanied by increased mitochondrial ROS, down-regulation of Bcl-2 protein, and up-regulation of Bim protein. The cleavage of autophagy-related protein ATG5 in cyathin Q (**27**) induced apoptosis (He et al. 2016).

Ten new polyoxygenated cyathane diterpenoids neocyathins A–H (**28–35**) (Figure 3) neocyathins I–J and (**36-37**) along with four known diterpenes cyathin I (**38**) (12R)-11α,14α-epoxy-13α,14β,15-trihydroxycyath-3-ene (**39**) cyathin O (**40**) (Figure 4) and allocyafrin B4 (**14**) were isolated from the *C. africanus* liquid culture. In the pathogenesis of different neurodegenerative diseases [Alzheimer's disease (AD)], neuroinflammation is implicated. Fourteen compounds (**28–40, 14**) were assessed for the anti-neuroinflammatory potential in BV2 microglia cells. Many compounds possess differential impacts on the expression of inducible nitric oxide synthase (iNOS), cyclooxygenase-2 (COX-2) in lipopolysaccharide (LPS)-stimulated and Aβ1–42-treated mouse microglia cell line BV-2. Based on molecular docking, bioactive compounds [e.g., (**38**)] could interact with iNOS protein other than the COX-2 protein. Collectively, the results indicated that this class of cyathane diterpenoids could serve as important lead compounds for drug discovery against neuroinflammation in AD (Wei et al. 2017).

Eight new polyoxygenated cyathane diterpenoids [neocyathins K–Q (**41–47**) (Figure 4) neocyathins R (**48**)] along with three known congeners [(12 S)-11α,14α-epoxy-13α,14β,15-trihydroxycyath-3-ene (**49**), allocyathin B2 (**50**) (Figure 5) and cyathin V (**25**)] were isolated from the *C. africanus* solid culture in cooked rice. Compounds (**41 and 42**) denote the first reported naturally occurring compounds with a 4,9-seco-cyathane carbon skeleton incorporating an unprecedented medium-sized 9/7 fused ring system, whereas the 3,4-seco-cyathane derivative (**43**) was isolated from *Cyathus* species for the first instance. All the isolates showed differential nerve growth factor (NGF)-induced neurite outgrowth-promoting activity in PC-12 cells at 1–25 μM, whereas one of the compounds [allocyathin B2 (**50**)] inhibited NO production in lipopolysaccharide (LPS)-stimulated microglia BV-2 cells. Besides, molecular docking showed that the compound (**50**) generated interactions with the iNOS protein (Wei et al. 2018).

Five terpenoids [possess two new cyathane diterpenoids neocyathin S (**51**) and neocyathin T (**52**) with three drimane sesquiterpenoids, 3β,6β-dihydroxycinnamolide (**53**), two new 3β,6α-dihydroxycinnamolide (**54**) and 2-keto-3β,6β-dihydroxycinnamolide (**55**) (Figure 5) were isolated from *Cyathus africanus*. Compounds (**51–55**) boosted the nerve growth factor (NGF)-mediated neurite outgrowth by rat pheochromocytoma (PC12) cells at 10 μM concentration (Kou et al. 2019).

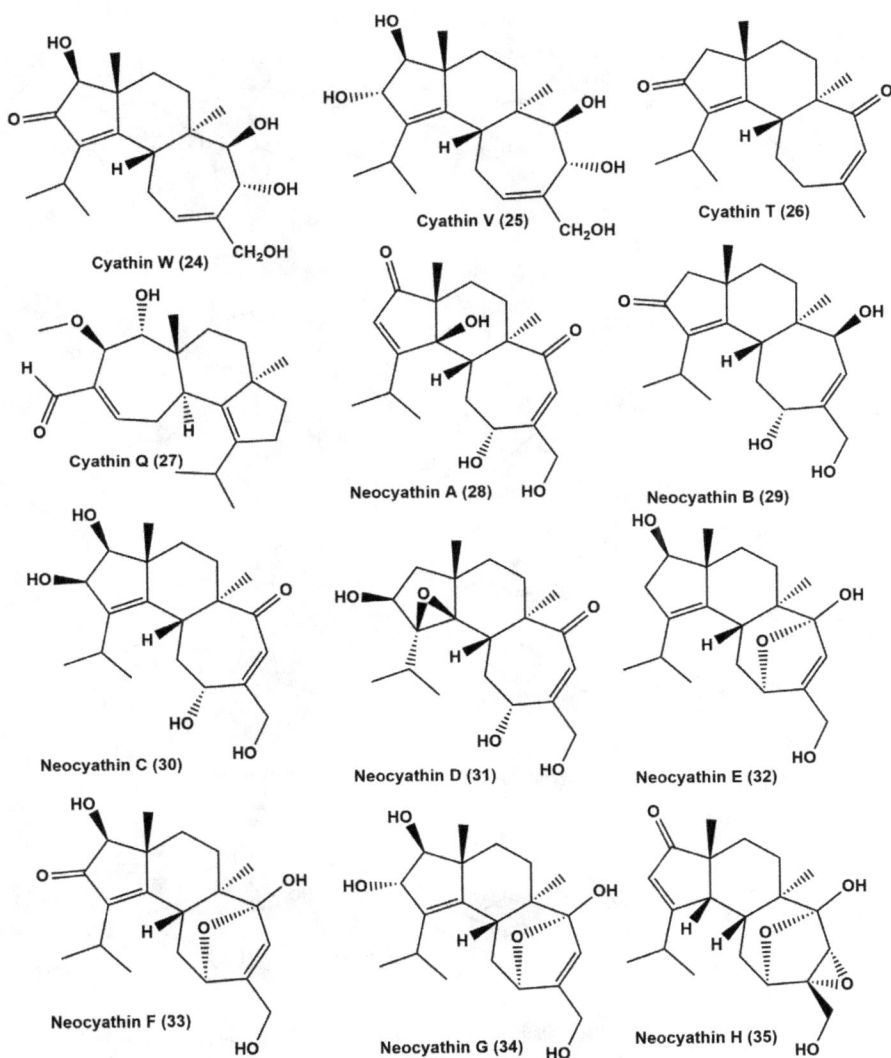

Figure 3: Structural details of metabolites isolated from *C. africanus* (24–35).

New 11 cyathane diterpenoids [designated as cyafricanins A–G (**56–62**) (Figure 5) **H-K** (**63–66**) (Figure 6), were isolated from the *C. africanus* culture broth. The cyafricanins A (**56**) was known to possess an unusual 3,4-secocarbon skeleton. All compounds were assessed for their neurotrophic activity against PC-12 cells and anti-neuroinflammatory activity in BV2 microglia cells. All the diterpenoids at a concentration of 20 μM showed nerve growth factor-induced neurite outgrowth-promoting activity. Among these cyafricanin B (**57**) and cyafricanin G (**62**) showed considerable neurotrophic activity, and cyafricanin A (**56**) possesses strong inhibitory

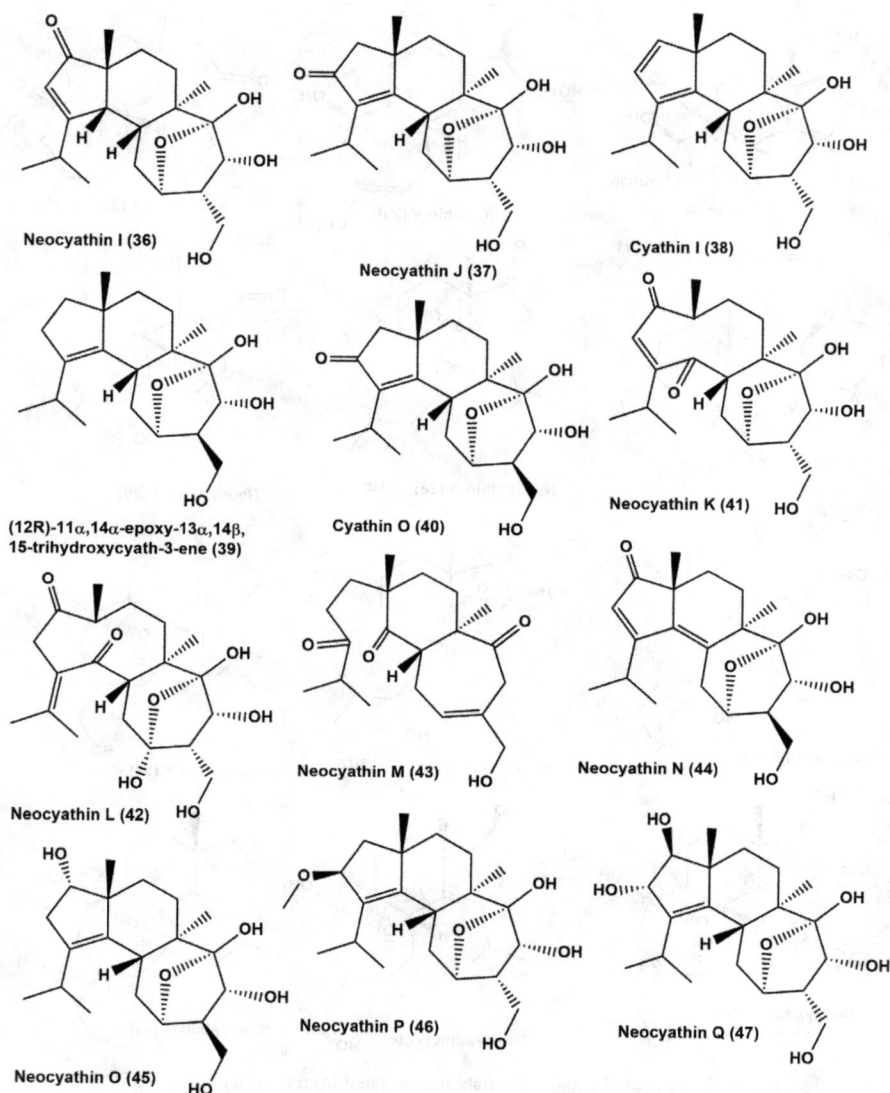

Figure 4: Structural details of metabolites isolated from *C. africanus* (36–47).

effects against both iNOS and cyclooxygenase-2 (COX-2) expression. Further, cyafricanin A (**56**) showed strong interactions with the iNOs protein in the active cavity based on molecular docking studies revealed that (Yin et al. 2019).

Eight undescribed cyathane diterpenoids, named cyathin S (**68**), cyathin U (**69**), cyathin X (**70**), (Figure 6), cyathin Q (**27**), cyathin R (**67**), cyathin T (**26**), cyathin V (**25**), cyathin W (**24**), along with five known congeners, neocyathin B (**29**), (12S)-11α,14α-epoxy-13α,14β,15-trihydroxycyath-3-ene (**49**), neocyathin J

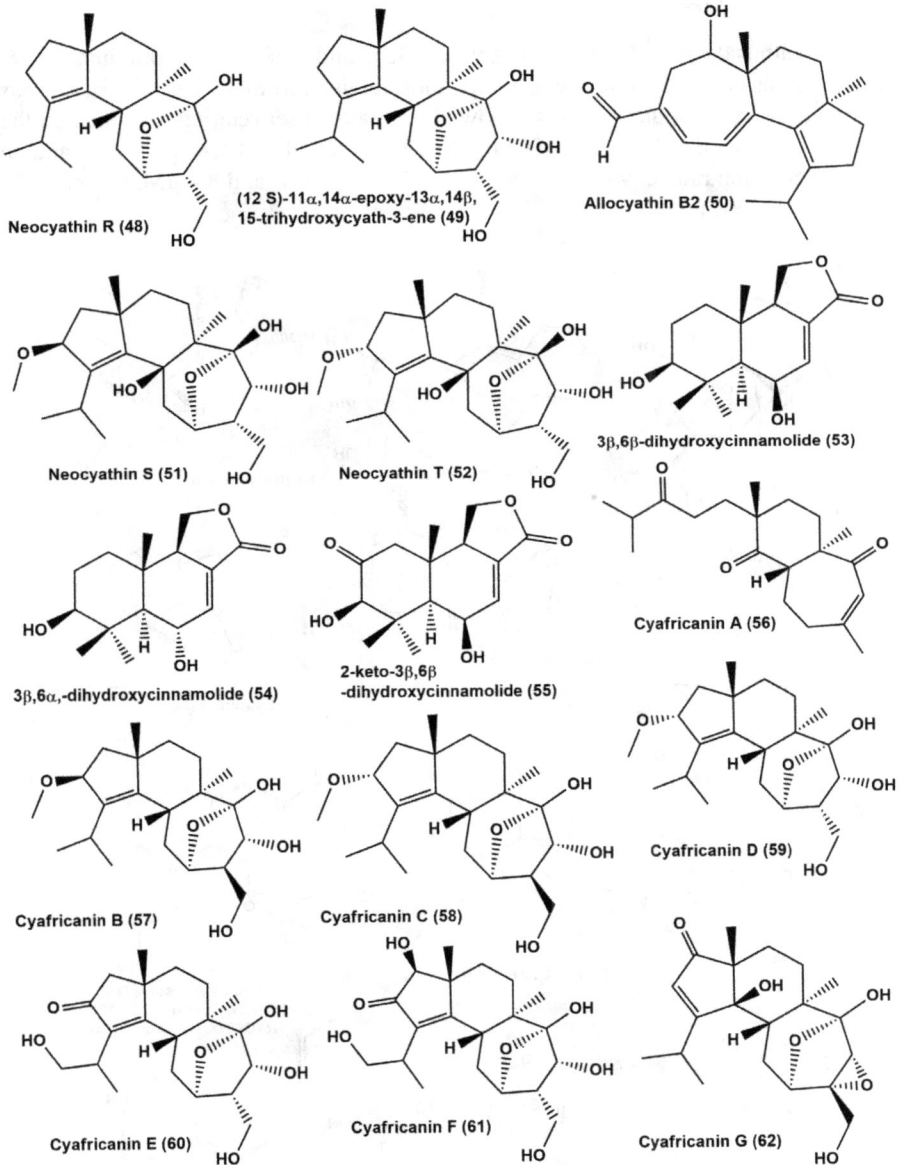

Neocyathin R (48)

(12 S)-11α,14α-epoxy-13α,14β,
15-trihydroxycyath-3-ene (49)

Allocyathin B2 (50)

Neocyathin S (51)

Neocyathin T (52)

3β,6β-dihydroxycinnamolide (53)

3β,6α,-dihydroxycinnamolide (54)

2-keto-3β,6β
-dihydroxycinnamolide (55)

Cyafricanin A (56)

Cyafricanin B (57)

Cyafricanin C (58)

Cyafricanin D (59)

Cyafricanin E (60)

Cyafricanin F (61)

Cyafricanin G (62)

Figure 5: Structural details of metabolites isolated from *C. africanus* (48–62).

(**37**), neocyathins K (**41**), and cyathin I (**38**), were isolated from the *C. africanus* liquid fermentation. The structural identification in this study revealed the number of candidates to allow for more bioactivity-screening choices. Among them, compound (12*S*)-11α,14α-epoxy-13α,14β,15-trihydroxycyath-3-ene (**49**), displayed significant acetylcholinesterase (AChE) inhibitory activity (IC$_{50}$ 4.60 μM) (Yu et al. 2022).

2.3 Cyathus stercoreus

Three cyathusals A (71), B (72), and C (73), and one known pulvinatal (74) (Figure 6), as new polyketide-type compounds obtained from the *Cyathus stercoreus* fermentation. Compounds (71–74) showed free radical scavenging activities on the DPPH radical with EC_{50} values of 41.6, 46.0, 26.6, and 28.6 µM, respectively, and on the ABTS cation radical with EC_{50} values of 7.9, 11.1, 9.1, and 8.4 µM, respectively (Kang et al. 2007).

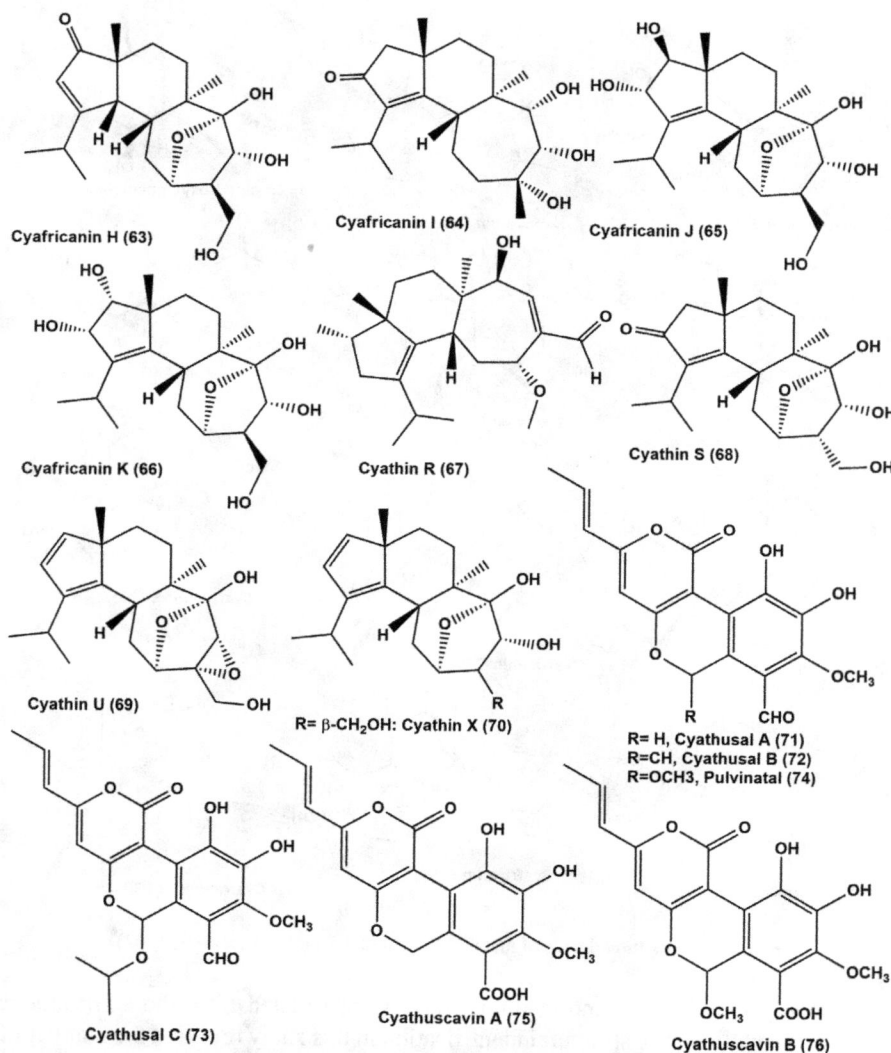

Figure 6: Structural details of metabolites isolated from *Cyathus stercoreus* (63–76).

Cyathuscavins A (**75**), B (**76**) (Figure 6), and C (**77**) together with the known 4-hydroxy-6-propenyl-pyran-2-one (**78**) (Figure 7), new polyketides were isolated from the *C. stercoreus*. Compounds (**75–77**) displayed antioxidant activity with EC_{50} values 19.31, 13.92, and 19.63 μM in the DPPH assay, respectively (Positive control BHA, EC_{50}, 99.22 μM) while these compounds exhibited antioxidant activity with EC_{50} values of 14.85, 9.84 and 7.08 μM in ABTS+ assay (Positive control EC_{50}, 9.929 μM). Compound (**75, 76**) displayed antioxidant activity with EC_{50} values of

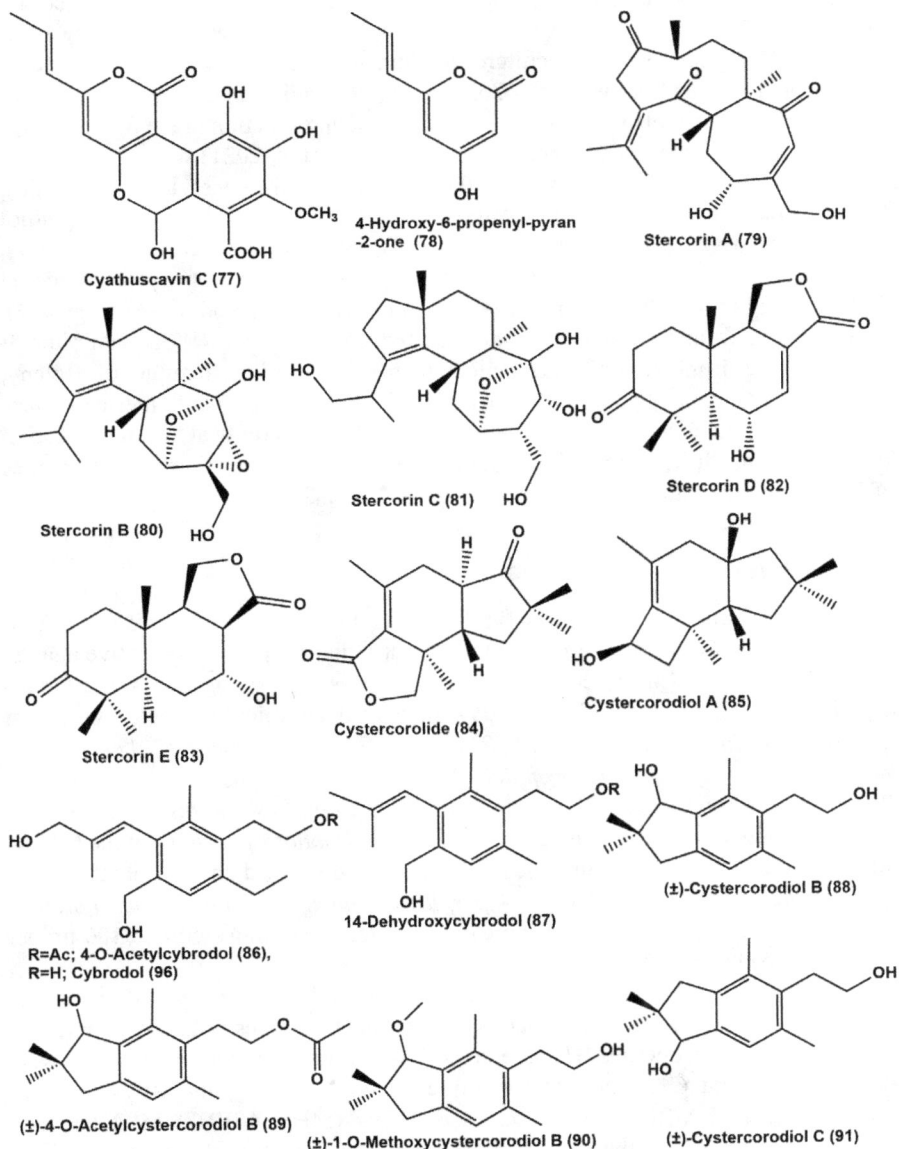

Figure 7: Structural details of metabolites isolated from *C. stercoreus* (77–91).

21.90 and 25.12 µM in d superoxide anion radicals' assay (Positive control BHA, EC_{50}, 82.56 µM). Another positive control trolox showed antioxidant activity with an EC_{50} value of 22.10 and 11.78 µM in DPPH and ABTS+ assay. Cyathuscavins A–C (**75–77**) prevented supercoiled plasmid DNA from Fe2+/H2O2-induced breakage (Kang et al. 2008).

Three new cyathane diterpenoids [stercorins A–C (**79–81**)], with two new drimane sesquiterpenoids stercorins D (**82**) and E (**83**) (Figure 7), isolated from the liquid cultures *C. stercoreus*. All cyathane diterpenoids possess neurotrophic activity against PC-12 cells at 10 µM. Compounds (**79–81**), suppressed significantly LPS-induced NO production in the culture medium in BV2 cells with IC_{50} 1.64 µM, 9.25 µM and 8.71 µM, respectively. Compound (**79**), having an uncommon medium-sized 9/7 ring system also possesses the most promising function in the prevention of activities-associated neurodegenerative diseases (Yin et al. 2021).

Cystercorolide (**84**), cystercorodiol A (**85**), 4-O-acetylcybrodol (**86**), 14-dehydroxycybrodol (**87**), (±)-cystercorodiol B (**88**), (±)-4-O-acetylcystercorodiol B (**89**), (±)-1-O-methoxycystercorodiol B (**90**), cystercorodiol C (**91**) (Figure 7), cystercorotone (**92**), and four already reported sesquiterpenes, epicoterpene D (**93**), russujaponol F (**94**), riparol B (**95**) (Figure 8), and cybrodol (**96**) (Figure 7), were purified from *C. stercoreus* when treated with SAHA a HDAC inhibitor. It was collected from Kunming Botanical Garden, Kunming Institute of Botany, Yunnan, China. Russujaponol F (**94**), was determined to be an enantiomeric mixture. Compounds (**85, 86, 92,** and **96**) showed weak inhibitory activity at the concentration 200 µM, against the growth of *Escherichia coli* ATCC25922 with the inhibition rates of 34.7, 33.0, 32.3, and 29.6%, respectively (Liu et al. 2022).

2.4 Cyathus striatus

Three new striatins, striatins A (**97**), B (**98**), and C (**99**) (Figure 8), were isolated from *Cyathus striatus* strain No. 12 mycelium. The striatins are potentially active against fungi imperfecti and a variety of Gram-positive as well as against some Gram-negative bacteria. The striatins (**97–99**), displayed antimicrobial activity against *Arthrobacter citreus, Bacillus brevis, B. subtilis, Escherichia coli K12, Leuconostoc mesenteroides, Mycobacterium phlei, Nocardia brasiliensis, Proteus vulgaris, Pseudomonas fluorescens, Sarcina lutea, Staphylococcus aureus, Streptomyces viridochromogenes, Saccharomyces cerevisiae* and *Rhodotorula rubra* at the MIC in the range 0.2–20 µg/ml. Compounds (**97–99**), also displayed antimicrobial activity against *Clostridium pasteurianum, Aspergillus panamensis, Penicillium notatum, Fusarium cubense,* 8–17 mm at Diameter inhibition zone (mm) with 20/100 µg per paper disc (Anke et al. 1977).

Striatin A, B, and C (**97–99**) (Figure 8), are the novel diterpenoid antibiotics obtained from *C. striatus*. The striatins are unique antibiotics with antibacterial as well as antifungal properties (Hecht et al. 1978). Striatin A, striatin B (**97, 98**), were also purified from *C. striatus* and exhibited in vitro activity against *Leishmania amazonensis* (IFLA/BR/67/PH8), *L. braziliensis* (MHOM/BR/75/M 2903) and *L. donovani* with IC_{90} in the range of 5–10 µg/mL (Inchausti et al. 1997).

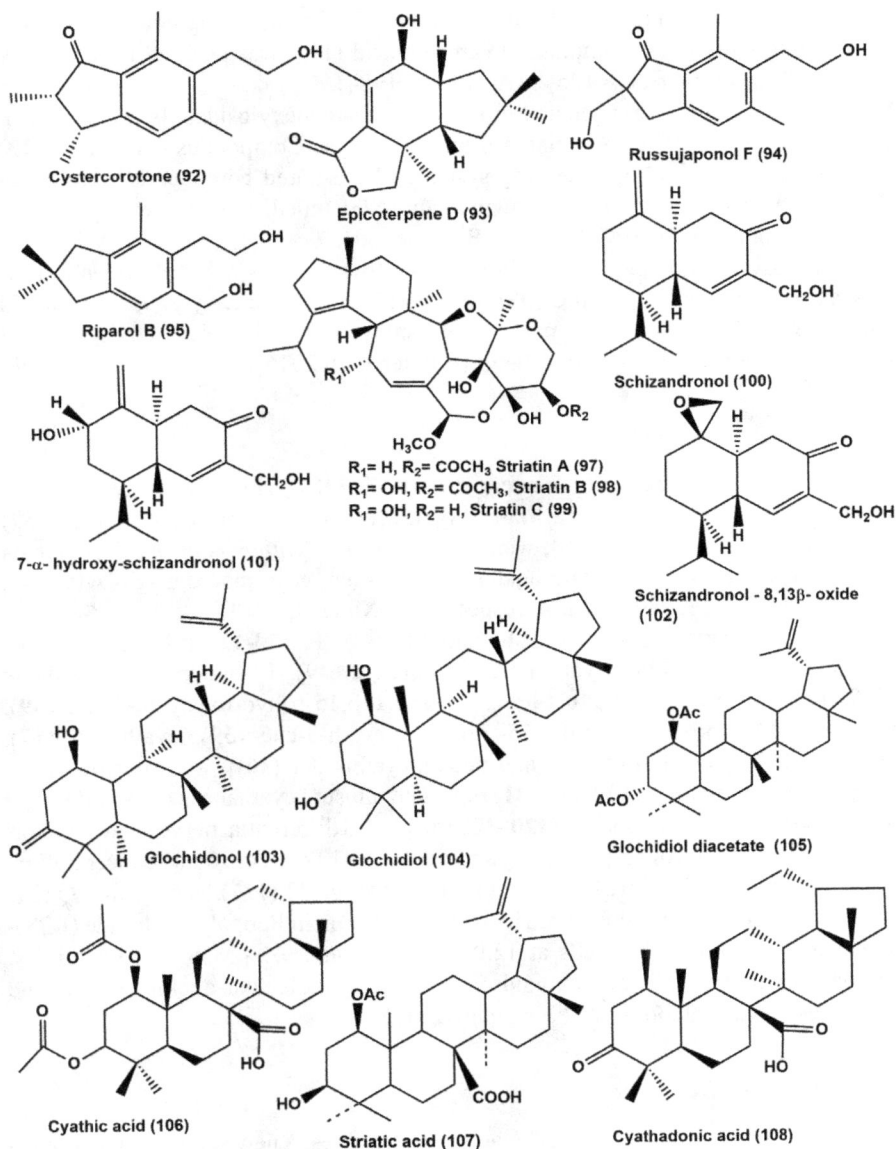

Figure 8: Structural details of metabolites isolated from *Cyathus* spp. (92–108).

Three sesquiterpenes were identified as schizandronol (**100**), 7-α-hydroxy-schizandronol (**101**) and schizandronol-,13β-oxide (**102**) (Figure 8), from the aerated liquid culture *C. striatus* grown (Ayer and Reffstrup 1982).

Known three triterpenes [glochidone (**11**), glochidonol (**103**), glochidiol (**104**), and glochidiol diacetate (**105**)], and four new triterpenoic acids [cyathic acid (**106**), striatic acid (**107**), cyathadonic acid (**108**) (Figure 8), and epistriatic acid (**109**)

(Figure 9)] have been purified from *C. striatus*. The known compound cyathic acid (**106**) along with a new compound pygmaeic acid (**110**) was purified from another species, *Cyathus pygmaeus* Lloyd (Ayer et al. 1984).

Six new highly oxygenated polycyclic cyathanexylosides [striatoids A–F (**111–116**) (Figure 9)] were isolated from *C. striatus*. Compounds (**112**) and (**113**) showed an unusual 15,4'-ether ring system. The isolated compounds were dose-dependently enhanced the nerve growth factor (NGF)-mediated neurite outgrowth in rat pheochromocytoma (PC-12) cells (Bai et al. 2015).

Two novel pyridino-cyathane diterpenoids [pyristriatins A and B (**117** and **118**) (Figure 9)] along with striatin C (**99**) were purified from the *C. cf. striatus* obtained from northern Thailand. The pyristriatins showed antimicrobial activity against Gram-positive bacteria as well as fungi (Richter et al. 2016).

2.5 Cyathus hookeri

A new cyathane diterpene, named cyathin I (**38**), along with two known cyathane diterpenes (12R)-11a,14a-epoxy-13a,14b,15-trihydroxycyath-3-ene (**39**) and erinacine I (**119**) (Figure 9), were isolated from *Cyathus hookeri*. Compounds (**38, 39, 119**) showed inhibition against NO production in macrophages with IC_{50} values of 15.5, 52.3, and 16.8 µM, respectively (Xu et al. 2013).

Six new cyathane diterpenoids [cyahookerins A–F (**120-125**)], and nine known analogues erinacol (**126**), cyathin B2 (**127**) (Figure 9), 14-oxo-cyatha-3,12-diene (**128**) (Figure 10), [(12S)-11α,14α-epoxy-13α,14β,15-trihydroxycyath-3-ene (**49**), (12R)-11α,14α-epoxy-13α, 14β, 15-trihydroxycyath-3-ene (**39**), cyathin E (**17**), cyathatriol (**22**), cyathin Q (**27**), and neoallocyathin A4 (**10**)], were purified from *C. hookeri*. Compounds (**120** and **121**) represent unusual cyathane acetals featuring a dioxolane ring. Five compounds (**120–125**) displayed differential nerve growth factor-induced neurite outgrowth-promoting activity in PC-12 cells at 10 µM concentration. Besides, cyahookerin B (**121**), cyathin E (**17**), cyathin B2 (**127**), and cyathin Q (**27**) possess significant inhibition of nitric oxide production in lipopolysaccharide (LPS)-activated BV-2 microglial cells at 12.0, 6.9, 10.9, and 9.1 µM IC_{50}, respectively. Similar activity of the four compounds was shown by molecular-docking studies and structure-activity relationships (Tang et al. 2019).

2.6 Other Cyathus Species

Other species also produce various bioactive metabolites. A new xanthone 1-hydroxy-6-methyl-8-hydroxymethylxanthone (**129**) (Figure 10), was purified from *Cyathus intermedius*, and has been confirmed by synthesis (Ayer and Taylor 1976).

Cyathin B2 (**127**) (Figure 9), Cyarharriol (**130**), 11,15-0,O-Diacetylcyathatriol (**131**), 15-0-acetylcyathatriol (**132**), and 11-0-acetylcyathatriol (**133**) (Figure 10), and allocyathin B2 (**50**), diterpenoid metabolites were purified from *Cyathus earlei* (Ayer and Lee 1979).

Seven new cyathane type diterpenes [cyathins J–P (**134–140**) (Figure 10)], with known two diterpenes [(12R)-11a,14a-epoxy-13a,14b,15-trihydroxycyath-3-

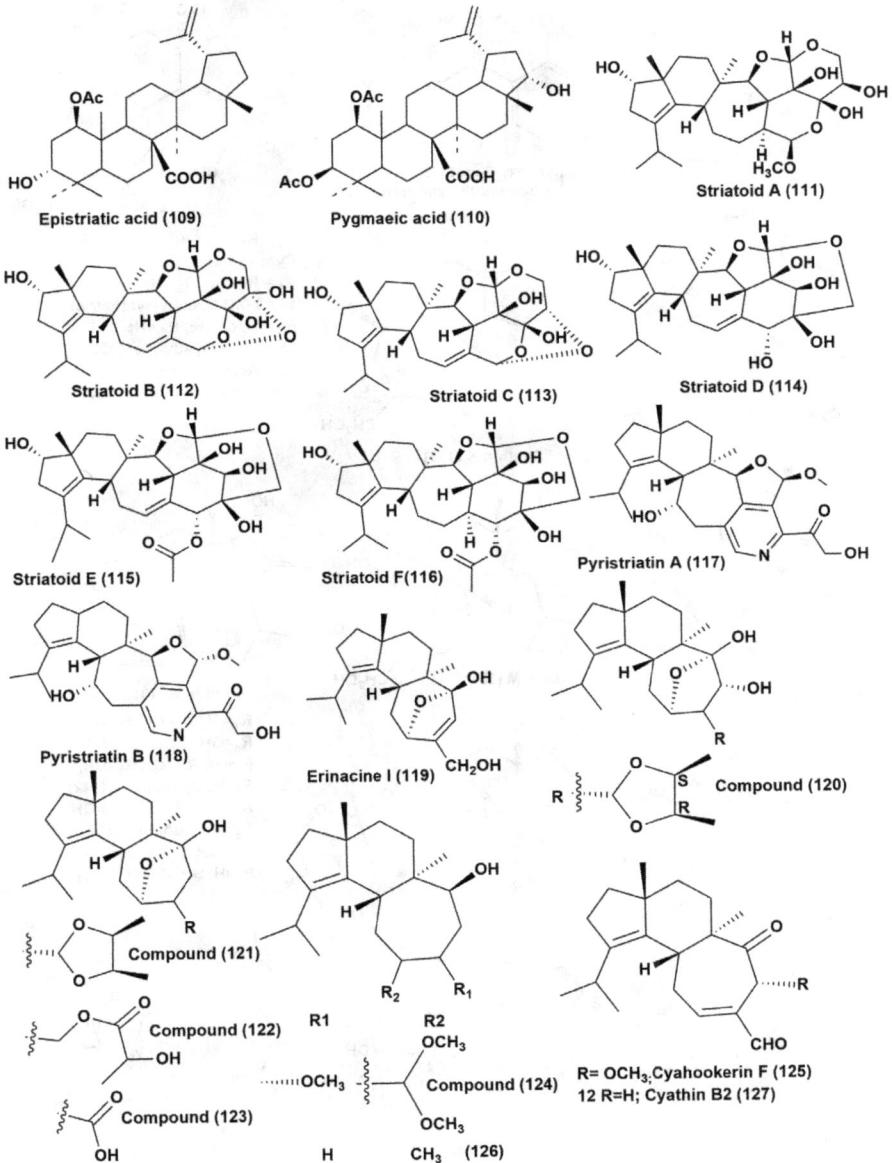

Figure 9: Structural details of metabolites isolated from *Cyathus* spp. (109–127).

ene (**39**)] and (12S)-11a,14a-epoxy-13a,14b,15-trihydroxycyath-3-ene (**49**) were isolated from the solid culture of *Cyathus gansuensis* on cooked rice. Compounds (**134, 135, 137** and **39**) showed moderate inhibitory activity against NO production in lipopolysaccharide-activated macrophages with IC_{50} values 42, 78, 80, and 16 µM, respectively (Wang et al. 2014).

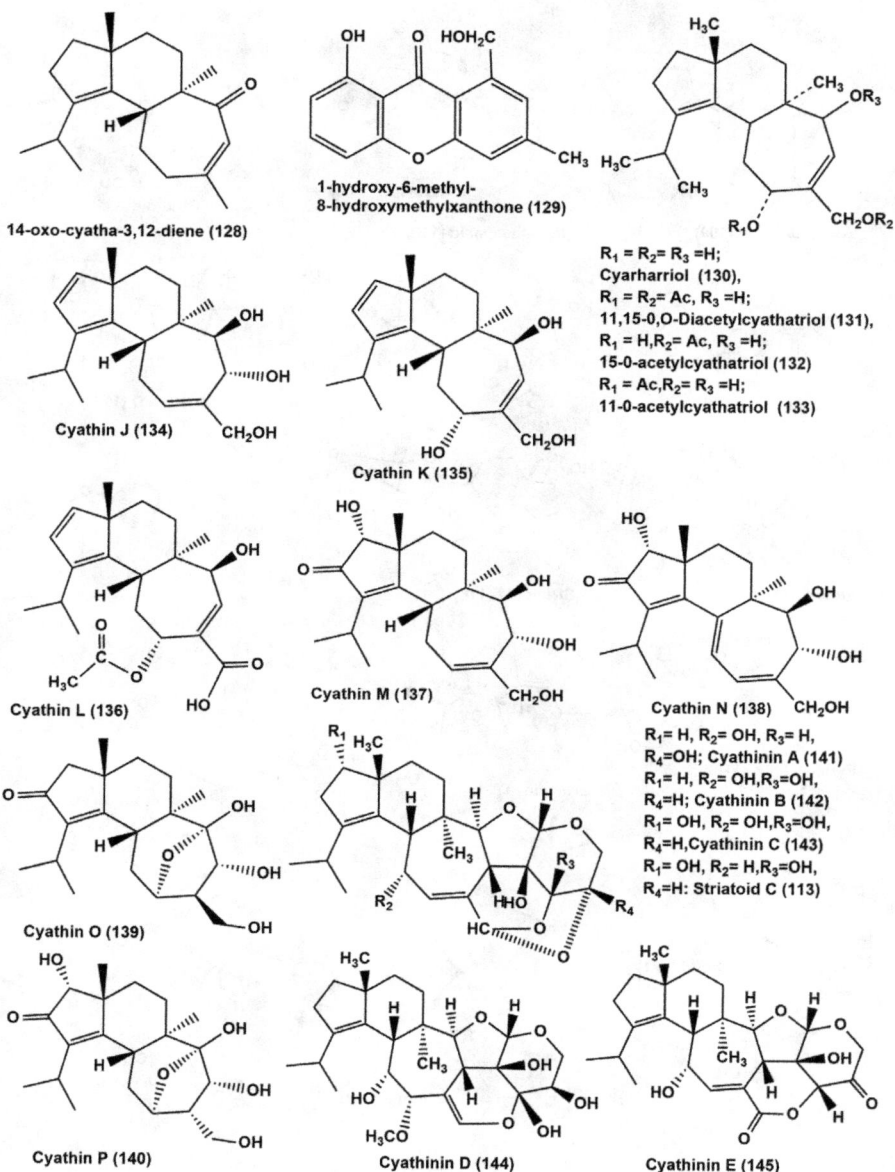

14-oxo-cyatha-3,12-diene (128)

1-hydroxy-6-methyl-8-hydroxymethylxanthone (129)

R₁ = R₂= R₃ =H; Cyarharriol (130),
R₁ = R₂= Ac, R₃ =H; 11,15-0,O-Diacetylcyathatriol (131),
R₁ = H,R₂= Ac, R₃ =H; 15-0-acetylcyathatriol (132)
R₁ = Ac,R₂= R₃ =H; 11-0-acetylcyathatriol (133)

Cyathin J (134)

Cyathin K (135)

Cyathin L (136)

Cyathin M (137)

Cyathin N (138)

R₁= H, R₂= OH, R₃= H, R₄=OH; Cyathinin A (141)
R₁= H, R₂= OH,R₃=OH, R₄=H; Cyathinin B (142)
R₁= OH, R₂= OH,R₃=OH, R₄=H,Cyathinin C (143)
R₁= OH, R₂= H,R₃=OH, R₄=H: Striatoid C (113)

Cyathin O (139)

Cyathin P (140)

Cyathinin D (144)

Cyathinin E (145)

Figure 10: Structural details of metabolites isolated from *Cyathus* spp. (128–145).

Six unknown cyathane diterpenes cyathinins A -E **(141–145)**, (Figure 10) and 10-hydroxyerinacine S **(146)**, and six previously known compounds striatal A **(147)**, striatal C **(148)**, striatal D **(149)** (Figure 11) striatoid C **(113)**, Striatin C **(99)** and Glochidone C **(11)** were isolated from *Cyathus subglobisporus* BCC43381. Compounds **(141, 113, 144, 99, 147–149)** exhibited antimalarial activity against *Plasmodium falciparum* with a range of 0.88–7.51 μM IC₅₀,

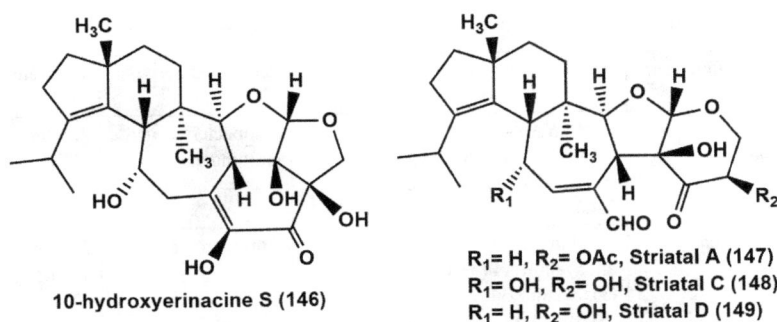

Figure 11: Structural details of metabolites isolated from *Cyathus* spp. (146–149).

while four compounds (**144, 99, 147, 148**) showed antifungal activity against *C. albicans* with a range of 8.6–80.3 μM IC_{50}. Four compounds (**141, 144, 99, 147, 148**) displayed anti-*M. tuberculosis* activity at a range of 25.0–50.0 μM/mL MIC. Compounds (**141, 144, 99, 147–149**) possess antibacterial activity against Gram-positive bacteria *Bacillus cereus*, and *Enterococcus faecium* with a range of 0.78–50.0 μM/mL MIC. Three compounds (**99, 147, 148**) showed antibacterial activity against three Gram-negative bacteria (*A. baumannii*, *E. coli* and *K. pneumoniae*) in a presence of PAβN (phenylalanine-arginine β-naphthylamide) with a range of 3.13–50 μM/mL MIC. Furthermore, three compounds (**141, 113, 144**) also showed antibacterial activity against *A. aumannii* as well as *E. coli* in a presence of PAβN in a range of 6.25–50.0 μM/mL MIC. However, all the tested compounds were not effective against Gram-negative bacteria without PAbN at a maximum tested concentration (50 μM/mL). Five compounds (**141, 144, 99, 147, 148**) displayed cytotoxicity against cell lines MCF-7, KB, NCI-H187, and Vero with a range of 0.13 to 28.32 μM MIC. Compound (**149**) displayed cytotoxicity against NCI-H187, and Vero cell lines with 6.21 and 0.51 μM MIC, respectively (Nitthithanasilp et al. 2018).

3. Activation of Biosynthetic Genes for Secondary Metabolite Production

Fungi have shown the promising source of bioactive compounds, but they secret many novel compounds only in their natural habitat and are unable to produce them in the axenic culture system. The expression of such secondary metabolites related gene clusters like polyketide synthase (PKS), non-ribosomal peptide synthase (NRPS), hybrid polyketide- non-ribosomal peptide synthase natural (HPN), etc. remain silent during the growth of microbial culture in laboratory conditions (Fisch et al. 2009). In their natural system, the production of bioactive compounds is the result of the activation of gene clusters belonging to bioactive compounds or the interaction of surrounding microbial communities (Marmann et al. 2014, Liu et al. 2017). Several methods have been developed to induce such silent gene clusters via modification of the genome or induction through chemicals, enzymes or changes in growth conditions in axenic culture (Cichewicz 2010, Winter et al. 2011, Goss

Table 1: Novel bioactive compounds reported from the genus *Cyathus*.

Sr. no.	Producing organism	Name of the compound	Biological activity	References
1.	*Cyathus helenae*	Cyathin A3 - A4 (**1 - 2**), cyathin B3 - B4 (**3 - 4**), cyathin C5 (**5**), allocyathin A4 (**6**), and 2,4,5-trihydroxybenzaldehyde (**7**)	Antibacterials and antifungals	Allbutt et al. 1971
2.	*C. helenae*	Cyathin A3 - A4 (**1 - 2**), allocyathin A4 (**6**), cyathin B3 (**3**), cyathin B4 (**4**), cyathin C5 (**5**), and chromocyathin (=2,4, 5-trihydroxybenzaldehyde) (**7**)	Antimicrobial	Johri et al. 1971
3.	*C. helenae*	Cyathin A3 - A4 (**1 - 2**) cyathin B3 - B4 (**3 - 4**), cyathin C5 (**5**), allocyathin A4 (**6**), allocyathin B3/B (**8**), cyathin C3 (**9**), neoallocyathin A4 (**10**), glochidone (**11**)	-	Ayer and Taube 1972, Ayer and Carstens, 1973, Ayer et al. 1978
4.	*Cyathus africanus*	Cyafrin A4 (**12**), cyafrin B4 (**13**), allocyafrin B4 (**14**), and cyafrin A5 (**15**), cyathin A3, (**1**) and allocyathin B3, (**8**)		Ayer et al. 1978
5.	*C. africanus*	Cyathins D–H (**16-20**), neosarcodonin O (**21**), cyathatriol (**22**), and 11-O-acetylcyathatriol (**23**)		Han et al. 2013
		Compounds (**18, 20, 21**, and **23**)	Inhibit NO production	
		Compounds (**21** and **23**)	Cytotoxic	
6.	*C. africanus*	Cyathin W (**24**), cyathin V (**25**), and cyathin T (**26**)		Han et al. 2015
		Compounds (**24** and **26**)	Inhibit NO production	
		Compound (**24**)		
7.	*C. africanus*	Cyathin Q (**27**)	Cytotoxic	He et al. 2016
8.	*C. africanus*	Neocyathins A–J (**28-37**), cyathin I (**38**) (12R)-11α,14α-epoxy-13α,14β,15-trihydroxycyath-3-ene (**39**) cyathin O (**40**) and allocyafrin B4 (**14**)	Anti-neuroinflammatory	Wei et al. 2017
9.	*C. africanus*	Neocyathins K–R (**41- 48**), cyathin V (**25**), (12 S)-11α,14α-epoxy-13α,14β,15-trihydroxycyath-3-ene (**10**) (**49**), and allocyathin B2 (**50**)	Anti-neuroinflammatory activity	Wei et al. 2018
		Allocyathin B2 (**50**)	Inhibited NO production	

Table 1 contd. ...

...Table 1 contd.

Sr. no.	Producing organism	Name of the compound	Biological activity	References
10.	*C. africanus*	Neocyathin S (**51**), neocyathin T (**52**), 3β,6β-dihydroxycinnamolide (**53**), 3β,6α-dihydroxycinnamolide (**54**) and 2-keto-3β,6β-dihydroxycinnamolide (**55**)	Anti-neuroinflammatory activity	Kou et al. 2019
11.	*C. africanus*	Cyafricanins A–K (**56-66**)	Anti-neuroinflammatory activity	Yin et al. 2019
12.	*C. africanus*	Cyathin Q (**27**), cyathin R (**67**), cyathin S (**68**), cyathin T (**26**), 4.3.5. cyathin U (**69**), cyathin V (**25**), cyathin W (**24**), cyathin X (**70**), neocyathin B (**29**), (12S)-11α,14α-epoxy-13α,14β,15-trihydroxycyath-3-ene (**49**), neocyathin J (**37**), neocyathins K (**41**), and cyathin I (**38**)		Yu et al. 2022
		(12S)-11α,14α-epoxy-13α,14β,15-trihydroxycyath-3-ene (**49**)	AChE inhibitory effect	
13.	*Cyathus stercoreus*	Cyathusals A-C (**71-73**), pulvinatal (**74**)	Antioxidant	Kang et al. 2007
14.	*C. stercoreus*	Cyathuscavins A-C (**75 -77**), 4-hydroxy-6-propenyl-pyran-2-one (**78**)	Antioxidant/ protected supercoiled plasmid DNA from Fe2+/2O2-induced breakage	Kang et al. 2008
15.	*C. stercoreus*	Stercorins A–E (**79-83**)		Yin et al. 2021
		Stercorins A–C (**79-81**)	Neurotrophic activity /Inhibited NO production	
16.	*C. stercoreus*	Cystercorolide (**84**), cystercorodiol A (**85**), 4-O-Acetylcybrodol (**86**), 14-Dehydroxycybrodol (**87**), (±)-cystercorodiol B (**88**), (±)-4-O-Acetylcystercorodiol B (**89**), (±)-1-O-Methoxycystercorodiol B (**90**), cystercorodiol C (**91**), cystercorotone (**92**), epicoterpene D (**93**), russujaponol F (**94**), riparol B (**95**), and cybrodol (**96**)		Liu et al. 2022
		Compounds (**85, 86, 92**, and **96**)	Antibacterial against *Escherichia coli* ATCC25922	

Table 1 contd. ...

...Table 1 contd.

Sr. no.	Producing organism	Name of the compound	Biological activity	References
17.	*Cyathus striatus*	Striatins A-C **(97-99)**	Antibacterial and Antifungal	Anke et al. 1977
18.	*C. striatus*	Striatin A- C **(97-99)**		Hecht et al. 1978
19.	*C. striatus*	Striatin A, B, **(97-98)**	Antileishmanial Activity	Inchausti et al. 1997
20.	*C. striatus*	Schizandronol **(100)**, 7-α-hydroxy - schizandronol **(101)** and schizandronol - 8,13β-oxide **(102)**		Ayer and Reffstrup 1982
21.	*C. striatus*	Glochidone **(11)**, glochidonol **(103)**, glochidiol **(104)**, glochidiol diacetate **(105)**, cyathic acid **(106)**, striatic acid **(107)**, cyathadonic acid **(108)**, and epistriatic acid **(109)**		Ayer et al. 1984
22.	*Cyathus pygmaeus*	Cyathic acid **(106)**, pygmaeic acid **(110)**		Ayer et al. 1984
23.	*C. striatus*	Striatoids A−F **(111-116)**	Anti-neuritogenic activity	Bai et al. 2015
24.	*C. striatus*	Pyristriatins A and B **(117-118)**, striatin C **(99)**		Richter et al. 2016
		Pyristriatins A and B **(117-118)**	Antimicrobial	
25.	*Cyathus hookeri*	Cyathin I **(38)**, (12R)-11a,14a-epoxy-13a,14b,15-trihydroxycyath-3-ene **(39)** and erinacine I **(119)**	Inhibited NO production	Xu et al. 2013
26.	*C. hookeri*	Cyahookerins A−F **(120-125)**, (12S)-11α,14α-epoxy-13α,14β,15-trihydroxycyath-3-ene **(49)**, (12R)-11α,14α-epoxy-13α,14β, 15-trihydroxycyath-3-ene **(39)**, cyathin E **(17)**, erinacol **(126)**, cyathatriol **(22)**, cyathin B2 **(127)**, cyathin Q **(27)**, 14-oxo-cyatha-3,12-diene **(128)**, and neoallocyathin A4 **(10)**		Tang et al. (2019)
		Compounds **(120-125)**	Neurotrophic Activity	
		Compounds **(121, 17, 127, 27)**	Inhibited NO production	
27.	*Cyathus intermedius*	1-Hydroxy-6-methyl-8-hydroxymethylxanthone **(129)**		Ayer and Taylor (1976)

Table 1 contd. ...

...Table 1 contd.

Sr. no.	Producing organism	Name of the compound	Biological activity	References
28.	*Cyathus earlei*	Cyathin B2 **(127)**, cyarharriol **(130)**, 11,15-0,O-Diacetylcyathatriol **(131)**, 15-0-acetylcyathatriol **(132)**, 11-0-acetylcyathatriol **(133)**, and allocyathin B2 **(50)**		Ayer and Lee (1979)
29.	*Cyathus gansuensis*	Cyathins J–P **(134-140)**, (12R)-11α,14α-epoxy-13α,14β,15-trihydroxycyath-3-ene **(39)**, and (12S)-11α,14α-epoxy-13α,14β,15-trihydroxycyath-3-ene **(49)**		Wang et al. (2014)
		Compounds **(134, 135, 137, 39)**	Inhibited NO production	
30.	*Cyathus subglobisporus*	Cyathinin A-E **(141-145)**, 10-hydroxyerinacine S **(146)**, striatoid C **(113)**, striatin C **(99)**, striatals A **(147)**, C **(148)**, D **(149)**, glochidone C **(11)**		Nitthithanasilp et al. (2018)
		Compounds **(141, 113, 144, 99,147, 148, 149)**	Antimalarial	
		Compounds **(144, 99, 147, 148)**	Anti-*Candida*	
		Compounds **(141, 144, 99, 147, 148)**	Anti-Mycobacterial	
		Compounds **(141, 144, 99,147, 148, 149)**	Antibacterial (Gram Positive)	
		Compounds **(99,147, 148)**	Antibacterial (Gram Negative)	
		Compound **(141, 113, 144,)**	Antibacterial activity against *E. coli* and *A. baumannii*	
		Compounds **(141, 144, 99,147, 148)**	Cytotoxicity against MCF-7, KB, NCI-H187, and Vero cell lines	
		Compound **(149)**	Cytotoxicity against NCI-H187, and Vero cell lines	

et al. 2012, Ochi and Hosaka 2012). The use of the chemical epi-genetic method is another approach to induce the cryptic silent biosynthetic genes. Along with these methods, various other methods of gene manipulation in the genome by gene knockout, promoter exchange and overexpression of transcription factor have been reported by (Bertrand et al. 2014). At a molecular level epigenetic modification of DNA histone proteins by methylation, acetylation or phosphorylation modulate the chromatin structure and induces the expression of silent biosynthetic genes (Bok et al. 2009). Similarly, SUMOylation is another technique at the transcriptional level to induce secondary metabolite production via changes in protein expression using small ubiquitin-related modifiers (Szewczyk et al. 2008).

One strain of many compounds (OSMAC) is also another method to induce secondary metabolites in microorganisms. In this method, changes in environmental conditions and substrate for microbial growth induce the production of various types of metabolite production. Changes in media substrate also support the selective enhancement in the production of interesting metabolites and suppress the production of other secondary metabolites. A co-culture technique is also one of the successful techniques to induce silent secondary metabolite production in laboratory conditions (Akone et al. 2016). Mimicking of the natural environment may lead to the production of a secondary metabolite which remains silent in the axenic culture system in the laboratory. Competition in local habitats for food and space between microbial communities induces the silent biogenetic gene cluster for the production of secondary metabolites which may be silent in axenic culture (Meyer and Nai 2018). The presence of the other microorganism in the culture medium may generate biotic stress, which stimulate the production of silent secondary metabolites or polyketide synthase (PKS) genes (Bertrand et al. 2013, De Roy et al. 2013, König et al. 2013). The co-existence of other microbes affects the availability of nutrition, generation of metabolites, and growth pattern as per their antagonistic or symbiotic nature. Use of above-discussed methods is helpful to generate silent secondary metabolites in laboratory conditions and provide various novel compounds to the drug industry.

4. Conclusion

Cyathus is an inedible, gasteromyceteous mushroom with limited traditional use. The present review showed that the *Cyatus* spp. possess diverse chemical compounds as well as prolific producers of a variety of bioactive metabolites of biotechnological potential. Among the 149 compounds reported so far, mainly belong to the cyathane diterpene class. The bioactive metabolites from *Cyathus* showed several biological activities such as anticancer, anti-neuroinflammatory, neurotrophic, NO production inhibition, NGF stimulation, antibacterial and antifungal. The metabolites with anti-neuroinflammatory and neurotrophic activities can be used to develop drugs for neuropathic pain and related brain disorders. Similarly, NO production inhibitors are useful to mitigate the propagation of neuroinflammation. NGF promotes nerve regeneration and therefore, metabolites which stimulate NGF can be a potential drug for Alzheimer's disease treatment. *Cyathus* metabolites with antibacterial activities against gram-positive as well as gram-negative bacteria are valuable scaffolds for developing broad-spectrum antibacterial agents. Fungi secret many novel compounds only in their natural habitat, therefore, specific techniques such as activation of biosynthetic genes for specific secondary metabolite production is required to produce them in sufficiently large quantity to facilitate their bioactivity evaluation and safety studies. The Bird's nest fungi are less explored for bioactive metabolites.

The titter of the metabolites produced are in small quantity or not detectible level needs optimization of media by the one strain many compounds (OSMAC) method, co-cultivation as well as application of epigenetic modifier. The whole-genome sequencing is necessary for the evaluation of their potential for production of the novel compounds. Molecular methods like the transfer of biosynthetic gene clusters to a suitable vector for large-scale production could be targeted. As opined by Ramm et al. (2017), additional avenues to produce novel compounds from *Cyathus* include the use of precursors of biosynthetic pathways in the culture medium or broth which may stimulate the biosynthetic pathways of secondary metabolites. Bioactive metabolites from *Cyathus* pave the ways to explore new compounds as well as new structural skeletons towards developing directed therapeutic agents.

References

Accioly, T., R.H. Cruz, N.M. Assis, N.K. Ishikawa, K. Hosaka, M.P. Martín and I.G. Baseia. 2018. Amazonian bird's nest fungi (Basidiomycota): Current knowledge and novelties on *Cyathus* species. *Mycoscience*, 59(5): 331–342.

Akondi, K.B. and V.V. Lakshmi. 2013. Emerging trends in genomic approaches for microbial bioprospecting. *OMICS*, 17: 61–70.

Allbutt, A.D., W.A. Ayer, H.J. Brodie, B.N. Johri and H. Taube. 1971. Cyathin, a new antibiotic complex produced by *Cyathus helenae*. *Can. J. Microbiol.*, 17(11): 1401–1407.

Anke, T., F. Oberwinkler, W. Steglich and G. Hofle. 1977. The striatins-new antibiotics from the basidiomycete *Cyathus striatus* (Huds. ex Pers.) Willd. *J. Antibiot.*, 30(3): 221–225.

Ayer, W.A. and H. Taube. 1972. Metabolites of *Cyathus helenae*, cyathin A3 and allocyathin B3, members of a new group of diterpenoids. *Tetrahedron Lett.*, 13(19): 1917–1920.

Ayer, W.A. and H. Taube. 1973. Metabolites of *Cyathus helenae*. A new class of diterpenoids. *Can. J. Chem.*, 51(23): 3842–3854.

Ayer, W.A. and L.L. Carstens. 1973. Diterpenoid metabolites of *Cyathus helenae*. Cyathin B3 and cyathin C3. *Can. J. Chem.*, 51(19): 3157–3160.

Ayer, W.A. and D.R. Taylor. 1976. Metabolites of bird's nest fungi. Part 5. The isolation of 1-hydroxy-6-methyl-8-hydroxymethylxanthone, a new xanthone, from *Cyathus intermedius*. Synthesis via photoenolisation. *Can. J. Chem.*, 54(11): 1703–1707.

Ayer, W.A., L.M. Browne, J.R. Mercer, D.R. Taylor and D.E. Ward. 1978. Metabolites of bird's nest fungi. Part 8. Some minor metabolites of *Cyathus helenae* and some correlations among the cyathins. *Can. J. Chem.*, 56(5): 717–721.

Ayer, W.A., T. Yoshida and D.M. van Schie. 1978. Metabolites of bird's nest fungi. Part 9. Diterpenoid metabolites of *Cyathus africanus* Brodie. *Can. J. Chem.*, 56(16): 2113–2120.

Ayer, W.A. and S.P. Lee. 1979. Metabolites of bird's nest fungi. Part 11. Diterpenoid metabolites of *Cyathus earlei* Lloyd. *Can. J. Chem.*, 57(24): 3332–3337.

Ayer, W.A. and T. Reffstrup.1982. Metabolites of bird's nest fungi. Part 18. new oxygenated cadinane derivatives from *Cyathus striatus*. *Tetrahedron*, 38(10): 1409–1412.

Ayer, W.A., R.J. Flanagan and T. Reffstrup. 1984. Metabolites of bird's nest fungi-19: New triterpenoid carboxylic acids from *Cyathus striatus* and *Cyathus pygmaeus*. *Tetrahedron*, 40(11): 2069–2082.

Bai, R., C.C. Zhang, X. Yin, J. Wei and J.M. Gao. 2015. Striatoids A–F, cyathane diterpenoids with neurotrophic activity from cultures of the fungus *Cyathus striatus*. *J. Nat. Prod.*, 78: 783–788.

Bertrand, S., C. Petit, L. Marcourt, R. Ho, K. Gindro, M. Monod and J.L. Wolfender. 2013. HPLC profiling with at-line microdilution assay for the early identification of antifungal compounds in plants from French Polynesia. *Phytochem. Anal.*, 25: 106–112.

Bertrand, S., N. Bohni, S. Schnee, O. Schumpp, K. Gindro and J.L. Wolfender. 2014. Metabolite induction via microorganism co-culture: A potential way to enhance chemical diversity for drug discovery. *Biotechnol. Adv.*, 32(6): 1180–1204.

Bok, J.W., Y.M. Chiang, E. Szewczyk, Y. Reyes-Dominguez, A.D. Davidson, J.F. Sanchez, H.C. Lo, K. Watanabe, J. Strauss, B.R. Oakley, C.C. Wang and N.P. Keller. 2009. Chromatin-level regulation of biosynthetic gene clusters. *Nat. Chem. Biol.*, 5: 462–464.

Brodie, H.J. 1975. The Bird's Nest Fungi. University of Toronto Press, Toronto, CA. 199 pp.

Cichewicz, R.H. 2010. Epigenome manipulation as a pathway to new natural product scaffolds and their congeners. *Nat. Prod. Rep.*, 27: 11–22.

De Roy, K., M. Marzorati, P. Van den Abbeele, T. Van de Wiele and N. Boon. 2014. Synthetic microbial ecosystems: an exciting tool to understand and apply microbial communities. *Environ. Microbiol.*, 16(6): 1472–1481.

Fisch, K.M., A.F. Gillaspy, M. Gipson, J.C. Henrikson, A.R. Hoover, L. Jackson, F.Z. Najar, H. Wägele and R.H. Cichewicz. 2009. Chemical induction of silent biosynthetic pathway transcription in *Aspergillus niger*. *J. Ind. Microbiol. Biotechnol.*, 36(9): 1199–1213.

Góis, J.S., R.H. da Cruz, P.H. Nascimento and I.G. Baseia. 2020. A new species and new records of *Cyathus* (Agaricales, Basidiomycota) from a National Park in Bahia, Brazil. *N. Z. J. Bot.*, 12: 1–12.

Goss, R.J.M., S. Shankar and A.A. Fayad. 2012. The generation of "unNatural" products: Synthetic biology meets synthetic chemistry. *Nat. Prod. Rep.*, 29: 870–889.

Han, J., Y. Chen, L. Bao, X. Yang, D. Liu, S. Li, F. Zhao and H. Liu. 2013. Anti-inflammatory and cytotoxic cyathane diterpenoids from the medicinal fungus *Cyathus africanus*. *Fitoterapia*, 84: 22–31.

Han, J.J., L. Zhang, J.K. Xu, L. Bao, F. Zhao, Y.H. Chen, W.K. Zhang and H.W. Liu. 2015. Three new cyathane diterpenoids from the medicinal fungus *Cyathus africanus*. *J. Asian Nat. Prod. Res.*, 17(5): 541–549.

He, L., J. Han, B. Li, L. Huang, K. Ma, Q. Chen, X. Liu, L. Bao and H. Liu. 2016. Identification of a new cyathane diterpene that induces mitochondrial and autophagy-dependent apoptosis and shows a potent *in vivo* anti-colorectal cancer activity. *Eur. J. Med. Chem.* 111: 183–192.

He, M.Q., R.L. Zhao, K.D. Hyde, D. Begerow, R.L. Zhao, K.D. Hyde, D. Begerow, M. Kemler, A. Yurkov, E.H. McKenzie, O. Raspe, M. Kakishima, S. Sanchez-Ramirez and E.C. Vellinga. 2019. Notes, outline and divergence times of Basidiomycota. *Fungal Divers.*, 99: 105–367.

Hecht, H.J., G. Höfle, W. Steglich, T. Anke and F. Oberwinkler. 1978. Striatin A, B, and C: Novel diterpenoid antibiotics from Cyathus striatus; X-ray crystal structure of striatin A. *J. Chem. Soc. Chem. Commun.*, (15): 665–666.

Inchausti, A., G. Yaluff, A. Rojas de Arias, S. Torres, M.E. Ferreira, H. Nakayama, A. Schinini, K. Lorenzen, T. Anke and A. Fournet. 1997. Leishmanicidal and trypanocidal activity of extracts and secondary metabolites from Basidiomycetes. *Phytother. Res.*, 11(3): 193–197.

Johri, B.N., H.J. Brodie, A.D. Allbutt, W.A. Ayer and H. Taube. 1971. A previously unknown antibiotic complex from the fungus *Cyathus helenae*. *Experientia*, 27(7): 853–853.

Kang, H.S., E.M. Jun, S.H. Park, S.J. Heo, T.S. Lee, I.D. Yoo and J.P. Kim. 2007. Cyathusals A, B, and C, antioxidants from the fermented mushroom *Cyathus stercoreus*. *J. Nat. Prod.*, 70: 1043–1045.

Kang, H.S., K.R. Kim, E.M. Jun, S.H. Park, T.S. Lee, J.W. Suh and J.P. Kim. 2008. Cyathuscavins A, B, and C, new free radical scavengers with DNA protection activity from the Basidiomycete *Cyathus stercoreus*. *Bioorg. Med. Chem. Lett.*, 18(14): 4047–4050.

König, C.C., K. Scherlach, V. Schroeckh, F. Horn, S. Nietzsche, A.A. Brakhage and C. Hertweck. 2013. Bacterium induces cryptic meroterpenoid pathway in the pathogenic fungus *Aspergillus fumigatus*. *Chembiochem.*, 14: 938–942.

Kou, R.W., S.T. Du, Y.X. Li, X.T. Yan, Q. Zhang, C.Y. Cao, X. Yin and J.M. Gao. 2019. Cyathane diterpenoids and drimane sesquiterpenoids with neurotrophic activity from cultures of the fungus *Cyathus africanus*. *J. Antbiot.*, 72: 15–21.

Liu, Y.J. and K.Q. Zhang. 2004. Antimicrobial activities of selected *Cyathus* species. *Mycopathologia*, 157: 185–189.

Liu, S.L., L. Zhou, H.P. Chen and J.K. Liu. 2022. Sesquiterpenes with diverse skeletons from histone deacetylase inhibitor modified cultures of the basidiomycete *Cyathus stercoreus* (Schwein.) De Toni HFG134. *Phytochemistry*, 195: 113048.

Liu, S., H. Dai, C. Heering, C. Janiak, W. Lin, Z. Liu and P. Proksch. 2017. Inducing new secondary metabolites through co-cultivation of the fungus *Pestalotiopsis* sp. with the bacterium *Bacillus subtilis*. *Tetrahedron. Lett.*, 58: 257–261.

Marmann, A., A. Aly, W. Lin, B. Wang and P. Proksch. 2014. Co-cultivation—A powerful emerging tool for enhancing the chemical diversity of microorganisms. *Mar. Drugs*, 12: 1043–1065.

Matheny, P.B., J.M. Curtis, V. Hofstetter, M.C. Aime, J.-C. Moncalvo, Z.-W. Ge, Z-L. Yang, J.C. Slort, J.F. Ammirati, T.J. Baroni, N.L. Bougher, K.W. Hughes, D.J. Lodge, R.W. Kerrigan, M.T. Seidl, D.K. Aanen, M. DeNitis, G.M. Daniele, D. Desjardin, B.R. Kropp, L.L. Norvell, A. Parker, E.C.Vellinga, R. Vilgalys and D. Hibbett. 2006. Major clades of Agaricales: A multilocus phylogenetic overview. *Mycologia*, 98(6): 982–995.

Meyer, V. and C. Nai. 2018. From axenic to mixed cultures: Technological advances accelerating a paradigm shift in microbiology. *Trends Microbiol.*, 26: 538–554.

Nitthithanasilp, S., C. Intaraudom, N. Boonyuen, R. Suvannakad and P. Pittayakhajonwut. 2018. Antimicrobial activity of cyathane derivatives from *Cyathus subglobisporus* BCC44381. *Tetrahedron*, 74(48): 6907–6916.

Ochi, K. and T. Hosaka. 2012. New strategies for drug discovery: Activation of silent or weakly expressed microbial gene clusters. *Appl. Microbiol. Biotechnol.*, 97: 87–98.

Rai, B.K., S.S. Ayachi and A. Rai. 1993. A note on ethno-myco-medicines from Central India. *Mycologist*, 7: 192–193.

Ramm, S., B. Krawczyk, A. Mühlenweg, A. Poch, E. Mösker and R.D. Süssmuth. 2017. A self-sacrificing n-methyltransferase is the precursor of the fungal natural product omphalotin. *Angew. Chem. Int. Ed.*, 56: 9994–9997. doi: 10.1002/anie.201703488.

Richter, C., S.E. Helaly, B. Thongbai, K.D. Hyde and M. Stadler. 2016. Pyristriatins A and B: pyridino-cyathane antibiotics from the basidiomycete *Cyathus cf. striatus*. *J. Nat. Prod.*, 79: 1684–1688.

Szewczyk, E., Y.M. Chiang, C.E. Oakley, A.D. Davidson, C.C.C. Wang and B.R. Oakley. 2008. Identification and characterization of the asperthecin gene cluster of *Aspergillus nidulans*. *Appl. Environ. Microbiol.*, 74: 7607–7612.

Tang, D., Y.Z. Xu, W.W. Wang, Z. Yang, B. Liu, M. Stadler, L.L. Liu and J.M. Gao. 2019. Cyathane diterpenes from cultures of the bird's nest fungus *Cyathus hookeri* and their neurotrophic and anti-neuroinflammatory activities. *J. Nat. Prod.*, 82(6): 1599–1608.

Wang, B., J. Han, W. Xu, Y. Chen and H. Liu. 2014. Production of bioactive cyathane diterpenes by a bird's nest fungus *Cyathus gansuensis* growing on cooked rice. *Food Chemistry*, 152: 169–176.

Wei, J., Y. Cheng, W.H. Guo, D.C. Wang, Q. Zhang, D. Li, J. Rong and J.M. Gao. 2017. Molecular diversity and potential anti-neuroinflammatory activities of cyathane diterpenoids from the basidiomycete *Cyathus africanus*. *Sci. Rep.*, 7(1): 1–14.

Wei, J., W.H. Guo, C.Y. Cao, R.W. Kou, Y.Z. Xu, M. Górecki, L. Di Bari, G. Pescitelli and J.M. Gao. 2018. Polyoxygenated cyathane diterpenoids from the mushroom *Cyathus africanus*, and their neurotrophic and anti-neuroinflammatory activities. *Sci. Rep.*, 8(1): 1–15.

Winter, J.M., S. Behnken and C. Hertweck. 2011. Genomics-inspired discovery of natural products. *Curr. Opin. Chem. Biol.*, 15: 22–31.

Xu, Z., S. Yan, K. Bi, J. Han, Y. Chen, Z. Wu and H. Liu. 2013. Isolation and identification of a new anti-inflammatory cyathane diterpenoid from the medicinal fungus *Cyathus hookeri* Berk. *Fitoterapia*, 86: 159–162.

Yin, X., J. Wei, W.W. Wang, Y.Q. Gao, M. Stadler, R.W. Kou and J.M. Gao. 2019. New cyathane diterpenoids with neurotrophic and anti-neuroinflammatory activity from the bird's nest fungus *Cyathus africanus*. *Fitoterapia*, 134: 201–209.

Yin, X., J. Qi, Y. Li, Z.A. Bao, P. Du, R. Kou, W. Wang and J.M. Gao. 2021. Terpenoids with neurotrophic and anti-neuroinflammatory activities from the cultures of the fungus *Cyathus stercoreus*. *Nat. Prod. Res.*, 35(22): 4524–4533.

Yu, M., X. Kang, Q. Li, Y. Liang, M. Zhang, Y. Gong, C. Chen, H. Zhu and Y. Zhang. 2022. Thirteen cyathane diterpenoids with acetylcholinesterase inhibitory effects from the fungus *Cyathus africanus*. *Phytochemistry*, 193: 112982.

8

Irpex lacteus — A Less Explored Source of Bioactive Metabolites

Sunil Kumar Deshmukh,[1,]* Manish K Gupta,[2] Kandikere R Sridhar[3] and Hemraj Chhipa[4]

1. Introduction

Wood-decaying fungus *Irpex lacteus* (Fr.) Fr. (Basidiomycota; Class, Agaricomycetes; Order, Polyporales; Family, Steccherinaceae) is a pathogenic mushroom. It causes a white rot and occurs on numerous hardwood trees, growing on dead branches of either dead or living trees as well as on dead-standing trunks or fallen trunks and branches. The host spectrum is broad including *Acer, Alnus, Betula, Cornus, Corylus, Fagus, Frangula, Juglans, Populus,* and *Prunus. Rosa, Sorbus* and *Tilia,* are rarely on conifers (e.g., *Juniperus*). The geographic distribution of *I. lacteus* is restricted to the Northern Hemisphere, including Europe, Asia, and North Africa, and is widely distributed in North America (Ryvarden and Gilbertson 1993, Bernicchia and Gorjón 2020, Gafforov et al. 2020, Gafforov and Ordynets 2022). *I. lacteus* was also reported from *Populus, Pyrus, Quercus, Salix* and *Ulmus* trees from Uzbekistan (Gafforov et al. 2020).

---

[1] R & D Division, Greenvention Biotech Pvt. Ltd., Uruli-Kanchan, Pune, 412202, Maharashtra, India.
[2] SGT College of Pharmacy, SGT University, Gurugram 122505, Haryana, India.
[3] Department of Biosciences, Mangalore University, Mangalagangotri, Mangalore, Karnataka, India.
[4] Agriculture University, Kota Rajasthan-324001, India.
* Corresponding author: sunil.deshmukh1958@gmail.com

Irpex lacteus is a common wood rot fungus used for the destruction of recalcitrant organics in contaminated soils as well as decolorizing synthetic dyes (Novotný et al. 2009, Svobodová et al. 2008). It is also reported from various plants as endophytic (Wang et al. 2013, Duan et al. 2019, Luo et al. 2022a, Luo et al. 2022b) and also known to cause infections in humans (Buzina et al. 2011, Kano et al. 2021, Saha et al. 2022). *Irpex lacteus* is a source of several classes of bioactive compounds such as sterol, terpenoid, sesquiterpene, tremulane, glucoside, furan, polyketide, indole, naphthalene, lactone and others possessing diverse biological activities. The present review highlights the bioactive metabolites reported from the fungus *I. lacteus* and their biological activities as shown in Table 1.

2. Biological Investigations on the Extracts of *Irpex lacteus*

Irpex lacteus is widely dispersed throughout temperate areas globally traditionally used medicinal fungus for several medicinal purposes (antibacterial, diuretic, and anti-inflammatory functions). The capsule Yishenkang made up of fermented polysaccharide fraction of *I. lacteus* is clinically used as a remedy for chronic glomerulonephritis in China (Chen et al. 2020). Firstly, the culture extract of *I. lacteus* was found active against *Bacillus cereus*, *Candida albicans*, *C. glabrata*, *C. parapsilosis*, *Escherichia coli*, *Staphylococcus aureus* and *S. typhimurium* (Rosa et al. 2003). Ethyl acetate extract of *I. lacteus* CCB 196, exhibited 100% growth of UACC-62, MCF-7, and TK-10 human cancer cell lines and Inhibition of PBMC proliferation. Ethyl acetate extract of *I. lacteus* CCB 196 also displayed 91% inhibition of the enzyme trypanothione reductase (TryR) and kill 87% of *Leishmania amazonensis* (Rosa et al. 2009).

The water-soluble polysaccharide (ILN III) isolated from *I. lacteus* by hot-water extraction, displayed antitumor activity against HepG2 as well as HeLa cell lines with IC 50 60.95 and 99.95 µg/mL, respectively. The ILA I showed significant inhibition of murine mesangial cells (HBZT-1) with IC_{50} of 185.06 µg/mL (Na et al. 2012).

The hot-water extract of this mushroom possesses seven water-soluble polysaccharide fractions. These fractions possess significant inhibition of HepG2, as well as HeLa tumour cells with IC_{50} of 60.95 and 99.95 µg/mL, respectively. These findings also suggest that the polysaccharide fractions of *I. lacteus* possess significant antitumor activities (Zhang et al. 2012). Polysaccharides in aqueous extracts have renoprotective activities via anti-inflammatory and inhibition of nuclear factor κB 65 heterodimer (NF-κB-p65) (Jiang et al. 2014). Methanolic extract of *I. lacteus* exhibited antibacterial activity against *Klebsiella pneumonia* (MTCC109), *S. aureus* (MTCC 737), *E. coli* (MTCC-739), with a zone of inhibition of 28, 27.33 and 19.33 mm respectively at the highest concentration of 500 mg/ml. (Chaudhary and Tripathy 2015). Methanolic extract of *I. lacteus* also showed antifungal activity against *C. albicans* (MTCC-227) and *Trichophyton mentagrophyte* (MTCC-8476) with a zone of inhibition 10.33 and 25.66 mm, respectively at the same concentration (Chaudhary and Tripathy 2015).

Table 1: Novel bioactive compounds reported from *Irpex lacteus*.

Sr. no.	Producing organism	Name of the compound	Biological target	Biological activity[a] ($MIC/IC_{50}/ID_{50}$)	References
1.	*I. lacteus* (IFO 5367)	5-pentyl-2-furaldehyde (1), and 5-(4-pentenyl)-2-furaldehyde (2)	Antinematicidal activity against *Aphelencoides besseyi*	50% mortality at 50 ppm	Hayashi et al. 1981
		3-*p*-anisoloxypropionate (3)		50% mortality at 25 ppm	
2.	*I. lacteus* strain E21, endophyte associated with the root of *Lycium ruthenicum*	Monacolin K (4)	Gram negative and Gram positive bacteria	Stronger inhibition	Wang et al. 2013
		Dehydromonacolin K (5)	*B. subtilis*	Moderate inhibition	
		Ergosterol peroxide (6)			
		Ergosterol (7)			
3.	*I. lacteus*	Irlactins A–D (8-11), irlactin E (12)			Ding, et al. 2013
4.	*I. lacteus*	Irlactins F-J (13-17, tremulenediol A (18), 1β,12-epoxy-14-hydroxy-2(11)-tremulene (19)			Ding et al. 2018
		Irlactin I (16)	Anticancer activity against HL-60, SMMC-7721, A-549, MCF-7, and SW480 cells	IC_{50} values of 16.23, 20.40, 25.55, 19.05, and 18.58 μM	
5.	*I. lacteus*	Irlactam A (20)	Isozymes of 11β-hydroxysteroid dehydrogenases (11β-HSD)	No significant activity against	Ding et al. 2019
6.	*I. lacteus* HFG1102	11,12-Epoxy-15-hydroxy-5,6-*seco*tremula-1,6(13)-dien-5,12-olide (21), 12b,15-dihydroxy-5,6-*seco*-tremula-1,6(13)-dien-5-oate (22), 11,12-epoxy-14-hydroxy-5,6-*seco*-tremula-1,6(13)-dien-5,12-olide (23), methyl 12β,14-dihydroxy-5,6-*seco*-tremula-1,6(13)-dien-5-oate (24), 6,11-dihydroxy5,6-*seco*-tremul-1-en-5,12-olide (25), 11,12-epoxy-6-hydroxy-5,6-*seco*-tremul-1-en-5,12-olide (26), conocenolide A (27), and conocenolide B (28)			Chen et al. 2018

No.	Source	Compounds	Activity	Values	References
7.	*I. lacteus* DR10-1 endophyte of *Distylium chinense*	Irpexlactes A–D (**29-32**)	*Pseudomonas aeruginosa*	MIC values ranging from 23.8 to 35.4 μM	Duan et al. 2019
		Irpexlacte A (**29**), and irpexlacte D (**32**)	Antioxidant activity DPPH assay	IC$_{50}$ values of 2.50 and 5.75 μM	
		Irlactin E (**12**) and 3β-hydroxycinnamolide (**33**)			
8.	*I. lacteus*	Tremutins A-H (**34–41**), irlactin I (**16**), (+)-(1R,6S,7S)-tremul-2-ene-12(11)-lactone (**42**), and ceriponol C (**43**)	Anticancer activity against MCF-7, SMMC-7721, HL-60, SW480, and A549 cell lines	Poor activity	Wang et al. 2020
		Tremutin F (**39**), tremutin G (**40**), (+)-(1R,6S,7S)-tremul-2-ene-12(11)-lactone (**42**), and ceriponol C (**43**)			
		Tremutin H (**41**)	Inhibitory effect on NO production	IC$_{50}$ value of 22.7 μM	
		Tremutin A (**34**)	LPS-induced proliferation of B lymphocyte cells	IC$_{50}$ value of 22.4 μM	
		Tremutin B (**35**)	Concanavalin A (Con A)-induced T cell proliferation and LPS-induced B lymphocyte cell proliferation	IC$_{50}$ values of 16.7 and 13.6 μM	
9.	*I. lacteus*	Irpexoates A (**44**), C (**46**)			Tang et al. 2018a
		Irpexoate B (**45**)	Anticancer activity against A-549, SMMC-7721, MCF-7, SW480 cell lines	IC$_{50}$ values varying from 22.9 to 34.0 μM	
		Irpexoate D (**47**)	Anticancer activity against SW480 cell lines	IC$_{50}$ value of 35.2 μM	
10.	*I. lacteus*	Irpeksins A–E (**48-52**)	Inhibitory activity against NO production in LPS-activated RAW 264.7 macrophage cells	IC$_{50}$ values in the range of 2.2 to 19.6 μM	Tang et al. 2018b

Table 1 contd. ...

...*Table 1 contd.*

Sr. no.	Producing organism	Name of the compound	Biological target	Biological activity[a] (MIC/IC$_{50}$/ID$_{50}$)	References
11.	*I. lacteus*	Irpexolidal (53), irpexolide A (54)			Tang et al. 2019a
12.	*I. lacteus*	Irpeksolactins A-J (55–64), asiatic acid (65), 15α-hydroxydehydrotumulosic acid (66), polyporenic acid C (67), 24-methylene-lanosta-7,9 (11)-diene-3-one (68), 29-hydroxypolyporenic acid C (69), 6α-hydroxypolyporenic acid C (70), de-hydrosulphurenic acid (71), polycarpol (72), dehydrotrametenonic acid (73), 3-oxo-6,16α-dihydroxylanosta-7,9(11),24(31)-trien-21-oic acid (74), ganoderol A (75), 13α,14β,17α-lanosta-7,9,24-triene-3β,16α-diol (76), 3α-hydroxy-24-methylene-23-oxolanost-8-en-26-carboxylic acid (77), 3α-carboxyacetoxyquercinic acid (78), inonotusane C (79), daedalol C (80), ganodermanondiol (81), hexatenuins B (82)			Tang et al. 2019b
		Irpeksolactin J (64)	Anticancer activity against A549 and SMMC-7721 cell line	54.6 and 50.0% inhibition rates at the concentration of 40 µM	
13.	*I. lacteus* HFG1102	Irlactin K (83), irpexolactins A-N (84–97), conocenol C (98), tremulenolide D (99), tremulenediol A (18), (+)-(3S,6R,7R)-tremulene-6,11,12-triol (100), conocenol B (101), (−)-(3S,6S,7S,10S)-tremulene-10,11,12-triol (102), irlactam A (20), 11,12-dihydroxy-1-tremulen-5-one (103), and irlactin A (8)			Chen et al. 2020
		Irlactin K (83), irpexolactin A (84), irpexolactin C (85), irpexolactin K (94), and irlactam A (20)	Vasorelaxant effect on KCl precontracted thoracic aorta rings	No significant vasorelaxant effect (Nifedipine was used as the positive control)	

No.	Source	Metabolites	Target	Activity	Reference
14.	*I. lacteus*	Irlactin K (1) (83)	Anticancer activity against SK-BR-3, SMMC-7721, HL-60, PANC-1, and A-549 cells	Inactive at IC$_{50}$ values >40 µM	Ding et al. 2020
			Isozymes of 11β-hydroxysteroid dehydrogenases (11β-HSD)	No significant activity against	
15.	*I. lacteus*	Irlactin L (104) and irlactin M (105), 6-hydroxy-2,6-dimethyloct-7-enoic acid (106)			Ding et al. 2021
16.	*I. lacteus* endophyte of *Gastrodia elata*	Irpexlactin A-D (107–110), Irpexlactin E (111)/11,12-epoxy-5,6-secotremula-1,6(13)-dien-5,12-olide (114), irpexlacin (112), irlactin G (14), conocenol C (98), conocenol B (101), 11,12-dihydroxy-1-tremulen-5-one (103), neroplofurol (113), irpexlacte B (30)	*P. polonicum, T. atroviride, Psathyrella subsingeri Armillaria* sp.	MIC values in the range of 4–64 µg/mL	Wang et al. 2021
		Positive control nystatin		MICs in the range of 4–16 µg/mL	
		Irpexlactin A (107), irpexlacin B (108), Irpexlactin E (111) / 11,12-epoxy-5,6-secotremula-1,6(13)-dien-5,12-olide (114), irlactin G (14), conocenol C (98), conocenol B (101),11,12-dihydroxy-1-tremulen-5-one (103), irpexlacte B (30)	*I. lacteus*	MIC values in the range of 32–64 µg/mL	
		Positive control nystatin		MIC, 16 µg/mL	
		2,3-Dihydroxydodacane-4,7-dione (115)		MIC values of 16, 32 and 16 µg/mL	
17.	*I. lacteus*	Lactedine (116), conocenol B (101), phellininginicisterol C (117), ceriponol A (118), nigrosirpexin A (119), 11,12-dihydroxy-1-termulen-5- one (120), conocenol A (121), 10β,11-dihydroxy-5,6-*seco1*,6(13)-termulacation-5,12-olide (122), irpenigirin A (123)		no obvious antibacterial and AChE inhibitory activity	Sun et al. 2022

Table 1 contd. ...

...*Table 1 contd.*

Sr. no.	Producing organism	Name of the compound	Biological target	Biological activity[a] (MIC/IC$_{50}$/ID$_{50}$)	References
18.	(*I. lacteus* OV38) endophyte of *Orychophragmus violaceus*	Irpexolaceus A (124), irpexolaceus B (125), irpexolaceus C (126), irpexolaceus D (127), irpexolaceus E (128), irpexolaceus F (129), irpexolaceus G (130), irpexonjust A (131), irpexonjust B (132), tremulenolide D (99), irpexlacte B(30), irpexlacte C (31), 5-(3-Oxopentyl)-2-furaldehyde (133), irpexolactin I (134), irpexolactin J (135), 2-furoic acid (136), and 5-(3-methoxy-3-oxopropyl)furan-2-carboxylic acid (137)			Luo et al. 2022a
		Irpexolaceus A (124), C (126), D (127), F (129), and G (130), irpexonjust B (132), and irpexlacte B (30)	Inhibitory effects of against NO release from LPS-induced RAW 264.7 cells	were higher than > 45%, at the concentration of 50 µg/mL	
		Irpexlacte C (31)		42.6% at the concentration of 50 µg/mL	
		Irpexolaceus B (125)		39.6% at the concentration of 50 µg/mL	
		Irpexonjust A (131)		43.7%at the concentration of 50 µg/mL	
		Irpexolaceus E (128)		33.6% at the concentration of 50 µg/mL	

19.	*I. lacteus, Orychophragmus violaceus.*	Irpexolaceus H (**138**) and I (**139**), irpexlactes C (**31**)		Luo et al. 2022b
		Irpexlactes B (**30**)	QS inhibitory activity at 50 mg/ml against biomarker strains of *Agrobacterium tumefaciens* A136 and *Chromobacterium violaceum* CV026	
20.	*I. lacteus* strain CMU-8413	Apotrichodiol (**140**), apotrichothecene (**141**), blennin D (**142**), collybial (**143**), cyclocalopin A (**144**), dehydroreadone (**145**), dictyoquinazol A (**146**), dihydromarasmone (**147**), frequentin (**148**), ganodermic acid Jb (**149**), geosmin (**150**), microdiplodiasol (**151**), pandangolide 1 (**152**), and piperdial (**153**)		Pineda-Suazo et al. 2021
21.	*I. lacteus*	Irpexins A–H (**154–161**), irpexins I (**162**) and J (**163**), irpexlacte B-D (**30–32**), 1-(5-(hydroxymethyl)-2-furanyl)-pentanone (**33**), 5-(3-oxopentyl)-2-furancarboxaldehyde (**34**), (1R,2R)-1-(5-(pent-4-en-1-yl)furan-2-yl)propane-1,2-diol (**35**), (1R,2R)-1-(5-pentylfuran-2-yl)propane-1,2-diol (**36**), 2,5,8-decanetrione (**37**), 2-butyl-3hydroxy-2-cyclopenten-1-one (**38**)	No cytotoxicity against MCF-7 and Hela cancer cell lines at the concentration of 10 µM	Wang et al. 2021
22.	*I. lacteus*	Irpexidines A (**164**) and B (**165**), irpexins K (**166**) and L (**167**), 5-carboxy-2-furanpropanoic (**168**), and 5-(methoxycarbonyl)-2-furanpropanoic acid (**169**)	No cytotoxicity against Hela cancer cell and inhibitory activity on NO production	Chen et al. 2021

Table 1 contd. ....

...Table 1 contd.

Sr. no.	Producing organism	Name of the compound	Biological target	Biological activity[a] (MIC/IC$_{50}$/ID$_{50}$)	References
	Coculture studies				
23.	*Nigrospora oryzae* co-cultured with *I. lacteus*, endophyte of *Dendrobium officinale*	Nigirpexin A-D (**170–173**), nigrosirpexin A (**119**)			Zhou et al. 2018
		Conocenol B (**101**)	Antifungal against *N. oryzae* and *I. lacteus*	MICs of 16 and 128 µg/mL	
		Nigrosirpexin A (**119**)		MICs of 64 and 128 µg/ml	
		Nigrosirpexin A (**119**)	AChE inhibition	Weak inhibition activity with an inhibition ratio of 35 % at a concentration of 50 µM	
24.	*Nigrospora oryzae* co-cultured with *I. lacteus, Dendrobium officinale*	Irpenigirin A (**123**), isonigirpexin C (**175**)			Wu et al. 2019
		Conocenol B (**101**) and 4-(4-dihydroxymethylphenoxy) benzaldehyde (**177**)	Antifungal activity against *N. oryzae*	MICs at 16 µg/mL	
		Irpenigirin B (**174**) and 5-Demethyl conocenol C (**176**)	Antifungal activities against *C. gloeosporioides*	MICs of 8 µg/mL	
		5-Demethyl conocenol C (**176**), conocenol B (**101**), conocenol C (**98**)	Anti-fungal activities against *Didymella glomerata*	MICs of 1, 8, and 4 µg/mL	
		Nigrosirpexin A (**119**)	anti-fungal activity against *N. oryzae*	MICs at 32 µg/mL	
			AChE inhibition	Weak inhibition activity with an inhibition ratio of 16% at a concentration of 50 µM	

		Conocenol B (101), and 4-(4-dihydroxymethylphenoxy)benzaldehyde (176)			
25.	*I. lacteus* and *Phaeosphaeria oryzae*	Irpexine (178), hypoxyxylerone (179)	Antifungal activity against *N. oryzae*	At a concentration of 100 µg/mL causal organism of *Cerasus cerasoides*	Sadahiro et al. 2020
26.	Phytopathogen–endophyte *N. oryzae* and *I. lacteus*	Nigrosirpexin B (180), 11,12-epoxy-5,6-secotremula-1,6(13)-dien-5,12-olide (114), 12-acetoxy-5,6-seco-1,6(13)-tremuladien-5,11-olide (115), conocenol B (101), 11,12-dihydroxy-1-tremulen-5-one (103), conocenolide A (27), and 11-acetoxy-5,6-*seco*-1,6(13)-tremuladien-5,12-olide (185)	Anti-phytopathogenic activity against *N. oryzae*	MICs of 1, 32, 256, 16, 128, 1, and 128 µg/mL, respectively	Shi et al. 2020
		Nigrosirpexin C (181), nigirpexin E (182), nigrosirpexin D (183), nigrosirpexin E (184)			
	phytopathogen–endophyte (nonhomologous *N. oryzae* and *I. lacteus*)	Nigrosirpexin F (186), 11,12-epoxy-5,6-*secotremula*-1,6(13)-dien-5,12-olide (114), 11,12-epoxy-12β-hydroxy-1-tremulen-5-one (187), and syringaresinol (188)			

Table 1 contd. ...

...Table 1 contd.

Sr. no.	Producing organism	Name of the compound	Biological target	Biological activity[a] (MIC/IC$_{50}$/ID$_{50}$)	References
27.	Co-culture of endophyte *I. lacteus*, phytopathogen *N. oryzae*, and entomopathogen *Beauveria bassiana*	Nigpexin B–D (190–192), p-hydroxybenzoic acid (196), tyrosol (197), β-sitosterol (200), and scytalone (201)	Anti-fungal activities against *I. lacteus*	MICs ≤ 8 μg/mL	Yin et al. 2021
		Tremulenediol A (18), 11-aldehyde-5,6-*seco*-1,6(13)-tremuladien- 5,12-olide (199), scytalone (201), 4,6,8-trihydroxy-3,4-dihydronaphthalen-1(2H)-one (202), and (3S,4R)-3,4-dihydroxypentanoic acid (203)	Anti-fungal activity against *N. oryzae*	MICs ≤ 4 μg/mL	
		Nigpexin B (190), nigpexin E (193), 11-aldehyde-5,6-*seco*-1,6(13)-tremuladien- 5,12-olide (199), (3S,4R)-3,4-dihydroxypentanoic acid (203)	Anti-fungal activity against *B. bassiana*	MICs ≤ 8 μg/mL	
		Nigpexin A–E (189–193), mevalonolactone (194), microsphaerophthalide F (195), p-hydroxybenzoic acid (196), tyrosol (197), 2-hydroxyphenylacetic acid (198), tremulenediol A (18), 11-aldehyde-5,6-*seco*-1,6(13)-tremuladien- 5,12-olide (199), conocenolide A (27), conocenolide B (28), β-sitosterol (200), scytalone (201), 4,6,8-trihydroxy-3,4-dihydronaphthalen-1(2H)-one (202), and (3S,4R)-3,4-dihydroxypentanoic acid (203)	Antifeedant activities against silkworm	% inhibition of 73–99%, at concentrations of 50 μg/cm^2	
		Tremulenediol A (18)		% inhibition of 93%, at concentrations of 6.25 μg/cm^2.	
		Positive drug avermectin inhibition percentage > 95%, at concentration of 50 μg/cm^2 or 6.25 μg/cm^2		% inhibition of > 95%, at concentration of 50 μg/cm^2 or 6.25 μg/cm^2	

No.	Source	Compounds	Activity	Results	Reference
28.	Co-culture of endophyte *I. lacteus* and pathogenic *N. oryzae*	Nigirpexin C (172), tremulenediol A (18), (+)-(3S,6R,7R)-tremulene-6,11,12-triol (100), Irperide (204), lactedine (116), and conocenol B (101)	Antifungal activity against *Aspergillus fumigatus*	MIC values of 1, 2 and 1 μg/mL, respectively	Wu Hayashi et al. 2022
Application in Biotransformation					
29.	*I. lacteus* CCTCC M 2017161 Biotransformation of huperzine A (205)	Compounds (206–207), 8α,15α-epoxyhuperzine A (208)		Neuroprotective activity by increasing the viability of U251 cell lines with EC50 = 35.3, 32.1, and 50.3 nM	Ying et al. (2019)

Polysaccharide ILN3A was purified from hot water extract of mutant (mutant ILN10 obtained via chemical mutagenesis) *I. lacteus*, grown in a liquid medium. The backbone of ILN3A (264 kDa) comprises (1→2) and (1→4) linkages and the ^{1}H NMR spectrum indicates the existence of α- and β-glycosidic anomeric carbon (Wang et al. 2016). Polysaccharide ILN3A was shown to inhibit the proliferation of mesangial cells. In MGN rats, ILN3A reverses structural changes in the kidney, suppresses the abnormally high level of urine protein and restores the concentration of serum albumin. The ILN3A also reduced total cholesterol, triglycerides, serum creatinine, and 6-keto-PGF in the kidney cortex. Added study shows that ILN3A restores serum Interleukin 2, Interleukin 2 receptor, Interleukin 6, tumor necrosis factor α, and renal cortical nuclear factor kappa B (Wang et al. 2016). Later, aqueous extract from the fruit body of *I. lacteus* has a preventive impact against adenine-induced chronic nephritis (Han et al. 2018). The *I. lacteus* extract increases endurance capacity by activating the AMPK-linked antioxidant pathway, which provides evidence of the clinical application of ILE (*I. lacteus* extract) is a potent agent against fatigue (Wang et al. 2019). *I. lacteus* extract displayed anti-fatigue activities in mice (Wang et al. 2019).

In coculture confrontation assays, *I. lacteus* efficiently antagonized *Colletotrichum* spp., *Fusarium* spp. and *Phytophthora* spp. phytopathogenic strains, with inhibition growth (between 16.7 and 46.3%). Assays of antibiosis showed the inhibitory effect of soluble extracellular metabolites of *I. lacteus* against phytopathogen growth (between 32.0 and 86.7%) (Pineda-Suazo et al. 2021).

3. Compounds Isolated from *Irpex lacteus*

Compounds 5-pentyl-2-furaldehyde (1), 5-(4-pentenyl)-2-furaldehyde (2), and 3-*p*-anisoloxypropionate (3) (Figure 1), were purified from the culture filtrate of *I. lacteus* (IFO 5367) (Hayashi et al. 1981).

Compounds monacolin K (4), dehydromonacolin K (5), Ergosterol peroxide (6) and ergosterol (7) (Figure 1), were extracted from *I. lacteus* strain E21 an endophyte associated with the root of *Lycium ruthenicum* (Wang et al. 2013).

Irlactins A–D (8–11) (Figure 1), four sesquiterpenoids with a rearranged 6/6 bicyclic system, with their expected biosynthetic precursor irlactin E (12), were isolated from cultures of the *I. lacteus*. Their structures were assessed based on spectroscopic methods as well as the absolute configurations of irlactins A–D (8–11), which were established by single crystal X-ray diffraction analysis (Ding et al. 2013).

Five new tremulane sesquiterpenes irlactins F-J (13–17), along with two known analogues tremulenediol A (18) (Figure 1), 1β,12-epoxy-14-hydroxy-2(11)-tremulene (19) (Figure 2), were isolated from cultures of the *I. lacteus*. Structures and relative configurations of compounds irlactins F-J (13–17), were determined by spectroscopic analysis (Ding et al. 2018). The new tremulane sesquiterpenoid irlactam A (20) (Figure 2) was purified from cultures of the *I. lacteus* and was elucidated by spectroscopic analysis (Ding et al. 2019).

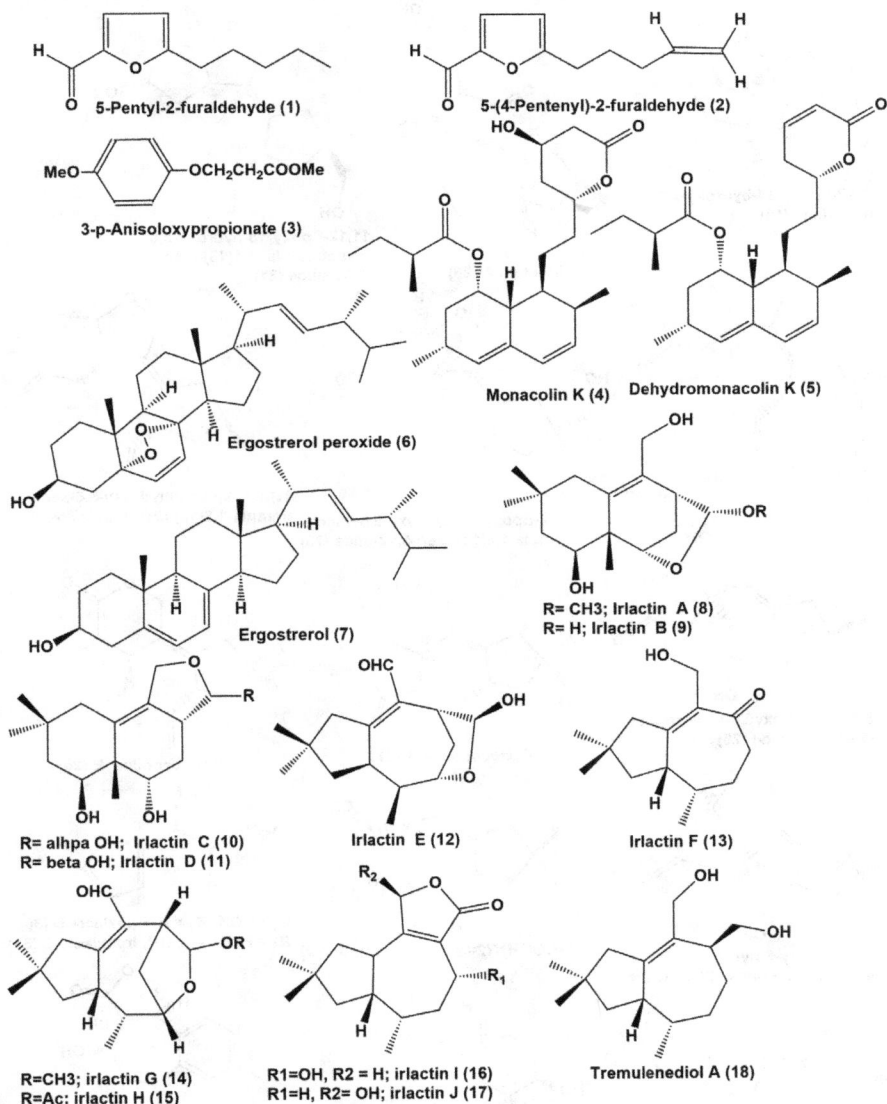

Figure 1: Metabolites isolated from *I. lacteus* (1–18).

Six previously unknown 5,6-*seco*-tremulane analogues 11,12-epoxy-15-hydroxy-5,6-*seco*-tremula-1,6(13)-dien-5,12-olide **(21)**, 12b,15-dihydroxy-5,6-*seco*-tremula-1,6(13)-dien-5-oate **(22)**, 11,12-epoxy-14-hydroxy-5,6-*seco*-tremula-1,6(13)-dien-5,12-olide **(23)**, methyl 12β,14-dihydroxy-5,6-*seco*-tremula-1,6(13)-dien-5-oate **(24)**, 6,11-dihydroxy5,6-*seco*-tremul-1-en-5,12-olide **(25)**, and 11,12-epoxy-6-hydroxy-5,6-*seco*-tremul-1-en-5,12-olide **(26)**, along with previously reported

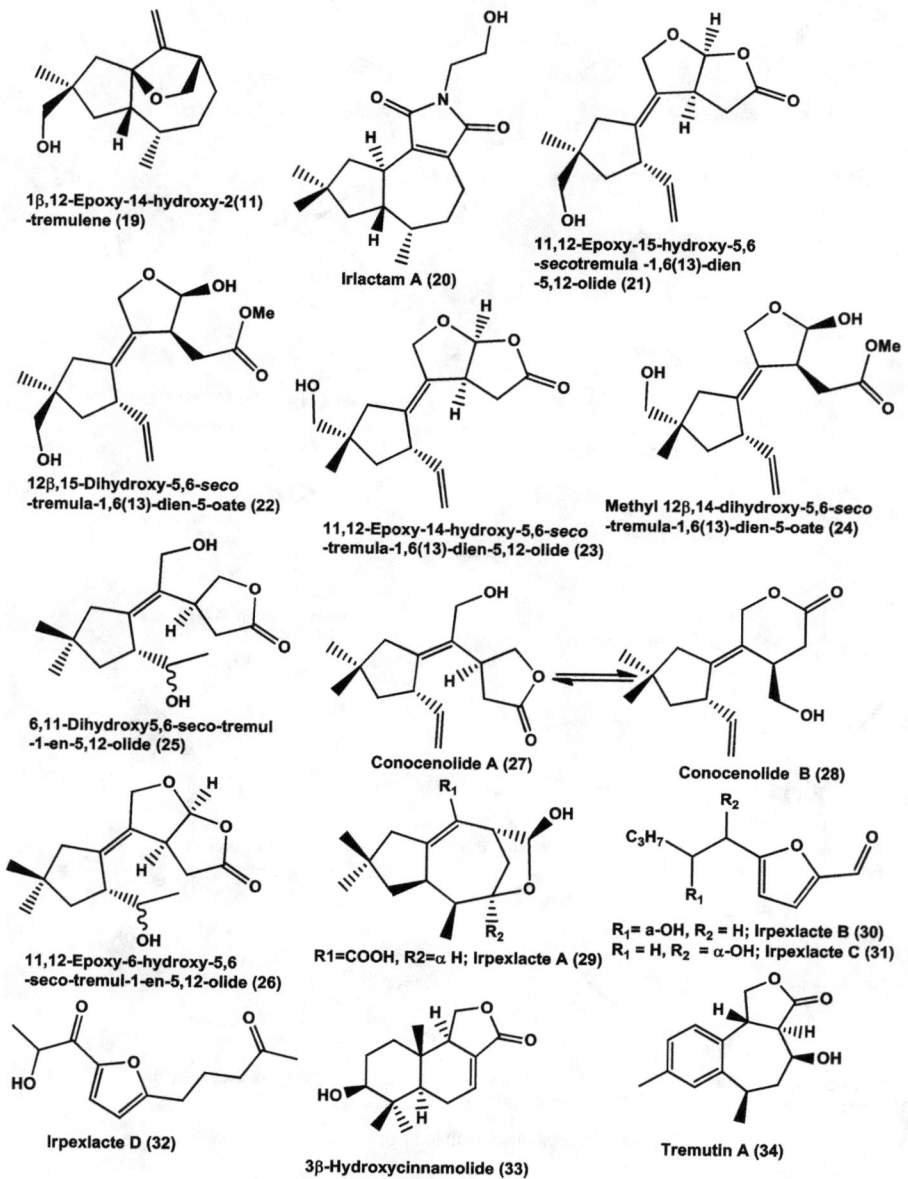

1β,12-Epoxy-14-hydroxy-2(11)
-tremulene (19)

Irlactam A (20)

11,12-Epoxy-15-hydroxy-5,6
-*seco*tremula -1,6(13)-dien
-5,12-olide (21)

12β,15-Dihydroxy-5,6-*seco*
-tremula-1,6(13)-dien-5-oate (22)

11,12-Epoxy-14-hydroxy-5,6-*seco*
-tremula-1,6(13)-dien-5,12-olide (23)

Methyl 12β,14-dihydroxy-5,6-*seco*
-tremula-1,6(13)-dien-5-oate (24)

6,11-Dihydroxy5,6-seco-tremul
-1-en-5,12-olide (25)

Conocenolide A (27)

Conocenolide B (28)

11,12-Epoxy-6-hydroxy-5,6
-seco-tremul-1-en-5,12-olide (26)

R1=COOH, R2=α H; Irpexlacte A (29)

R1= a-OH, R2 = H; Irpexlacte B (30)
R1 = H, R2 = α-OH; Irpexlacte C (31)

Irpexlacte D (32)

3β-Hydroxycinnamolide (33)

Tremutin A (34)

Figure 2: Metabolites isolated from *I. lacteus* (19–34).

compound conocenolide A **(27),** and conocenolide B **(28)** (Figure 2), were isolated from the culture of *I. lacteus* (HFG1102). The structures of these compounds were elucidated by extensive spectroscopic analysis including NMR and HRMS spectroscopic analyses (Chen et al. 2018).

The new tremulane sesquiterpene, irpexlacte A **(29)**, and three new furan derivatives, irpexlacte B–D **(30-32)** (Figure 2), were obtained from the endophytic fungus *I. lacteus* DR10-1 waterlogging tolerant plant *Distylium chinense*, along with two known compound irlactin E **(12)** and 3β-hydroxycinnamolide **(33)** (Figure 2). The structures of these metabolites were assessed by HRESIMS and NMR spectroscopic analysis (Duan et al. 2019).

Cultures of *I. lacteus* were the source of eight undescribed sesquiterpenoids, tremutin A **(34)** (Figure 2) B-H **(35–41)**, along with three known compounds irlactin I **(16)**, (+)-(1R,6S,7S)-tremul-2-ene-12(11)-lactone **(42)**, and ceriponol C **(43)** (Figure 3). The structures of the new compounds along with absolute configurations were evaluated based on extensive spectroscopic methods, single-crystal X-ray diffractions and equivalent circulating density calculations. Compounds tremutin A **(34)** and B **(35)** possess an unusual 6/7-fused ring system that might be derived from a tremulane framework. Compounds tremutins C-G **(36–40)** and irlactin I **(16)**, (+)-(1R,6S,7S)-tremul-2-ene-12(11)-lactone **(42)**, and ceriponol C **(43)**, are tremulane sesquiterpenoids of which tremutin D **(37)** and E **(38)** are the first tremulane examples with a 1,2-epoxy moiety to be reported (Wang et al. 2020).

4. Tremulane Sesquiterpenoids

4.1 Eburicane Triterpenoids

Four eburicane-type triterpenoids along with malonyl modifications [irpexoates A-D **(44–47)**] (Figure 3), were isolated from the fruit bodies of the *I. lacteus*. The structures of the new compounds were established by extensive spectroscopic methods along with ^{1}D and ^{2}D NMR, HRESIMS spectroscopic analyses (Tang et al. 2018a).

Five new triterpenoids, irpeksins A-C **(48–50)** (Figure 3), and **D-E (51–52)** (Figure 4), were purified from fruit bodies of *I. lacteus*. The structures and absolute configurations of the new compounds were established through extensive spectroscopic analysis, computational methods, and Cotton effects. Compounds irpeksins A-D **(48–51)** are featured by a scaffold of 1,10-*seco*- and ring B aromatic eburicane (24-methyllanostane), and compound irpeksins E **(52)** are characterized by a scaffold of 1,10-9,11- di*seco*- as well as ring B aromatic eburicane, which represents unprecedented cleavage patterns in the lanostane family (Tang et al. 2018b).

Irpexolidal **(53)**, a triterpenoid with an unprecedented carbon skeleton with its biogenetic-related compound irpexolide A **(54)** (Figure 4) isolated from the fruit bodies of *I. lacteus*. The irpexolidal **(53)**, features a 6/5/6/5/6/5-fused polycyclic skeletal system, which arises from the eburicane-type triterpene by a 6,7-*seco*-6,8-cyclo pattern. The structures of irpexolidal **(53)**, as well as irpexolide A (54), were established by extensive spectroscopic techniques, ECD calculation, and DP4+ probability based on GIAO NMR chemical shift calculations (Tang et al. 2019a).

Ten previously unknown triterpenoid congeners [irpeksolactins A-F **(55–60)**] (Figure 4) and G-J **(61–64)** with eighteen known compounds asiatic acid **(65)**,

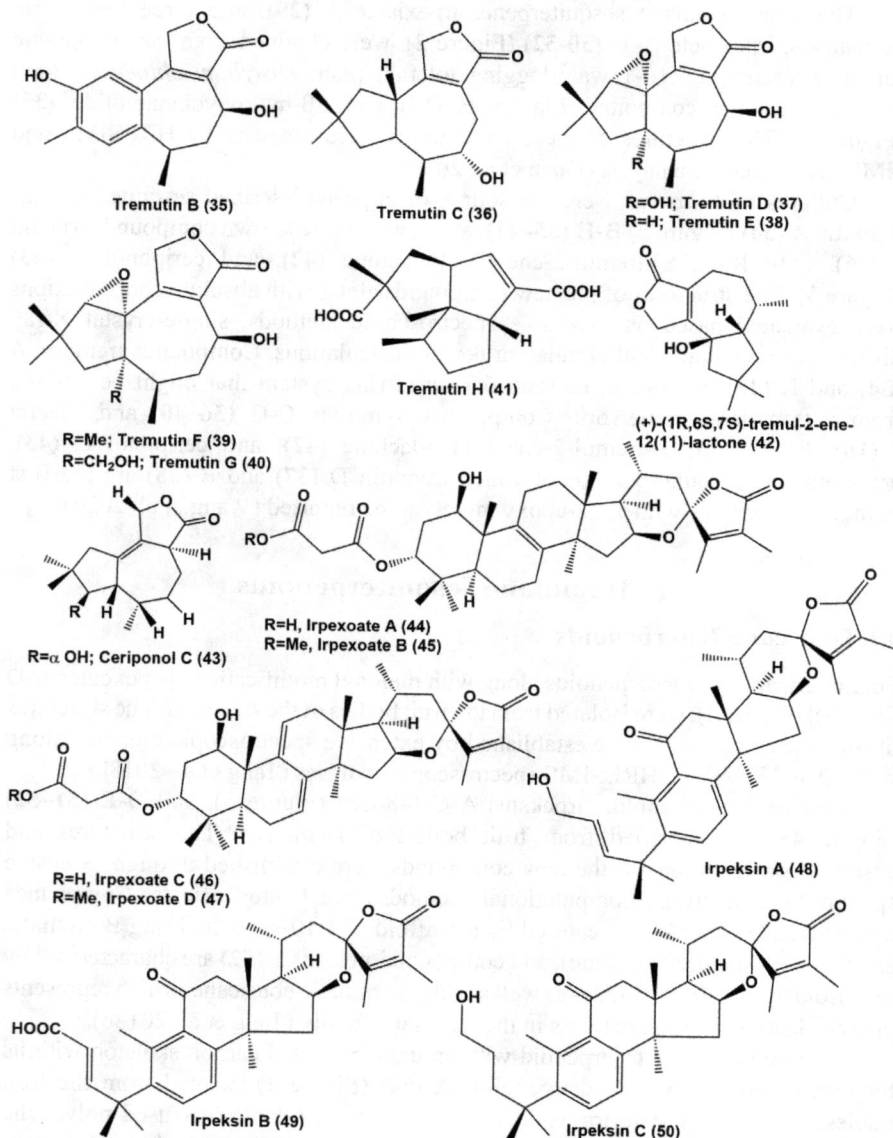

Figure 3: Metabolites isolated from *I. lacteus* (35–50).

15α-hydroxydehydrotumulosic acid **(66)**, polyporenic acid C **(67)**, 24-methylene-lanosta-7,9 (11)-diene-3-one **(68)** (Figure 5), 29-hydroxypolyporenic acid C **(69)**, 6α-hydroxypolyporenic acid C **(70)**, de-hydrosulphurenic acid **(71)**, polycarpol **(72)**, dehydrotrametenonic acid **(73)**, 3-oxo-6,16α-dihydroxylanosta-7,9(11),24(31)-trien-21-oic acid **(74)**, ganoderol A **(75)**, 13α,14β,17α-lanosta-7,9,24-triene-3β,16α-

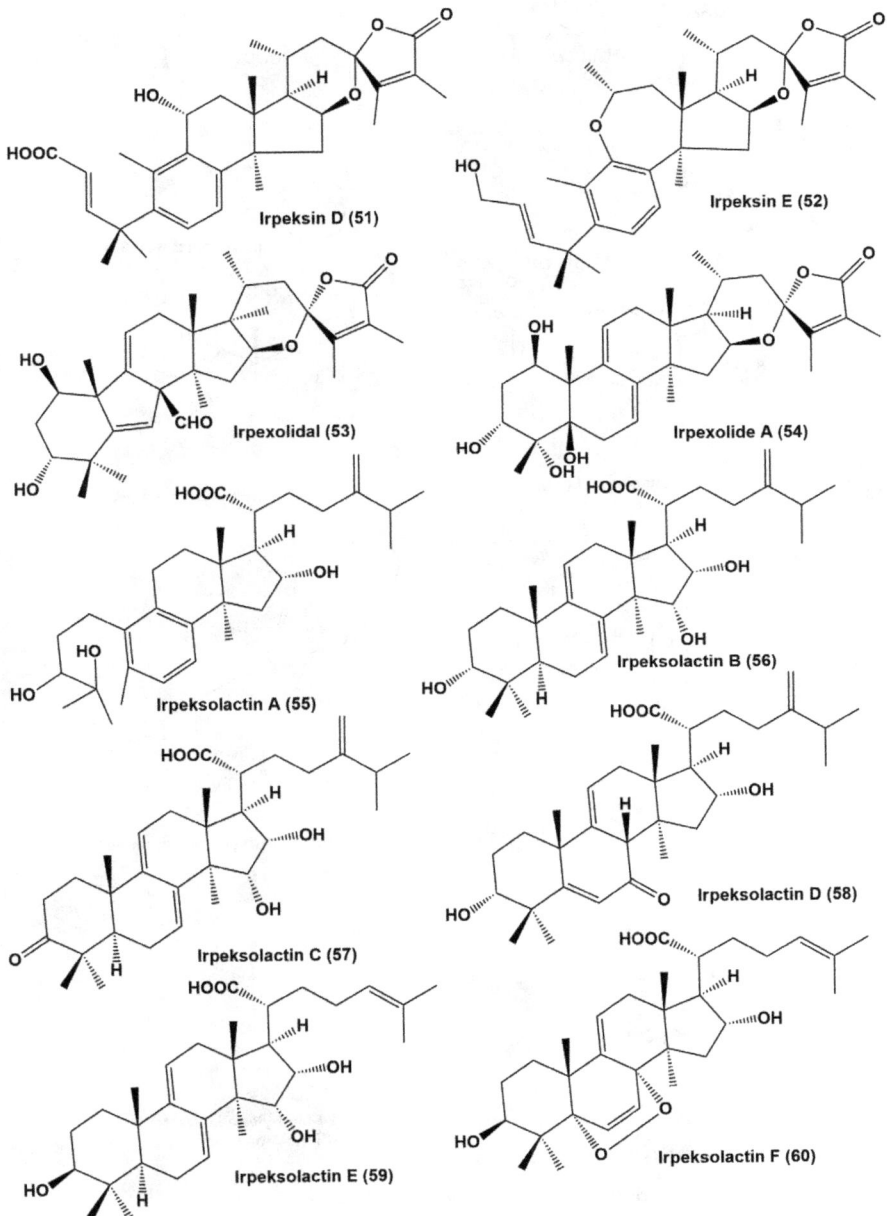

Figure 4: Metabolites isolated from *I. lacteus* (51–60).

diol **(76)**, 3α-hydroxy-24-methylene-23-oxolanost-8-en-26-carboxylic acid **(77)** (Figure 6), 3α-carboxyacetoxyquercinic acid **(78)**, inonotusane C **(79)**, daedalol C **(80)**, ganodermanondiol **(81)**, hexatenuins B **(82)** (Figure 7) were isolated from the

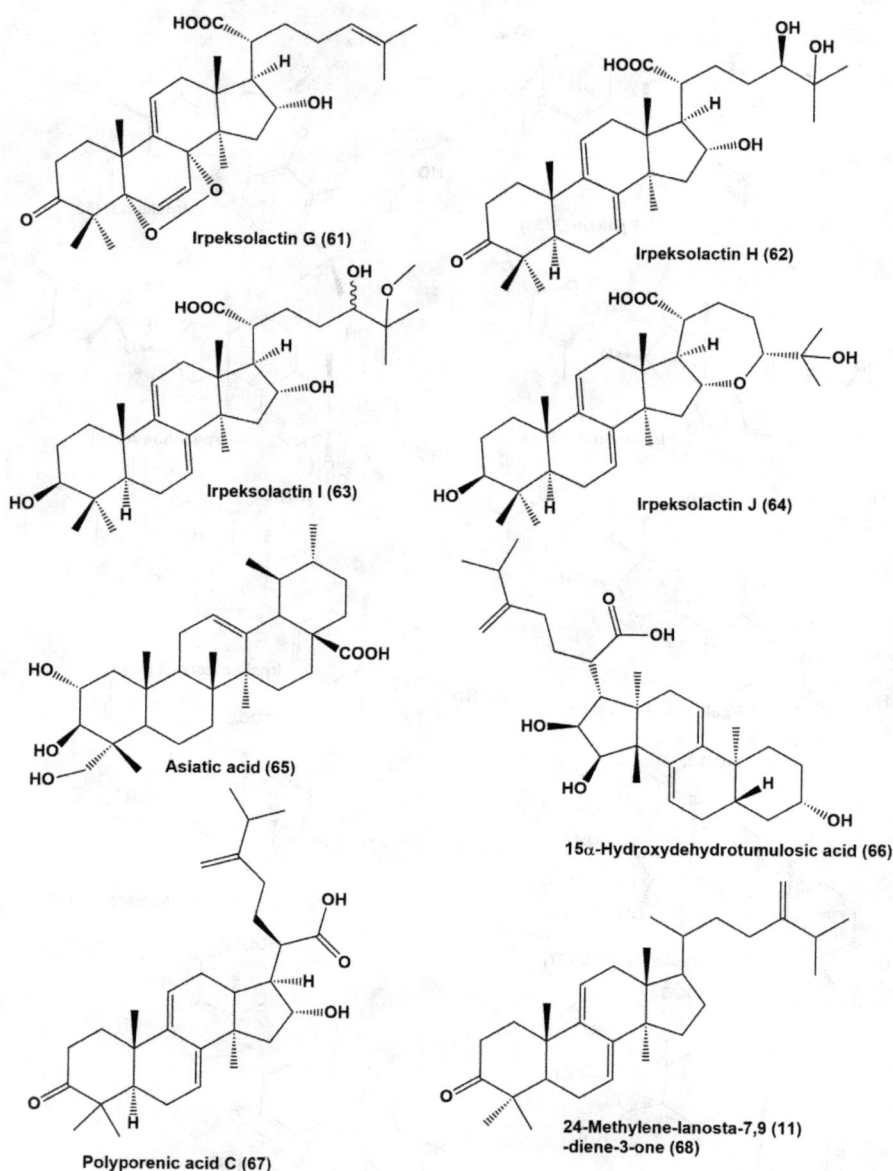

Figure 5: Metabolites isolated from *I. lacteus* (61–68).

fruit bodies of the rainforest-dwelling *I. lacteus*. The structures of all the compounds were characterized by extensive spectroscopic analysis including 1D and 2D NMR and MS spectroscopic techniques (Tang et al. 2019b).

A new irlactane-type, irlactin K **(83)**, and 22 tremulane-type sesquiterpenes along with 14 previously undescribed ones [irpexolactins A–N **(84–97)**], and known irlactane-

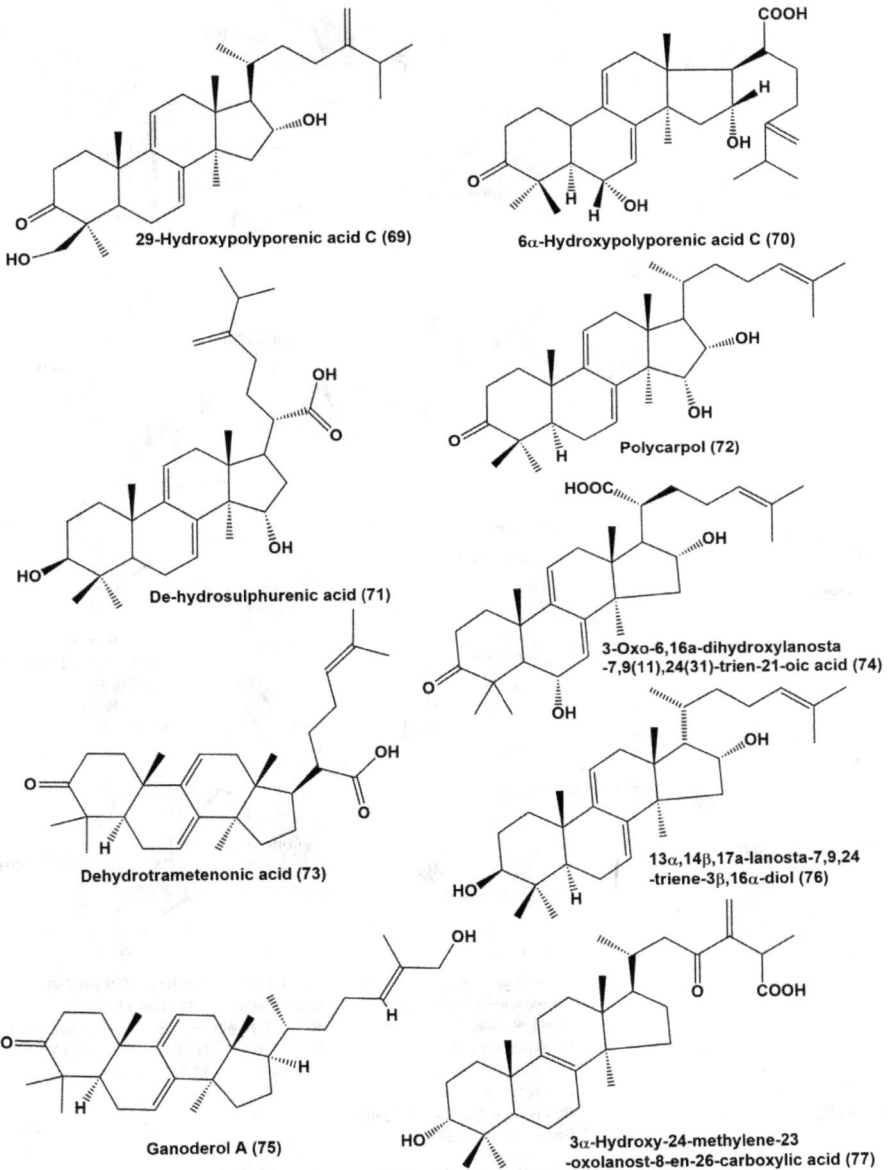

29-Hydroxypolyporenic acid C (69)

6α-Hydroxypolyporenic acid C (70)

De-hydrosulphurenic acid (71)

Polycarpol (72)

3-Oxo-6,16a-dihydroxylanosta-7,9(11),24(31)-trien-21-oic acid (74)

Dehydrotrametenonic acid (73)

13α,14β,17a-lanosta-7,9,24-triene-3β,16α-diol (76)

Ganoderol A (75)

3α-Hydroxy-24-methylene-23-oxolanost-8-en-26-carboxylic acid (77)

Figure 6: Metabolites isolated from *I. lacteus* (69–77).

type sesquiterpenoid, conocenol C **(98)**, tremulenolide D **(99)**, (+)-(3S,6R,7R)-tremulene-6,11,12-triol **(100)**, conocenol B **(101)**, (−)-(3S,6S,7S,10S)-tremulene-10,11,12-triol **(102)** (Figure 7), 11,12-dihydroxy-1-tremulen-5-one **(103)** (Figure 8), tremulenediol A **(18)**, irlactam A **(20)**, and irlactin A **(8)** were isolated

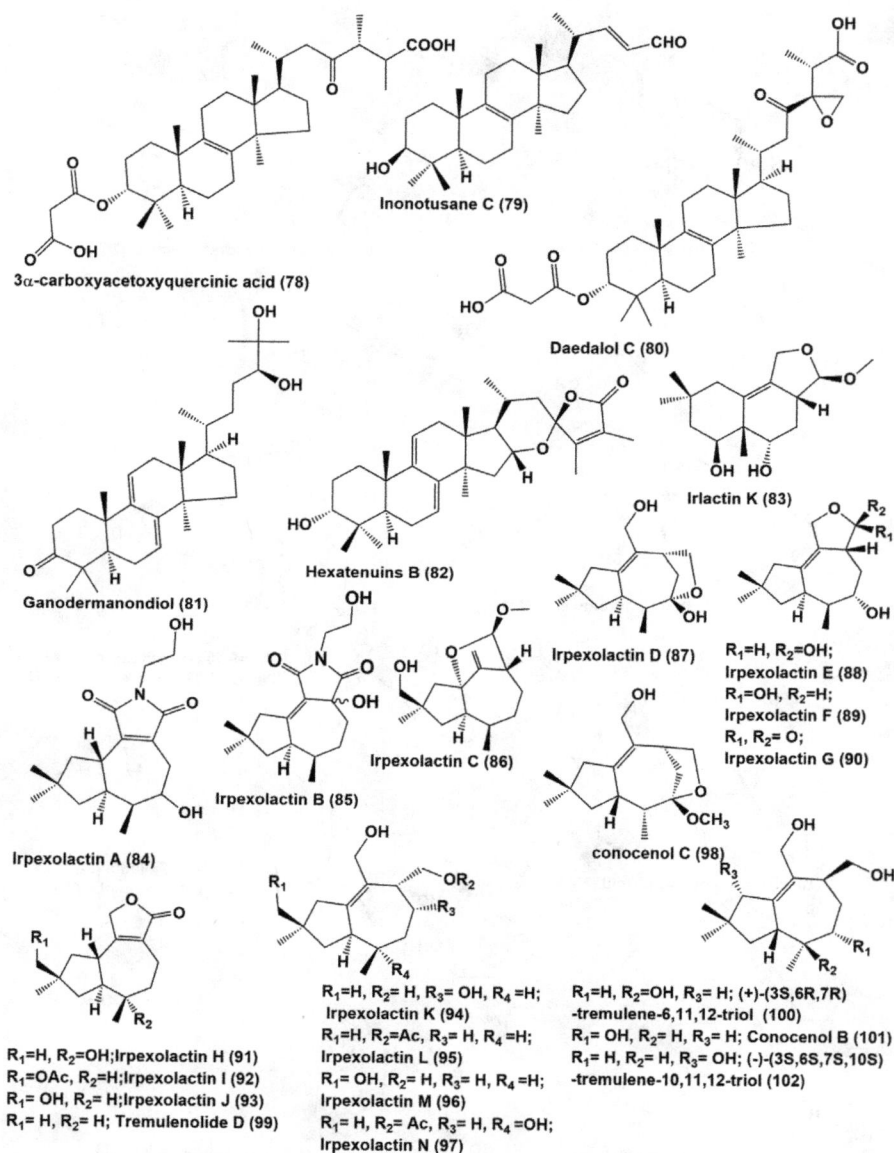

Figure 7: Metabolites isolated from *I. lacteus* (78–102).

from the fermentation broth of *I. lacteus* (HFG1102). Their structures were characterized by extensive spectroscopic methods along with 1D and 2D NMR and MS spectroscopic analysis. The absolute configurations of irlactin K **(83)**, as well as the known compound conocenol B (**(101)** (Figure 7), were established by single-crystal X-ray diffraction analysis (Chen et al. 2020). Compounds irlactin

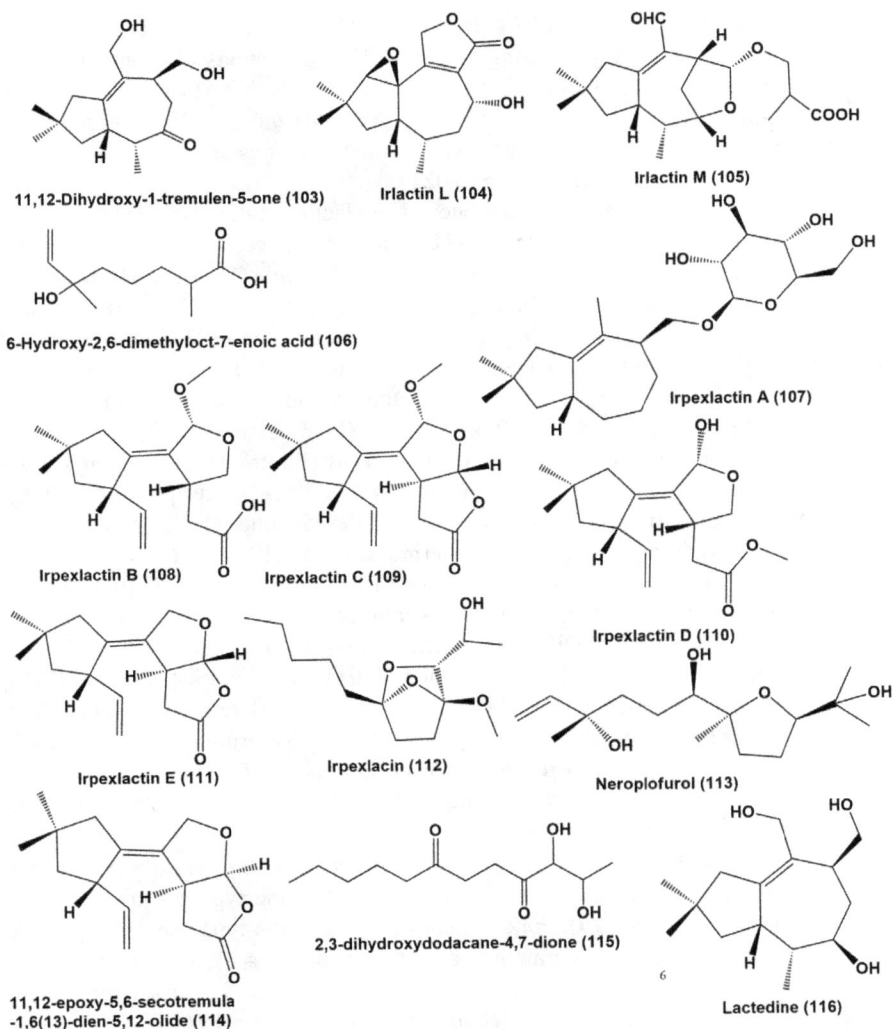

Figure 8: Metabolites isolated from *I. lacteus* (103–116).

K **(83)**, irpexolactins A **(84)**, C **(85)**, K **(94)** (Figure 7), and irlactam A **(20)** were evaluated on vasorelaxant effect on KCl pre-contracted thoracic aorta rings with no significant vasorelaxant effect. The nifedipine was used as the positive control (Chen et al. 2020). Irlactin K **(83)** (Figure 7), was also isolated from the culture broth of *I. lacteus*. Its structure was established through extensive spectroscopic analyses (Ding et al. 2020).

Two new tremulane-type sesquiterpenes irlactin L **(104)** and irlactin M **(105)** were purified from cultures of *I. lacteus*, together with a known compound ([6-hydroxy-

2,6-dimethyloct-7-enoic acid **(106)**] (Figure 8). Their structures were determined by spectroscopic data analysis (Ding et al. 2021). Compounds **(104 and 105)** did not show any significant cytotoxicities in human SK-BR-3, SMMC-7721, HL-60, PANC-1, A-549 cell lines using the MTT (IC_{50}, > 40 lM) and in addition, compounds **(104 and 105)** showed no significant activity against isozymes of 11β-hydroxysteroid dehydrogenases (11β-HSD) (Ding et al. 2021).

Five new tremulane sesquiterpenes irpexlactin A-E **(107–111)**, a new tetrahydrofuran derivative irpexlacin **(112)**, and known compounds neroplofurol **(113)**, 11,12-epoxy-5,6-*seco*tremula-1,6(13)-dien-5,12-olide **(114)**, irpexlacte B **(30)**, and 2,3-dihydroxydodacane-4,7-dione **(115)** (Figure 8), irlactin G **(14)**, conocenol C **(98)**, conocenol B **(101)**, 11,12-dihydroxy-1-tremulen-5-one **(103)**, were extracted from endophyte *I. lacteus* associated with the plant *Gastrodia elata*. Compound irpexlactin A **(107)** was the first tremulane glucoside, and irpexlacin **(112)** consists of a rare tetrahydropyran-tetrahydrofuran scaffold (Wang et al. 2021a).

A new tremulane sesquiterpene, lactedine **(116)** (Figure 8) with seven known tremulane sesquiterpenes as phellinignincisterol C **(117)**, ceriponol A **(118)**, nigrosirpexin A **(119)**, 11,12-dihydroxy-1-termulen-5- one **(120)**, conocenol A **(121)**, 10β,11-dihydroxy-5,6-*seco*1,6(13)-termulacation-5,12-olide **(122)** (Figure 9), conocenol B (2) **(101)**, and a known triterpene irpenigirin A **(123)** (Figure 9) were isolated from *I. lacteus* and their structures were elucidated on the basis of extensive spectroscopic data as well as DP4+ probability analyses (Sun et al. 2022).

Compounds lactedine **(116)**, conocenol B **(101)**, phellinignincisterol C **(117)**, ceriponol A **(118)**, nigrosirpexin A **(119)**, 11,12-dihydroxy-1-termulen-5-one **(120)**, conocenol A **(121)**, 10β,11-dihydroxy-5,6-*seco*1,6(13)-termulacation-5,12-olide **(122)**, and irpenigirin A **(123)** did not showed antibacterial and AChE inhibitory activities at a concentration of 64 μg/mL and 50.00 μM, respectively (Sun et al. 2022).

Irpex lacteus (OV38) endophytic with *Orychophragmus violaceus* was the source of nine novel compounds along with six tremulane-type sesquiterpenoids, Irpexolaceus A-F **(124–129)**, one phenolic bisabolane-type sesquiterpenoid, Irpexolaceus G **(130)** and two furan derivatives, irpexonjust A **(131)** and irpexonjust B **(132)**, along with eight known compounds tremulenolide D **(99)**, irpexlacte B **(30)** and C **(31)**, 5-(3-Oxopentyl)-2-furaldehyde **(133)**, irpexolactin I **(134)**, irpexolactin J **(135)**, 2-furoic acid **(136)**, and 5-(3-methoxy-3-oxopropyl)furan-2-carboxylic acid **(137)** (Figure 9). The structural elucidation of these natural compounds was evaluated based on NMR, HRESIMS, single-crystal X-ray diffraction, and ECD spectroscopic data (Luo et al. 2022a).

Two new tremulane-type sesquiterpenoids [irpexolaceus H **(138)** and I **(139)**] (Figure 10) together with two known furan compounds [irpexlacte B **(30)** and C **(31)**] were isolated from endophytic *I. lacteus* with the plant *Orychophragmus violaceus*. Their structures were determined by spectroscopic data (NMR, HRESIMS, IR, and UV), single-crystal X-ray diffraction, and electronic circular dichroism (ECD) assessment (Luo et al. 2022b).

R$_1$ = R$_2$=R$_3$ = R$_5$ =R$_6$ = H, R$_4$ = OH; Phellinignincisterol C (117)
R$_1$ = R$_2$=R$_3$ = R$_4$ =R$_6$ = H, R$_5$= OH; Ceriponol A (118)
R$_2$=R$_3$ = R$_4$ =R$_5$=R$_6$ = H, R$_1$ = OH;Nigrosirpexin A (119)
R$_1$=R$_4$ = R$_5$ =R$_6$= H, R$_2$ = R$_3$=O; 11,12-Dihydroxy-1-termulen-5- one (120)
R$_1$=R$_2$ = R$_3$ =R$_4$= R$_5$ = H, R$_6$ = OH; Conocenol A (121)

10b,11-dihydroxy-5,6-seco1,6(13)
-termulacation-5,12-olide (122)

Irpenigirin A (123)

Irpexolaceu A (124)

Irpexolaceus B (125)

Irpexonjust B (126)

R$_1$ = OH; R$_2$ =H; Irpexolaceus D (127)
R$_1$ = H; R$_2$ =OH; irpexolaceus E (128)

Irpexolaceus F (129)

irpexolaceus G (130)

Irpexonjust A (131)

Irpexonjust B (132)

5-(3-Oxopentyl)-2-furaldehyde (133)

R=OAc; Irpexolactin I (134)
R= OH; Irpexolactin J (135)

2-Furoic acid (136)

5-(3-methoxy-3-oxopropyl)
furan-2-carboxylic acid (137)

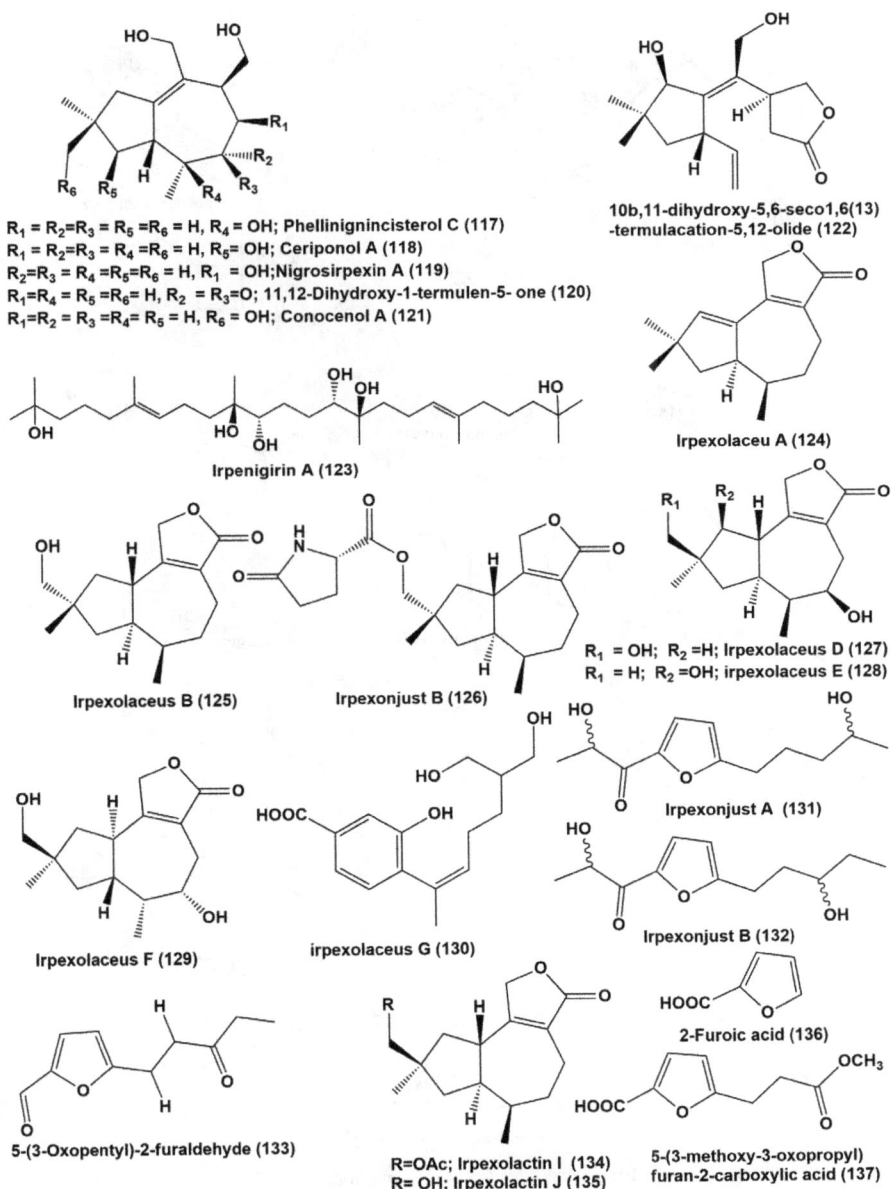

Figure 9: Metabolites isolated from *I. lacteus* (117–137).

4.2 Other Secondary Metabolites

Known compounds apotrichodiol **(140)**, apotrichothecene **(141)**, blennin D **(142)**, collybial **(143)**, cyclocalopin A **(144)**, dehydrooreadone **(145)**, dictyoquinazol A **(146)**, dihydromarasmone **(147)**, frequentin **(148)**, ganodermic acid Jb **(149)**

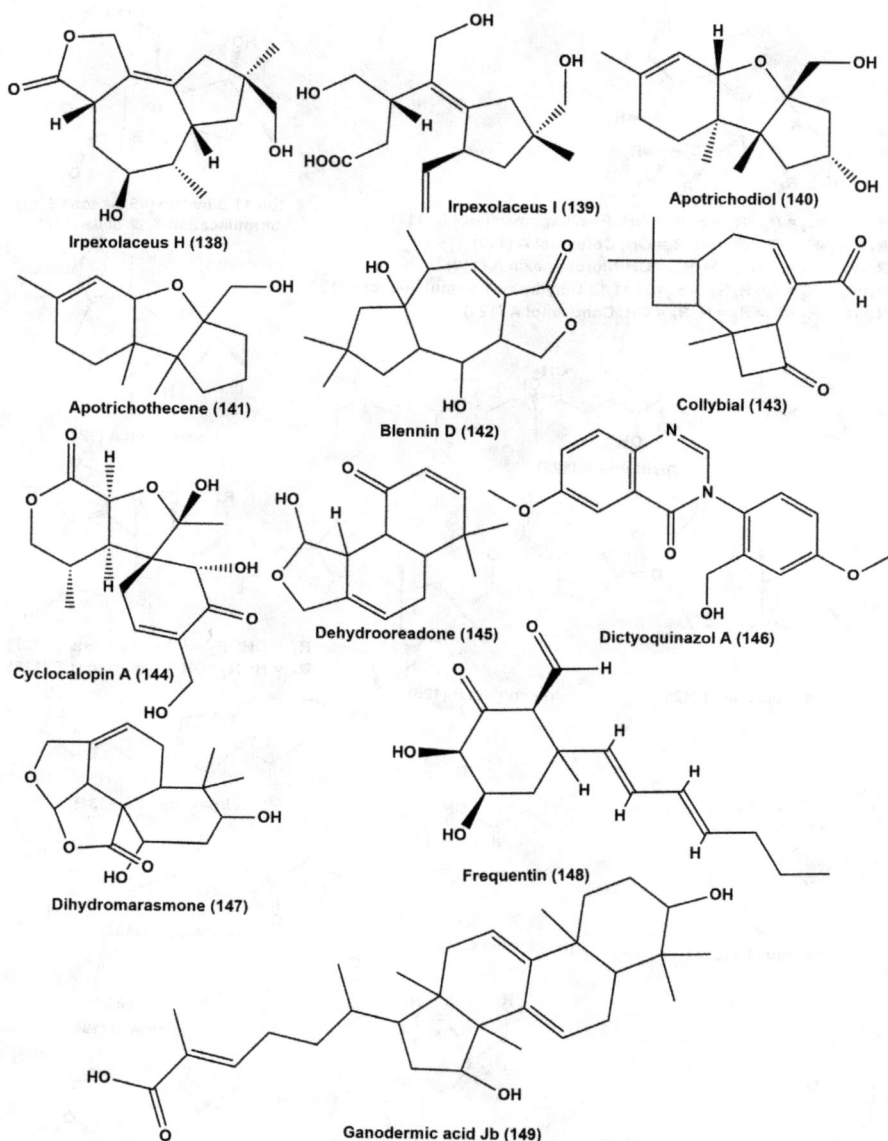

Irpexolaceus H (138)

Irpexolaceus I (139)

Apotrichodiol (140)

Apotrichothecene (141)

Blennin D (142)

Collybial (143)

Cyclocalopin A (144)

Dehydrooreadone (145)

Dictyoquinazol A (146)

Dihydromarasmone (147)

Frequentin (148)

Ganodermic acid Jb (149)

Figure 10: Metabolites isolated from *I. lacteus* (138–149).

(Figure 10), geosmin **(150)**, microdiplodiasol **(151)**, pandangolide 1 **(152)**, and piperdial **(153)** (Figure 11), were purified from *I. lacteus* (CMU-8413) basidiocarps (Pineda-Suazo et al. 2021).

Eight new furan derivatives [irpexins A–H **(154–161)**], two new polyketides [irpexins I **(162)** and J **(163)** (Figure 11), together with nine known compounds

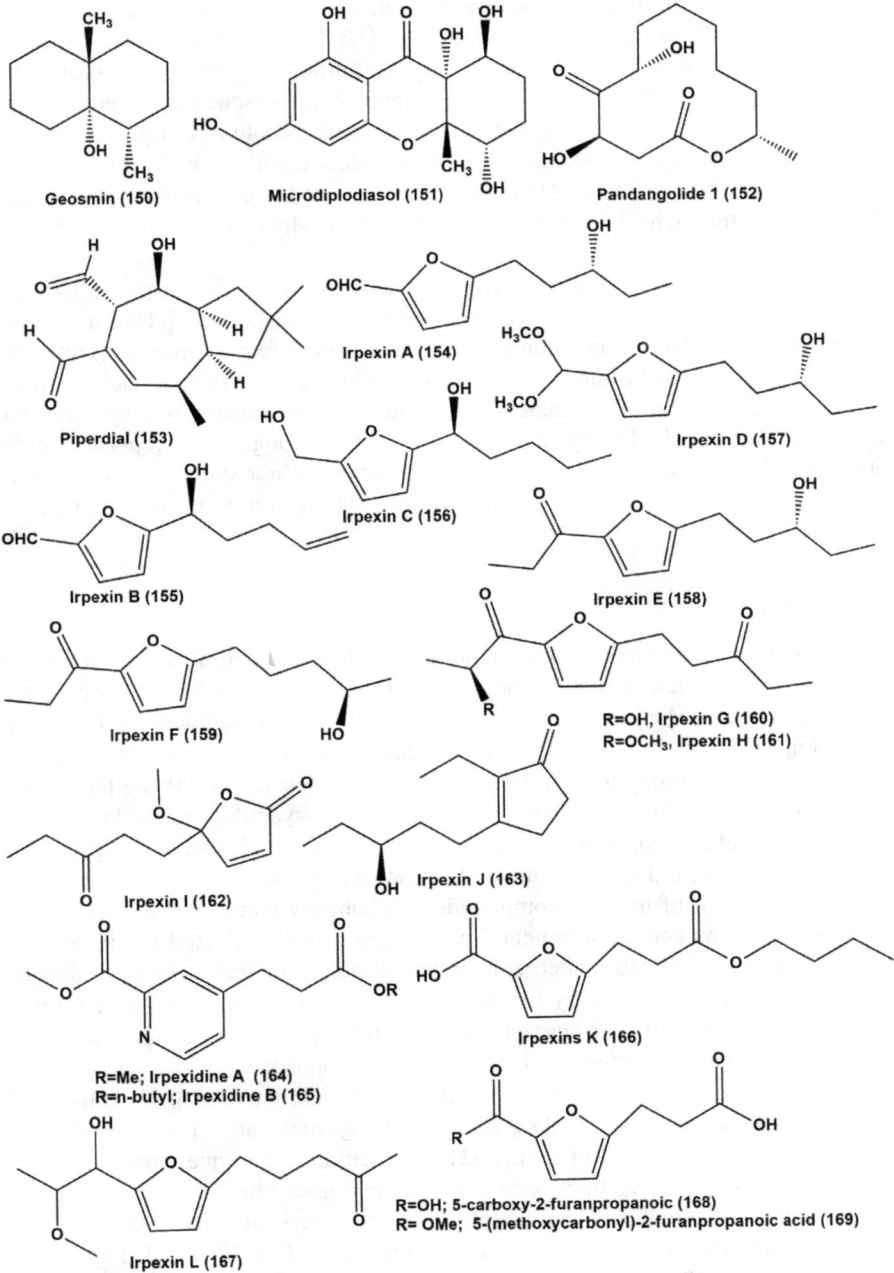

Geosmin (150)

Microdiplodiasol (151)

Pandangolide 1 (152)

Piperdial (153)

Irpexin A (154)

Irpexin B (155)

Irpexin C (156)

Irpexin D (157)

Irpexin E (158)

Irpexin F (159)

R=OH, Irpexin G (160)
R=OCH₃, Irpexin H (161)

Irpexin I (162)

Irpexin J (163)

R=Me; Irpexidine A (164)
R=n-butyl; Irpexidine B (165)

Irpexins K (166)

R=OH; 5-carboxy-2-furanpropanoic (168)
R= OMe; 5-(methoxycarbonyl)-2-furanpropanoic acid (169)

Irpexin L (167)

Figure 11: Metabolites isolated from *I. lacteus* (150–169).

[irpexlacte   B-D   **(30–32)**,   1-(5-(hydroxymethyl)-2-furanyl)-pentanone   **(33)**, 5-(3-oxopentyl)-2-furancarboxaldehyde **(34)**, (1R,2R)-1-(5-(pent-4-en-1-yl)furan-2-yl)propane-1,2-diol **(35)**, (1R,2R)-1-(5-pentylfuran-2-yl)propane-1,2-diol **(36)**, 2,5,8-decanetrione **(37)**, 2-butyl-3hydroxy-2-cyclopenten-1-one **(38)**] were isolated from the fermentation of *I. lacteus*. The structures and absolute configurations were determined on the basis of extensive spectroscopic methods as well as the Mosher ester reaction (Wang et al. 2021b). All compounds at the concentration of 10 μM showed no cytotoxicity to human MCF-7 as well as Hela cancer cell lines (Wang et al. 2021b).

Two undescribed disubstituted pyridine derivatives [irpexidines A [(164) and B **(165)**] and two undescribed alkylfuran derivatives [irpexins K [(166) and L (4) **(167)**] along with two known compounds [5-carboxy-2-furanpropanoic **(168),** and 5-(methoxycarbonyl)-2-furanpropanoic acid **(169)**] (Figure 11) were isolated from the fermentation broth of *I. lacteus*. Their structures were determined by extensive spectroscopic analysis. The pyridine derivatives of this fungus were reported for the first time (Chen et al. 2021). The new compounds did not show any cytotoxicity against Hela cancer cells as well as inhibitory activity on NO production (Chen et al. 2021).

4.3 Co-culture

In the co-culture technique, fungi are cultivated with other microorganisms to induce the production of silent bioactive compounds (Bertrand et al. 2014). The inoculated co-cultured microorganism induces the cryptic biosynthetic pathways in *I. lacteus*, which remain silent in the axenic culture under laboratory culture conditions. In co-culture, one microorganism produces signaling molecules to activate the cryptic genes of other co-culture organisms (Okada and Seyedsayamdost 2017). In addition, the production of secondary metabolites by one co-culture strain can act as a precursor molecule of the desired active compound. This technique supports the induction and yield enhancement of induced compounds in laboratory culture (Zhu et al. 2011). Growth conditions and environmental parameters directly affected the interaction of both constituted microbes of co-culture, which shows the impact on the induction of cryptic genes. It is necessary to maintain or provide suitable growth conditions for both strains to grow and interact with each other properly (Zuhang and Zhang 2021). On the basis of interaction, the mode of action of stimulating silent gene clusters has been categorized as (1) cell-to-cell physical interaction (2) cell-to-cell communication through chemical signaling and (3) genomic alteration via horizontal gene transfer or mutation (Kim et al., 2021) in co-culture technique. Some of the co-culture studies of *I. lacteus* have been discussed in this section.

Five new metabolites belonging to two backbones of pulvilloric acid-type azaphilone and tremulane sesquiterpenes [nigirpexin A-D **(170–173)**] (Figure 12), nigrosirpexin A **(119)** and conocenol B **(101)** were isolated from *Nigrospora oryzae* co-cultured along with *I. lacteus*, The tremulane sesquiterpene conocenol B **(101)** production by *I. lacteus* was induced via the induction of *Nigrospora oryzae* (Zhou et al. 2018).

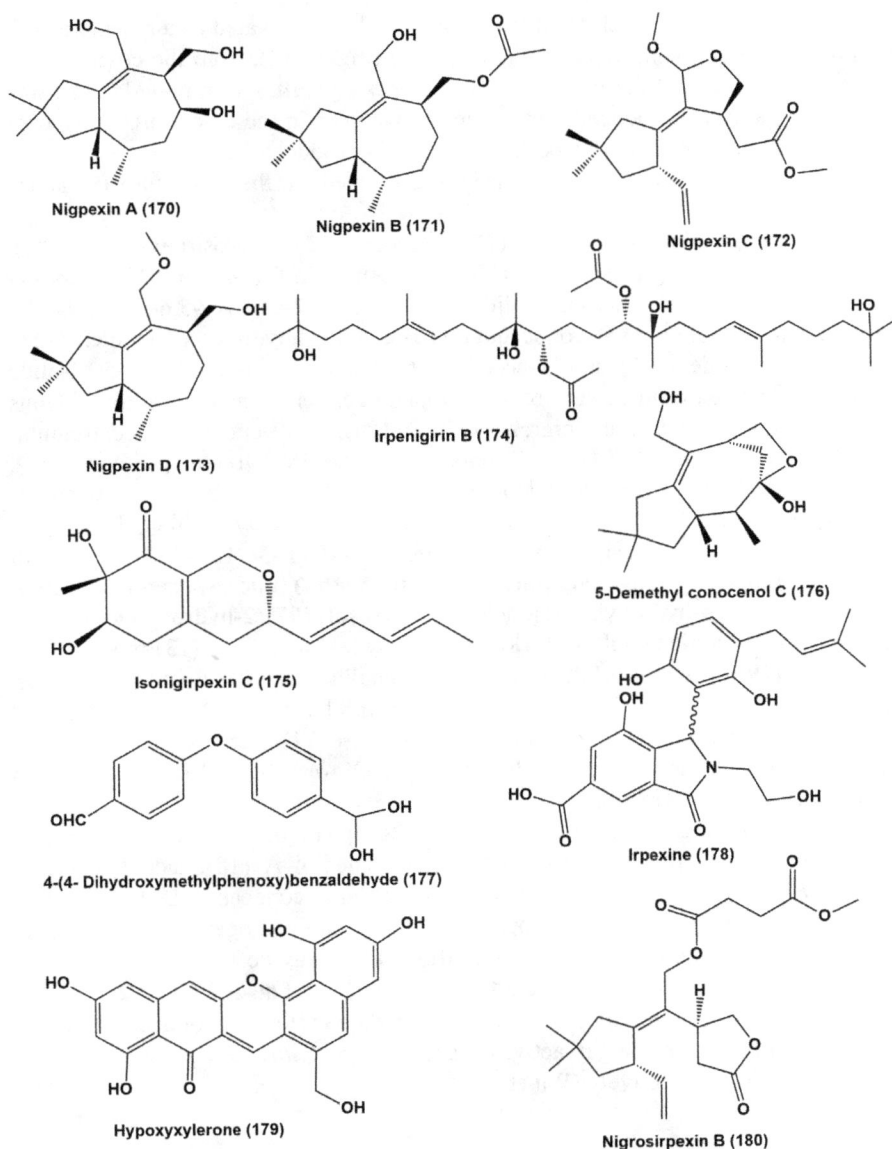

Figure 12: Metabolites isolated from *I. lacteus* (170–180).

The co-culture of the phytopathogenic *N. oryzae* with endophytic *I. lacteus* from the host orchid *Dendrobium officinale* afforded two new squalenes Irpenigirin A (123), irpenigirin B (174), one new azaphilone isonigirpexin C (175), two new tremulane sesquiterpenes 5-demethyl conocenol C (176) (Figure 12), nigrosirpexin A (119), and known compounds conocenol B (101) and C (98), and 4-(4-dihydroxymethylphenoxy)benzaldehyde (177) (Wu et al. 2019).

A new isoindolinone alkaloid [irpexine **(178)**], was isolated as a racemate with a known green pigment, hypoxyxylerone **(179)** (Figure 12), from the co-culture of two endophytic fungi (*I. lacteus* and *Phaeosphaeria oryzae*). Compound **(178)**, was found to be a newly produced metabolite of *I. lacteus* in the co-culture along with *P. oryzae*. Even though hypoxyxylerone **(179)** is produced by the monoculture of *I. lacteus*, its production was significantly enhanced by the co-culture (Sadahiro et al. 2020).

Compounds nigrosirpexin B **(180)** (Figure 12), nigrosirpexin C **(181)**, nigirpexin E **(182)**, nigrosirpexin D **(183)**, nigrosirpexin E **(184)** and 11,12-epoxy-5,6-*seco*tremula-1,6(13)-dien-5,12-olide **(114)**, 12-acetoxy-5,6-*seco*-1,6(13)-tremuladien-5,11-olide **(115)**, conocenol B **(101)**, 11,12-dihydroxy-1-tremulen-5-one **(103)**, conocenolide A **(27)**, and 11-acetoxy-5,6-*seco*-1,6(13)-tremuladien-5,12-olide **(185)** were isolated from co-culture of phytopathogen–endophyte (non-homologous *N. oryzae* and *I. lacteus*), and nigrosirpexin F **(186)**, 11,12-epoxy-5,6-*seco*tremula-1,6(13)-dien-5,12-olide **(114)**, 11,12-epoxy-12β-hydroxy-1-tremulen-5-one **(187)**, and syringaresinol **(188)** (Figure 13) were isolated from co-culture of endophyte-host (*N. oryzae* and *I. lacteus* from the host orchid *D. officinale*) (Shi et al. 2020).

Five new tremulane sesquiterpenoids [nigpexin A-E **(1–5)** **(189–193)**] along with previously reported compounds [mevalonolactone **(194)**, microsphaerophthalide F **(195)** (Figure 13), *p*-hydroxybenzoic acid **(196)**, tyrosol **(197)**, 2-hydroxyphenylacetic acid **(198)**, tremulenediol A **(18)**, 11-aldehyde-5,6-*seco*-1,6(13)-tremuladien-5,12-olide **(199)**, conocenolide A **(27)** conocenolide B **(28)**, β-sitosterol **(200)**, scytalone **(201)**, 4,6,8-trihydroxy-3,4-dihydronaphthalen-1(2H)-one **(202)**, and (3S,4R)-3,4-dihydroxypentanoic acid **(203)**] (Figure 14) were isolated from co-culture of endophytic *I. lacteus*, phytopathogen *N. oryzae*, and entomopathogen *Beauveria bassiana* (Yin et al. 2021).

A new antifungal butenolide [irperide **(204)**] (Figure 14) along with five known compounds [nigirpexin C **(172)**, tremulenediol A **(18)**, lactedine **(116)**, (+)-(3S,6R,7R)-tremulene-6,11,12-triol (5) **(100)** and conocenol B **(101)**] were isolated from the co-culture of endophyte *I. lacteus* and pathogenic *N. oryzae*. The structure of 1, including the absolute configuration, was determined by analyses based on NMR, HR-ESI-MS data and ECD spectra (Wu et al. 2022)

Compounds irperide **(204)** (Figure 14), lactedine **(116)**, and conocenol B **(101)** showed significant antifungal activity against *Aspergillus fumigatus* with MIC 1, 2 and 1 μg/mL, respectively (Wu et al. 2022).

4.4 *Application in Biotransformation*

The biotransformation of huperzine A (hupA) **(205)** (one of the characteristic bioactive constituents of the medicinal plant *Huperzia serrate*) by the endophytic *I. lacteus* (CCTCC M 2017161) yielded two previously unknown compounds **(206-207)** (Figure 14), featuring a hupA-butanone structure, along with a known

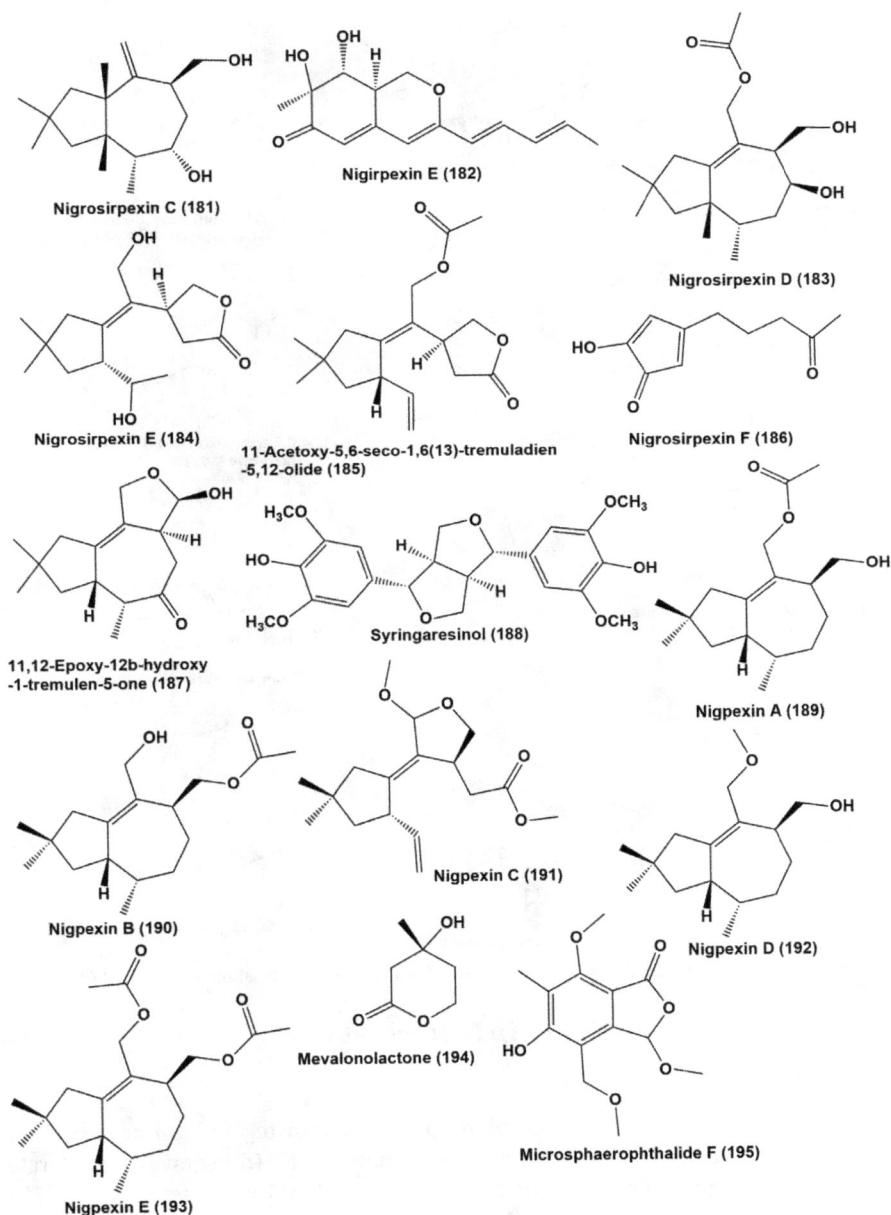

Nigrosirpexin C (181)

Nigirpexin E (182)

Nigrosirpexin D (183)

Nigrosirpexin E (184)

11-Acetoxy-5,6-seco-1,6(13)-tremuladien-5,12-olide (185)

Nigrosirpexin F (186)

11,12-Epoxy-12b-hydroxy-1-tremulen-5-one (187)

Syringaresinol (188)

Nigpexin A (189)

Nigpexin B (190)

Nigpexin C (191)

Nigpexin D (192)

Mevalonolactone (194)

Microsphaerophthalide F (195)

Nigpexin E (193)

Figure 13: Metabolites isolated from *I. lacteus* (181–195).

analog 8α,15α-epoxyhuperzine A (**208**). The structures of all the isolates were evaluated by spectroscopic methods including NMR, MS, IR, and UV spectra (Ying et al. 2019).

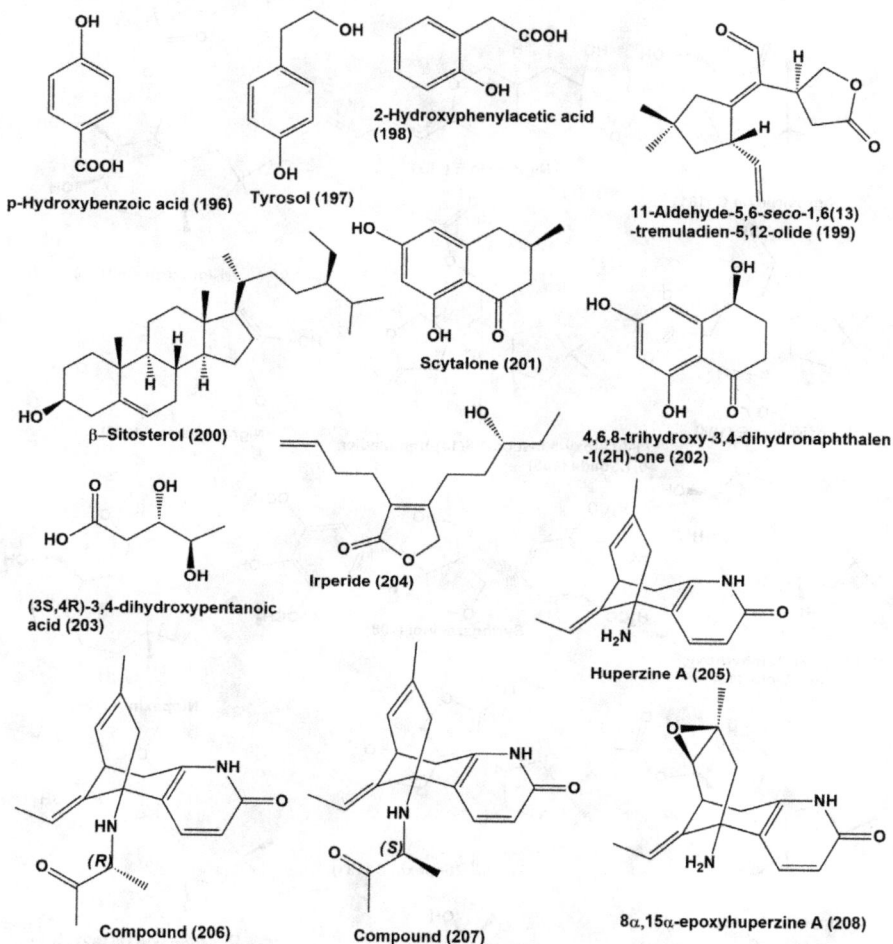

Figure 14: Metabolites isolated from *I. lacteus* (**196–204**) and biotransformation of huperzine (**205-208**).

5. Modern Medicinal Uses

5.1 Antimicrobial Activities

Monacolin K (**4**) shows a stronger inhibition of the growth of Gram-negative and Gram-positive bacteria; however, dehydromonacolin K (**5**) shows a moderate inhibition of the growth of *B. subtilis* (Wang et al. 2013). Irpexlactes A–D (**29–32**) displayed average antibacterial activity against *Pseudomonas aeruginosa* with MIC ranging between 23.8 and 35.4 μM (Duan et al. 2019).

The tremulane-type sesquiterpene conocenol B (**101**) displayed antifungal activity against *N. oryzae* and *I. lacteus* with MIC 16 and 128 μg/mL, respectively (Zhou et al. 2018). Nigrosirpexin A (**119**) displayed poor activity against *N. oryzae* and *I. lacteus* with MIC 64 and 128 μg/ml, respectively (Zhou et al. 2018).

The metabolites conocenol B **(101)** and 4-(4-dihydroxymethylphenoxy) benzaldehyde **(177)** revealed selectivity of antifungal activity against *N. oryzae* with MIC 16 μg/mL. The irpenigirin B **(174)** and 5-Demethyl conocenol C **(176)** exhibited antifungal activity against *C. gloeosporioides* with MIC 8 μg/mL; 5-Demethyl conocenol C **(176)**, conocenol B **(101)**, conocenol C **(98)**, showed anti-fungal activities against *Didymella glomerata* with MIC 1, 8, and 4 μg/mL, respectively; and nigrosirpexin A **(119)** showed anti-fungal activity against *N. oryzae* with MIC 32 μg/mL. Compound conocenol B **(101)**, and 4-(4-dihydroxymethylphenoxy) benzaldehyde **(177)** displayed antifungal activity against *N. oryzae* at a concentration of 100 μg/mL causal organism of *Cerasus cerasoides* (Wu et al. 2019).

Metabolites nigrosirpexin B **(180)**, 11,12-epoxy-5,6 *seco*tremula-1,6(13)-dien-5,12-olide **(114)**, 12-acetoxy-5,6-*seco*-1,6(13)-tremuladien-5,11-olide **(115)**, conocenol B **(101)**, 11,12-dihydroxy-1-tremulen-5-one **(103)**, conocenolide A **(27)**, and 11-acetoxy-5,6-*seco*-1,6(13)-tremuladien-5,12-olide **(185)** from *I. lacteus* showed selective anti-phytopathogenic activity against *N. oryzae* with MIC 1, 32, 256, 16, 128, 1, and 128 μg/mL, respectively, while no obvious anti-phytopathogenic activity against *I. lacteus* was seen (MIC of 512 μg/mL) (Shi et al. 2020).

Compound 2,3-dihydroxydodacane-4,7-dione **(115)**, displayed antifungal activity against *Penicillium polonicum, Trichoderma atroviride, Psathyrella subsingeri* with MIC 16, 32 and 16 μg/mL, respectively. Compounds irpexlactin A-D **(107-110)**, Irpexlactin E **(111)**/11,12-epoxy-5,6-secotremula-1,6(13)-dien-5,12-olide **(114)**, irpexlacin **(112)** and known compound irlactin G **(14)**, conocenol C **(98)**, conocenol B **(101)**, 11,12-dihydroxy-1-tremulen-5-one (10) **(103)**, neroplofurol **(113)**, irpexlacte B **(30)**, were found active against *P. polonicum, T. atroviride, Psathyrella subsingeri Armillaria* sp. with MIC range between 4 and 64 μg/mL.

Compounds irpexlactin A **(107)**, irpexlactin B **(108)**, Irpexlactin E **(111)**/ 11,12-epoxy-5,6-secotremula-1,6(13)-dien-5,12-olide (12) **(114)**, irlactin G **(14)**, conocenol C **(98)**, conocenol B **(101)**, 11,12-dihydroxy-1-tremulen-5-one (10) **(103)**, irpexlacte B **(30)**, were found active against *I. lacteus* with MIC range from 32–64 μg/mL. The positive control nystatin displayed antifungal activity against *P. polonicum, T. atroviride, P. subsingeri, I. lacteus* with MIC range from 4–16 μg/mL (Wang et al. 2021a).

Compounds nigpexin B-D **(2–4) (190–192)**, p-hydroxybenzoic acid **(196)**, tyrosol **(197)**, β-sitosterol **(200)**, and scytalone **(201)**, displayed anti-fungal activities against *I. lacteus* with MIC ≤ 8 μg/mL. Compounds tremulenediol A **(18)**, 11-aldehyde-5,6-*seco*-1,6(13)-tremuladien- 5,12-olide **(199)**, scytalone **(201)**, 4,6,8-trihydroxy-3,4-dihydronaphthalen-1(2H)-one **(202)**, and (3S,4R)-3,4-dihydroxypentanoic acid **(203)** displayed significant anti-fungal activity against *N. oryzae* with MICs ≤ 4 μg/mL, and compounds nigpexin B **(190)**, nigpexin E **(5) (193)**, 11-aldehyde-5,6-*seco*-1,6(13)-tremuladien- 5,12-olide **(199)**, from *I. lacteus*, and (3S,4R)-3,4-dihydroxypentanoic acid **(203)** displayed significant anti-fungal activity against *B. bassiana* with MIC ≤ 8 μg/mL (Yin et al. 2021).

Compounds nigpexin A–E **(189–193)**, mevalonolactone **(194)**, microsphaerophthalide F **(195)**, p-hydroxybenzoic acid **(196)**, tyrosol **(197)**,

2-hydroxyphenylacetic acid **(198)**, tremulenediol A **(18)**, 11-aldehyde-5,6-*seco*-1,6(13)-tremuladien-5,12-olide **(199)**, conocenolide A **(27)**, conocenolide B (14) **(28)**, β-sitosterol **(200)**, scytalone **(201)**, 4,6,8-trihydroxy-3,4-dihydronaphthalen-1(2H)-one **(202)**, and (3S,4R)-3,4-dihydroxypentanoic acid **(203)** displayed antifeedant potential against silkworm with inhibition percentages ranging from 73-99% at concentrations of 50 μg/cm². Tremulenediol A **(18)**, indicated notable antifeedant potential with an inhibition percentage of 93% at the concentration of 6.25 μg/cm². Avermectin is used as a positive drug with an inhibition percentage > 95% at a concentration 50 μg/cm² or 6.25 μg/cm² (Yin et al. 2021).

5.2 Antitumor Activity

Irpexoate B **(45)** displayed poor cytotoxicity against human cancer cell lines (A-549, SMMC-7721, MCF-7, SW480) with IC_{50} varying from 22.9 to 34.0 μM, and irpexoate D **(47),** showed poor cytotoxicity against the human cancer cell line SW480 with an IC_{50} 35.2 μM (Tang et al. 2018a). Compounds irpeksins A-E **(48–52)**, showed potent inhibitory activity against NO production in LPS-activated RAW 264.7 macrophage cells with IC_{50} varying from 2.2 to 19.6 μM (Tang et al. 2018b).

Compound irlactin I **(16)**, exhibited average cytotoxic activity against HL-60, SMMC-7721, A-549, MCF-7, and SW480 cells with IC_{50} 16.23, 20.40, 25.55, 19.05, and 18.58 μM, respectively (Ding et al. 2018).

Irpeksolactin J (10) **(64)** showed selective and weak cytotoxicity against the A549 and SMMC-7721 cancer cell lines with inhibition rates of 54.6 and 50% at the concentration of 40 μM, respectively (Tang et al. 2019b).

The compound irlactin K **(83)**, was inactive against breast cancer (SK-BR-3), hepatocellular carcinoma (SMMC-7721), human myeloid leukaemia (HL-60), pancreatic cancer (PANC-1), and lung cancer cells (A-549) (IC_{50} > 40 μM) (Ding et al. 2020).

5.3 Anti-Inflammatory Activity

Compounds irpeksins A–E **(48–52)**, revealed significant inhibitory potential against NO production in LPS-activated RAW 264.7 macrophage cells with IC_{50} varying between 2.2 and 19.6 μM (Tang et al. 2018b).

At the concentration (50 μg/mL), the inhibitory effects of irpexolaceus A **(124)**, C **(126)**, D **(127)**, F **(129)**, and G **(130)**, irpexonjust B **(132)**, and irpexlacte B **(30)** against NO release from LPS-induced RAW 264.7 cells were more than 45%, while irpexlacte C **(31)** (42.6%), irpexolaceus B **(125)**, (39.6%), irpexonjust A **(131)** (43.7%), and irpexolaceus E **(128)**, (33.6%) exhibited weaker inhibitory effects on the release of NO (Luo et al. 2022a).

Compounds tremutin F **(39)**, tremutin G **(40)**, (+)-(1R,6S,7S)-tremul-2-ene-12(11)-lactone **(42)**, and ceriponol C **(43)**, exhibited poor activities against MCF-

7, SMMC-7721, HL-60, SW480, and A549 cancer cell lines. Compound tremutin H **(41)**, showed a weak inhibitory effect on NO production with a half maximal inhibitory concentration (IC$_{50}$ 22.7 μM). Compound tremutin A **(34)**, inhibited the lipopolysaccharide (LPS)-induced proliferation of B lymphocyte cells with an IC$_{50}$ 22.4 μM, while tremutin B **(35)**, inhibits concanavalin A (Con A)-induced T cell proliferation and LPS-induced B lymphocyte cell proliferation with IC$_{50}$ 16.7 and 13.6 μM, respectively (Wang et al. 2020).

5.4 *Other Biological Activities*

In the LPS-induced neuro-inflammation injury detection, compounds 1–2 **(206–207)**, and 8α,15α-epoxyhuperzine A (3) **(207)**, showed moderate neuroprotective activity through an increase of viability of U251 cell lines with EC50 of 35.3, 32.1, and 50.3 nM, respectively (Ying et al. 2019).

The compound 3-p-anisoloxypropionate **(3)**, showed 50% mortality of nematode *Aphelencoides besseyi* in the solution of 25 ppm, and compound 5-pentyl-2-furaldehyde **(1)**, 5-(4-pentenyl)-2-furaldehyde **(2)**, showed the similar activity at 50 ppm (Hayashi et al. 1981).

Nigrosirpexin A **(119)** was active against acetylcholinesterase (AChE) with a ratio of 35% at 50 μM (Zhou et al. 2018). Compound nigrosirpexin A **(119)**, from *I. lacteus*, revealed weak activity against AChE, with an inhibition ratio of 16% at 50 μM (Wu et al. 2019).

Irlactam A **(20)**, and irlactin K **(83)**, showed no significant activity against isozymes of 11β-hydroxysteroid dehydrogenases (11β-HSD) (Ding et al. 2019, 2020). Compounds irpexlacte A **(29)**, and irpexlacte D **(32)**, revealed remarkable antioxidant activity with IC$_{50}$ 2.50 and 5.75 μM, respectively (Duan et al. 2019).

Compounds irlactin K **(83)**, irpexolactin A **(84)**, irpexolactin C **(85)**, irpexolactin K **(94)**, and irlactam A **(20)**, were assessed on vasorelaxant effect on KCl pre-contracted thoracic aorta rings with no significant vasorelaxant effect. Nifedipine was used as a positive control (Chen et al. 2020).

Compounds nigpexin A–E **(189–193)**, mevalonolactone **(194)**, microsphaerophthalide F **(195)**, p-hydroxybenzoic acid **(196)**, tyrosol **(197)**, 2-hydroxyphenylacetic acid **(198)**, tremulenediol A **(18)**, 11-aldehyde-5,6-*seco*-1,6(13)-tremuladien-5,12-olide **(199)**, conocenolide A **(27)**, B **(28)**, β-sitosterol **(200)**, scytalone **(201)**, 4,6,8-trihydroxy-3,4-dihydronaphthalen-1(2*H*)-one **(202)**, and (3*S*,4*R*)-3,4-dihydroxypentanoic acid **(203)** displayed antifeedant activities against silkworm with an inhibition percentages of 73–99% at 50 μg/cm^2. Tremulenediol A **(18)**, showed notable antifeedant activity with inhibition percentage of 93% at 6.25 μg/cm^2 among them. Avermectin was used as positive drug with inhibition percentage > 95% at 50 μg/cm^2 or 6.25 μg/cm^2 (Yin et al. 2021).

Recently, irpexlacte B **(30)** displayed potential QS inhibitory activity at 50 mg/ml against biomarker strains of *Agrobacterium tumefaciens* A136 and *Chromobacterium violaceum* CV026 (Luo et al. 2022b).

6. Concluding Remarks

Irpex lacteus belongs to basidiomycetes and 208 compounds are reported from this fungus. These compounds belong to sterol, terpenoid, sesquiterpene, tremulane, glucoside, furan, polyketide, indole, naphthalene, lactone and others. classes and displayed antibacterial, antifungal, anticancer, anti-inflammatory, antioxidant, antifeedant, nematicidal, AChE inhibitory, and vasorelaxant activities. Co-culture technique was also used to produce a new bioactive compound using *Nigrospora oryzae, Phaeosphaeria oryzae, C. gloeosporioides,* and *Beauveria bassiana* as co-culture partners. Co-culture stimulated the cryptic biosynthetic gene cluster to trigger novel bioactive compound production by partner microorganism *Irpex lacteus.* The whole-genome sequencing is necessary for the evaluation of their potential for production of the novel compounds. Molecular methods like the transfer of biosynthetic gene clusters to a suitable vector for large-scale production could be targeted. The diverse metabolites from *I. lacteus* linked with various biological activities have shown the avenue for researchers to develop new compounds with drug-like potential. The novel compounds could be evaluated for their pharmacological activities in the treatment of various diseases. Besides, a combination of existing drugs with metabolites of *I. lacteus* may be evaluated for their synergistic effects in the treatment of specific diseases.

References

Bau, T. and Y.C. Dai. 2004. Diversity and conservation of main wood-rotting fungi in Changbai Mountains. *J. Fungal Res.,* 2: 26–30.

Bernicchia, A. and S.P. Gorjón. 2020. Polypores of the Mediterranean Region. Romar: Segrate, Italy, p. 903.

Bertrand, S., N. Bohni, S. Schnee, O. Schumpp, K. Gindro and J.L. Wolfender. 2014. Metabolite induction via microorganism co-culture: A potential way to enhance chemical diversity for drug discovery. *Biotechnol. Adv.,* 32: 1180–1204.

Buzina, W., C. Lass-Flörl, G. Kropshofer, M.C. Freund and E. Marth. 2005. The polypore, mushroom *Irpex lacteus,* a new causative agent of fungal infections. *J. Clin. Microbiol.,* 43: 2009–2011.

Chaudhary, R. and A. Tripathy. 2015. Isolation and identification of bioactive compounds from *Irpex lacteus* wild fleshy fungi. *J. Pharm. Sci. Res.,* 7(7): 424.

Chen, H.P., X. Ji, Z.H. Li, T. Feng and J.K. Liu. 2020. Irlactane and tremulane sesquiterpenes from the cultures of the medicinal fungus *Irpex lacteus* HFG1102. *Nat. Prod. Bioprospect.,* 10(2): 89–100.

Chen, H.P., Z.Z. Zhao, Z.H. Li, T. Feng and J.K. Liu. 2018. *Seco*-tremulane Sesquiterpenoids from the cultures of the medicinal Fungus *Irpex lacteus* HFG1102. *Nat. Prod. Bioprospect.,* 8(2): 113–119.

Chen, Q., M. Wang, X.W. Yi, Z.H. Li, T. Feng and J.K. Liu. 2021. Two new pyridine derivatives and two new furan derivatives from *Irpex lacteus. Nat. Prod. Res.,* 18: 1–7.

Ding, J.H., T. Feng, B.K. Cui, K. Wei, Z.H. Li and J.K. Liu. 2013. Novel sesquiterpenoids from cultures of the basidiomycete *Irpex lacteus. Tetrahedron Lett.,* 54(21): 2651–2654.

Ding, J.H., Z.H. Li, T. Feng and J.K. Liu. 2018. Tremulane sesquiterpenes from cultures of the basidiomycete *Irpex lacteus. Fitoterapia,* 125: 245–248.

Ding, J.H., Z.H. Li, T. Feng and J.K. Liu. 2019. A new tremulane sesquiterpenoid from the fungus *Irpex lacteus. Nat. Prod. Res.,* 33(3): 316–320.

Ding, J.H., Z.H. Li, T. Feng and J.K. Liu. 2020. A sesquiterpene lactone from *Irpex lacteus. Chem. Nat. Compd.,* 56(3): 403–405.

Ding, J.H., Z.H. Li, T. Feng and J.K. Liu. 2021. Two new sesquiterpenes from cultures of the fungus *Irpex lacteus. J. Asian Nat. Prod. Res.*, 23(4): 348–352.

Dong, X.M., X.H. Song, K.B. Liu and C.H. Dong. 2017. Prospect and current research status of medicinal fungus *Irpex lacteus. Mycosystema*, 36: 28–34.

Duan, X.X., D. Qin, H.C. Song, T.C. Gao, S.H. Zuo, X. Ding, Y.T. Di and J.Y. Dong. 2019. *Irpex lacte* AD, four new bioactive metabolites of endophytic fungus *Irpex lacteus* DR10-1 from the waterlogging tolerant plant *Distylium chinense. Phytochemistry Letters*, 32: 151–156.

Gafforov, Y. and A. Ordynets. 2022. Aphyllophoroid fungi of Uzbekistan. Institute of Botany of the Academy of Sciences of the Republic of Uzbekistan. Occurrence dataset https://doi.org/10.15468/vsru5z (accessed via GBIF.org on 2022-05-25).

Gafforov, Y., A. Ordynets, E. Langer, M. Yarasheva, A.M. Gugliotta, D. Schigel, L. Pecoraro, Y. Zhou, L. Cai and L.W. Zhou. 2020. Species diversity with comprehensive annotations of wood-inhabiting poroid and corticioid fungi in Uzbekistan. *Front. Microbiol.*, 11: 598321. doi.org/10.3389/fmicb.2020.598321.

Gilbertson, R.L. and L. Ryvarden. 1987. North American Polypores, vol. 1. Fungiflora: Oslo, Norway. 433 p.

Han, Y., H.Y. Bao, L. Ma, W.J. Chen and T. Bau. 2018. Prevention and treatment of chronic glomerulonephritis in mice by administrating *Irpex lacteus* fruiting body extract. *Mycosystema*, 38(3): 428–439.

Hayashi, M., K. Wada and K. Munakata. 1981. New nematicidal metabolites from a fungus, *Irpex lacteus. Agric. Biol. Chem.*, 45(6): 1527–1529.

Jiang, X., X. Zhao, H. Luo and K. Zhu. 2014. Therapeutic effect of polysaccharide of large yellow croaker swim bladder on lupus nephritis of mice. *Nutrients*, 6(3): 1223–1235.

Kano, Y., Y. Yamagishi, K. Kamei, Y. Mutoh, H. Yuasa, K. Yamada and N. Matsukawa. 2021. First case report of fungal meningitis due to a polypore mushroom *Irpex lacteus. Neurol. Sci.*, 42(5): 2075–2078.

Luo, H., H. Jiang, X. Huang and A. JIA. 2022b. New Sesquiterpenoids from plant-associated *Irpex lacteus. Front. Chem.*, 461.

Luo, H.Z., H. Jiang, B. Sun, Z.N. Wang and A.Q. Jia. 2022a. Sesquiterpenoids and furan derivatives from the *Orychophragmus violaceus* (L.) OE Schulz endophytic fungus *Irpex lacteus* OV38. *Phytochemistry*, 194: 112996. doi: 10.1016/j.phytochem.2021.112996.

Na, Z., L. Yan, L. Jia-hui, W. Juan, Y. Shuang, Z. Nan, M. Qing-fan and T. Li-rong. 2012. Isolation, purification and bioactivities of polysaccharides from *Irpex lacteus. Chem. Res. Chin. Univ.*, 28(2): 249–254.

Novotný, C., P. Erbanová, T. Cajthaml, N. Rothschild, C. Dosoretz and V. Sasek. 2000. *Irpex lacteus*, a white rot fungus applicable to water and soil bioremediation. *Appl. Microbiol. Biotechnol.*, 54(6): 850–853.

Novotný, Č., T. Cajthaml, K. Svobodová, M. Šušla and V. Šašek. 2009. *Irpex lacteus*, a white-rot fungus with biotechnological potential—review. *Folia Microbiol.*, 54: 375–390.

Okada, B.K. and M.R. Seyedsayamdost. 2017. Antibiotic dialogues: Induction of silent biosynthetic gene clusters by exogenous small molecules. *FEMS Microbiol. Rev.*, 41: 19–33.

Pineda-Suazo, D., J.M. Montero-Vargas, J.J. Ordaz-Ortiz and G. Vázquez-Marrufo. 2021. Growth inhibition of phytopathogenic fungi and oomycetes by basidiomycete *Irpex lacteus* and identification of its antimicrobial extracellular metabolites. *Pol. J. Microbiol.*, 70(1): 131–136. doi: 10.33073/pjm-2021-014.

Rosa, L.H., K.M.G. Machado, C.C. Jacob, M. Capelari, C.A. Rosa and C.L. Zani. 2003. Screening of Brazilian basidiomycetes for antimicrobial activity. *Mem. Inst. Oswaldo Cruz*, 98(7): 967–974.

Rosa, L.H., K.M. Machado, A.L. Rabello, E.M. Souza-Fagundes, R. Correa-Oliveira, C.A. Rosa and C.L. Zani. 2009. Cytotoxic, immunosuppressive, trypanocidal and antileishmanial activities of basidiomycota fungi present in Atlantic Rainforest in Brazil. *Antonie Leeuwenhoek*, 95(3): 227–237.

Ryvarden, L. and R.L. Gilbertson. 1993. European polypores: Part 1: Abortiporus-Lindtneria. Fungiflora A/S.

Sadahiro, Y., H. Kato, R.M. Williams and S. Tsukamoto. 2020. Irpexine, an isoindolinone alkaloid produced by coculture of endophytic fungi, *Irpex lacteus* and *Phaeosphaeria oryzae. J. Nat. Prod.*, 83(5): 1368–1373. doi: 10.1021/acs.jnatprod.0c00047.

Saha, S., B. Goueli and Y. Perry. 2022. First adult pulmonary case of prior infection of polypore mushroom *Irpex lacteus. Chest*, A109–A109.

Shi, L.J., Y.M. Wu, X.Q. Yang, T.T. Xu, S. Yang, X.Y. Wang, Y.B. Yang and Z.T. Ding. 2020. The cocultured *Nigrospora oryzae* and *Collectotrichum gloeosporioides*, *Irpex lacteus*, and the plant host *Dendrobium officinale* bidirectionally regulate the production of phytotoxins by anti-phytopathogenic metabolites. *J. Nat. Prod.*, 83(5): 1374–1382.

Sun, C.T., J.P. Wang, Y. Shu, X.Y. Cai, J.T. Hu, S.Q. Zhang, L. Cai and Z.T. Ding. 2022. A new tremulane sesquiterpene from *Irpex lacteus* by solid-state fermentation. *Nat. Prod. Res.*, 36(3): 862–867.

Svobodová, K., A. Majcherczyk, C. Novotný and U. Kües. 2008. Implication of mycelium-associated laccase from Irpex lacteus in the decolorization of synthetic dyes. *Bioresour Technol.*, 99(3): 463–471.

Tang, Y., Z.Z. Zhao, K. Hu, T. Feng, Z.H. Li, H.P. Chen and J.K. Liu. 2019a. Irpexolidal represents a class of triterpenoid from the fruiting bodies of the medicinal fungus *Irpex lacteus. J. Org. Chem.*, 84(4): 1845–1852.

Tang, Y., Z.Z. Zhao, T. Feng, Z.H. Li, H.P. Chen and J.K. Liu. 2019b. Triterpenes with unusual modifications from the fruiting bodies of the medicinal fungus *Irpex lacteus. Phytochemistry*, 162: 21–28.

Tang, Y., Z.Z. Zhao, Z.H. Li, T. Feng, H.P. Chen and J.K. Liu. 2018a. Irpexoates A–D, four triterpenoids with malonyl modifications from the fruiting bodies of the medicinal fungus *Irpex lacteus. Nat. Prod. Bioprospect.*, 8(3): 171–176.

Tang, Y., Z.Z. Zhao, J.N. Yao, T. Feng, Z.H. Li, H.P. Chen and J.K. Liu. 2018b. Irpeksins A–E, 1, 10-*seco*-eburicane-type triterpenoids from the medicinal fungus *Irpex lacteus* and their anti-NO activity. *J. Nat. Prod.*, 81(10): 2163–2168.

Wang, D.L., X.Q. Yang, W.Z. Shi, R.H. Cen, Y.B. Yang and Z.T. Ding. 2021a. The selective anti-fungal metabolites from *Irpex lacteus* and applications in the chemical interaction of *Gastrodia elata*, *Armillaria* sp. and endophytes. *Fitoterapia*, 155: 105035. doi: 10.1016/j.fitote.2021.105035.

Wang, M., Z.H. Li, M. Isaka, J.K. Liu and T. Feng. 2021b. Furan derivatives and polyketides from the fungus *Irpex lacteus. Nat. Prod. Bioprospect.*, 11(2): 215–222.

Wang, M., J.X. Du, Y. Hui-Xiang, Q. Dai, Y.P. Liu, J. He, Y. Wang, Z.H. Li, T. Feng and J.K. Liu. 2020. Sesquiterpenoids from cultures of the basidiomycetes *Irpex lacteus. J. Nat. Prod.*, 83(5): 1524–1531. doi: 10.1021/acs.jnatprod.9b01177.

Wang, J., C. Li, W. Hu, L. Li, G. Cai, Y. Liu and D. Wang. 2019. Studies on the anti-fatigue activities of *Irpex lacteus* polysaccharide-enriched extract in mouse model. *Pak. J. Pharm. Sci.*, 32(3): 1011–1018.

Wang, J., J. Song, D. Wang, N. Zhang, J. Lu, Q. Meng, Y. Zhou, N. Wang, Y. Liu and L. Teng. 2016. The anti-membranous glomerulonephritic activity of purified polysaccharides from *Irpex lacteus* Fr. *Int. J. Biol. Macromol.*, 84: 87–93.

Wang, W., Y.M. Ma, H.C. Zhang, G. Liu and X. Zhang. 2013. Metabolites from the strain E21, an endophytic fungus in *Lycium ruthenicum. Chin. J. New Drugs*, 22.

Wu, Y.M., Q.Y. Zhou, X.Q. Yang, Y.J. Luo, J.J. Qian, S.X. Liu, Y.B. Yang and Z.T. Ding. 2019. Induction of antiphytopathogenic metabolite and squalene production and phytotoxin elimination by adjustment of the mode of fermentation in cocultures of phytopathogenic *Nigrospora oryzae* and *Irpex lacteus. J. Agric. Food Chem.*, 67(43): 11877–11882.

Yang, Y.B. and Z.T. Ding. 2022. A new butenolide with antifungal activity from solid co-cultivation of *Irpex lacteus* and *Nigrospora oryzae. Nat. Prod. Res.*, 11: 1–5. doi: 10.1080/14786419.2022.2037589.

Yin, H.Y., X.Q. Yang, D.L. Wang, T.D. Zhao, C.F. Wang, Y.B. Yang and Z.T. Ding. 2021. Antifeedant and antiphytopathogenic metabolites from co-culture of endophyte Irpex lacteus, phytopathogen *Nigrospora oryzae*, and entomopathogen *Beauveria bassiana. Fitoterapia*, 148: 104781. doi: 10.1016/j.fitote.2020.104781.

Ying, Y.M., Y.L. Xu, H.F. Yu, C.X. Zhang, W. Mao, C.P. Tong, Z.D. Zhang, Q.Y. Tang, Y. Zhang, W.G. Shan and Z.J. Zhan. 2019. Biotransformation of Huperzine A by *Irpex lacteus*—A fungal endophyte of *Huperzia serrata. Fitoterapia*, 138: 104341. doi.org/10.1016/j.fitote.2019.104341.

Zhang, N., Y. Liu, J.H. Lu, J. Wang, S. Yang, N. Zhang, Q.F. Meng and L.R. Teng. 2012. Isolation, purification and bioactivities of polysaccharides from *Irpex lacteus*. *Chem. Res. Chin. Univ.*, 28(2): 249–254.

Zhou, Q.Y., X.Q. Yang, Z.X. Zhang, B.Y. Wang, M. Hu, Y.B. Yang, H. Zhou and Z.T. Ding. 2018. New azaphilones and tremulane sesquiterpene from endophytic *Nigrospora oryzae* cocultured with *Irpex lacteus*. *Fitoterapia*, 130: 26–30.

Zhu, F., G. Chen, X. Chen, M. Huang and X. Wan. 2011. Aspergicin, a new antibacterial alkaloid produced by mixed fermentation of two marine-derived mangrove epiphytic fungi. *Chem. Nat. Compd.*, 47(5): 767–769.

Zhuang, L. and H. Zhang. 2021. Utilizing cross-species co-cultures for discovery of novel natural products. *Curr. Opin. Biotechnol.*, 69: 252–262.

9

Laetiporus sulphureus
A Rich Source of Diverse Bioactive Metabolites

Sunil K Deshmukh,[1] *Manish K Gupta,*[2]
Hemraj Chhipa[3] *and Kandikere R Sridhar*[4]

1. Introduction

Laetiporus sulphureus basidiomycete mushroom (Class, Agaricomycetes; Family, Polyporaceae; Order, Polyporales) grew on rotting wood of many species of trees in Africa, Asia, Europe and South America (Vasaitis et al. 2009, Khatua et al. 2017). Its fruit bodies were obtained in culture using a substrate encompassing a blend of oak sawdust with sawdust of deciduous trees (Pleszczyńska et al. 2013). The *L. sulphureus* has been used in Asia for a long time as a source of nutrition and medicine. Media like malt extract and potato dextrose agars are usually preferred media for the vegetative growth of *Laetiporus*. The necessary optimum requirements for mycelial production were a temperature of 25–30°C and a pH of 6–8 (Luangharn et al. 2014). The fruit bodies of *L. sulphureus* have golden-yellow tinges and develop on the dead tree trunks and branches. The old fruit bodies attain pale-grey or pale-beige color.

[1] R & D Division, Greenvention Biotech Pvt. Ltd., Uruli-Kanchan, Pune 412202, Maharashtra, India.
[2] SGT College of Pharmacy, SGT University, Gurugram 122505, Haryana, India.
[3] College of Horticulture and Forestry, Jhalawar, Agriculture University Kota-326001
[4] Department of Biosciences, Mangalore University, Mangalagangotri, Mangalore, Karnataka, India.
* Corresponding author: sunil.deshmukh1958@gmail.com

Laetiporus sulphureus was described for the first time by the French mycologist Pierre Bulliard in 1789 and it was designated as *Boletus sulphureus*. Tender fruit bodies of *L. sulphureus* have exceptional morphology, hence it has been defined as "chicken polypore" as well as "chicken of the woods" owing to its texture and taste almost similar to poultry. It is one of the delicacies of the vegetarian diet as a substitute for chicken in different regions of Germany as well as North America. Sometimes consumption of this mushroom leads to gastrointestinal disturbances, vomiting, fever and allergic responses (Jordan 1995, Watling 1997). Its consumption is also known to cause hallucinations owing to the presence of alkaloids like those known in psychoactive plant species (Appleton et al. 1988). The fruit bodies of this mushroom are capable to regulate the human system by improving health as well as defend the body from illnesses (Ying et al. 1987). Fruit bodies are used in Europe to treat gastric cancer, rheumatism, pyretic diseases and coughs (Matt 1947, Rios et al. 2012). Fruit bodies burn to oust out midges and mosquitoes (Ying et al. 1987). *L. sulphureus* is known to produce polysaccharides, melanin, lectins, phenolic acids, saponin, sterols, terpenes and tocopherols (Olennikov et al. 2011, Petrović et al. 2014b). This mushroom also possesses much pharmaceutical significance such as antimicrobial, antiviral, antioxidant, anticoagulant, antitumor, anti-inflammatory, hypoglycemic, immunostimulant and cytotoxic potential (Grienke et al. 2014, Sułkowska-Ziaja et al. 2018, Gründemann et al. 2020). The present chapter highlights the bioactive metabolites obtained from *L. sulphureus* and their biological functions. The details of *L. sulphureus* bioactive compounds as well as their biological properties are given in Table 1.

2. Bioactives from *Laetiporus sulphureus*

Laetiporus sulphureus produces a large number of bioactive metabolites with various biological activities and a few of them have been discussed.

2.1 Polysaccharides

Polysaccharides derived from fungi usually consist of β-1,3-glucans (Olennikov et al. 2009). The lentinan and ganoderan of mushrooms have a major influence on the improvement of immunity, in to treat cancer and are useful in pharmaceutical industries (Chihara et al. 1970, Feng et al. 2014). Polysaccharides present in the cell walls of *L. sulphureus* consist of α-1,3-glucan as a major source (~ 78%) (Jelsma and Kreger 1978). Monosaccharides present in polysaccharides of *L. sulphureus* include glucose, fructose, galactose, rhamnose, mannose, arabinose and xylose (Hwang et al. 2008, Olennikov et al. 2009, Min et al. 2010, Seo et al. 2010). The polysaccharides of *Laetiporus* have diverse structures as well as possess high medicinal importance (see Table 1).

The polysaccharides of *L. sulphureus* are derived from the fruit bodies as well as from submerged mycelial fermentation. They could also be obtained as water-soluble or alkali-soluble portions. The molecular weights of such polysaccharides

Table 1: Novel bioactive compounds reported from *Laetiporus sulphurous.*

Sr. no.	Producing organism	Name of the compound	Biological activity	References
1.	Fruit body of *Laetiporus sulphureus*	ergost-7-en-3-ol (**1**), ergost-7,22-dien-3-ol (**2**), (majority), ergost-5,7,22-trien-3-ol (ergosterol) (**3**), ergost-5,7,9(11),22-tetraen-3-ol (dehydroergosterol) (**4**), ergost-3,5,7,9(11),22-pentaen (**5**), 24-methylenelanost-7,9-dien3-ol (**6**), 24-methylenelanost-8-en-3-ol (obtusifoldienol) (**7**), 4,4-dimethylergost-24-en-3-ol (**8**), 4-methylergost-5,7,25-trien-3-ol (**9**), 4-methylergost-7,14,25-trien-3-ol (**10**), ergosterol peroxide (**11**), and ergosta-7,22-dien-3,5,6- triol (cerevisterol) (**12**)		Ericsson and Ivonne 2009
2.	Fruit body of *L. sulphureus*	lanost-22-en-3β-ol (**13**), 4,4-dimethylergost-22-en3β-ol (**14**), lanost-8,25-dien-3β-ol (**15**), 24-ethylcholest5,7,22-trien-3β-ol (**16**), ergost-5,7,9(11)-trien-3-ona (**17**), ergost-4,6,8(14),15-tetraen-3-one (**18**), ergost4,6,8(14),22-tetraen-3-one (**19**), ergost-4,6,15,22-tetraen-3-one (**20**), 4,4,14,20-tetramethylpregn-8-en3β,22-diol (**21**), 28-norlanost-8,22-dien-3β,24-diol (**22**), lanost-7,9(11),24-triene-3β,15α-diol (polycarpol) (**23**), 4,4,14,20-tetramethylpregn-8-en-3β,7,15,22-tetraol (**24**), 30-norlanost-9(11)-en-3β,14,15-triol (**25**), ergost-7,9(11),22-trien-3β,5,14-triol (**26**), ergosta7,22-dien-3β,5,6-triol (**27**), 3β,16α-dihydroxy-24-methylenelanost-7,9(11)-dien-21-oic acid (**28**), 3β-hydroxy-24-methylenelanost-8-en-21-oic acid (eburicoic acid) (**29**), 3β,15α-dihydroxy-24-methylenelanost-8-en-21-oic acid (sulphurenic acid) (**30**)		Barrera and Ramírez 2011.
3.	Fruit body of *L. sulphureus*	Sulphurenolide A (**31**)		Khalilov et al. 2019
		Sulphurenolide B (**32**), sulphurenolide C (**33**)	Anti-inflammatory activity	
4.	Fruit bodies of *L. sulphureus*	3-oxosulfurenic acid (**34**), eburicoic acid (**29**), sulfurenic acid (**35**), 15α-hydroxytrametenolic acid (**36**), versisponic acid C (**37**), acetyl eburicoic acid (**38**), acetyl trametenolic acid (**39**)	Antitumor agents.	León et al. 2004
5.	Fruit body of *L. sulphureus* var. *miniatus*	Acetyl eburicoic acid (**38**)	Anti-inflammatory	Saba et al. 2015

Table 1 contd. ...

...Table 1 contd.

Sr. no.	Producing organism	Name of the compound	Biological activity	References
6.	Fruiting bodies of *L. sulphureus* var. *miniatus*	15α-hydroxy-3,4-seco-lanosta-4(28),8,24-triene-3,21-dioic acid **(40)**, 5α-hydroxy-3,4-seco-lanosta-4(28),8,24-triene-3,21-dioic acid 3-methyl ester **(41)**, 15α-acetoxylhydroxytrametenolic acid **(42)**, versisponic acid D **(43)**		Yin et al. 2015
7.	Fruit body of *L. sulphureus*	Laetiporins C **(44)**	Antifungal activity	Hassan et al. 2021
		Laetiporins C **(44)**, and D **(45)**,	Antiproliferative activity	
		Fomefficinic acid **(46)**, eburicoicacid **(47)**, 15 α-hydroxytrametenolic acid **(36)**, trametenolic acid **(48)**		
8.	Fruits of *L. sulphureus*	Sulphurenoids A **(49)**, 15α-hydroxy-3-oxolanosta-8,24-dien-21-oic acid **(53)**, 3-ketodehydrosulfurenic acid **(54)**, 3β-hydroxylanosta-8,24-dien-21-oic acid **(55)**, eburicoic acid **(29)**, 3-oxolanosta-8,24-dien-21-oic acid **(56)**, laricinolic acid **(57)**, 3beta, 15α-dihydroxy-24-keto-double bond 8-lanostene-21-oic acid **(58)**, 15α-hydroxytrametenolic acid **(36)**, sulphurenic acid **(59)**, 3-oxosulfurenic acid **(60)**, dehydrosulphurenic acid **(61)**, and trametenolic acid **(48)**		Khalilov et al. 2022
		Sulphurenoids B–D **(50-52)**	Anti-inflammatory activity	
9.	Cultures of *L. sulphureus*	Sulphureuines B–H **(63-69)**, agripilol A **(70)**, 3β-hydroxy-11,12-O-isopropyldrimene **(71)**, and sulphurenic acid **(59)**,		He et al. 2015
		eburicoic acid **(29)**,	Cytotoxic activity	
10.	Fruit body of *L. sulphurous*	laetiporic acid A **(72)**, 2-dehydro-3-deoxylaetiporic acid A **(73)**	antioxidant activities	Davoli et al. 2005
11.	Cultures of *L. sulphureus*	6-((2E, 6E)-3, 7-dimethyldeca-2, 6-dienyl)-7-hydroxy-5-methoxy-4-methylphtanlan-1-one **(74)**, 6-((2E, 6E)-3, 7, 11-trimethyldedoca-2, 6, 10-trienyl)-5,7-dihydroxy-4-methylphtanlan-1-one **(75)**	Cytotoxic activity	Fan et al. 2014
		6-((2E, 6E)-3, 7, 11-trimethyldedoca-2, 6, 10-trienyl)-7-hydroxy-5-methoxy-4-methylphtanlan-1-one **(76)**		

Table 1 contd. ...

...Table 1 contd.

Sr. no.	Producing organism	Name of the compound	Biological activity	References
12.	Cultures of *L. sulphureus*	3-Methyl-dihydrofuran-2-one **(77)**, 3-methyl-5H-furan-2-one **(78)**, 4-methyl-dihydrofuran-2-one **(79)**, 4,5-dimethyl-dihydro-furan-2-one **(80)**, 6-methylmaltol **(81)**, 4-methyl-5H-furan-2-one; **(82)**, 5-isopropyl-3-methyl-5H-furan-2-one **(83)**, pantolactone **(84)**, sotolon **(85)**, 4,5-dimethyl-3-methoxy-5H-furan-2-one **(86)**		Krings et al. 2011
13.	Fruit body of the fungus *L. sulphureus* var. *miniatus*	Masutakeside I **(87)**, masutakic acid A **(88)**, egonol **(89)**, demethoxyegonol **(90)**, egonol glucoside **(91)**, egonol gentiobioside **(92)**	Cytotoxic activity	Yoshikawa et al. 2001
14.	*L. sulphureus*	(±)-Laetirobin (1) **(93)**	Anticancer	Lear et al. 2009
15.	Mycelium of *L. sulphureus*	Beauvericin **(94)**, enniatin Al **(95)**, enniatin B1 **(96)**, enniatin B **(97)** and enniatin A **(98)**		Deol et al. 1978
16.	Fruiting bodies of *L. sulphureus*	N-Phenethylhexadecanamide **(99)** eburicoic acid **(29)**		Shiono et al. 2005

reach up to 2.8×10^4 kDa and a minimum of 0.6 kDa (Alquini et al. 2004, Hwang et al. 2008). Min et al. (2010) established that mycelial growth is related to the pH of the environment and the production of polysaccharides. Acidified conditions with a pH range between 2 and 4 are favorable for the accumulation of mycelium, thus such increased mycelium results in the production of exopolysaccharides (EPS). Polysaccharides are also linked with proteins and phenols, which enhance their biological potential (Klaus et al. 2013). Methods of extraction in an alkaline medium, infrared spectroscopy and nuclear magnetic resonance (NMR) could also be employed to assess the structural properties of polysaccharides of *L. sulphureus*, which provide a basis to elucidate their biological properties.

Hot water extract of fruit bodies and mycelia of *L. sulphureus* was a rich source of α-(1→3)-glucan with glucose as a monosaccharide composition (Wiater et al. 2012). The polysaccharides of the cell wall are also a potential source of α-1,3-glucan (~ 78%) (Jelsma and Kreger 1978). Alkali-soluble polysaccharides (ASPS) contain glucans and laetiglucan I [a linear β-(1→3)-glucan]. Separation of the ASPS complex possesses LSA-1b-1 (MW, 180 kDa) and LSA-1b-2 (MW, 150 kDa) (Olennikov et al. 2010). The bioactive metabolites of mycelium as well as fruit bodies of *L. sulphureus* includes polysaccharides, phenolic acids, saponin, melanin, lectins, sterols, terpenes and tocopherols (Olennikov et al. 2011a, Petrović et al. 2014b).

The oligosaccharides derived from the fruit bodies of *L. sulphurous* through acid hydrolysis of α-(1→3)-glucan have been evaluated by HPLC. Fermentation of α-(1→3)-GOS as well as the model prebiotics was matched *in vitro* (e.g., *Bifidobacterium*, *Lactobacillus* and enteric bacteria). A blend of α-(1→3)-GOS could be obtained by polymerization between 2 and 9. This hydrolysate showed a robust bifidogenic impact on lactobacilli, however, it failed to stimulate the growth of *Enterococcus faecalis* as well as *Escherichia coli*. The α-(1→3)-GOS also proved to be an active compound to selectively stimulate many beneficial bacteria (Wiater et al. 2020).

The intracellular, as well as extracellular polysaccharides (EPS) of *L. sulphureus* obtained from the mycelia and its filtrates through submerged cultures, exhibited strong antioxidant activities by reducing ability as well as scavenging superoxide anion with EC50 < 5 mg/ml (Lung and Huang 2011, 2012). The principal compounds that possess antioxidant potential in mushrooms include phenols as well as polysaccharides (Mwangi et al. 2022). In comparison to phenolics of *L. sulphureus*, the extract of polysaccharide also showed a major role to induce antioxidant activity. Polysaccharides of this mushroom possess high quantities of α-glucan as well as β-glucan. The EC_{50} of antioxidant assays was highly correlated with the quantity of α-glucan. The hot water extract, partly purified lipopolysaccharides (LP) and hot alkali-derived polysaccharides (LNa) of the fruit bodies of *L. sulphureus* were also examined for antioxidant activities. The LNa possesses the highest quantity of α-glucan (17.3 g/100 g), while the LP possesses mainly β-glucans (66.8 g/100 g). Klaus et al. (2013) demonstrated that the LNa serves as the most potent antioxidant with effective concentrations (EC_{50}), reducing power (4.0 mg/ml), the DPPH radical-scavenging potential (0.5 mg/ml) and ferrous ion-chelation capacity (1.5 mg/ml). A significant relation was established between the total phenolics, the DPPH radical-scavenging potential and the reducing power. Klaus et al. (2013) also showed that the three extracts studied (at concentrations 0.1–10 mg/ml) were also non-toxic to the trophoblast cell line (HTR-8/SVneo).

In addition to polysaccharides and lectin, there are many compounds of *L. sulphureus* that possess anticancer activity. The EPS treatment activates the Bax and Bad proteins, which are necessary to trigger the apoptosis of human leukaemia cells (U937) (Seo et al. 2011). The EPS of *L. sulphureus* stimulated insulin secretion as well as insulin sensitivity possibly by regulation of lipid metabolism in the diabetic mouse (Cho et al. 2007, Hwang et al. 2008).

2.2 Lectins

Lectins are the most assessed glycoproteins of mushrooms, which possess a variety of biological functions. It is known that about 82% of lectins are derived mainly from mushrooms (Singh et al. 2011). Several investigations were carried out by Nikitina et al. (2017) on the lectins obtained from basidiomycetes (e.g., *Flammulina velutipes*, *Lentinula edodes* and *Pleurotus ostreatus*).

Lectins of *L. sulphureus* (LSL) are known for the residues of N-acetyllactosamine, which is responsible for hemagglutination and hemolysis. The LSL has varied structural components as well as molecular weight. Konska et al. (1994) isolated the LSL (190 kDa) for the first time as a heterotetrameric protein (A2B2 type 36) employing affinity chromatography with Sepharose. Similarly, the lectin was also isolated from the *Agaricus bisporus* as well as *A. campestris*. The LSL is composed of two discrete modules such as N-terminal lectin as well as pore-forming modules with 35 kDa, which shows a β-trefoil scaffold structure, which is similar to the bacterial toxins (Tateno and Goldstein 2003, Mancheño et al. 2005). Mancheño et al. (2004) crystallized the LSL (140 kDa) at 291 K by the hanging-drop vapour-diffusion technique, which was made up of four subunits of 35 kDa associated with non-covalent bonds. The LSL with 52 kDa as well as LSL with 76 kDa contained large quantities of glutamic acid, aspartic acid, threonine and leucine, while the sulfur amino acids (Cystine and Methionine) contents were low (Wang et al. 2018, 2019). Thus, LSLs have the potential for immunopotentiation in functional foods as well as pharmacology.

2.3 Sterols

Ericsson and Ivonne (2009) from *L. sulphureus* fruit bodies purified 12 sterols: ergost-7-en-3-ol **(1)**, ergost-7,22-dien-3-ol **(2)**, ergost-5,7,22-trien-3-ol (ergosterol) **(3)**, ergost-5,7,9(11),22-tetraen-3-ol (dehydroergosterol) **(4)**, ergost-3,5,7,9(11),22-pentaen **(5)**, 24-methylenelanost-7,9-dien3-ol **(6)**, 24-methylenelanost-8-en-3-ol (obtusifoldienol) **(7)**, 4,4-dimethylergost-24-en-3-ol **(8)**, 4-methylergost-5,7,25-trien-3-ol **(9)**, 4-methylergost-7,14,25-trien-3-ol **(10)**, ergosterol peroxide **(11)**, and ergosta-7,22-dien-3,5,6- triol (cerevisterol) **(12)** (Figure 1). The ergost-7,22-dien-3-ol **(2)** is the major sterol that has been found in several polyporous fungi (Yokoyama et al. 1975). Subsequently, Barrera et al. (2011) isolated 18 steroid-related compounds from the *L. sulphureus* fruit bodies based on the GC-MS analyses: lanost-22-en-3β-ol **(13)**, 4,4-dimethylergost-22-en3β-ol **(14)**, lanost-8,25-dien-3β-ol **(15)**, 24-ethylcholest5,7,22-trien-3β-ol **(16)**, ergost-5,7,9(11)-trien-3-ona **(17)**, ergost-4,6,8(14),15-tetraen-3-one **(18)**, ergost4,6,8(14),22-tetraen-3-one **(19)**, ergost-4,6,15,22-tetraen-3-one **(20)**, 4,4,14,20-tetramethylpregn-8-en3β,22-diol **(21)**, 28-norlanost-8,22-dien-3β,24-diol **(22)**, lanost-7,9(11),24-triene-3β,15α-diol (polycarpol) **(23)**, 4,4,14,20-tetramethylpregn-8-en-3β,7,15,22-tetraol **(24)**, 30-norlanost-9(11)-en-3β,14,15-triol **(25)**, ergost-7,9(11),22-trien-3β,5,14-triol **(26)**, ergosta7,22-dien-3β,5,6-triol **(27)**, 3β,16α-dihydroxy-24-methylenelanost-7,9(11)-dien-21-oic acid **(28)**, 3β-hydroxy-24-methylenelanost-8-en-21-oic acid (eburicoic acid) **(29)**, 3β,15α-dihydroxy-24-methylenelanost-8-en-21-oic acid (sulphurenic acid) **(30)** (Figure 1).

From the fruit bodies of *L. sulphurous*, three new analogs of brassinosteroids such as sulphurenolide A **(31)** (Figure 1), sulphurenolide B **(32)**, and sulphurenolide C **(33)** (Figure 2) were obtained. Sulphurenolide A as well as Sulphurenolide B, are a pair of C-20 epimer, while sulphurenolide B signifies the first natural 20R-brassinosteroid. Besides, the sulphurenolides A–C **(31–33)** were reported for the first time by Khalilov et al. (2019) as 5-hydroxylation as well as homo-6-oxa derivatives of brassinosteroids.

Ergost-3,5,7,9(11),22-pentaen (5)

R_1	R_2	R_3	R_4	Δ	
CH₃	CH₃	H	H	7	Compound (1)
CH₃	CH₃	H	H	5, 22	Compound (2)
CH₃	CH₃	H	H	7,5,22	Compound (3)
CH₃	CH₃	H	H	5,7,9,(11),22	Compound (4)
CH₂	H₃	CH₃	CH₃	24	Compound (8)
CH₃	CH₂	H	CH₃	5,7,25	Compound (9)
CH₃	CH₂	H	CH₃	7,14,25	Compound (10)

R_1	R_2	R_3	R_4	Δ	
CH₂	CH₃	CH₃	CH₃	7,9,24	Compound (6)
CH₂	CH₃	CH₃	CH₃	8, 24	Compound (7)

Ergosterol peroxide (11)

Ergosta-7,22-dien-3,5,6-triol (12)

R_1	R_2	R_3	R_4	Δ	
H	H	H	CH₃	22	Compound (13)
CH3	H	H	H	22	Compound (14)
H	H	H	CH₃	8.25	Compound (15)
CH₂CH₃	OH	H	CH₃	5,7,22	Compound (16)
H	H	OH	H	7,9,(11)24	Compound (23)
H	OH	OH	H	9(11)	Compound (25)

R1	R2	R3	R4	R5	R6	Δ	
H	H	H	H	CH3	C=O	5.7.9(11)	Compound (17)
H	H	H	H	CH3	C=O	4,6,8(14),15	Compound (18)
H	H	H	H	CH3	C=O	4,6,8(14),22	Compound (19)
H	H	H	H	CH3	C=O	4,6,15,22	Compound (20)
CH3	H	H	CH3	OH	βOH	8,22	Compound (22)
H	OH	H	OH	CH3	βOH	7,9(11),22	Compound (26)
H	OH	OH	H	CH3	βOH	7, 22	Compound (27)

R_1	R_2	Δ	
H	H	8	Compound (21)
OH	OH	8	Compound (24)

R_1	R_2	Δ	
H	OH	7,9(11)	Compound (28)
H	H	8	Compound (29)
OH	H	8	Compound (30)

Sulphurenolides A (31)

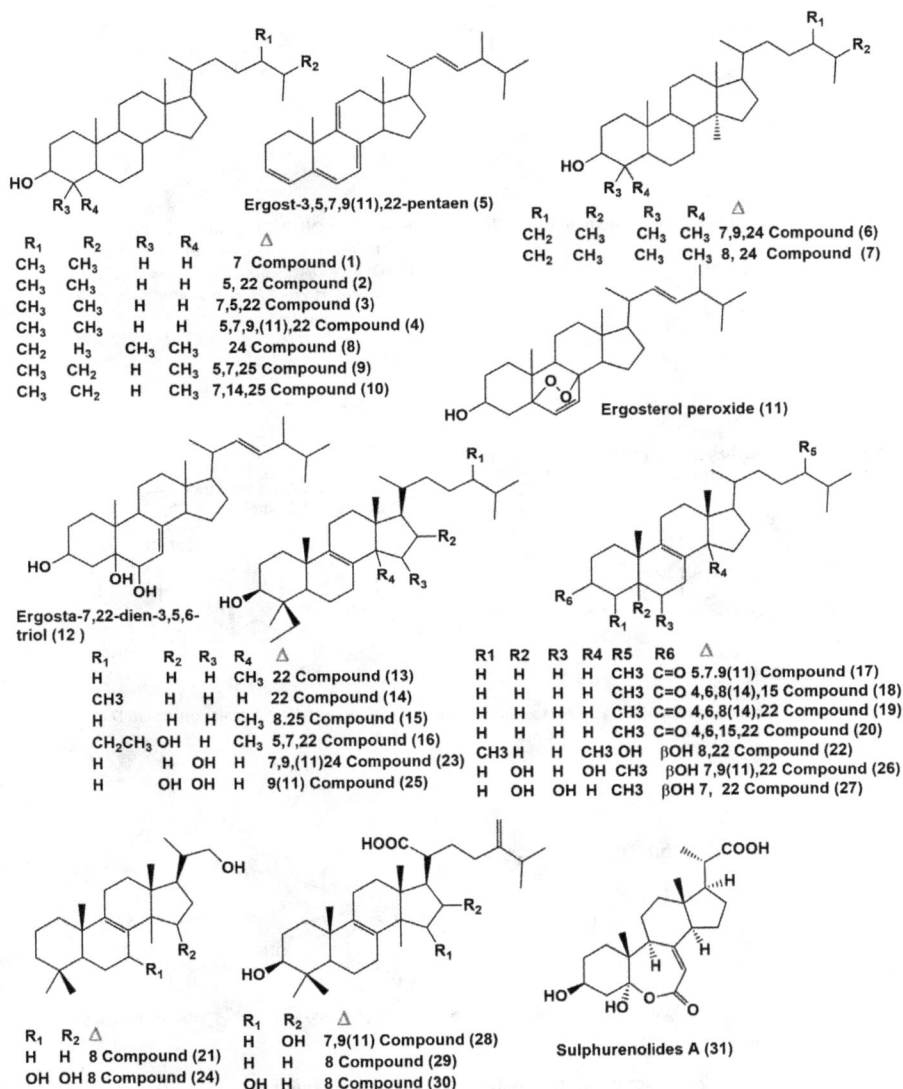

Figure 1: Structures of metabolites isolated from *L. sulphureus* (1–31).

2.4 Terpenes and Terpenoids

Lanostanoids obtained from the *L. sulphurous* possess many bioactive potentials (e.g., anti-inflammatory, immunomodulating, antioxidant, protease-inhibition, antibacterial and antiviral). León et al. (2004) from the fruit bodies of *L. sulphurous* isolated a new lanostanoid triterpene, 3-oxosulfurenic acid **(34)** along with three known triterpenes sulfurenic acid **(35)**, 15α-hydroxytrametenolic acid **(36)**

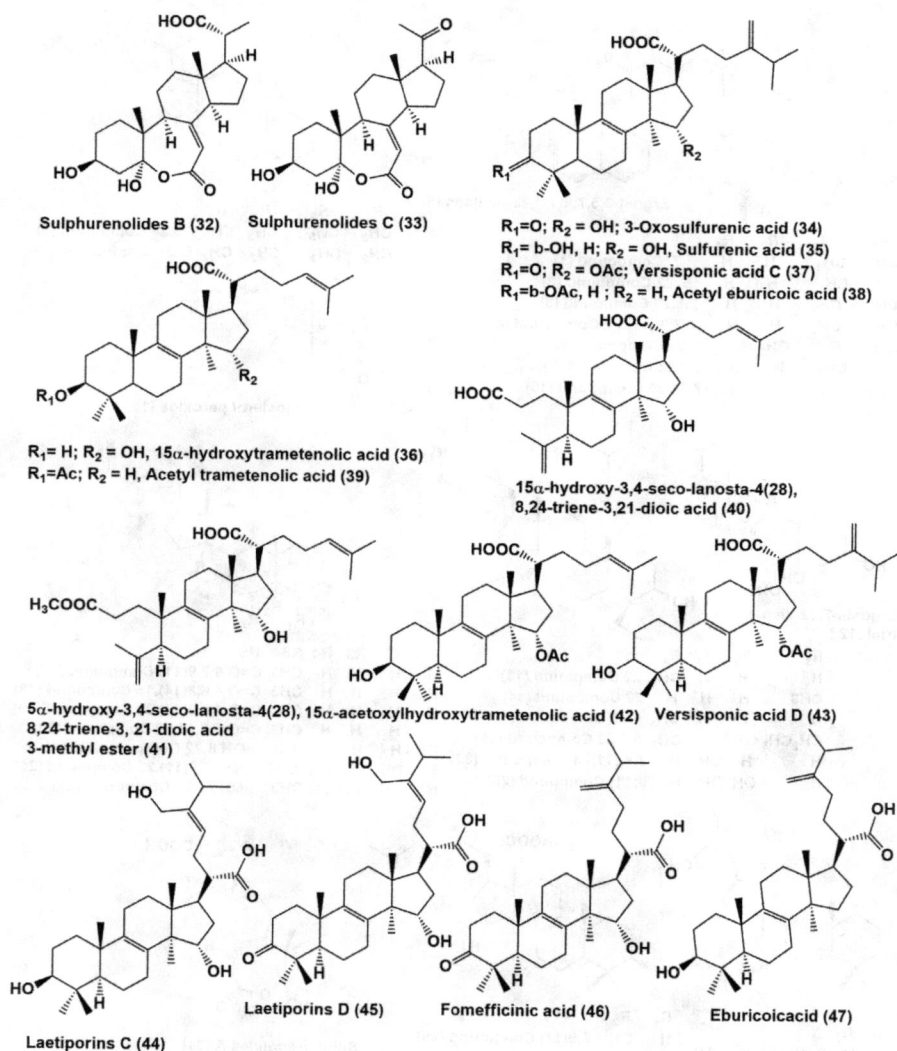

Sulphurenolides B (32)　　Sulphurenolides C (33)

R_1=O; R_2 = OH; 3-Oxosulfurenic acid (34)
R_1= b-OH, H; R_2 = OH, Sulfurenic acid (35)
R_1=O; R_2 = OAc; Versisponic acid C (37)
R_1=b-OAc, H ; R_2 = H, Acetyl eburicoic acid (38)

R_1= H; R_2 = OH, 15α-hydroxytrametenolic acid (36)
R_1=Ac; R_2 = H, Acetyl trametenolic acid (39)

15α-hydroxy-3,4-seco-lanosta-4(28),
8,24-triene-3,21-dioic acid (40)

5α-hydroxy-3,4-seco-lanosta-4(28),
8,24-triene-3, 21-dioic acid
3-methyl ester (41)

15α-acetoxylhydroxytrametenolic acid (42)　Versisponic acid D (43)

Laetiporins D (45)　　Fomefficinic acid (46)　　　Eburicoicacid (47)

Laetiporins C (44)

Figure 2: Structures of metabolites isolated from *L. sulphureus* (32–47).

(Figure 2), eburicoic acid **(29)**, acetyl derivative versisponic acid C **(37)**, acetyl eburicoic acid **(38)**, acetyl trametenolic acid **(39)** (Figure 2). Further, from the fruit bodies, *L. sulphureus* var. *miniatus*, two new 3,4-seco-lanostane-type triterpenes [15α-hydroxy-3,4-seco-lanosta-4(28),8,24-triene-3,21-dioic　　acid　　**(40)**　　and 5α-hydroxy-3,4-seco-lanosta-4(28),8,24-triene-3,21-dioic acid 3-methyl ester **(41)** and a new lanostane triterpene [15α-acetoxylhydroxytrametenolic acid **(42)** along with versisponic acid D **(43)** (Figure 2) were isolated. Compounds **(40–43)**, did not exhibit any toxicity against many cell lines (MCF-7, SMMC7721, HL-60, SW480, and A549) (Yin et al. 2015).

Two new triterpenoids [laetiporins C **(44)** and D **(45)** along with four triterpenes [fomefficinic acid **(46)**, eburicoicacid **(47)** (Figure 2), 15α-hydroxytrametenolic acid **(36)** and trametenolic acid **(48)** (Figure 3) obtained from the *L. sulphureus* fruit bodies. Laetiporin C **(44)** showed a weak antifungal activity against the fungus *Mucor hiemalis* with MIC 66 µg/mL (positive control nystatin MIC 8.3 µg/mL). Furthermore, Hassan et al. (2021) isolated compounds **(44 and 45)** and exhibited weak antiproliferative activity against fibroblast L929 of mice as well as five human cancer cell lines (KB-3-1, A431, MCF-7, PC-3 and A549).

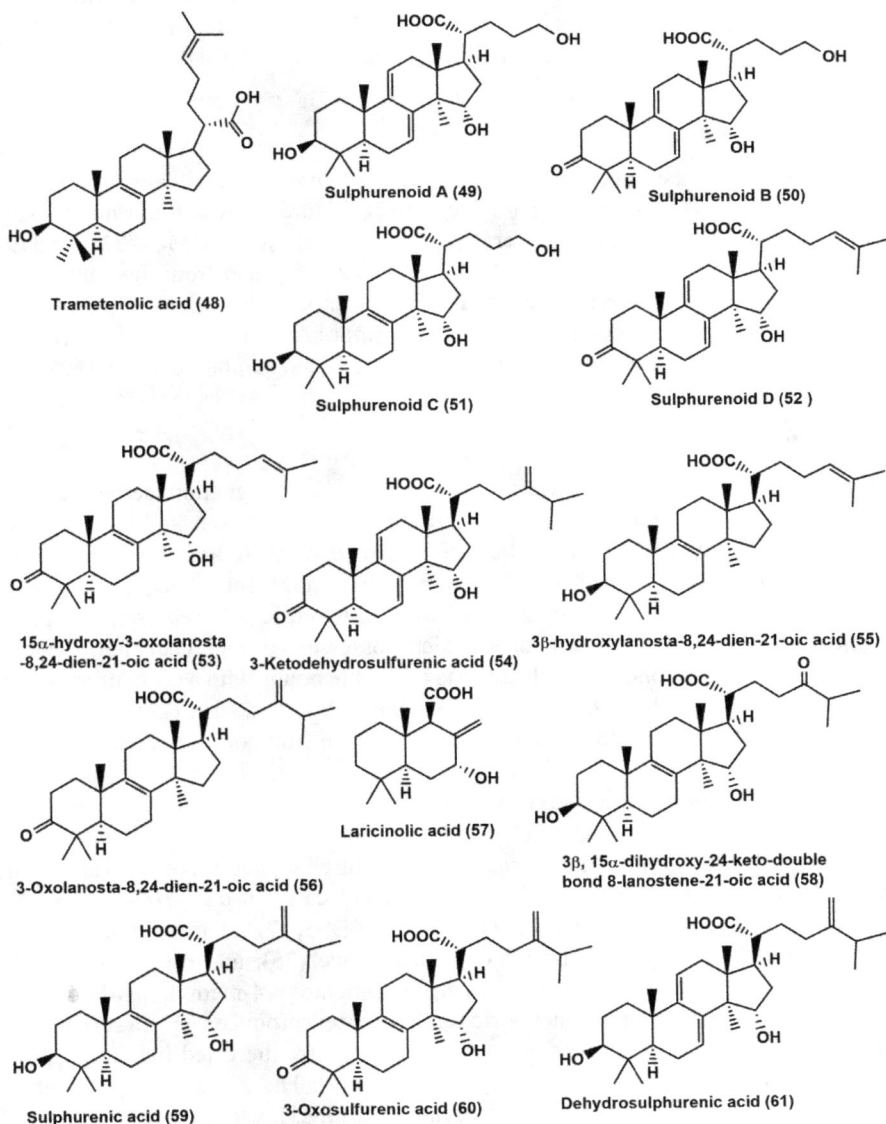

Sulphurenoid A (49)

Sulphurenoid B (50)

Trametenolic acid (48)

Sulphurenoid C (51)

Sulphurenoid D (52)

15α-hydroxy-3-oxolanosta-8,24-dien-21-oic acid (53)

3-Ketodehydrosulfurenic acid (54)

3β-hydroxylanosta-8,24-dien-21-oic acid (55)

3-Oxolanosta-8,24-dien-21-oic acid (56)

Laricinolic acid (57)

3β, 15α-dihydroxy-24-keto-double bond 8-lanostene-21-oic acid (58)

Sulphurenic acid (59)

3-Oxosulfurenic acid (60)

Dehydrosulphurenic acid (61)

Figure 3: Structures of metabolites isolated from *Laetiporus sulphureus* (48–61).

Four new triterpenoids were obtained from the *L. sulphureus* fruit bodies include sulphurenoids A–D **(49–52)** with 12 identified analogues such as 15α-hydroxy-3-oxolanosta-8,24-dien-21-oic acid **(53)**, 3-ketodehydrosulfurenic acid **(54)**, 3β-hydroxylanosta-8,24-dien-21-oic acid **(55)**, 3-oxolanosta-8,24-dien-21-oic acid **(56)**, laricinolic acid **(57)**, 3β,15α-dihydroxy-24-keto-double bond 8-lanostene-21-oic acid **(58)**, 15α-hydroxytrametenolic acid **(36)**, sulphurenic acid **(59)**, 3-oxosulfurenic acid **(60)**, dehydrosulphurenic acid **(61)** (Figure 3) eburicoic acid **(29)**, and trametenolic acid **(48)**. Khalilov et al. (2022) showed that the compounds **(50–52)** exhibited a significant anti-inflammatory potential by inhibiting the release of NO in LPS-stimulated cell line (RAW 264.7) with IC_{50} 14.3–42.3 μM in comparison with the positive control minocycline (IC_{50}, 73.0 μM).

2.5 Sesquiterpenoids

He et al. (2015a) assessed the cultures of *L. sulphureus* for isolation of a new illudin-type sesquiterpenoid sulphureuine A **(62)** (Figure 4), which did not display significant inhibition against several cell lines (HL-60, SMMC-7721, A549, MCF-7 and SW480) (IC_{50}, > 40 mM). He et al., (2015b) also from the cultures of *L. sulphurous* obtained seven novel drimane sesquiterpenoids such as sulphureuines B–H **(63–69)** along with compounds such as agripilol A **(70)**, 3β-hydroxy-11,12-O-isopropyldrimene **(71)** (Figure 4), eburicoic acid **(29)** and sulphurenic acid **(59)**.

2.6 Polyenes

An innovative carotenoid carboxylic acid with the trivial designation laetiporxanthin has been identified from *L. sulphureus* growing on cherry, oak, and elm trees in the vicinity of Royal Holloway College, UK. However, there are no reports on its structural elucidation (Valadon and Mummery 1969). A pigment called laetiporic acid was obtained from the fruit body *L. sulphureus*. Laetiporic acid represents a major pigment of *L. sulphureus* fruit bodies, which possesses an unprecedented decaene skeleton in its chromophore, which consists of double bonds with a cis conformation (Weber et al. 2004). The non-isoprenoid polyene laetiporic acid A **(72)**, 2-dehydro-3-deoxylaetiporic acid A **(73)** (Figure 4), purified from fruit bodies of *L. sulphurous*. Laetiporic acid A (1) **(72)** was the major orange pigment. The pigments laetiporic acid A **(72)** and 2-dehydro-3-deoxylaetiporic acid A **(73)** were stable in the presence of oxygen as well as light (Davoli et al. 2005).

Fan et al. (2014) from the *L. sulphurous* culture isolated three derivatives of mycophenolic acid [6-((2E, 6E)-3, 7-dimethyldeca-2, 6-dienyl)-7-hydroxy-5-methoxy-4-methylphtanlan-1-one **(74)**, 6-((2E, 6E)-3, 7, 11-trimethyldedoca-2, 6, 10-trienyl)-5, 7-dihydroxy-4-methylphtanlan-1-one **(75)**, and 6-((2E, 6E)-3, 7, 11-trimethyldedoca-2, 6, 10-trienyl)-7-hydroxy-5-methoxy-4-methylphtanlan-1-one **(76)** (Figure 4). The pigment laetiporic acid has been produced by *L. sulphureus* during varied growth conditions as well as in media by the dried biomass (up to 10 g/l with a fermenter of 7 L capacity after 17 days). The extractions obtained at 70°C with 15 min incubation revealed optimum results. Freshly obtained biomass

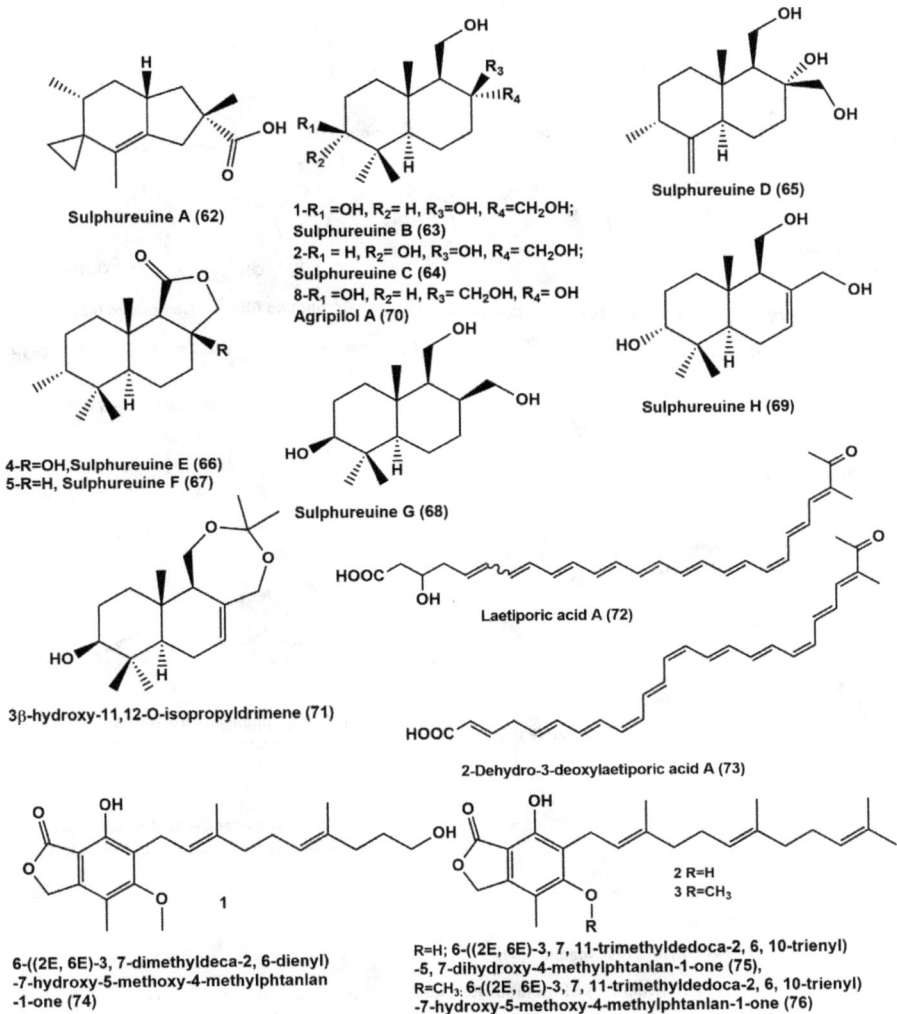

Sulphureuine A (62)

1-R_1 =OH, R_2= H, R_3=OH, R_4=CH$_2$OH;
Sulphureuine B (63)
2-R_1 = H, R_2= OH, R_3=OH, R_4= CH$_2$OH;
Sulphureuine C (64)
8-R_1 =OH, R_2= H, R_3= CH$_2$OH, R_4= OH
Agripilol A (70)

Sulphureuine D (65)

Sulphureuine H (69)

4-R=OH,**Sulphureuine E (66)**
5-R=H, **Sulphureuine F (67)**

Sulphureuine G (68)

Laetiporic acid A (72)

3β-hydroxy-11,12-O-isopropyldrimene (71)

2-Dehydro-3-deoxylaetiporic acid A (73)

6-((2E, 6E)-3, 7-dimethyldeca-2, 6-dienyl)
-7-hydroxy-5-methoxy-4-methylphtanlan
-1-one (74)

2 R=H
3 R=CH$_3$

R=H; **6-((2E, 6E)-3, 7, 11-trimethyldedoca-2, 6, 10-trienyl)**
-5, 7-dihydroxy-4-methylphtanlan-1-one (75),
R=CH$_3$: **6-((2E, 6E)-3, 7, 11-trimethyldedoca-2, 6, 10-trienyl)**
-7-hydroxy-5-methoxy-4-methylphtanlan-1-one (76)

Figure 4: Structures of metabolites isolated from *L. sulphureus* (62–76).

was diluted using ethanol (f.c. 50%; v/v) to obtain considerable biomass (50 or 200 g/l). Such a blend was employed for dyeing up to 60 min with increased color with increasing biomass. Elevated temperature (90°C than 70°C) also increased the red value for high biomass. According to Zschätzsch et al. (2021), the brilliant-hue color of the silk is not seen often with natural pigments.

2.7 *Volatile Compounds*

The mushroom *L. sulphureus* possesses a robust seasoning aroma. The volatile compounds were obtained using gluten of wheat as a substrate to culture

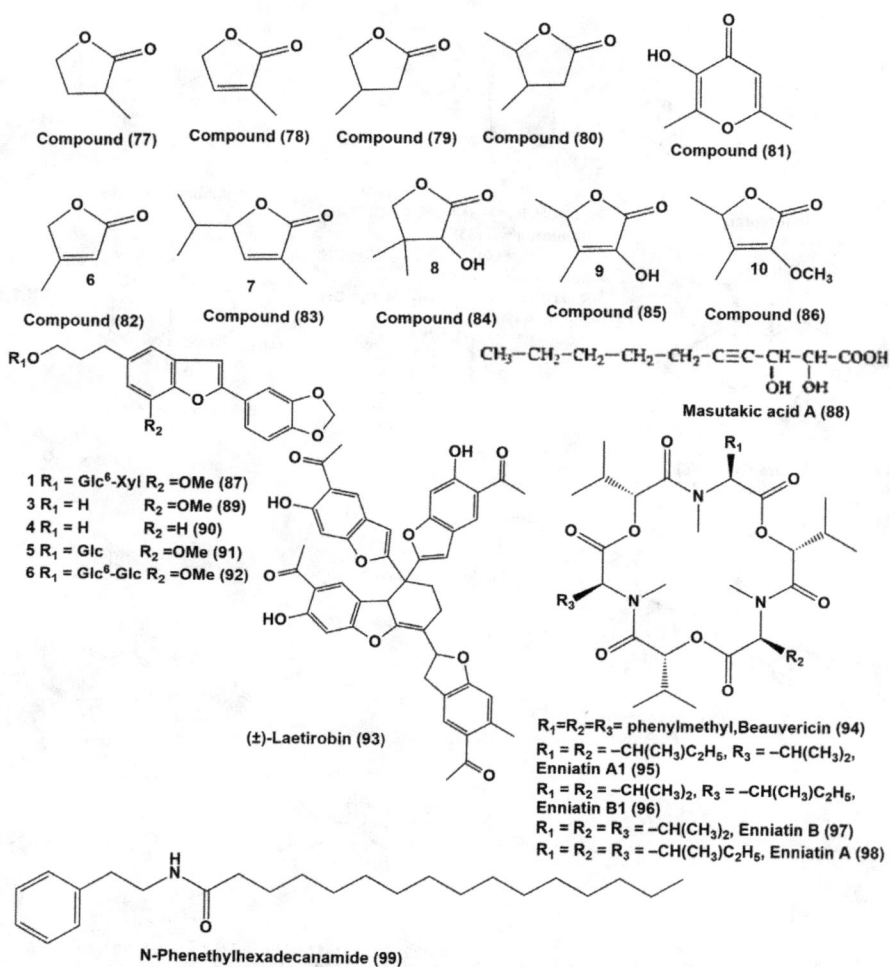

Compound (77)   Compound (78)   Compound (79)   Compound (80)

Compound (81)

6   7   8   OH   9   OH   10   OCH₃

Compound (82)   Compound (83)   Compound (84)   Compound (85)   Compound (86)

1 R₁ = Glc⁶-Xyl R₂ =OMe (87)
3 R₁ = H   R₂ =OMe (89)
4 R₁ = H   R₂ =H (90)
5 R₁ = Glc   R₂ =OMe (91)
6 R₁ = Glc⁶-Glc R₂ =OMe (92)

$CH_3-CH_2-CH_2-CH_2-CH_2-C\equiv C-CH-CH-COOH$
                                      $\underset{OH}{|}\ \underset{OH}{|}$

Masutakic acid A (88)

(±)-Laetirobin (93)

R₁=R₂=R₃= phenylmethyl,Beauvericin (94)
R₁ = R₂ = –CH(CH₃)C₂H₅, R₃ = –CH(CH₃)₂,
Enniatin A1 (95)
R₁ = R₂ = –CH(CH₃)₂, R₃ = –CH(CH₃)C₂H₅,
Enniatin B1 (96)
R₁ = R₂ = R₃ = –CH(CH₃)₂, Enniatin B (97)
R₁ = R₂ = R₃ = –CH(CH₃)C₂H₅, Enniatin A (98)

N-Phenethylhexadecanamide (99)

Figure 5: Structures of metabolites isolated from *L. sulphureus* (77–99).

L. sulphurous. About 10 heterocyclic compounds were identified by Krings et al. (2011), which include 3-methyl-dihydrofuran-2-one (**77**), 3-methyl-5H-furan-2-one (**78**), 4-methyl-dihydrofuran-2-one (**79**), 4,5-dimethyl-dihydro-furan-2-one (**80**), 6-methylmaltol (**81**), 4-methyl-5H-furan-2-one (**82**), 5-isopropyl-3-methyl-5H-furan-2-one (**83**), pantolactone (**84**), sotolon (**85**), 4,5-dimethyl-3-methoxy-5H-furan-2-one (**86**) (Figure 5).

2.8 Other Compounds

Yoshikawa et al. (2001) from fruit bodies of *L. sulphureus* var. *miniatus* obtained masutakeside I (**87**), a new benzofuran glycoside as well as another novel C10 acetylenic acid, masutakic acid A **(88)** along with the compounds egonol **(89)**, demethoxyegonol **(90)**, egonol glucoside **(91)** and egonol gentiobioside **(92)**. (±)-Laetirobin **(93)** (Figure 5) was derived as a cytostatic lead from parasitically growing *L. sulphureus* with black locust tree species (*Robinia pseudoacacia*) by the reverse-immunoaffinity technique. Preliminary studies by Lear et al. (2009) and Simon et al. (2010) showed that (±)-laetirobin **(93)** quickly passes into the tumor cells, preventing cell division during the late stage of mitosis, which leads to apoptosis.

From the lipid extract of fruit bodies of *L. sulphurous*, an insecticide, cyclodepsipeptide beauvericin **(94)** as well as a mixture of the cyclodepsipeptides enniatin Al **(95)**, enniatin B1 **(96)**, enniatin B **(97)** and enniatin A **(98)** (Figure 5) were derived. The lipids extracted from the mycelium of *L. sulphureus* were split into neutral as well as polar lipids by chromatographic techniques, beauvericin and a mixture of cyclodepsipeptides were isolated from the fraction of neutral lipids (Deol et al. 1978). Shiono et al. (2005) obtained N-phenethylhexadecanamide **(99)** (Figure 5) from the methanol extract of *L. sulphureus* fruit bodies along with eburicoic acid **(29)**.

3. Pharmacological Activities

3.1 Antimicrobial Activity

The ethanolic extract of *L. sulphureus* displayed antibacterial activity against *Bacillus cereus*, *B. subtilis*, *E. coli*, *Micrococcus flavus*, *M. luteus*, *Morganella morganii*, *Proteus vulgaris*, *Pseudomonas aeruginosa*, *Salmonella enteritidis*, *Staphylococcus aureus* and *Yersinia enterecolitica* with inhibition zone ranging between 4.5 and 23 mm. Turkoglu et al. (2007) exhibited considerable antifungal activity against *Candida albicans* by the ethanolic extract *L. sulphureus*.

Sinanoglou et al. (2015) demonstrated that fruit bodies of *L. sulfureus* n-hexane extract is active against *Salmonella typhimurium* (MIC, 0.15 mg/ml) with a positive control of streptomycin (MIC, 0.2 mg/ml) and *Trichoderma viride* (MIC, 1.5 mg/ml) with a positive control of ketoconazole (MIC, 1.0 mg/ml). The compounds extracted from *L. sulphureus* showed higher fungal inhibitory activity compared to bacteria (Petrović et al. 2014a). For example, the MIC of the compounds against the fungi *Penicillium funiculosum* and *Trichoderma viride* were 0.023 and 0.006 mg/ml, respectively, while for the bacteria *E. coli*, *Micrococcus flavus* and *Pseudomonas aeruginosa* with MIC up to 0.40 mg/ml.

L. sulphureus is also known to produce huge quantities of citric, oxalic and fumaric acids, which possess antibacterial as well as antifungal (e.g., *Aspergillus* and *Trichoderma*) (Petrović et al. 2014b). The polysaccharides of *L. sulphureus* showed antitumor activity *in vitro* against HCT15 (GI50, 68.96 µg/ml) as well as HeLa (GI50, 72.26 µg/ml). The methanolic extract employed to study the in situ potential against *Aspergillus flavus* showed considerable preserving potential.

The solvent extracts obtained from *L. sulphureus* mycelia (ethanol, petroleum ether and ethyl acetate) showed antibacterial activity for many bacteria (*B. subtilis*,

E. coli, P. aeruginosa and *S. aureus*). Zhang et al. (2018) opined that such activity may be owing to the presence of compounds like oleic acid, palmitic acid, lactic acid, succinic acid, glycerol, glycerol α-monopalmitate, ergosterol, uracil, D-mannitol, ring-(proline-glycine), cyclo-L-Hyp-Gly and L-methyl L-glutamate compounds in the mycelial extracts of *L. sulphureus*. Another study by Petrović et al. (2013) showed that 15 mg/ml methanol extract of *L. sulphureus* resulted in total inhibition of the growth of *Aspergillus flavus* in tomato paste (room temperature, 25°C) for up to 15 days. Seibold et al. (2020) cloned the lapA of *L. sulphureus* (GenBank accession # MT304701) to express in strains of *Aspergillus* for the purpose of biosynthesis of laetiporic acids. Compounds such as laetiporic acid A **(72)** and 2-dehydro-3-deoxylaetiporic acid A **(73)** of *L. sulphureus* revealed a noticeable antifungal activity against the protoplasts of *Aspergillus* (4 mg/ml), while undiluted as well as 1:5 dilution of polyenes significantly reduced the viability of *A. nidulans* protoplast. Popa et al. (2016) presented that the ethanol extracts of fruit bodies of *L. sulphureus* as well as the dried biomass were antimicrobial against *E. coli* (ATCC8739), *Enterococcus faecalis*, *S. aureus* (ATCC6538), *S. epidermidis* (ATCC12228), *Candida albicans* (ATCC10321) and *C. parapsilopsis* (CBS604).

3.2 *Antioxidant Activity*

Ethanol extract of *L. sulphureus* possessing polyphenols and flavonoid-type compounds increased the antioxidant ability up to 63.8 and 14.2 μg/mg, respectively. At the concentration of 80 μg/ml, the *L. sulphureus* extract, BHA and α-tocopherol revealed 57.4, 88.2 and 93.3% inhibition, respectively. At 160 μg/ml concentration these compounds showed 82.2, 96.4, and 98.6% DPPH radical-scavenging activity, respectively. At concentrations of 100, 200, 400 and 800 μg/ml, the DPPH radical-scavenging potential was up to 14, 26, 55 and 86%, respectively. The total phenolic content of fruit bodies was 283.9 mg GAE/100g (Popa et al. 2016). A positive correlation was found between the total phenolics and their antioxidant capabilities in mushroom extracts (Turkoglu et al. 2007). The ethanol extract of fruit bodies and dried biomass of *L. sulphureus* displayed antioxidant activity like the DPPH radical-scavenging with EC50 of 35.45 and 46.67 mg/ml, respectively. Lin and Lee (2021) demonstrated that the fermented extract of *L. sulphureus* improved the antioxidant activity of broilers through the enhancement of expression of the GSH-Px as well as CAT.

According to Olennikov et al. (2011b), melanin in *L. sulphureus* possesses antiradical activity (IC 50 up to 57.8 μg/ml) with marked radical-scavenging potential by melanin-containing phenolic compounds. The hot alkali extract of *L. sulphureus* (LNa) showed antioxidant potential such as the DPPH radical-scavenging potential (0.5 mg/ml) and reducing ability (4.0 mg/ml) including ferrous ion-chelating capacity (1.5 mg/ml). The LNa is known to possess the highest quantity of α-glucan (17.3 g/100), which is a ubiquitous monosaccharide in all extracts. According to Klaus et al. (2013), the LNa at concentrations between 0.1 and 10 mg/ml was non-toxic to the trophoblast cell line (HTR-8/SVneo). Lung

(2012) reported that the intracellular as well as extracellular polysaccharides (EPS) obtained from the mycelia and filtrates of submerged culture *L. sulphureus* exhibited a strong antioxidant capacity especially the reducing power as well as radical-scavenging superoxide (EC50, < 5 mg/ml). The hot-water extract of polysaccharides (H-SMPS), as well as enzyme-extractable polysaccharides (E-SMPS), derived from the spent residues of *L. sulphurous* exhibited hepatoprotective as well as antioxidant activities. Activities of glutathione peroxidase (GSH-Px), superoxide dismutase and catalase (CAT) were higher compared to the model control (MC) group with H-SMPS group up to 30.6, 36.8 and 44.6%, respectively. The H-SMPS with a high dose inhibited the malondialdehyde (MDA) and lipid peroxidation (LPO) up to 27.9 and 41.0%, respectively. Zhao et al. (2017) showed that the E-SMPS inhibits MDA and LPO up to 40.7 and 50.6, respectively than the MC group.

4. Anticancer and Cytotoxic Activities

Eburicoic acid (29) showed cytotoxic potential against several cell lines (HL-60, SW-480, A-549 and SMMC-721) with IC_{50}, 37.5, 36.1, 15.6 and 14.8 µM, respectively with a positive control cisplatin cytotoxic activity (IC_{50}, 12.8 µM) (He et al. 2015b). Petrović et al. (2020) demonstrated that the LSL has highly significant inhibitory activity against the development of tumor cells HCT-116 as well as tumor cell B16-F10 in xenografts with melanoma with the highest reduction up to 57.22% at a concentration of 100 µg/ml.

The sulfurenic acid (35), 15α-hydroxytrametenolic acid (36) and acetyl eburicoic acid (38) were potent inhibitors of propagation of human myeloid leukaemia cells HL-60 with IC_{50}, 14, 12, and 15 µM, respectively. The versisponic acid C (2) (37), eburicoic (3) (29), and 15α-hydroxytrametenolic (7) (36) acids showed moderate activity with IC_{50} ranging between 25 and 31 µM, while 3-oxosulfurenic acid (1) (34) showed very poor activity of IC_{50} 407 µM with a positive control ursolic acid IC_{50} 21 µM. Programmed cell death was seen by the stimulation of caspase-3 as well as the disintegration of poly(ADP-ribose) polymerase-1 (PARP-1), while it was also found an early release of mitochondrial cytochrome c. Such an event delivers the signal for the commencement of the apoptosome complex assembly that employs sequential progressions as well as activates caspase-9 as an initiator caspase that eventually triggers downstream of caspases. Caspase-3 is one of the key components of apoptosis, which is liable for the proteolytic cleavage of many proteins (e.g., nuclear enzyme PARP-1). The acetyl eburicoic acid (38) was one of the most potent inducers of apoptosis. Such apoptosis will go along with the activation of caspase-3 as well as PARP-1 fragmentation, which was related to an initial discharge of mitochondrial cytochrome c (León et al. 2004).

From the *L. sulphurous*, drimane kind of sesquiterpenoids sulphureuine B (63) was also obtained. The sulphureuine B (63) displayed cytotoxicity against glioma cells (U-87MG) with IC_{50} 16 µM. Zhang et al. (2015) demonstrated that the sulphureuine B-induced death of U-87MG cells as well as the mechanisms include endoplasmic reticulum (ER) stress with mitochondrial death receptor-mediated pathways. The ER stress was recognized by high cytoplasmic vacuolation, CHOP elevation as well as caspase-12 cleavage. The PERK, IRE1α and ATF6α activations were induced by the treatment of sulphureuine B. Besides, the sulphureuine B-induced Bcl-2 down-regulation, caspase-8 activation and cleavage of PARP were hampered.

Compounds such as egonol (**89**), demethoxyegonol (**90**), egonol glucoside (**91**) exhibited *in vitro* cytotoxicity against Kato III cells with IC_{50} 28.8, 27.5, and 24.9 mg/ml, respectively with positive control of hinokitiol 0.6 mg/ml (Yoshikawa et al. 2001). Mycophenolic acid derivatives such as 6-((2E, 6E)-3, 7-dimethyldeca-2, 6-dienyl)-7-hydroxy-5-methoxy-4-methylphtanlan-1-one (**74**), 6-((2E, 6E)-3, 7, 11-trimethyldedoca-2, 6, 10-trienyl)-5, 7-dihydroxy-4-methylphtanlan-1-one (**75**) showed modest cytotoxicity against the many cell lines (HL-60, SMMC-7721, A-549 and MCF-7) with IC_{50} 39.1, 31.1, 27.4, and 35.7 μM/l, respectively (Fan et al. 2014). (±)-Laetirobin (**93**) was obtained as a cytostatic lead from *L. sulphureus*. Based on the studies by Lear et al. (2009), the preliminary cellular investigations showed that (±)-laetirobin (**93**) rapidly moves into the tumor cells and blocks the cell division at a later stage of mitosis, which provokes apoptosis.

5. Other Pharmacological Properties

The compounds sulphurenolide B (**32**), and sulphurenolide C (**33**) exhibited substantial inhibition of NO production in LPS-induced RAW 264.7 cells, such inhibition ratios at 50 μM were 68.2 and 87.3%, respectively, while at the same concentration, the positive control minocycline exhibited inhibition up to 73% (Khalilov et al. 2019). Eburicoic acid (EA) (**29**), di9d does not cause cytotoxicity on RAW264.7 cells at concentrations between 0.02 and 0.08 μM. The Eburicoic acid has significantly repressed the release of inflammatory mediators such as nitrite oxide (NO) as well as prostaglandin E2 (PGE2); blocked the mRNA as well as protein expression of inducible nitrite oxide synthase (iNOS), cyclooxygenase-2 (COX-2) and pro-inflammatory cytokines TNF-α, IL-6 and IL-1β; reduced the contents of phosphorylated PI3K, Akt, mTOR and NF-κBp65 in LPS-induced RAW264.7 cells in a dose- as well as time-dependent way. These results showed that the Eburicoic acid performed anti-inflammatory effects on LPS-induced RAW264.7 cells, such effects could be attained through inhibiting the PI3K/Akt/mTOR/NF-κB signaling pathway as well as inhibiting the LPS-induced production of inflammatory intermediaries including pro-inflammatory cytokines (Wang et al. 2015). The anti-inflammatory triterpene, and acetyl eburicoic acid (LSM-H7) (**38**) were obtained from *L. sulphureus* var. *miniatus* fruit bodies. The LSM-H7 (**38**) inhibited the NO production in RAW 264.7 cells dose-dependent way devoid of cytotoxicity. Furthermore, Saba et al. (2015) also showed the prevention of the production of pro-inflammatory cytokines, cyclooxygenase-2, inducible nitric oxide synthase, interleukin (IL)-1β, IL-6 and the tumor necrosis factor α with glyceraldehyde 3-phosphate dehydrogenase.

The EPS obtained from *L. sulphureus* upsurge the propagation, and insulin secretion events of rat insulinoma cells (RINm5F) and possesses an antidiabetic impact. Applying the streptozotocin-induced diabetic rat model, significantly increased (68%) cell viability by EPS treatment (2 mg/ml). Early treatment with EPS prior to treatment with streptozotocin enhanced the cell viability (86%) (Hwang et al. 2008, Hwang and Yun 2010).

Okamura reported that the crude extracts of fruit bodies of *L. sulphureus* possess anti-coagulation activity duration with thrombin (more than 600 s), hence, these could be utilized in foodstuffs to prevent thrombosis (Okamura et al. 2000). The alanine aminotransferase (ALT), as well as aspartate aminotransferase (AST), serve as indicators of membrane damage of liver cells as they are the most sensitive. Damage to hepatocyte membranes leads to the spillage of enormous amounts of ALT and AST into the bloodstream, which leads to elevating in serum contents dramatically (Schindhelm et al. 2006). Sun et al. (2014) demonstrated that by treating with 50 mg/kg of *L. sulphurous*, the serum AST and ALT contents in rats significantly reduced (13.4 and 9.02 U/L to 3.51 and 1.67 U/L, respectively) than the model group.

Ethanol extract of *L. sulphureus* (LSE) is non-toxic at high doses (400–500 µg/ml), while efficiently inhibiting melanogenesis in a dose-dependent way in the zebrafish (*Danio rerio*) as a pre-clinical fish model. At the depigmenting doses, the explored extracts revealed no adverse effects on the embryos and melanocytes of zebrafish. Further, they did not provoke inflammation or neutropenia on the application of the highest dose ensuring complete cell depigmentation. Since the LSE has shown significantly higher therapeutic potential compared to kojic acid and hydroquinone, two familiar depigmenting agents, overall results of this study strongly indicate that the studied mushroom extracts could be used as efficient as well as safe topical agents to treat the disorders like skin hyperpigmentation (Pavic et al. 2021). *Laetiporus* also generate organic acids (e.g., oxalic acid) to progress Fenton-type reactions for carbohydrate depletion during its growth as well as development (Iii and Highley 1997, Kartal et al. 2004).

To identify the sesquiterpene synthases in *L. sulphurous*, *Coprinopsis cinerea* as well as *Stereum hirsutum* are used. The common sesquiterpene synthases could be divided into four clades as Clade I and III possess enzymes that utilize 1, 10 and 1,11-cyclization of (2E,6E)-farnesyl diphosphate to produce sesquiterpenes derived from an E, E-germacradienyl cation. The Clade II and IV possess enzymes that shared a 1,10 and 1,6-cyclization of (3R)-nerolidyl diphosphate mechanism, producing sesquiterpenes derived from a Z, E-germacradienyl cation (Schmidt-Dannert 2015, Nagamine et al. 2019).

The *L. sulphureus* fermented product (FL) is capable to improve the Feed Conversion Ratio (FCR) in broilers through the improvement of antioxidative status by increasing the gastrointestinal TJ mRNA expression and reducing the negative impacts of dietary NSP. Besides, 5% FL addition showed the top results in increasing the activity of serum SOD as well as antioxidant gene expression (in jejunum, ileum and liver) through the improvement of expression of gastrointestinal TJ mRNA (Lin and Lee 2021).

6. Activation of Biosynthetic Genes for Secondary Metabolite Production

Mushrooms are getting more recognition as an alternative source of food supplements due to the presence of proteins and bioactive compounds. The edible or non-edible nature of mushroom mainly depends on the presence of harmful or beneficial natural products or secondary metabolites so, identification of all chemical compounds are more important in mushroom (Lee et al. 2022). The production of such beneficial compounds for living beings in microorganisms has required the availability of native habitats and surrounding environments for microbial and environmental interactions. Production of such compounds is silent during culturing of microbes at laboratory conditions and biosynthetic gene clusters remain silent or downregulated due to changes in natural growth conditions (Scherlach and Hertweck 2021). Different methods have been discovered to activate such silent bioactive gene clusters to induce the production of bioactive compounds including ribosomal engineering, data mining, use of chemical inducers, epigenetic modifications, and co-culturing (Deshmukh et al. 2022). In ribosomal engineering, the resistant microbial strains are selected using different antibiotics containing media and further such strains are used in the activation of silent biosynthetic genes (Scherlach and Hertweck 2021). In genetic engineering, the target or silent biosynthetic gene site is identified and manipulated by the gene expression with the introduction of promoters or changes in regulatory genes (Amos et al. 2017, Zhang et al. 2017). The co-culture method is another approach to induce the silent biosynthetic gene cluster. In this method, the interactions of co-culture microbes play a significant role via chemical mediators to induce the expression of silent genes (Scherlach and Hertweck 2020). Co-culture of microbes not only induce silent gene but also supports the production yield of metabolites. The production of emericellamides A and emericellamides B increased up to 100-fold in the fungus *Emericella parvathecia* after co-culturing with actinomycete *Salinispora arenicola* (Oh et al. 2007). Similarly, the co-culture of *Thalassopia* sp. and *Libertella* sp. induced the production of libertellenones A-D compounds which are unexpressed in the mono-culture of *Libertella* sp. (Chiang et al. 2008). One strain of many compounds (OSMAC) is another approach to induce silent biosynthetic genes. In this method alternation in culture conditions also changes the compound profiling of target strains. Changes in media composition, growth temperature, aeration, use of inhibitors and use of inducer compounds manipulate the cultures condition and production of secondary metabolites. In addition to them, changes in light intensity and fermentation volume also affect the production profile of secondary metabolites (Daletos et al. 2017). Various chemical epigenetic modifiers such as histone deacetylases inhibitors and DNA methyltransferases inhibitors also play a significant role in the induction or suppression of silent bioactive gene clusters via chromatin modification or changes in DNA- post-translation (Magotra et al. 2017). All such explained methods have been used in the activation of silent biosynthetic gene clusters in microbes. Still uses of such method in mushrooms for activation of the silent gene have wide potential and more research is required to understand their suitability.

7. Conclusion

Laetiporus sulphureus is commonly called the "chicken of the woods" as it involves wood rot in Africa, Asia, Europe and South America. Various components and metabolites are produced by *L. sulphureus* such as polysaccharides, phenolic acids, saponin, melanin, lectins, sterols, terpenes and tocopherols. These compounds possess different biological functions (e.g., antimicrobial, antiviral, anti-inflammatory, anticoagulant, antioxidant, hypoglycaemic, immunostimulant, antitumor, anticancer and cytotoxic activities). The *L. sulphurous* being edible will be extremely useful in the production of nutraceutical products with nutritional as well as health-promoting prospects.

References

Alquini, G., E.r. Carbonero, F.R. Rosado, C. Cosentino and M. Iacomini. 2004. Polysaccharides from the fruit bodies of the basidiomycete *Laetiporus sulphureus* (Bull.: Fr. Murr). *FEMS Microbiol. Lett.*, 230(1): 47–52.

Amos, G.C., T. Awakawa, R.N. Tuttle, A.C. Letzel, M.C. Kim, Y. Kudo, W.S. Fenical, B. Moore and P.r. Jensen. 2017. Comparative transcriptomics as a guide to natural product discovery and biosynthetic gene cluster functionality. *Proceedings of the National Academy of Sciences*, 114(52): E11121–E11130.

Appleton, R.E., J.E. Jan and P.D. Kroeger. 1988. *Laetiporus sulphureus* causing visual hallucinations and ataxiaina child. *Can. Med. Assoc. J.*, 139: 48–49.

Barrera, E.D.C. and I.J.N. Ramírez. 2011. Steroid-related metabolites from edible mushroom *Laetiporus sulphureus*. *Revista Facultad De Ciencias Básicas*, 7(2): 298–307.

Chiang, Y.-M., E. Szewczyk, T. Nayak, A.D. Davidson, J.F. Sanchez, H.C. Lo, W.Y. Ho, H. Simityan, E. Kuo, A. Praseuth and K. Watanabe. 2008. Molecular genetic mining of theAspergillus secondary metabolome: Discovery of the emericellamide biosynthetic pathway. *Chem. Biol.*, 15: 527–532. doi: 10.1016/j.chembiol.2008.05.010.

Chihara, G., J. Hamuro, Y. Maeda, Y. Arai and F. Fukuoka. 1970. Fractionation and purification of the polysaccharides with marked antitumor activity, especially lentinan, from *Lentinus edodes* (Berk.) Sing. (an edible mushroom). *Cancer Res.*, 30(11): 2776–2781.

Cho, E.J., H.J. Hwang, S.W. Kim, J.Y. Oh, Y.M. Baek, J.W. Choi, S.H. Bae and J.W. Yun. 2007. Hypoglycemic effects of exopolysaccharides produced by mycelial cultures of two different mushrooms *Tremella fuciformis* and *Phellinus baumii* in ob/ob mice. *Appl. Microbiol. Biotechnol.*, 75: 1257–1265.

Daletos, G., W. Ebrahim, E. Ancheeva, M. El-Neketi, W. Lin, P. Proksch, D.J. Newman, G.M. Cragg and P.G. Grothaus. 2017. Microbial coculture and OSMAC approach as strategies to induce cryptic fungal biogenetic gene clusters. *In* Chemical biology of natural products (pp. 233–284). CRC Press.

Davoli, P., A. Mucci, L. Schenetti and R.W. Weber. 2005. Laetiporic acids, a family of non-carotenoid polyene pigments from fruit-bodies and liquid cultures of *Laetiporus sulphureus* (Polyporales, Fungi). *Phytochemistry*, 66(7): 817–823.

Deol, B.S., D.D. Ridley and P. Singh. 1978. Isolation of cyclodepsipeptides from plant pathogenic fungi. *Aust. J. Chem.*, 31: 1397–1399.

Deshmukh, S.K., L. Dufossé, H. Chhipa, S. Saxena, G.B. Mahajan and M.K. Gupta. 2022. Fungal endophytes: A potential source of antibacterial compounds. *J. Fungi*, 8(2): 164.

Duan, Y., J. Qi, J.M. Gao and C. Liu. 2022. Bioactive components of *Laetiporus* species and their pharmacological effects. *Appl. Microbiol. Biotechnol.*, 106(18): 5929–5944.

Ericsson, D. and J. Ivonne. 2009. Sterol composition of the macromycete fungus *Laetiporus sulphureus*. *Chem. Nat. Compd.*, 45(2): 193–196.

Fan, Q.Y., X. Yin, Z.H. Li, Y. Li, J.K. Liu, T. Feng and B.H. Zhao. 2014. Mycophenolic acid derivatives from cultures of the mushroom *Laetiporus sulphureu. Chin. J. Nat. Med.*, 12(9): 685–688.

Grienke, U., M. Zöll, U. Peintner and J.M. Rollinger. 2014. European medicinal polypores—A modern view on traditional uses. *J. Ethnopharmacol.*, 154(3): 564–583.

Gründemann, C., J.K. Reinhardt and U. Lindequist. 2020. European medicinal mushrooms: Do they have potential for modern medicine?–An update. *Phytomedicine*, 66: 53131. doi: 10.1016/j. phymed.2019.153131.

Hassan, K., B. Matio Kemkuignou and M. Stadler. 2021. Two new triterpenes from basidiomata of the medicinal and edible mushroom, *Laetiporus sulphureus. Molecules*, 26(23): 7090. doi: 10.3390/ molecules26237090. 26(23):7090.

He, J.B., J. Tao, X.S. Miao, Y.P. Feng, W. Bu, Z.J. Dong, Z.H. Li, T. Feng and J.K. Liu. 2015a. Two new illudin type sesquiterpenoids from cultures of *Phellinus tuberculosus* and *Laetiporus sulphureus. J. Asian Nat. Prod. Res.*, 17(11): 1054–1058.

He, J.B., J. Tao, X.S. Miao, W. Bu, S. Zhang, Z.J. Dong, Z.H. Li, T. Feng and J.K. Liu. 2015b. Seven new drimane-type sesquiterpenoids from cultures of fungus *Laetiporus sulphureus. Fitoterapia*, 102: 1–6.

Hwang, H.S., S.H. Lee, Y.M. Baek, S.W. Kim, Y.K. Jeong and J.W. Yun. 2008. Production of extracellular polysaccharides by submerged mycelial culture of *Laetiporus sulphureus* var. *miniatus* and their insulinotropic properties. *Appl. Microbiol. Biotechnol.*, 78: 419–429.

Hwang, H.S. and J.W. Yun. 2010. Hypoglycemic effect of polysaccharides produced by submerged mycelial culture of *Laetiporus sulphureus* on streptozotoc ininduced diabetic rats. *Biotechnol. Bioprocess Eng.*, 15(1): 173–181.

Iii, F.G. and T.L. Highley. 1997. Mechanism of brown-rot decay: Paradigm or paradox. *Int. Biodeterior. Biodegrad.*, 39(2-3): 113–124.

Jelsma, J. and D.R. Kreger. 1978. Observations on the cell-wall compositions of the bracket fungi *Laetiporus sulphureus* and *Piptoporus betulinus. Arch. Microbiol.*, 119(3): 249–255.

Jordan, M. 1995. Evidence of severe allergic reactions to *Laetiporus sulphureus. Mycologist*, 9: 157–158.

Kartal, S.N., E. Munir, T. Kakitani and Y. Imamura. 2004. Bioremediation of CCA-treated wood by brown-rot fungi *Fomitopsis palustris, Coniophora puteana,* and *Laetiporus sulphureus. J. Wood Sci.*, 50(2): 182–188.

Khalilov, Q., L. Li, Y. Liu, W. Liu, S. Numonov, H.A. Aisa and T. Yuan. 2019. Brassinosteroid analogues from the fruiting bodies of *Laetiporus sulphureus* and their anti-inflammatory activity. *Steroids*, 151: 108468.

Khalilov, Q., S. Numonov, P. Sukhrobov, K. Bobakulov, F. Sharopov, M. Habasi, J. Zhao, T. Yuan and H.A. Aisa. 2022. New triterpenoids from the fruiting bodies of *Laetiporus sulphureus* and their anti-inflammatory activity. *ACS Omega*, 7(31): 27272–27277.

Khatua, S., S. Ghosh and K. Acharya. 2017. *Laetiporus sulphureus* (Bull.: Fr.) Murr. as food as medicine. *Pharmacogn. J.*, 9: s1–s15.

Klaus, A., M. Kozarski, M. Niksic, D. Jakovljevic, N. Todorovic, I. Stefanoska and L.J. Van Griensven. 2013. The edible mushroom *Laetiporus sulphureus* as potential source of natural antioxidants. *Int. J. Food Sci. Nutr.*, 64(5): 599–610.

Konska, G., J. Guillot, M. Dusser, M. Damez and B. Botton. 1994. Isolation and characterization of an N-acetyllactosamine-binding lectin from the mushroom *Laetiporus sulfureus. J. Biochem.*, 116(3): 519–523.

Krings, U., A. Grimrath, S. Schindler and R.G. Berger. 2011. Volatiles responsible for the seasoning-like flavour of cell cultures of *Laetiporus sulphureus. Flavour Frag. J.*, 26(3): 174–179.

Lear, M.J., O. Simon, T.L. Foley, M.D. Burkart, T.J. Baiga, J.P. Noel, A.G. DiPasquale, A.L. Rheingold and J.J. La Clair. 2009. Laetirobin from the parasitic growth of *Laetiporus sulphureus* on *Robinia pseudoacacia. J. Nat. Prod.*, 72(11): 1980–1987.

Lee, S., J.S. Yu, S.R. Lee and K.H. Kim. 2022. Non-peptide secondary metabolites from poisonous mushrooms: Overview of chemistry, bioactivity, and biosynthesis. *Nat. Prod. Rep.*, 39(3): 512–559.

León, F., J. Quintana, A. Rivera, F. Estévez and J. Bermejo. 2004. Lanostanoid triterpenes from *Laetiporus sulphureus* and apoptosis induction on HL-60 human myeloid leukemia cells. *J. Nat. Prod.*, 67(12): 2008–2011.

Luangharn, T., S.C. Karunarathna, K.D. Hyde and E. Chukeatirote. 2014. Optimal conditions of mycelia growth of *Laetiporus sulphureus* sensu lato. *Mycology*, 5(4): 221–227.

Lung, M.Y. 2012. Antioxidant properties of polysaccharides from *Laetiporus sulphureus* in submerged cultures. *Afr. J. Biotechnol.*, 11(23): 6350–6358.

Lung, M.Y. and W.Z. Huang. 2011. Production, purification and tumor necrosis factor-α (TNF-α) release capability of exopolysaccharide from *Laetiporus sulphureus* (Bulliard: Fries) Bondartsev & Singer in submerged cultures. *Process Biochem.*, 46(2): 433–439.

Lung, M.Y. and W.Z. Huang. 2012. Antioxidant properties of polysaccharides from *Laetiporus sulphureus* in submerged cultures. *Afr. J. Biotechnol.*, 11(23): 6350–6358.

Mancheño, J.M., H. Tateno, I.J. Goldstein and J.A. Hermoso. 2004. Crystallization and preliminary crystallographic analysis of a novel haemolytic lectin from the mushroom *L. sulphureus*. *Acta Crystallogr.*, D60: 1139–1141.

Mancheño, J.M., H. Tateno, I.J. Goldstein, M. Martínez-Ripoll and J.A. Hermoso. 2005. Structural analysis of the *Laetiporus sulphureus* hemolytic pore-forming lectin in complex with sugars. *J. Biol. Chem.*, 280(17): 17251–17259.

Magotra, A., M. Kumar, M. Kushwaha, P. Awasthi, C. Raina, A.P. Gupta, B.A. Shah, S.G. Gandhi and A. Chaubey. 2017. Epigenetic modifier induced enhancement of fumiquinazoline C production in *Aspergillus fumigatus* (GA-L7): An endophytic fungus from Grewia asiatica L. *AMB Express*, 7: 1–10.

Matt, G. 1947. Pilze heilen Rheumatismus. *Schweizerische Zeitschrift für Pilzkunde*, 25: 167–168.

Min, J.S., J.K. Min, H.H. Lee, S.R. Kim, B.W. Kang, J.U. Park, E.J. Rhu, Y.H. Choi and K.J. Yong. 2010. Initial acidic pH is critical for mycelial cultures and functional exopolysaccharide production of an edible mushroom, *Laetiporus sulphureus* var *miniatus* JM 27. *J. Microbiol.*, 48(6): 881–884.

Mwangi, R.W., J.M. Macharia, I.N. Wagara and R.L. Bence. 2022. The antioxidant potential of different edible and medicinal mushrooms. *Biomed. Pharmacother.*, 147: 112621. doi.: 10.1016/j.biopha.2022.112621.

Nagamine, S., C. Liu, J. Nishishita, T. Kozaki, K. Sogahata, Y. Sato, A. Minami, T. Ozaki, C. Schmidt-Dannert, J.I. Maruyama and H. Oikawa. 2019. Ascomycete *Aspergillus oryzae* is an efficient expression host for production of basidiomycete terpenes by using genomic DNA sequences. *Appl. Environ. Microbiol.*, 85(15): e00409–19. doi: 10.1128/AEM.00409-19.

Nikitina, V.E., E.A. Loshchinina and E.P. Vetchinkina. 2017. Lectins from mycelia of basidiomycetes. *Int. J. Mol. Sci.*, 18(7): 1334. doi: 10.3390/ijms18071334.

Oh, D.C., C.A. Kauffman, P.R. Jensen and W. Fenical. 2007. Induced production of emericellamides A and B from the marine-derived fungus *Emericella* sp. in competing co-culture. *J. Nat. Prod.*, 70: 515–520. doi: 10.1021/np060381f

Okamura, T., T. Takeno, M. Dohi, I. Yasumasa, T. Hayashi, M. Toyoda, H. Noda, S. Fukuda, N. Horie and M. Ohsugi. 2000. Development of mushrooms for thrombosis prevention by protoplast fusion. *J. Biosci. Bioeng.*, 89(5): 474–478.

Olennikov, D.N., S.V. Agafonova, G.B. Borovskii, T.A. Penzina and A.V. Rokhin. 2009. Alkali-soluble polysaccharides of *Laetiporus sulphureus* BullFr Murr fruit bodies. *Appl. Biochem. Microbiol.*, 45(6): 626–630.

Olennikov, D.N., S.V. Agafonova, A.V. Stolbikova and A.V. Rokhin. 2010. Minor glucans from *Laetiporus sulphureus* basidiocarps. *Chem. Nat. Compd.*, 46(3): 444–445.

Olennikov, D.N., L.M. Tankhaeva and S.V. Agafonova. 2011a. Antioxidant components of *Laetiporus sulphureus* (Bull.: Fr.) Murr. fruit bodies. *Appl. Biochem. Microbiol.*, 47: 419–425.

Olennikov, D.N., S.V. Agafonova, A.V. Stolbikova and A.V. Rokhin. 2011b. Melanin of *Laetiporus sulphureus* (Bull: Fr) Murr sterile form. *Appl. Biochem. Microbiol.*, 47(3): 298–303.

Pavic, A., T. Ilic-Tomic and J. Glamočlija. 2021. Unravelling anti-melanogenic potency of edible mushrooms *Laetiporus sulphureus* and *Agaricus silvaticus in vivo* using the zebrafish model. *J. Fungi*, 7(10): 834. doi: 10.3390/jof7100834.

Petrović, J., J. Glamočlija, D.S. Stojković, A. Ćirić, M. Nikolić, D. Bukvički, M.E. Guerzoni and M.D. Soković. 2013. *Laetiporus sulphureus*, edible mushroom from Serbia: Investigation on volatile compounds, *in vitro* antimicrobial activity and in situ control of *Aspergillus flavus* in tomato paste. *Food Chem. Toxicol.*, 59: 297–302.

Petrović, J., J. Glamočlija, T. Ilić-Tomić, M. Soković, D. Robajac, O. Nedić and A. Pavić 2020. Lectin from *Laetiporus sulphureus* effectively inhibits angiogenesis and tumor development in the zebrafish xenograft models of colorectal carcinoma and melanoma. *Int. J. Biol. Macromol.*, 148: 129–139.

Petrović, J., M. Papandreou, J. Glamočlija, A. Ćirić, C. Baskakis, C. Proestos, F. Lamari, P. Zoumpoulakis and M. Soković. 2014a. Different extraction methodologies and their influence on the bioactivity of the wild edible mushroom *Laetiporus sulphureus* (Bull) Murrill. *Food Funct.*, 5(11): 2948–2960.

Petrović, J., D. Stojković, F.S. Reis, L. Barros, J. Glamočlija, A. Ćirić, I.C. Ferreira and M. Soković. 2014b. Study on chemical, bioactive and food preserving properties of *Laetiporus sulphureus* (Bull.: Fr.) Murr. *Food Funct.*, 5(7): 1441–1451.

Pleszczyńska, M., A. Wiater, M. Siwulski and J. Szczodrak. 2013. Successful large-scale production of fruiting bodies of *Laetiporus sulphureus* (Bull.: Fr.) Murrill on an artificial substrate. *World J. Microbiol. Biotechnol.*, 29(4): 753–758.

Popa, G., C.P. Cornea, G. Luta, E. Gherghina, F. Israel-Roming, C. Bubueanu and R. Toma. 2016. Antioxidant and antimicrobial properties of *Laetiporus sulphureus* (Bull.) Murrill. *AgroLife Sci. J.*, 5(1): 168–173.

Rios, J., I. Andujar, M.C. Recio and R.M. Giner. 2012. Lanostanoids from fungi: A Group of potential anticancer compounds. *J. Nat. Prod.*, 75: 2016–2044.

Saba, E., Y. Son, B.R. Jeon, S.E. Kim, I.K. Lee, B.S. Yun and M.H. Rhee. 2015. Acetyl eburicoic acid from *Laetiporus sulphureus* var. *miniatus* suppresses inflammation in murine macrophage RAW 264.7 cells. *Mycobiology*, 43(2): 131–136.

Scherlach, K. and C. Hertweck. 2020. Chemical mediators at the bacterial-fungal interface. *Annu. Rev. Microbiol.*, 74: 267–290.

Scherlach, K. and C. Hertweck. 2021. Mining and unearthing hidden biosynthetic potential. *Nat. Commun.*, 12(1): 1–12.

Schindhelm, R.K., M. Diamant, J.M. Dekker, M.E. Tushuizen, T. Teerlink and R.J. Heine. 2006. Alanine aminotransferase as a marker of non-alcoholic fatty liver disease in relation to type 2 diabetes mellitus and cardiovascular disease. *Diabetes-Metab. Res.*, 22(6): 437–443.

Schmidt-Dannert, C. 2015. Biosynthesis of terpenoid natural products in fungi. *Adv. Biochem. Eng. Biot.*, 148: 19–61.

Seibold, P.S., C. Lenz, M. Gressler and D. Hoffmeister. 2020. The *Laetiporus* polyketide synthase LpaA produces a series of antifungal polyenes. *J. Antibiot.*, 73(10): 711–720. doi: 10.1038/s41429-020-00362-6.

Seo, M.J., B.W. Kang, J.U. Park, M.J. Kim, H.H. Lee, Y.H. Choi and Y.K. Jeong. 2011. Biochemical characterization of the exopolysaccharide purified from *Laetiporus sulphureus* mycelia. *J. Microbiol. Biotechnol.*, 21(12): 1287–1293.

Seo, M.J., M.J. Kim, H.H. Lee, S.R. Kim, B.W. Kang, J.U. Park, E.J. Rhu, Y.H. Choi and Y.K. Jeong. 2010. Initial acidic pH is critical for mycelial cultures and functional exopolysaccharide production of an edible mushroom, *Laetiporus sulphureus* var. *miniatus* JM 27. *J. Microbiol.*, 48(6): 881–884.

Shiono, Y., Y. Tamesada, Y.D. Muravayev, T. Murayama and M. Ikeda. 2005. N-Phenethylhexadecanamide from the edible mushroom *Laetiporus sulphureus*. *Nat. Prod. Res.*, 19(4): 363–366.

Simon, O., B. Reux, J.J. La Clair and M.J. Lear. 2010. Total synthesis confirms laetirobin as a formal Diels–Alder adduct. *Chem. Asian J.*, 5(2): 342–351.

Sinanoglou, V.J., P. Zoumpoulakis, G. Heropoulos, C. Proestos, A. Ćirić, J. Petrovic, J. Glamoclija and M. Sokovic. 2015. Lipid and fatty acid profile of the edible fungus *Laetiporus sulphurous*. Antifungal and antibacterial properties. *J. Food Sci. Technol.*, 52(6): 3264–3272.

Singh, R.S., R. Bhari and H.P. Kaur. 2011. Current trends of lectins from microfungi. *Crit. Rev. Biotechnol.*, 31(3): 193–210.

Singh, S.S., H. Wang, Y.S. Chan, W. Pan, X. Dan, C.M. Yin, O. Akkouh and T.B. Ng. 2014. Lectins from edible mushrooms. *Molecules*, 20(1): 446–469.

Sułkowska-Ziaja, K., B. Muszyńska, A. Gawalska and K. Sałaciak. 2018. *Laetiporus sulphureus*: Chemical composition and medicinal value. *Acta Sci. Pol. Hortorum Cultus.*, 17(1): 87–96.

Sun, W.J., H.B. He, J.Z. Wang, L. Wu, F. Cheng and Z.S. Deng. 2014. The main components analysis of *Laetiporus sulphureu* crude extract and its hepatoprotective effect on carbon tetrachloride-induced hepatic fibrosis in rats. *Appl. Mech. Mater.*, 568-570: 1934–1939.

Tateno, H. and I.J. Goldstein. 2003. Molecular cloning, expression, and characterization of novel hemolytic lectins from the mushroom *Laetiporus sulphureus*, which show homology to bacterial toxins. *J. Biol. Chem.*, 278(42): 40455–40463.

Turkoglu, A., M.E. Duru, N. Mercan, I. Kivrak and K. Gezer. 2007. Antioxidant and antimicrobial activities of *Laetiporus sulphureus* (Bull.) Murrill. *Food Chem.*, 101(1): 267–273.

Valadon, L.R.G. and R.S. Mummery. 1969. A new carotenoid from *Laetiporus sulphureus*. *Bot. Ann.*, 5: 879–882.

Vasaitis, R., A. Menkis, Y.W. Lim, S. Seok, M. Tomsovsky, L. Jankovsky, V. Lygis, B. Slippers and J. Stenlid. 2009. Genetic variation and relationships in *Laetiporus sulphureus* s lat, as determined by ITS rDNA sequences and *in vitro* growth rate. *Mycol. Res.*, 113(3): 326–336.

Wang, J., W. Sun, H. Luo, H. He, W.Q. Deng, K. Zou, C. Liu, J. Song and W. Huang 2015. Protective effect of eburicoic acid of the chicken of the woods mushroom, *Laetiporus sulphureus* (higher Basidiomycetes), against gastric ulcers in mice. *Int. J. Med. Mushrooms*, 17(7): 619–626.

Wang, Y., B. Wu, J. Shao, J. Jia, Y. Tian, X. Shu, X. Ren and Y. Guan. 2018. Extraction, purification and physicochemical properties of a novel lectin from *Laetiporus sulphureus* mushroom. *Lwt*, 91: 151–159.

Wang, Y., Y. Zhang, J. Shao, B. Wu and B. Li. 2019. Potential immunomodulatory activities of a lectin from the mushroom *Latiporus sulphureus*. *Int. J. Biol. Macromol.*, 130: 399–406.

Watling, R. 1997. Poisoning by fungi: Interesting cases. *Mycologist*, 11: 101.

Weber, R.W.S., A. Mucci and P. Davoli. 2004. Laetiporic acid, a new polyene pigment from the wood-rotting basidiomycete *Laetiporus sulphureus* (Polyporales, Fungi). *Tetrahedron Lett.*, 45(5): 1075–1078.

Wiater, A., A. Waśko, P. Adamczyk, K. Gustaw, M. Pleszczyńska, K. Wlizło, M. Skowronek, M. Tomczyk and J. Szczodrak. 2020. Prebiotic potential of oligosaccharides obtained by acid hydrolysis of α-(1→3)-Glucan from *Laetiporus sulphureus*: A pilot study. *Molecules*, 25(23): 5542. doi: 10.3390/molecules25235542.

Wiater, A., M. Pleszczyńska, J. Szczodrak and G. Janusz. 2012. Comparative studies on the induction of *Trichoderma harzianum* mutanase by α-(1→3)-glucan-rich fruiting bodies and mycelia of *Laetiporus sulphureus*. *Int. J. Mol. Sci.*, 13(8): 9584–9598.

Yin, X., Z.H. Li, Y. Li, T. Feng and J.K. Liu. 2015. Four lanostane-type triterpenes from the fruiting bodies of mushroom *Laetiporus sulphureus* var. *miniatus*. *J. Asian Nat. Prod. Res.*, 17(8): 793–799.

Ying, J., X. Mao, Q. Ma, Y. Zong and H. Wen. 1987. Icones of Medical Fungi from China. Science Press, Beijing, China.

Yokoyama, A., S. Natori and K. Aoshima. 1975. Distribution of tetracyclic triterpenoids of lanostane group and sterols in the higher fungi especially of the polyporaceae and related families. *Phytochemistry*, 14(2): 487–497.

Yoshikawa, K., S. Bando, S. Arihara, E. Matsumura and S. Katayama. 2001. A benzofuran glycoside and an acetylenic acid from the fungus *Laetiporus sulphureus* var. *miniatus*. *Chem. Pharm. Bull.*, 49(3): 327–329.

Zhang, J., J. Lv, L. Zhao, X. Shui and L.A. Wang. 2018. Antioxidant and antimicrobial activities and chemical composition of submerged cultivated mycelia of *Laetiporus sulphureus*. *Chem. Nat. Compd.*, 54(6): 1187–1188.

Zhang, J.W., G.L. Wen, L. Zhang, D.M. Duan and Z.H. Ren. 2015. Sulphureuine B, a drimane type sesquiterpenoid isolated from *Laetiporus sulphureus* induces apoptosis in glioma cells. *Bangladesh J. Pharmacol.*, 10(4): 896–902.

Zhang, M.M., F.T. Wong, Y. Wang, S. Luo, Y.H. Lim, E. Heng, W.L. Yeo, R.E. Cobb, B. Enghiad, E.L. Ang and H. Zhao. 2017.CRISPR-Cas9 strategy for activation of silent Streptomyces biosynthetic gene clusters. *Nat. Chem. Biol.*, 13(6): 607–609.

Zhao, H., Y. Lan, L. Hui, Y. Zhu, W. Liu, J. Zhang and L. Jia. 2017. Antioxidant and hepatoprotective activities of polysaccharides from spent mushroom substrates (*Laetiporus sulphureus*) in acute alcohol-induced mice. *Oxid. Med. Cell Longev.*, 2017: 5863523. doi: 10.1155/2017/5863523.

Zschätzsch, M., S. Steudler, O. Reinhardt, P. Bergmann, F. Ersoy, S. Stange, A. Wagenführ, T. Walther, R.G. Berger and A. Werner. 2021. Production of natural colorants by liquid fermentation with *Chlorociboria aeruginascens* and *Laetiporus sulphureus* and prospective applications. *Eng. Life Sci.*, 21(3-4): 270–282.

Biocomposites

10

Experimental Analysis of the Mechanics of Mycelium-based Biocomposites

Libin Yang[1,3] and *Zhao Qin*[1,2,3,]*

1. Introduction

Due to the restricted methods of generating construction materials and the rising demand by the worldwide population, the construction sector has been under intense strain during the past ten years (Madurwar et al. 2013, Pheng and Hou 2019). Traditional building materials like steel and concrete require a lot of energy. It pollutes our environment in ways embodied carbon can measure and trace, restricting its widespread production and utilization (Madurwar et al. 2013, Maraveas 2020). In addition, as the world's population is growing quickly, more agricultural products are consumed annually, producing more byproducts (e.g., rice husks, cotton stalks, and straw). Most of these byproducts are tracked as purely agricultural wastes that are primarily discarded or burned, producing carbon dioxide, atmospheric particulates, and other greenhouse gases (Defonseka 2014, Bhuvaneshwari et al. 2019, Maraveas 2020). Besides being used in low-quality building materials for infrastructure (e.g.,

[1] Department of Civil and Environmental Engineering, Syracuse University, Syracuse, NY 13244, United States of America.
[2] The BioInspired Institute, Syracuse University, Syracuse, NY 13244, United States of America.
[3] Laboratory for Multiscale Material Modeling, Syracuse University, Syracuse, NY, 13244, United States of America.
* Corresponding author: zqin02@syr.edu

brick elements and green concrete for low-rise buildings, insulation materials particleboard for non-structural applications, and fillings for road construction), they also have been used in part as an additive to fertilizers, animal bedding, and low-quality building materials for infrastructure (e.g., The local bitumen road that contains rice hull ash can bear a higher load and have water resistance) (Defonseka 2014).

Due to its low energy requirement for development, lack of byproducts, and wide range of potential applications, mycelium has recently attracted increasing interest in academic and commercial investigations (Holt et al. 2012, Pelletier et al. 2013, Jones et al. 2017, Nawawi et al. 2020)(Figure 1). The vegetative portion of a fungus is called mycelium, and it is made up of a network of tiny, white filaments that range in diameter from 1 to 30 μm and extend out from a single spore into every crevice of the substrate (Fricker et al. 2007, Islam et al. 2017). Proteins, glucans, and chitin are the various layers that make up each mycelium filament. The chemical compositions of different layers may differ from different species of mycelium (Haneef et al. 2017). The mycelium network's growth is nourished by the substrate made of organic material. These organic substances originate from the remains of living things like plants and animals and their waste products generated in the environment (Swift 1996, Steigerwald 2014). Their elemental composition comprises numerous proteins, lipids, and carbohydrates in addition to cellulose, tannin, lignin, and cutin. (Sejian et al. 2015). The general procedure used to grow the

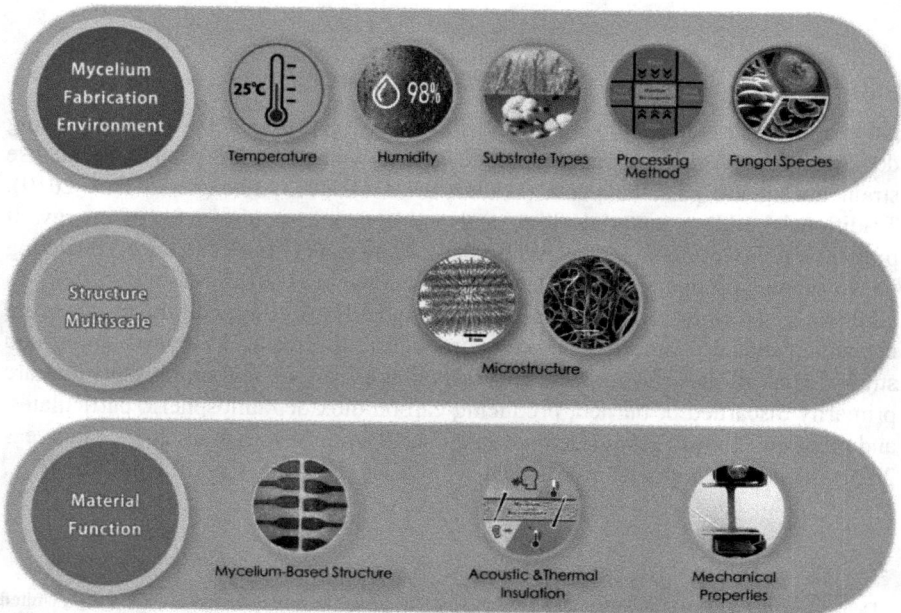

Figure 1: The mycelium study, including its multiscale structure, material function, and how environmental factors define these characteristics. The mycelium study, including its multiscale structure, material function, and how environmental factors define these characteristics. Using different substrate to create the dog bone sample to test the mechanical properties (Yang et al. 2021).

mycelium composite is similar to the standard protocol of raising mushrooms, which includes: (1) inoculating the culturing dish with mushroom spores and sufficient nutrients and water. The incubation time for the mycelium to completely cover the dish is about 7–14 days. (2) Prepare the sterilized growing substrate composed of various organic matters (e.g., brown rice, roasted buckwheat, wheat, and straw) and transfer a small piece of mycelium sample cut from the culturing dish into the growing substrate for further incubation. (3) When the substrate is full of mycelium, it is dried at a high temperature for several hours to inactivate the hyphae and stop the growth process before gaining the mycelium composite. Humidity and temperature are two important factors that can affect mycelium growth during the second stage. High humidity (relatively humidity 98%) and warm room temperature (24–25 degrees) with fresh air provide an excellent environment for growing mycelium (Hoa and Wang 2015).

The mycelium-based material can achieve certain structural and material properties by adjusting the substrate and processing technique. Mycelium combines with organic matter from industrial and agricultural wastes mix to create a bio-composite that may be utilized to create low-value materials (e.g., packaging and gap fillers) and high-value composite materials for structural purposes (Holt et al. 2012, Pelletier et al. 2013, Haneef et al. 2017, Islam et al. 2017, Jones et al. 2017). Unlike metal alloy or polymer composite that require energy or complex equipment to melt the raw material and mix different parts, one can uniformly mix various components in the form of small pieces to form the substrate before growing mycelium, which naturally binds and integrates the elements during its growth. Various substrates can perform particular behaviors by cultivating different mycelium species to form bio composites (e.g., structural support, fire resistance, and acoustic insulation).

Two different mycelium-based composite materials have been studied and produced for the construction: mycelium-based foam and mycelium-based sandwich composites are shown in Figure 2 (Girometta et al. 2019). Mycelium-based foam (MBF) is made by growing fungi homogenously in agricultural wastes in small pieces (Appels et al. 2019). As the mycelium network grows, it produces fibers that bind these pieces together to form a porous material (Bartnicki-Garcia 1968, Jiang et al. 2017, Karana et al. 2018). Mycelium-based sandwich composites (MBSC) add natural fiber fabric (e.g., jute, hemp, and cellulose) as the top and bottom layers aside from the central core as the agricultural wastes combined with mycelium form a sandwich structure of higher bending rigidity (Jiang et al. 2017). Both MBF and MBSC as the "mycelium bricks" or "panels" have shown mechanical strength, lightweight and environmental advantages in packing, building insulation, and interior design in comparison to expanded polystyrene (EPS) foams (Holt et al. 2012, Jones et al. 2017, 2018, Xing et al. 2018, Girometta et al. 2019).

As one of the main building blocks of mycelium, chitin is a natural polymer abundantly found in both fungal cell walls and exoskeletons of crustaceans (Jones et al. 2020a). It has been applied to biomedical applications (Morganti and Morganti 2008, Danti et al. 2019, Azimi et al. 2020, Jones et al. 2020a). Chitin and its derivative chitosan are both long linear macromolecules that can be used to make fibers for wound dressing by electrospinning (Naseri et al. 2014, Danti et al. 2019, Morganti

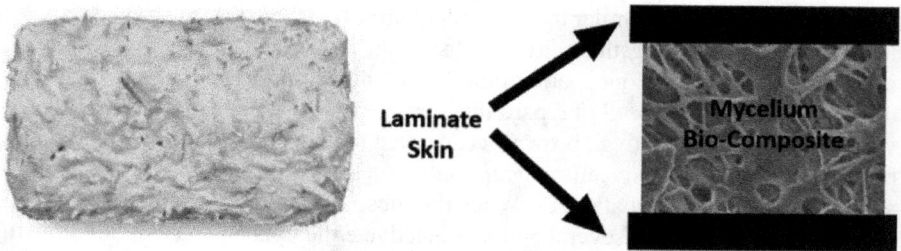

Figure 2: A snapshot of the mycelium-based foam (left) (Karana et al. 2018) (reused under a Creative Commons Attribution License) and a schematic of the mycelium-based sandwich composite (right).

et al. 2019, Azimi et al. 2020), which is a fiber production method that uses the force from an electric field to draw charged polymer chains from solutions to form a continuous fiber as a bundle of aligned chains (Wang et al. 2019). Chitin has been used to produce nonwoven cloths and gels to cover the wound and interact with the open tissue for healing, making it necessary to look into their multiscale structures at the interface with biological tissues (Muzzarelli et al. 2007, Jayakumar et al. 2011, Muzzarelli 2012). Both crustacean chitin and fungal chitin are applied in wound-dressing research, but there are significant differences in the structure, properties, and processing between them (Morin-Crini et al. 2019, Jones et al. 2020a). Both need to be extracted from the compound, as crustacean chitin often binds with sclerotized proteins and minerals, while fungal chitin binds to other polysaccharides (e.g., glucans) (Muzzarelli 2011). Highly purified crustacean-derived chitin and chitosan have been used more widely. Still, less research has been done on fungi-derived chitin, even though the extraction process of fungal chitin is more straightforward (Di Mario et al. 2008, Hassainia et al. 2018, Jones et al. 2020a, Azimi et al. 2021). Also, fungi-derived chitin has advantages in quantity and availability, as the growth of mycelium is not subjected to seasonal and regional restrictions as for that of crustaceans (Di Mario et al. 2008, Hassainia et al. 2018).

2. Mycelium Growth and Multiscale Structural Features

2.1 Substrate Types

The cotton, grains, wheat, coffee granules, and flax residues can be used as the substrate to grow the mycelium, a mixture consists of agricultural as lignocellulosic wastes (Jiang et al. 2017, Appels et al. 2019, Girometta et al. 2019). Due to fungi's ability for degrading cellulose or lignin in plant bio-mass, lignocellulosic waste is frequently used as the substrate for mycelium-based foams (Girometta et al. 2017). More effectively than brown-rot and soft-rot fungi, the white-rot fungi have capability to break down the lignin, cellulose, and hemicellulose's intricate molecular structures. Short sugars and starches are digested before complex polysaccharides because they are more available (Couturier and Berrin 2013, Xu et al. 2013). The fungi type and carbohydrate composition of each plant materials determine the degradation rates of short sugars and complex polysaccharides (Hori et al. 2013). Moreover, the final bio-composite can also be influenced by the type, dimension, and processing method of substrate particles (Elsacker et al. 2019). Compared with wood particles, using grains as the substrate for inoculum or nutrition will increase the overall density by up to twofold (Girometta et al. 2019).

2.2 Humidity and Temperature for Mycelium Growth

Mycelium growth can be impacted by several crucial elements, including humidity and temperature. A rather high-humidity environment should be maintained while mycelium grows. Therefore, for mycelium growth, humidifiers or sprinkler systems are typically employed. For instance, Jiang et al., used a semi-permeable polypropylene bag to provide a high humidity (up to 98 percent relative humidity) and a sterile environment for the respiration of mycelium fungi growing (Jiang et al. 2017). Room temperature (24–25°C) is ideal for the growth of mycelium (Hoa and Wang 2015).

2.3 Fabrication Process

Various fabrication techniques can achieve different functions of mycelium-based bio-composites. The oven can be used to dry the mycelium-based bio-composite. It is the most common method to remove the remaining water within the mycelium and substrate. Moreover, the sandwich structure, MBSC, can be created by consolidating with natural fiber on both sides of the mycelium-based foam to develop lightweight and high-strength foams (Attias et al. 2020). The mycelium not only links the small particle in the substrate together but also associate the core material with the fiber fabrics (through the interface created during mycelium growth). Since the mycelium connected the core foam with fiber layers, the structure is remodeled as a strong composite board with high bending stiffness. When the sandwich plate is subjected to shear forces, it can prevent delamination at the material interface (Jiang et al. 2017). To increase the bio-composites adhesion with mycelium foam, other natural glues can be added, for example, bio-resin, before combining the fabrics to the core part. The large cohesive zone in adhesion can prevent the fabric from easily separating from the foam part through a sharp single crack from defects. Moreover, the large cohesive zone cannot prevent fungus from spreading through additional fiber fabric layers, which is critical (Jiang et al. 2019).

2.4 Structural Features of Mycelium at Different Scales

Mycelium's intricate network structure is responsible for its diverse material functions at various scales. Branching filaments and the network structure's topology control the mycelium's mechanical characteristics (Islam et al. 2017). From the macro to the nanoscale, Figure 3 depicts the general organization of the mycelium. The contact area with the complex porous substrate can be increased because the mycelium has a symbiosis relationship with the substrate to produce the branch fiber network structure. From a spore, the mycelium network spreads outward through the cell wall and membrane at the tip of a mycelium fiber. Every single mycelium fiber is composed of an array of slender cells separated by cross walls, the septum, and enclosed within the same cell wall. The septum acts as a cross wall between the array of slender cells that make up the single mycelium fiber and are enclosed within the same cell wall. Rapid movement of nutrients, water, and other small molecules between cells and mycelium fiber is made possible by tiny holes in the septum. A layer of chitin, a layer of glucans, and a layer of proteins (e.g., mannoproteins and hydrophobins) on the cell membrane make up the cell wall, which shields the mycelium and gives mechanical strength (Haneef et al. 2017). Chitin is a complex polysaccharide, a polymer of N-acetylglucosamine that locates on the cell membrane and is crucial in giving the fungus cell walls their structural integrity and strength. We can observe the pure mycelium network without other substrate fibers by using the agar plate as the inoculated substrate for 7 to 14 days. Using the scanning electron microscope, we can observe that the mycelium network comprises many mycelium fibers with a diameter of about 2 μm. Then, to create a mycelium composite and mushrooms (e.g., the king oyster mushroom, as seen in the image in Figure 3's lower right corner), we migrate the mycelium network to the substrate in the lab. This enables us to conduct mechanical tests and capture microscopic images of the mycelium network at a large scale. We perform the tensile test of the mycelium-based bio composites dog bone sample to obtain mechanical properties. The rectangular figure in the upper left corner shows the nanostructure of the α-chitin. Two main polymorphic forms of chitin exist which are α and β. The α-chitin is the most common polymorph for both crustacean and fungal chitin. The β-chitin occurs only in squid pens, sea tube worms, and some algae (centric diatom) (Rinaudo 2006).

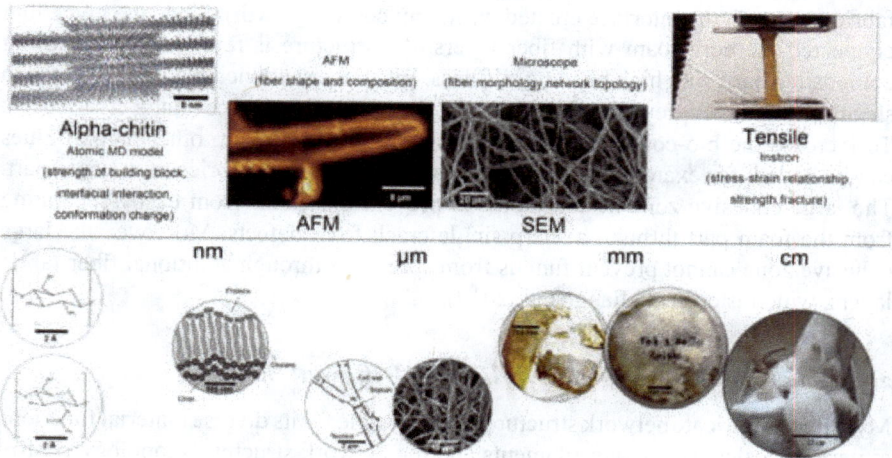

Figure 3: Multiscale structure of the mycelium (Yang et al. 2021). From the bottom left, the figure shows the molecular formula of chitin and chitosan, followed by two figures showing the structure of the single mycelium and mycelium cell wall and the SEM image of the mycelium network, as well as two figures showing that the wet and dry mycelium samples in the culturing disk. The figure in the right-up corner shows the cultured king oyster mushroom. Next to it is the snapshot of the tensile test of the mycelium-based dog bone sample. We can use different research methods to study the structure-function relationship of the mycelium network at different scale levels, as noted at the axis of the plot. (The AFM figure is reproduced under a Creative Commons Attribution License (Haneef et al. 2017).

The fungi's cell wall's innermost layer, chitin, can serve as reinforcement and strength. [(1-4) connected N-acetyl-2-amino-2-deoxy-D-glucose] units make up the biopolymer known as chitin (Dhillon et al. 2013). Smaller monomers combine to generate chitin, a structural polymer that helps create strong fibers. The fibers can form weak bonds to increase the strength of the integrated structure when the cells secrete them from inside or outside in an organized way (Karana et al. 2018). Chitosan application development has grown significantly in recent years, particularly in wound healing (Jones et al. 2020a). The advantages of obtaining the chitin from fungus outweigh that from the shells of the crustaceans; however, still, so many suppliers obtain it from crustacean shells (Dhillon et al. 2013). The most important advantage of getting chitin from the fungus is they are not limited to season and location. The benefits and drawbacks of getting chitosan from fungus and crustacean shells are listed in Table 1.

Chitin from crustaceans often binds with sclerotized proteins and minerals and contains minimal residual protein. Isolate the chitin from fungus becomes more accessible due to the difference between the fungus and crustaceans. The proteins can be efficiently removed using a kitchen blender for a brief period of mechanical agitation and following light, alkaline treatment (Fazli Wan Nawawi et al. 2019, Nawawi et al. 2020). However, the quantities of chitin can be exceeded by the glucan associated with fungal chitin (Hackman 1960, Attwood and Zola 1967, Kramer et al. 1995, Percot et al. 2003, Muzzarelli 2011). Moreover, the extracted α, β, and γ chitin can have different secondary structures, except for the most common polymorph α-chitin, squid pen, sea tube worms, and some algae contain the β-chitin (Rinaudo 2006). The molecular structure of α chitin and β chitin are shown in Figure 4. The figure shows that the neighboring chains of the α chitin are in antiparallel directions. In opposition, the chains are parallel in the β chitin (Figure 6), which means the primary difference between these two chitins is in the secondary structure (Rinaudo 2006). Additionally, parallel and antiparallel chains are present in the γ chitin (Rinaudo 2006). Due to this structural difference, the adjacent amide groups between nearby chains in the α chitin parallel, which do not in the β chitin, associating with the flexibility of the β chitin (Elieh-Ali-Komi and Hamblin 2016).

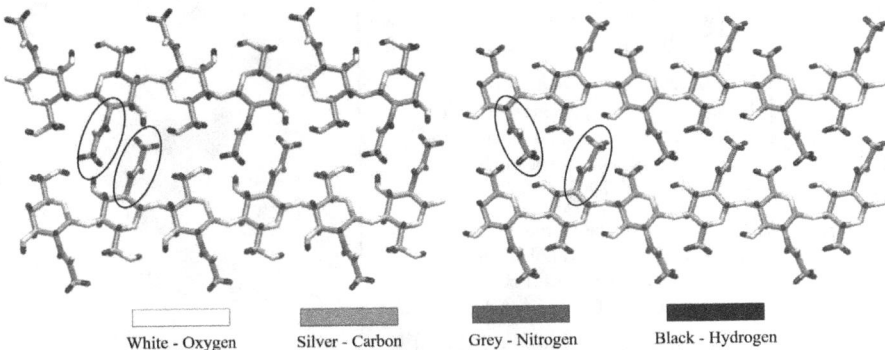

White - Oxygen   Silver - Carbon   Grey - Nitrogen   Black - Hydrogen

Figure 4: The molecular structure of α chitin (left) and β chitin (right). Each atom is colored according to its type, from light to dark, with white for oxygen, silver for carbon, grey for nitrogen, and black for hydrogen.

Table 1: Advantages and disadvantages of chitosan from crustacean shells and fungi.

Source	Advantage	Disadvantage
Crustacean Shells	Well-established method for industrial production of chitosan (Kilavan Packiam et al. 2011, Aranaz et al. 2012).	Seasonal and limited supply, high cost and laborious process and not environmentally friendly (Pochanavanich and Suntornsuk 2002, Streit et al. 2009, Kilavan Packiam et al. 2011).
		Large quantities of chemicals, such as alkali and acids, higher temperatures, and long processing time are required for extraction. Generally, alkali concentration of 30–50% w/v and temperature 100°C is required (Pochanavanich and Suntornsuk 2002, Streit et al. 2009, Aranaz et al. 2012).
		Demineralization treatment is required to remove calcium carbonate, which accounts for 30–50% of crustacean shells (Pochanavanich and Suntornsuk 2002, Streit et al. 2009, Aranaz et al. 2012).
		It possesses high molecular weight and protein contamination, limiting its applications in biomedicines (Pochanavanich and Suntornsuk, 2002, Streit et al. 2009).
Fungi	The medium-low molecular weight is suitable for many biomedical applications (Pochanavanich and Suntornsuk 2002).	Processes not scaled up to industrial level (Kilavan Packiam et al. 2011).
	A higher degree of deacetylation can be achieved (Pochanavanich and Suntornsuk 2002).	
	Free of allergenic shrimp protein (Pochanavanich and Suntornsuk 2002, Streit et al. 2009).	
	The molecular weight and degree of deacetylation of fungal chitosan can be controlled by varying the fermentation conditions (Pochanavanich and Suntornsuk 2002, Streit et al. 2009).	
	The supply of fungal biomass is infinite, mainly from the biotechnological and pharmaceutical industries (Pochanavanich and Suntornsuk 2002, Streit et al. 2009).	
	Cheap biowastes can be used as economic substrates for culturing fungi (Pochanavanich and Suntornsuk, 2002, Streit et al. 2009).	

3. Mycelium Mechanics and Effect of Water

3.1 Mycelium Water Content and Effective for Mechanical Properties

After naturally growing, mycelium contains a lot of water (over 60% (Elsacker et al. 2019)). Most of the water must be removed to inactivate its growth to gain a high and reliable mechanical performance. To achieve the high and dependable mechanical properties of the final mycelium-based biocomposite materials, most of the water must be removed to inactivate its growth. The ultimate residual moisture percentage in MBF or MBSC is not mentioned in the current literature. However, it must be sufficiently dry to stop fungal development (Girometta et al. 2019). The final water content of the mycelium was determined by the substrate and the type of fungi. For example, a hemp pulp substrate absorbs more water than a cotton wool substrate (Ziegler et al. 2016). Additionally, the way that moisture absorbs may vary depending on the coating. The moisture content before deactivation is typically between 59 and 80% (Deacon 2013, Velasco et al. 2014). However, researchers have found that the residual proportion in the finished product is only about 10 to 15 percent (Deacon 2013). Therefore, the significant influence on the mechanical properties of the mycelium samples is the water content of the final mycelium-based bio-composites.

According to prior studies, the manufacturing procedure affects the water content in the mycelium network, which substantially impacts its mechanical properties. According to Appels et al., pressing can significantly influence the water content, hence mycelium composites' mechanical characteristics (Appels et al. 2019). This outcome is anticipated since pressing can remove water and air from the porous mycelium network, reduce porosity, increase density, and increase Young's modulus and strength of the material (Gibson and Ashby 1982, Dai et al. 2007, Qin et al. 2017). Additionally, pressing encourages the horizontal reorientation of fibers in the panel plane (Butterfield et al. 1992) and reduces plate thickness, which improves fiber connection between the walls of the fibers at overlap spots (Carvalho and Costa 1998). Large voids that are structural flaws in the mycelium composite may be reduced by pressing to prevent the crack from forming after loading (Girometta et al. 2019, Jones et al. 2020b). According to an early study, hot pressing can enhance mechanical qualities compared to cold pressing by applying pressure and high temperature from two hot plates (Appels et al. 2019).

3.2 King Oyster Mushroom Water Content and Tensile Test

Since the water content is an important factor that can affect the mycelium-based bio-composite material, we perform our tests to understand the water loss of the pure mycelium network within the mushroom samples after baking for a certain amount of time, as shown in the Figure 5. We use the king oyster (*Pleurotus eryngii*) mushrooms. The reason that we used this fungal is based on its character and shape, it can be easily used in the tensile test, and based on the literature review it is more commonly used in the experiment than the other species. Preparing groups of samples with a total weight of $M_0 \approx 100$ g for each group, keep the temperature at an elevated level of constant 80°C in an oven and bake the samples for different amounts of time (t) before measuring and recording the weight of the residue materials ($M(t)$). We intentionally take this temperature to avoid breaking the molecular structure of mycelium. The percentage of water loss is thereafter defined by $P_{wat} = [M_0 - M(t)]/M_0$. Every 15 minutes of baking, the two sets of samples (skin and core of the mushroom) were taken out and weighted. Repeat the measurement until baking for 4 hours, when the P_{wat} curve starts to converge without further changing. The curve in Figure 5 shows that the total water loss of the samples near the skin of the mushroom goes up to $P_{wat} = 90\%$ after 4 hours of baking and the sample near the core has $P_{wat} = 91\%$, which is not so different from the skin samples. Also, we notice that baking the samples for more than 4 hours will not generate more water loss, suggesting the water content in the natural mycelium of this king oyster mushroom is ~ 90%, which is even higher than many of the hydrogels (Wu et al. 2019, Liu et al. 2020). By using the scanning electron microscope (SEM), it is shown that the natural mycelium within the mushroom is represented by a fully connected network of tubes partially filled by water, which become an array of flatten ribbons once lose water (Figure 5).

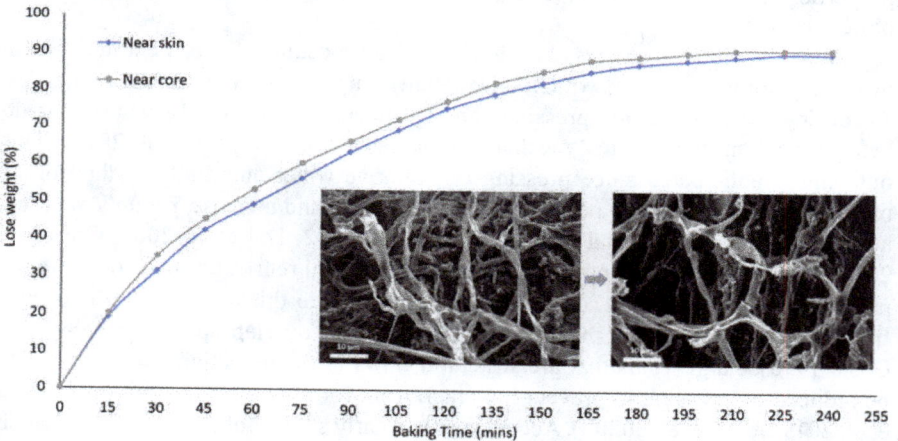

Figure 5: The amount of water loss percentage compared to the original weight (Yang et al. 2021). Inserted figure (left): the natural mycelium fiber from the skin of king oyster mushroom. Inserted figure (right): baked 30 minutes mycelium fiber from the skin of king oyster mushroom. It is shown that the natural fibers represent tubes with naturally bended overall conformation while the dry ones become a flat ribbon curved up in the radial direction with a straight overall conformation, suggesting the larger bending stiffness of the fiber.

After we test the water content, we perform our mechanical test to samples taken from the skin and middle parts of *Pleurotus Eryngii* mushrooms (king oyster, as shown in Figure 6A) to better understand the mechanics of mycelium with different water content and thus material density. We use an Instron 5966 machine (10 KN static load cell, 1 KN pneumatic grips with 90 psi holding pressure) to stretch all the material samples to get their stress-strain curves in tension. We measure the initial sample length as the distance between the edges of the two grips as L_0, zero the force before clamping and zero the displacement before the test. The lower grips during the test are fixed by a pin and the upper grips move at a constant displacement speed of $v = 2$ *mm/min*. Our mechanical testing results, as summarized in Figures 6B and C for the critical stress and strain of these samples after baking for different amounts of time, corresponds to a certain amount of water loss (Figure 5).

Table 2: The main mechanical properties result of different conditions of *Pleurotus eryngii* (Yang et al. 2021).

Sample conditions	Natural		Dry 30 min		Dry 40 min		Dry 50 min	
	skin	core	skin	core	skin	core	skin	core
Percentage of water loss (%)	0	0	31	35	38.4	42	45	47.5
Young's Modulus (MPa)	2.65	1.19	5.91	15.11	57.99	54.32	64.42	57.79
Ultimate stress (MPa)	0.28	0.18	2.09	3.32	5.11	4.58	6.29	7.73
Critical strain (%)	8.55	11.65	49.88	46.60	18.41	21.94	15.84	25.48
Critical stress (MPa)	0.18 ± 0.07	0.11 ± 0.04	1.56 ± 0.41	2.49 ± 0.68	4.14 ± 0.98	4.26 ± 0.30	4.07 ± 1.35	5.80 ± 1.36

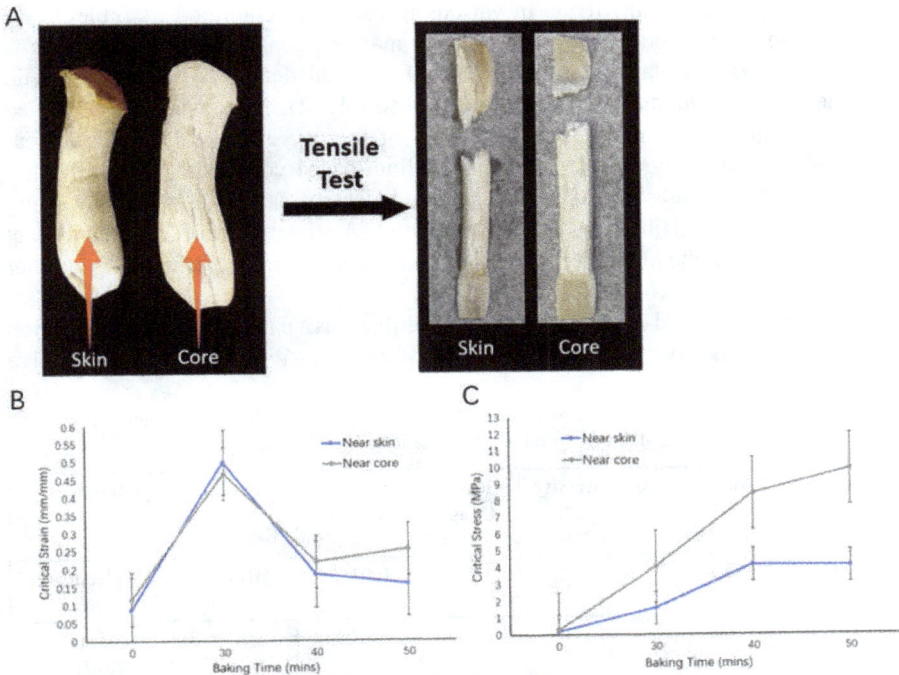

Figure 6: Tensile tests of *Pleurotus eryngii* samples after low-temperature baking (Yang et al. 2021). (A). snapshots of *Pleurotus eryngii* mushroom and the location where we obtain the skin and core samples. The natural *Pleurotus eryngii* (without baking and water loss) near the skin (left) and core (right) samples after the tensile test. (B). Critical stress of king oyster mushrooms as a function of baking time for samples at skin and core. (C). The critical strain of king oyster mushrooms as a function of baking time for samples at skin and core.

It is shown that the samples in tensile loading fail by generating zigzag surfaces at the breaking point after necking takes place, suggesting the ductile failure of the natural samples, govern by the sliding failure between mycelium fibers (Figure 6A). It is shown in Figure 6C that while the critical stress monotonically increases with the bake time, as well as the water loss (Figure 5), of the skin and core samples, the critical strain of the mushroom sample after baking for 30 minutes with 31% and 35% of water loss is larger than the other samples (Figure 6B). It is not clear why the critical stress keeps increasing for drier samples, but the critical strain increases up to 30% of water loss and then decreases afterwards. The interaction between chitin and water may strongly attribute to this phenomenon, as water can play a key role to turn a biological interface from ductile to brittle in mechanical loading, as what has been observed in collagen and wood materials (Qin et al. 2012, Jin et al. 2015).

4. Mycelium Composite and Mechanical Features

Mycelium-based bio-composites mechanical characteristics are essential for their use in engineering applications. By comparing various research, it can be seen that the network topology of the mycelium within the composited is greatly influenced by the fungal species and the substrate utilized to produce mycelium. The test results for the MBF and MBSC in various research are compiled in Table 3. We can observe a considerable difference in the material density between tests. As is evident in most cellular materials, increased material density typically results in a higher Young's modulus and strength (Gibson 2012). Moreover, when studied across multiple experiments, the mechanics of the mycelium composites appear extremely diverse. The density of the mycelium-based composite is significantly influenced by the substrate. Usually, a higher density can be achieved by using higher grain content (fibers, husks, or wood pulp) (Arifin and Yusuf 2013, Dhillon et al. 2013). The diverse mycelium species employed in various experiments is another factor.

A mycelium-based composite that can be utilized as a packaging and construction material needs to have the essential ability of compressive strength. Compressive

Table 3: The mechanical properties of mycelium-based composites. MBF = mycelium-based foam; MBSC = mycelium-based sandwich composite.

Density (g/cm³)	Young's modulus (MPa)	Compressive strength (kPa)	Flexural strength (kPa)	Tensile strength (kPa)	Material	References
0.10–0.14	66.14–71.77	670–1180	–	100–200	MBSC	(Ziegler et al. 2016)
0.10–0.24	2–97	–	50–860	10–240	MBF	(Appels et al. 2019)
0.183 ± 0.015	–	41.72 ± 13.49	10.91 ± 4.41	49.90 ± 20.00	MBF	(López Nava et al. 2016)
0.29–0.35	–	156–340	–	–	MBF	(Fallis 2013)
0.029–0.045	0.6–2	40–83	–	180–300	MBF	(Islam et al. 2017)

strength is the ability of a material or structure to endure loads that tend to compress the material. The mechanical characteristics of MBF are the main emphasis of López-Nava et al. focus. (Substrate: common wheat stalks; fungi: *Pleurotus* sp.). They claimed that because water absorption substantially impacts compressive strength and the substrate and mycelium absorb large amounts of water, MBFs always have lower compressive strengths than synthetic polymer foam of the same density (López Nava et al. 2016). Additionally, according to Silverman, different mushrooms can be used to create MBF with a higher compressive strength using fiber (e.g., psyllium husk) in the substrate. Moreover, they utilize chicken feathers to improve the compressive strength of the substrate. The durability of the keratin protein in the feathers can make them resistant to degradation during mycelium growth. Moreover, they can support the composite's structural integrity and aid in its mechanics' properties since they are lightweight and hydrophobic. They claimed that for the same density, the compressive strength of the composite significantly increases (Fallis 2013).

Flexural strength, also known as bend strength, transverse rupture strength, or modulus of rupture, is the stress at the sample's fracture site when it is bending. López-Nava et al. investigated that the range of the flexural strength of MBF is lower than the synthetic polymer foams with the same density, while the tensile strength is much larger than the synthetic ones (López Nava et al. 2016). According to research by López-Nava et al., MBF's tensile strength is substantially more than synthetic ones. However, the flexural strength range is less than that of synthetic polymer foams of the same density. However, the outcomes from Appels et al., are different. The mycelium-based bio-composite made by *Trametes multicolor* and *Pleurotus ostreatus* grows on rapeseed straw and beech sawdust is put to the test by the authors. The findings demonstrate that it has more flexural strength than synthetic polymer foam (Appels et al. 2019). According to the authors, the difference in contrasting mechanics between the substrate and the mycelium fibers is responsible for the impact. The mycelium fibers are more elastic than the colonized substrate particles, adding to the composite's flexibility and making them more likely to rupture under intense strain. Moreover, the authors also believe that the pressing method can impact tensile strength, especially heat pressing can greatly boost tensile strength, whereas substrate and fungi species have less or no impact (Appels et al. 2019).

One of the key elements that can influence mechanical properties is density, as is specified. Tension and compression tests are used by Islam et al. to test the mycelium samples. The mycelium exhibits linear elastic behavior at low strain and occurs yields. The samples undergo strain hardening before rupturing in the tensile test. The mycelium exhibits behaviors similar to open-cell foam during the compression test. The linear-elastic behavior is the initial appearance in the stress-strain curve, followed by a plateau with a softened response. When effect by repeated loading and unloading cycles, the mycelium displays strain-related hysteresis and stress softening effects (with their mechanical features summarized in Table 2) (Islam et al. 2017).

An MBSC with a core constructed of hemp pith and a cotton mat was described by Ziegler et al. The fabric used for the surface binding is a common cloth manufactured from natural fibers, like burlap. Similar to Jiang et al., the authors are following this methodology (Jiang et al. 2017). They covered both sides of the pre-inoculated active mycelium-based biocomposite foam with natural fiber fabric. As an organic adhesive, the mycelium will keep growing to join the two sides of the fabric made of natural fibers. The fiber surface gives MBSC a high tensile strength and enhances its compressive strength. Young's modulus, however, does not achieve the greatest value as MBF (Ziegler et al., 2016) (Table 2). Jiang et al., stated that creating the MBSC by trying to use various fibers as the MBF surface. The fungus cemented the fabric layers with the MBF core by creating a compact mycelium net. The findings indicate that flax, as opposed to jute or cellulose, is more effective for colonization and produces more mycelium. Compared to samples made with cellulose (16 kPa and 15 kPa, respectively) or jute (20 kPa and 12 kPa, respectively) surface layers, the ultimate strength and yield stress of the samples made with flax surface layers (35 kPa and 27 kPa, respectively) are nearly twice as high (Jiang et al. 2017).

4.1 Mycelium-Based Bio Composite Tensile Test

4.1.1 Mold Design

We design the mold in the shape of a dog bone to make the mycelium-based bio-composite plate. A dog-bone sample is designed to ensure the highest probability that the sample will fail due to maximum tensile loading. Dog bone sample is primarily used in tensile tests. The sample has a shoulder at each end and a gauge section. The shoulders are more comprehensive than the gauge section, which causes a stress concentration to occur in the middle when the sample is loaded with a tensile force. This stress concentration ensures a higher probability that the sample will rupture away from the ends. When the rupture of a sample occurs in the midsection, the material reaches its maximum tensile strength.

Figure 7A shows the design drawings. There are two main parts of the mold: the top and bottom parts are all made of aluminum. For the top part, the length of the dog bone shape is 114 mm, the width at the two ends is 18 mm, and the width of the narrow part in the middle is 5 mm. The effective pressure area of the top part is 1370.11 mm^2 (2.12 in^2). To take out the upper part more efficiently, the length and width of the bottom groove part are all longer than the top part, about 1.7 mm. So, the length is about 144–115.7 mm, with the end part width at 18–19.7 mm, and the middle part width measures 5–5.7 mm of the final mycelium-based plate. Except for this, we polished the side of the top of the dog bone shape mold about 1 degree from top to bottom. We also opened four holes at the bottom part and made an aluminum plate that could be put into the bottom part. Then, we can push out the sample and the aluminum plate by using the aluminum base with four steel columns. To prevent the mycelium-based material adhesive from sticking to the top part of the aluminum plate after the high-temperature baking, we cut aluminum foil of the dog bone shape. We place it inside the mold before putting the material in. After placing the material in the aluminum foil-lined mold, we placed another piece of foil on top of the material. Then we cover the upper part of the mold tightly and use it for the next step of the experiment. Moreover, we designed the corners and the connecting parts into a curved transition because of the mechanical requirements. Figure 7A shows the finished model.

Figure 7: A. Mold design drawing (left) and model (right) B. Three different materials: stalk, stalk with mycelium, and hardwood mixed with coffee grounds with mycelium (from left to right) C. The general process of the experiment. The dog bone samples shown in the Figure are made with three different materials based on the 80 centigrade and 6.75 MPa pressure combined with five different baking times, which are 1 hour, 2 hours, 4 hours, 8 hours, and 16 hours. (material → mix with water → mushy material → hot press machine → test samples → Instron machine → broken samples) D. The SEM image of mycelium-based bio-composite material at room temperature dried (left) and after the hot processing method of 90 centigrade, baking time 16 hours, and 20.27 MPa pressure (right).

4.1.2 *Material and Sample Preparation*

To ascertain whether mycelium can or cannot improve the material's mechanical properties, we prepared three different materials for the tensile test: stalk, stalk with mycelium, and hardwood mixed with coffee grounds and mycelium. Figure 7B shows the three different materials. We split the larger wood into small pieces and use a blender to crush it into the pelleted wood stalk powder. The mycelium-based bio composite sample preparation is generally divided into three main parts: substrate preparation, inoculate fungi, and the final mycelium-based composite material preparation. To avoid the mushroom infecting the mold and spot, we use 75% isopropyl alcohol to sterilize all the tools and polypropylene bags for the growth of the mycelium. We culture the king oyster mushroom spore into the stalk substrate for seven days. We chose the king oyster mushroom to grow the mycelium because this species is the most commonly used in the experiment. It is easier to grow successfully than the other species. Put the stalk in the mushroom culture bag at room temperature (around 25°C) and use the ultrasonic humidifier to generate the water mist to keep the growth environment at a higher humidity (relatively 98%). The mushroom species we chose for the last material, hardwood with mycelium with coffee grounds, is still the king oyster mushroom. We buy the mushroom grow kit from the online supplier as the control group, and after we receive it, we culture the mushroom for seven days. We use these three different materials to see how the mycelium can improve the mechanical properties of the material and how different substrates with mycelium can affect the mechanical properties of the material. The first three steps in Figure 7C show how we prepare the raw material to make the dog bone samples. We take 200 grams of each material into a blender and add 200 grams of water. Blend until the material becomes mushy. We use the blender to mix the material with water since we want to make the mycelium as a role of glue to discuss how this method may improve the mechanical properties or not. We take the material out and put it into a sterilized plastic box. The materials are stocked in the refrigerator to keep it at a lower temperature and avoid the temperature and environmental effects.

We use the 10-ton hot press machine with the hydraulic hand pump, as shown in Figure 7C step 4, performing the hot press to make the dog bone samples. The press machine connected with the PID temperature controller box allows us to set different temperatures. We selected three different temperatures for each material which are 176, 194, and 212 Fahrenheit (80, 90, and 100 Centigrade), and three different applied press forces, which are 1, 2, and 3 metric tons. Since the pressure gauge only shows the pressure, we convert the applied force to the pressure based on the website's equation. The equations are:

$$\sigma_{press} = \frac{F_{plate}}{A} \tag{1}$$

$$F_{plate} = A_{tank}P \tag{2}$$

where σ_{press} and F_{plate} are the normal stress and force applied on the top surface of the sample, respectively, $A = 2.12\ in^2$ is the area of top surface of the dog bone sample,

$A_{tank} = 2.25$ in^2 is the internal cross-section area of the hydraulic compressor and P is the pressure applied on the gauge. It is noted that the σ_{press} and A used here are for the sample preparation and they refer to mechanical and geometric features in a different direction from the mechanical tensile test. Table 4 shows the example pressures we applied to the material to obtain our mechanical samples and the pressure we read on the pressure gauge.

Table 4: The pressure applied to the material and the pressure read on the gauge.

Pressure type	Design applied force F_{plate} (LB)	Pressure on the material σ_{press} (psi)	Pressure on the gauge P (psi)
Low	2204.62	1038.10	979.83
Mid	4409.25	2076.21	1959.67
High	6613.87	3114.31	2939.49

For each material, we have three groups based on different temperatures combined with different pressures for each material. We set the five baking times for each combination: 1, 2, 4, 8, and 16 hours. So, for each material in total, we have 45 samples for the mechanical test. For each sample, we set the temperature first. When the temperature reaches the one designated, we put the mold with material on the bottom plate of the machine and begin to apply the pressure. We use the paper towel to wipe off excess water squeezed out after applying the target load. We allow the machine to bake the sample until the time that we are setting and then take the mold off the machine. Usually, the press machine will drop the pressure automatically, so we need to apply the pressure so that the machine will stay at the aim load that we set. Figure 7C shows the samples we did for each material. Moreover, we take a piece from the sample that we processed hot press treatment to do the SEM image to compare with mycelium-based bio composites without hot processing, as shown in Figure 7D. The left figure is the dried mycelium-based bio-composite material at room temperature for about one month. We can see that under slowly losing water, all the mycelium becomes flat strips. Also, it has a much smaller size than wood fiber, which allows us to identify it. The right figure is the mycelium-based bio composite sample after the hot processing. Since we use the blender to mix the material with water and then make the sample, it is hard to find the mycelium. The clamp connection becomes the only property that allows us to distinguish the mycelium from wood fiber since it is the unique structure that the mycelium has. Because the sample we chose to do the SEM image only baked under 90 centigrade, the mycelium still contained the water. The structure of mycelium is not like under the slow losing water condition; it is still maintained a thicker structure.

4.1.3 Tensile Test and Results

We perform the tensile test for each dog bone samples to understand how the mycelium can affect the mechanical properties of the bio-composite materials. Moreover, we want to find a relatively good combination of temperature, pressure, and time conditions that can show good mechanical properties. We use the same method that we test the natural mushroom samples; the only difference is that the upper grips move at a constant displacement speed of $v = 0.5$ *mm/min*. The last step of Figure 7C shows that the tensile test breaks the samples of the Instron machine. The software with the Instron machine returns the σ – ε curve as well as Young's modulus, yield stress, and breaking strain during the test. Before we performed the tensile test, we calculated the density to see if the different treatment conditions could affect it. After taking the sample out of the mold, we measure the weight. The equation $\rho = \dfrac{m}{v}$ calculates the density. The volume is calculated using an equation $v = At$, and the A area of each sample is the same and fixed by the mold, which is 1370.11 mm²; the t is the thickness which I directly measured by the sample. Moreover, based on the data obtained from the Instron, we calculate the ultimate stress, 0.1% Young's modulus, and toughness for further analysis. The results file provides the change of displacement of the sample for every 0.02 seconds. The equation $\epsilon = \dfrac{\Delta L}{L}$ is used to calculate the strain; the ΔL is the displacement of the last 0.02 seconds minus the displacement of the first 0.02 seconds. L is measured from the machine when we set the sample to the proper location, the distance from the metal piece connected by the upper clamp to the metal piece connected by the lower clamp. The equation $\sigma = \dfrac{F}{A}$ calculates the stress. The F is the force obtained from the results file. The A is the test area since our sample is dog bone shaped; the test area is the narrower part in the middle. We use the thickness times the width of the middle part to calculate the test area. By observing the stress-strain curve shown on the machine, we set the strain limitation to do the linear fitting, which is 0.1% (0.001), to calculate Young's modulus. The toughness is the area under the stress-strain curve which we use the integral to calculate.

Figures 8–10 shows the stress-strain curve for all three different material samples based on different pressure, temperature, and baking time. We summarized the maximum density and mechanical properties of three different materials to get the general results as shown in Table 5 From the table, we can see that the maximum stress, which is 12.985 MPa, occurs in the material that uses hardwood with coffee grounds with mycelium. Compared with the maximum stress of 10.89 MPa of corn straw bio-board (Wu et al. 2015), the mycelium-based bio composite shows higher stress, which means the mycelium can help improve the mechanical properties of the biocomposite material. However, the toughness is decreased with the mycelium involved in the biocomposite material. Toughness is the ability of a material to absorb energy and plastically deform without fracturing. Compared to bio-composite material with mycelium and without, bio-composite material is more brittle than the stalk-based dog bone sample. One reason is when we make the stalk-based sample, and we use the water to mix with the stalk to create the same experiment requirement as the way we make the mycelium-based samples. However, the structure of stalk-based samples is still loose even though we performed hot pressing. The mycelium acts as a filler and adhesive to union the substrate material into a whole part under hot pressing. Flexible biocomposites have higher toughness because they have the

Table 5. The maximum density and mechanical properties of each material.

Mechanical properties and density	Stalk	Stalk with mycelium	Hardwood with coffee grounds with mycelium
Density (g/cm³)	1.5101	1.55	1.542
Ultimate Stress (MPa)	8.8538	9.7043	12.985
0.1% Young's Modulus (MPa)	2669.5	2841.4	3664.8
Toughness (MJ/m³)	0.0591	0.0529	0.0357

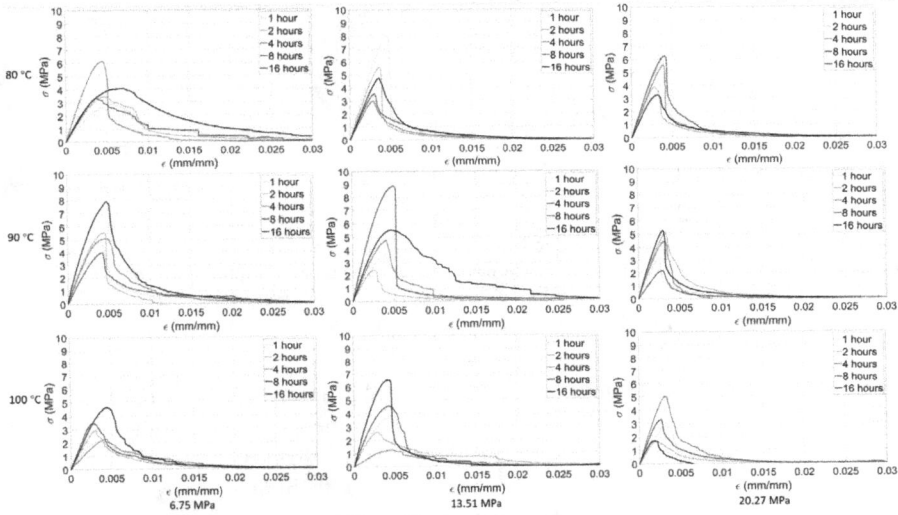

Figure 8: The stress-strain curve for dog bone samples tests of the composites made of pure stalk. Tests are based on samples prepared under different temperatures, pressure, and baking time. (Different color in the stress-strain curve means the different baking times, from light to darker colors, which are 1 hour, 2 hours, 4 hours, 8 hours, and 16 hours).

highest strain and compressive strength. However, an increase in temperature made the mycelial network less flexible, reducing the strain and absorption of mechanical stress, and this change in mechanical behavior affected toughness, reducing it (Santos et al. 2021). Moreover, the toughness can be affected by the original mechanical properties of the substrate. So, based on different manufacturing requirements, different kinds of substrates are needed to inoculate mycelium.

The material properties and composition are important to the mechanical properties of the biocomposite material. Moreover, the different treatment conditions also can affect the mechanical properties. Since we want to analyze how the temperature, pressure, and baking time affect the mechanical properties, we calculate the specific mechanical properties for each sample. We use each mechanical property to divide by the density to calculate the specific mechanical properties to involve the density effect on the mechanical properties. Moreover, the treatment conditions

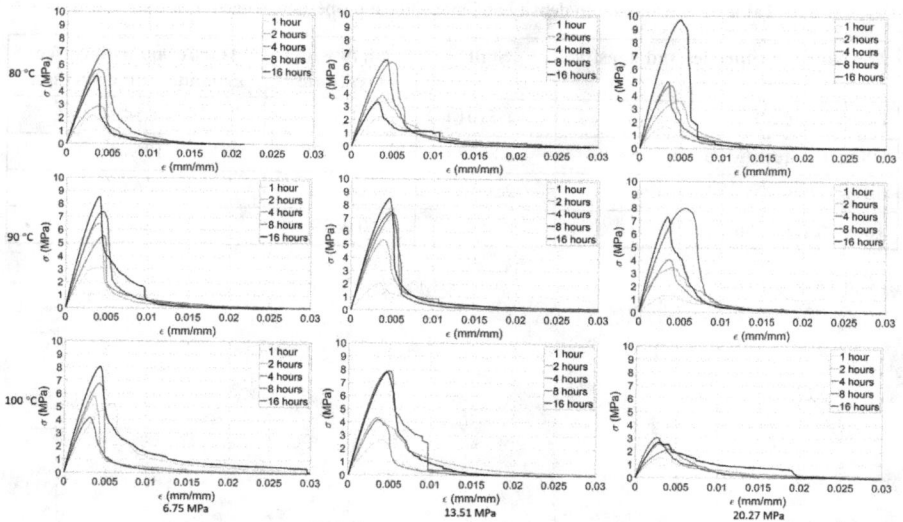

Figure 9: The stress-strain curve for dog bone samples tests of the composites made of the substrate of stalk with mycelium grown in it. Tests are based on samples prepared under different temperatures, pressure, and baking time. (Different color in the stress-strain curve means the different baking times, from light to darker colors, which are 1 hour, 2 hours, 4 hours, 8 hours, and 16 hours).

that can obtain the maximum specific mechanical property results cannot stand for all situations. So, we also plot the 3D plot for all the experiment data to observe the results, as shown in Figure 11. The figures show that when the temperature is 90 centigrade, the baking time is 16 hours, and with high pressure, the specific mechanical properties always show good results, around the average or better. The high pressure decreases the porosity and density of the mycelium-based biocomposites and makes the samples denser than the stalk-based samples. Moreover, even though the stalk-based samples are made under the same treatment conditions, the mycelium plays an essential role in combining the single stalk fiber. It is not only an adhesive to link the stalk fiber together, but also fills up the interspace between the stalk fibers. Based on the function of mycelium, it makes the mycelium-based samples achieve higher mechanical strength than the stalk-based samples. Since water can be an essential factor in affecting the mechanical properties, to avoid it has a high content in the samples. The water evaporation occurs at temperatures above 90 centigrade, so the relatively good mechanical properties always need 90 or 100 centigrade to be achieved.

5. Discussion

Less research is given to microorganisms and their multiscale structures than biological materials that include proteins (e.g., silk, collagen, the cytoskeleton, etc.). Mycelium and its composite materials can be studied to understand the fungus network's mechanics, biological function, and application to create green composite

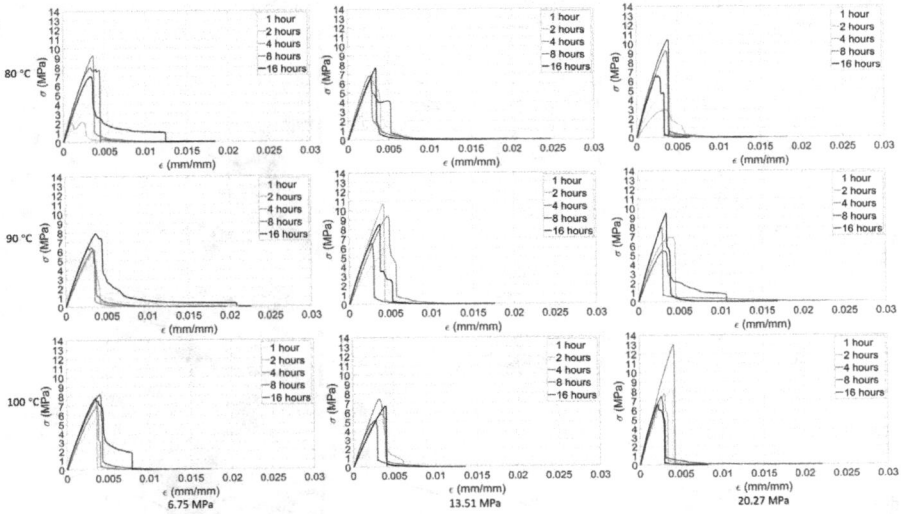

Figure 10: The stress-strain curve for dog bone samples tests of the composites made of the hardwood mixed with coffee grounds and mycelium grown in it. Tests are based on samples prepared under different temperatures, pressure, and baking time. (Different color in the stress-strain curve means the different baking times, from light to darker colors, which are 1 hour, 2 hours, 4 hours, 8 hours, and 16 hours).

materials with good mechanics and lightweight properties in both simulation and corresponding experiments for synthesis (Holt et al. 2012, Haneef et al. 2017, Jiang et al. 2017, Girometta et al. 2019, Attias et al. 2020). Method to grow and process mycelium-based composites can lead to a promising and innovative way to produce building materials from using the agricultural method (Attias et al. 2020). Investigating the molecular make-up and biological function of the mycelium network can aid in identifying novel drugs produced by fungi with particular biological functions or serve as an inspiration for the topology of the internet of things that uses little energy and responds quickly to pests and diseases (Muzzarelli et al. 2007, Naseri et al. 2014, Morin-Crini et al. 2019, Wang et al. 2019, Jones et al. 2020a).

Mycelium composite, a more environmentally friendly material than synthetic foams, has advantages in many engineering applications (e.g., packaging materials and boards for acoustic and thermal insulation) and is gaining popularity. The development of such materials is still in its infancy, and the standardized procedure that will result in optimum material properties has not yet been discovered. This bio-composite material has the potential to be extensively employed in the biomedical industry, furniture, agriculture, and civil engineering. In terms of mechanical properties, mycelium composites are generally distinct from natural cellular materials or synthetic polymer foams. Their mechanics are not simply defined by the processing method at the end of its production but as the collective result of the fungi species, their substrate, and related environments during the growth. Their mechanics' results are not commonly defined by the interaction between the fungal species and their substrate but by associated conditions during growth. Moreover,

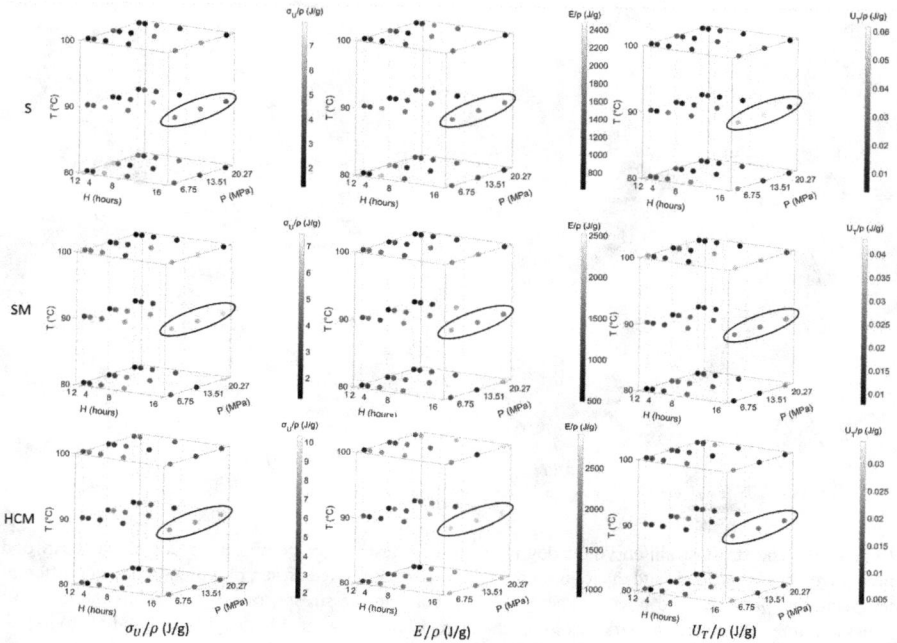

Figure 11: 3D scatter plots based on the treatment conditions and the color bar based on all tested samples' specific mechanical properties. Stalk (S); Stalk with mycelium (SM); Hardwood with coffee grounds with mycelium (HCM).

it is also determined by the processing technique used to create the biocomposite. The substrate's characteristics determine the matrix material's mechanics within the composite. The compositions and structure of the substrate have an impact on the mycelium network itself. Additionally, the final composite's water content is significant since both the mycelium and substrate can absorb water. The way that removes the water from the biocomposite should be considered. Mycelium can typically be inactivated and removed by a hot press technique, effectively stopping it from growing while applying to different applications. However, results from several studies are frequently incomparable because of the various available parameters in different studies. For instance, the mycelium-based bio-composite has not demonstrated many benefits over the most significant rival, traditional material (EPS).

Even mycelium-based composites have advantages in terms of mechanics, lightweight construction, and several eco-friendly qualities, but their use on a larger scale is constrained and fraught with difficulties. For example, because it is a biomaterial, its production is less standardized than traditional engineering materials like steel, cement, and polymer. It is also unclear how to tailor the substrates used by specific fungi species to increase mycelium's yield and improve the composite's mechanical properties. It may be necessary to investigate the structure-mechanics relationship of various classes of fungi (by type of rot, type of hyphae, gene, etc.) to identify the most promising species of yielding composite with the best mechanics because it is very challenging to test the microstructure and mechanics for all the mycelium species (Blackwell 2011). Furthermore, unlike polymer foams, machines cannot quickly and in large quantities generate mycelium-based bio composites since it takes two weeks or more to develop the mycelium-based bio-composite material. In an incubation environment, it's critical to automatically control the growing elements like temperature, humidity, nourishment, and light without directly involving human labor. It is also unknown how each component that makes up the composite influences the integrity of the fibrous network by contributing to its contact with wood fibers. These restrictions are essential before using the material for architectural and more extensive industrial applications.

6. Conclusion

Mycelium research is generally not limited to material usage. Mycelium, a fungus' vegetative portion, has the unique ability to utilize discrete agricultural wastes as substrates for the growth of its network, integrating wastes into continuous composites without the need for energy and not the creation of further waste (Jones et al. 2018, Appels et al. 2019, Girometta et al. 2019). In addition to repairing soil fragments, mycelium serves a more crucial role in nature as an information highway that facilitates interactions among various plants (Fricker et al. 2017) (Gorzelak et al. 2015) (Simard et al. 2012). It enables isolated plants to communicate and exchange information to successfully protect themselves against pests and illnesses (Babikova et al. 2014). Beyond the mechanical properties of material study, the study of mycelium-based composites, as it integrates various discrete building blocks and accomplishes material functions that none of the building blocks alone can accomplish, has become the primary motivation and reason for our desire to learn more about the mycelium network and its biological functions. Primary ecologists are currently interested in the functions of the mycelium network, but it is unclear exactly how the chemical signals are carried out in the network's hierarchical structure, how their effectiveness relates to the geometry and topology of the network, or how such knowledge may contribute to our understanding of the topology of the internet and the internet of things, or of innovative internet media with low energy consumption. Most of these inquiries require interdisciplinary work, and some might be resolved by creating a multiscale model of the mycelium network and applying it to pertinent simulations. The mycelium bio-composite material cannot carry much weight as the construction material. However, construction requires many parts that don't have to take much weight, for example, insulation material, inside walls, and facade cladding. Testing the thermal and acoustic insulation properties of mycelium-based bio-composite material properties to compare with the current material used in the construction building. Moreover, mycelium-based bio-composite material is durable and naturally fire-resistant and can be easily formed into any shape. They are eco-friendly and carbon neutral. It will break down if exposed to living organisms. Mycelium as a building material presents an excellent opportunity to upgrade agricultural waste into a low-cost, sustainable, and biodegradable material alternative. The manufacturing process of the mycelium-based bio composite material is not like cement needs heat and energy-consuming ingredients. In addition to researching its use in creating green composite materials, we will also gather expertise for designing an information network system about mycelium-based bio composite materials.

Acknowledgments

The authors acknowledge the National Science Foundation CAREER Grant (Award #: 2145392) and Collaboration for Unprecedented Success and Excellence (CUSE) Grants at Syracuse University for supporting the research work.

References

Appels, F.V.W., S. Camere, M. Montalti, E. Karana, K.M.B. Jansen, J. Dijksterhuis, P. Krijgsheld and H.A.B. Wösten. 2019. Fabrication factors influencing mechanical, moisture- and water-related properties of mycelium-based composites. *Mater. Des.*, 161: 64–71. doi: 10.1016/j.matdes.2018.11.027.

Aranaz, I., M. Mengibar, R. Harris, I. Panos, B. Miralles, N. Acosta, G. Galed and A. Heras. 2012. Functional characterization of chitin and chitosan. *Curr. Chem. Biol.*, 3: 203–230(28). doi: 10.2174/2212796810903020203.

Arifin, Y.H. and Y. Yusuf. 2013. Mycelium fibers as new resource for environmental sustainability. *Procedia Eng.*, 53: 504–508. doi: 10.1016/j.proeng.2013.02.065.

Attias, N., O. Danai, T. Abitbol, E. Tarazi, N. Ezov, I. Pereman and Y.J. Grobman. 2020. Mycelium bio-composites in industrial design and architecture: Comparative review and experimental analysis. *J. Clean. Prod.*, 246: 119037. doi: 10.1016/j.jclepro.2019.119037.

Attwood, M.M. and H. Zola. 1967. The association between chitin and protein in some chitinous tissues. *Comp. Biochem. Physiol.*, 20: 993–998. doi: 10.1016/0010-406X(67)90069-2.

Azimi, B., C. Ricci, A. Fusco, L. Zavagna, S. Linari, G. Donnarumma, A. Hadrich, P. Cinelli, M.-B. Coltelli, S. Danti and A. Lazzeri. 2021. Electrosprayed shrimp and mushroom nanochitins on cellulose tissue for skin contact application. *Molecules*, 26L 4374. doi: 10.3390/molecules26144374.

Azimi, B., L. Thomas, A. Fusco, O.I. Kalaoglu-Altan, P. Basnett, P. Cinelli, K. de Clerck, I. Roy, G. Donnarumma, M.B. Coltelli, S. Danti and A. Lazzeri. 2020. Electrosprayed chitin nanofibril/electrospun polyhydroxyalkanoate fiber mesh as functional nonwoven for skin application. *J. Funct. Biomater.*, 11: 62. doi: 10.3390/jfb11030062.

Babikova, Z., D. Johnson, T. Bruce, J. Pickett and L. Gilbert. 2014. Underground allies: How and why do mycelial networks help plants defend themselves? What are the fitness, regulatory, and practical implications of defence-related signaling between common plants via common mycelial networks? Insights & Perspectives Z. Babikova. *BioEssays*, 36: 21–26. doi: 10.1002/bies.201300092.

Bartnicki-Garcia, S. 1968. Cell wall chemistry, morphogenesis, and taxonomy of fungi. *Annu. Rev. Microbiol.*, 22: 87–108. doi: 10.1146/annurev.mi.22.100168.000511.

Bhuvaneshwari, S., H. Hettiarachchi and J.N. Meegoda. 2019. Crop residue burning in India: Policy challenges and potential solutions. *Int. J. Environ. Res. Public Health*, 16: 832. doi: 10.3390/ijerph16050832.

Blackwell, M. 2011. The fungi: 1, 2, 3 ... 5.1 million species? *Am. J. Bot.*, 98: 426–438. doi: 10.3732/ajb.1000298.

Butterfield, B., K. Chapman, L. Christie and A. Dickson. 1992. Ultrastructural characteristics of failure surfaces in medium density fiberboard. *For. Prod. J. (USA)*, 42: 55–60.

Carvalho, L.M.H. and C.A.V. Costa. 1998. Modeling and simulation of the hot-pressing process in the production of medium density fiberboard (MDF). *Chem. Eng. Commun.*, 170: 1–21.

Couturier, M. and J.G. Berrin. 2013. The saccharification step: The main enzymatic components. In Lignocellulose Conversion: Enzymatic and Microbial Tools for Bioethanol Production (Berlin, Heidelberg: Springer), pp. 93–110. doi: 10.1007/978-3-642-37861-4_5.

Dai, C., C. Yu and X. Zhou. 2007. Heat and mass transfer in wood composite panels during hot pressing. Part II. Modeling void formation and mat permeability. *Wood and Fiber Sci.*, 37: 242–257.

Danti, S., L. Trombi, A. Fusco, B. Azimi, A. Lazzeri, P. Morganti, M.B. Coltelli and G. Donnarumma. 2019. Chitin nanofibrils and nanolignin as functional agents in skin regeneration. *Int. J. Mol. Sci.*, 20: 2669. doi: 10.3390/ijms20112669.

Deacon, J. 2013. Fungal Biology: 4th Edition. doi: 10.1002/9781118685068.

Defonseka, C. 2014. Introduction to polymeric composites with rice hulls. Smithers Rapra. Technology, pp. 208.

Dhillon, G.S., S. Kaur, S.K. Brar and M. Verma.2013. Green synthesis approach: Extraction of chitosan from fungus mycelia. *Crit. Rev. Biotechnol.*, 33: 379–403. doi: 10.3109/07388551.2012.717217.

Di Mario, F., P. Rapanà, U. Tomati and E. Galli. 2008. Chitin and chitosan from Basidiomycetes. *Int. J. Biol. Macromol.*, 43: 8–12. doi: 10.1016/j.ijbiomac.2007.10.005.

Elieh-Ali-Komi, D. and M.R. Hamblin. 2016. Chitin and chitosan: Production and application of versatile biomedical nanomaterials. *Int. J. Adv. Res. (Indore)*, 4: 411–427.

Elsacker, E., S. Vandelook, J. Brancart, E. Peeters and L. de Laet. 2019. Mechanical, physical and chemical characterisation of mycelium-based composites with different types of lignocellulosic substrates. *PLoS ONE*, 14: e0213954. doi: 10.1371/journal.pone.0213954.

Fallis, A.G. 2013. Development and testing of mycelium-based composite materials for shoe sole applications. *Journal of Chemical Information and Modeling*.

Fazli Wan Nawawi, W.M., K.Y. Lee, E. Kontturi, R.J. Murphy and A. Bismarck. 2019. Chitin nanopaper from mushroom extract: Natural composite of nanofibers and glucan from a single biobased source. *ACS Sustainable Chemistry and Engineering*, 7: 6492–6496. doi: 10.1021/acssuschemeng.9b00721.

Fricker, M., L. Boddy and D. Bebber. 2007. Network organisation of mycelial fungi. In Biology of the Fungal Cell. doi: 10.1007/978-3-540-70618-2_13.

Fricker, M.D., L.L.M. Heaton, N.S. Jones and L. Boddy. 2017. The mycelium as a network. *The Fungal Kingdom*, 335–367. doi: 10.1128/9781555819583.ch15.

Gibson, L.J. 2012. The hierarchical structure and mechanics of plant materials. *Journal of the Royal Society Interface*, 9: 2749–2766. doi: 10.1098/rsif.2012.0341.

Gibson, L.J. and M.F. Ashby. 1982. The mechanics cellular materials of three-dimensional cellular materials. *Proc. R. Soc. Lond. A*, 382: 43–59.

Girometta, C., A.M. Picco, R.M. Baiguera, D. Dondi, S. Babbini, M. Cartabia, M. Pellegrini and E. Savino. 2019. Physico-mechanical and thermodynamic properties of mycelium-based biocomposites: A review. *Sustainability (Switzerland)*, 11: 281. doi: 10.3390/su11010281.

Girometta, C., A. Zeffiro, M. Malagodi, E. Savino, E. Doria, E. Nielsen, A. Buttafava and D. Dondi. 2017. Pretreatment of alfalfa stems by wood decay fungus Perenniporia meridionalis improves cellulose degradation and minimizes the use of chemicals. *Cellulose*, 24: 3803–3813. doi: 10.1007/s10570-017-1395-6.

Gorzelak, M.A., A.K. Asay, B.J. Pickles and S.W. Simard. 2015. Inter-plant communication through mycorrhizal networks mediates complex adaptive behaviour in plant communities. *AoB Plants*, 7: plv050. doi: 10.1093/aobpla/plv050.

Hackman, R. 1960. Studies on Chitin IV. The occurrence of complexes in which chitin and protein are covalently linked. *Aust. J. Biol. Sci.*, 13: 568–577. doi: 10.1071/bi9600568.

Haneef, M., L. Ceseracciu, C. Canale, I.S. Bayer, J.A. Heredia-Guerrero and A. Athanassiou. 2017. Advanced materials from fungal mycelium: Fabrication and tuning of physical properties. *Sci. Rep.*, 7: 41292. doi: 10.1038/srep41292.

Hassainia, A., H. Satha and S. Boufi. 2018. Chitin from Agaricus bisporus: Extraction and characterization. *Int. J. Biol. Macromol.*, 117: 1334–1342. doi: 10.1016/j.ijbiomac.2017.11.172.

Hoa, H.T. and C.L. Wang. 2015. The effects of temperature and nutritional conditions on mycelium growth of two oyster mushrooms (*Pleurotus ostreatus* and *Pleurotus cystidiosus*). *Mycobiology*, 43: 14–23. doi: 10.5941/MYCO.2015.43.1.14.

Holt, G.A., G. McIntyre, D. Flagg, E. Bayer, J.D. Wanjura and M.G. Pelletier. 2012. Fungal mycelium and cotton plant materials in the manufacture of biodegradable molded packaging material: Evaluation study of select blends of cotton byproducts. *J. Biobased Mater Bioenergy*, 6: 431–439. doi: 10.1166/jbmb.2012.1241.

Hori, C., J. Gaskell, K. Igarashi, M. Samejima, D. Hibbett, B. Henrissat and D. Cullen. 2013. Genomewide analysis of polysaccharides degrading enzymes in 11 white- and brown-rot Polyporales provides insight into mechanisms of wood decay. *Mycologia*, 105: doi: 10.3852/13-072.

Islam, M.R., G. Tudryn, R. Bucinell, L. Schadler and R.C. Picu. 2017. Morphology and mechanics of fungal mycelium. *Sci. Rep.*, 7: 13070. doi: 10.1038/s41598-017-13295-2.

Jayakumar, R., M. Prabaharan, P.T. Sudheesh Kumar, V.S.T. Furuike and H. Tamur. 2011. Novel chitin and chitosan materials in wound dressing. In Biomedical Engineering, Trends in Materials Science, 3–24. doi: 10.5772/13509.

Jiang, L., D. Walczyk, G. McIntyre, R. Bucinell and B. Li. 2019. Bioresin infused then cured mycelium-based sandwich-structure biocomposites: Resin transfer molding (RTM) process, flexural properties, and simulation. *J. Clean. Prod.*, 207: 123–135. doi: 10.1016/j.jclepro.2018.09.255.

Jiang, L., D. Walczyk, G. McIntyre, R. Bucinell and G. Tudryn. 2017. Manufacturing of biocomposite sandwich structures using mycelium-bound cores and preforms. *J. Manuf. Process.*, 28: 50–59. doi: 10.1016/j.jmapro.2017.04.029.

Jin, K., Z. Qin and M.J. Buehler. 2015. Molecular deformation mechanisms of the wood cell wall material. *J. Mech. Behav. Biomed. Mater.*, 42: 198–206. doi: 10.1016/j.jmbbm.2014.11.010.

Jones, M., T. Bhat, T. Huynh, E. Kandare, R. Yuen, C.H. Wang and S. John. 2018. Waste-derived low-cost mycelium composite construction materials with improved fire safety. *Fire Mater.*, 42: 816–825. doi: 10.1002/fam.2637.

Jones, M., T. Huynh, C. Dekiwadia, F. Daver and S. John. 2017. Mycelium composites: A review of engineering characteristics and growth kinetics. *J. Bionanosci.*, 11: 241–257. doi: 10.1166/jbns.2017.1440.

Jones, M., M. Kujundzic, S. John and A. Bismarck. 2020a. Crab vs. Mushroom: A review of crustacean and fungal chitin in wound treatment. *Mar. Drugs*, 18: 64. doi: 10.3390/md18010064.

Jones, M., A. Mautner, S. Luenco, A. Bismarck and S. John. 2020b. Engineered mycelium composite construction materials from fungal biorefineries: A critical review. *Mater. Des.*, 187: 108397. doi: 10.1016/j.matdes.2019.108397.

Karana, E., D. Blauwhoff, E.J. Hultink and S. Camere. 2018. When the material grows: A case study on designing (with) mycelium-based materials. *Int. J. Des.*, 12: 119–136.

Kilavan Packiam, K., T.S. George, S. Kulanthaivel and N.S. Vasanthi. 2011. Extraction, purification and characterization of chitosan from endophytic fungi isolated from medicinal plants. *World J. Sci. Technol.*, 1: 43–48.

Kramer, K.J., T.L. Hopkins and J. Schaefer. 1995. Applications of solids NMR to the analysis of insect sclerotized structures. *Insect Biochem. Mol. Biol.*, 25: 1067–1080. doi: 10.1016/0965-1748(95)00053-4.

Liu, J., S. Lin, X. Liu, Z. Qin, Y. Yang, J. Zang and X. Zhao. 2020. Fatigue-resistant adhesion of hydrogels. *Nat. Commun.*, 11: 1071. doi: 10.1038/s41467-020-14871-3.

López Nava, J.A., J. Méndez González, X. Ruelas Chacón and J.A. Nájera Luna. 2016. Assessment of edible fungi and films bio-based material simulating expanded polystyrene. *Mater. Manuf. Process.*, 31: 1085–1090. doi: 10.1080/10426914.2015.1070420.

Madurwar, M.V., R.V. Ralegaonkar and S.A. Mandavgane. 2013. Application of agro-waste for sustainable construction materials: A review. *Constr. Build. Mater.*, 38: 872–878. doi: 10.1016/j.conbuildmat.2012.09.011.

Maraveas, C. 2020. Production of sustainable construction materials using agro-wastes. *Materials*, 13: 262. doi: 10.3390/ma13020262.

Morganti, P. and G. Morganti. 2008. Chitin nanofibrils for advanced cosmeceuticals. *Clin. Dermatol.*, 26: 334–340. doi: 10.1016/j.clindermatol.2008.01.003.

Morganti, P., G. Morganti and M.B. Coltelli. 2019. Chitin nanomaterials and nanocomposites for tissue repair. *In*: Choi, A. and Ben-Nissan, B. (eds.). Marine-Derived Biomaterials for Tissue Engineering Applications. Springer Series in Biomaterials Science and Engineering, vol 14. Springer, Singapore. https://doi.org/10.1007/978-981-13-8855-2_21.

Morin-Crini, N., E. Lichtfouse, G. Torri and G. Crini. 2019. Applications of chitosan in food, pharmaceuticals, medicine, cosmetics, agriculture, textiles, pulp and paper, biotechnology, and environmental chemistry. *Environ. Chem. Lett.*, 17: 1667–1692. doi: 10.1007/s10311-019-00904-x.

Muzzarelli, R.A.A. 2011. Chitin nanostructures in living organisms. *In*: Gupta, N. (ed.). Chitin. Topics in Geobiology, vol 34. Springer, Dordrecht. https://doi.org/10.1007/978-90-481-9684-5_1.

Muzzarelli, R.A.A. 2012. Nanochitins and nanochitosans, paving the way to eco-friendly and energy-saving exploitation of marine resources. pp. 153–164. *In*: Matyjaszewski, K. and Möller, M. (eds.). Polymer Science: A Comprehensive Reference. doi: 10.1016/B978-0-444-53349-4.00257-0.

Muzzarelli, R.A.A., P. Morganti, G. Morganti, P. Palombo, M. Palombo, G. Biagini, M. Mattioli Belmonte, F. Giantomassi, F. Orlandi and C. Muzzarelli. 2007. Chitin nanofibrils/chitosan glycolate composites as wound medicaments. *Carbohydr. Polym.*, 70: 274–284. doi: 10.1016/j.carbpol.2007.04.008.

Naseri, N., C. Algan, V. Jacobs, M. John, K. Oksman and A.P. Mathew. 2014. Electrospun chitosan-based nanocomposite mats reinforced with chitin nanocrystals for wound dressing. *Carbohydr. Polym.*, 109: 7–15. doi: 10.1016/j.carbpol.2014.03.031.

Nawawi, W.M.F.B.W., M. Jones, R.J. Murphy, K.Y.E. Lee, Kontturi and A. Bismarck. 2020. Nanomaterials derived from fungal sources-Is it the new hype? *Biomacromolecules*, 21: 30–55. doi: 10.1021/acs.biomac.9b01141.

Pelletier, M.G., G.A. Holt, J.D. Wanjura, E. Bayer and G. McIntyre. 2013. An evaluation study of mycelium based acoustic absorbers grown on agricultural by-product substrates. *Ind. Crops Prod.*, 51: 480–485. doi: 10.1016/j.indcrop.2013.09.008.

Percot, A., C. Viton and A. Domard. 2003. Characterization of shrimp shell deproteinization. *Biomacromolecules*, 4: 1380–1385. doi: 10.1021/bm034115h.

Pheng, L.S. and L.S. Hou. 2019. The economy and the construction industry. *In*: Construction Quality and the Economy. Management in the Built Environment. Springer, Singapore. https://doi.org/10.1007/978-981-13-5847-0_2.

Pochanavanich, P. and W. Suntornsuk. 2002. Fungal chitosan production and its characterization. *Lett. Appl. Microbiol.*, 35: 17–21. doi: 10.1046/j.1472-765X.2002.01118.x.

Qin, Z., A. Gautieri, A.K. Nair, H. Inbar and M.J. Buehler. 2012. Thickness of hydroxyapatite nanocrystal controls mechanical properties of the collagen-hydroxyapatite interface. *Langmuir*, 28: 1982–1992. doi: 10.1021/la204052a.

Qin, Z., G.S. Jung, M.J. Kang and M.J. Buehler. 2017. The mechanics and design of a lightweight three-dimensional graphene assembly. *Sci. Adv.*, 3: e1601536. doi: UNSP e1601536 10.1126/sciadv.1601536.

Rinaudo, M. 2006. Chitin and chitosan: Properties and applications. *Prog. Polym. Sci.*, 31: 603–632. doi: 10.1016/j.progpolymsci.2006.06.001.

Santos, I.S., B.L. Nascimento, R.H. Marino, E.M. Sussuchi, M.P. Matos and S. Griza. 2021. Influence of drying heat treatments on the mechanical behavior and physico-chemical properties of mycelial biocomposite. *Compos. B. Eng.*, 217: doi: 10.1016/j.compositesb.2021.108870.

Sejian, V., J. Gaughan, L. Baumgard and C. Prasad. 2015. Climate change impact on livestock: Adaptation and mitigation. doi: 10.1007/978-81-322-2265-1.

Simard, S.W., K.J. Beiler, M.A. Bingham, J.R. Deslippe, L.J. Philip and F.P. Teste. 2012. Mycorrhizal networks: Mechanisms, ecology and modelling. *Fungal Biol. Rev.*, 26: 39–60. doi: 10.1016/j.fbr.2012.01.001.

Steigerwald, B. 2014. NASA Goddard Instrument Makes First Detection of Organic Matter on Mars. NASA. Available at: https://www.nasa.gov/content/goddard/mars-organic-matter.

Streit, F., F. Koch, M.C.M. Laranjeira and J.L. Ninow. 2009. Production of fungal chitosan in liquid cultivation using apple pomace as substrate. *Braz. J. Microbiol.*, 40: 20–25. doi: 10.1590/S1517-83822009000100003.

Swift, R.S. 1996. Organic matter characterization. *Methods of Soil Analysis: Part 3 Chemical Methods*, 5: 1011–1069.

Velasco, P.M., M.P.M. Ortiz, M.A.M. Giro, M.C.J. Castelló and L.M. Velasco. 2014. Development of better insulation bricks by adding mushroom compost wastes. *Energy Build.*, 80: 17–22. doi: 10.1016/j.enbuild.2014.05.005.

Wang, C., J. Wang, L. Zeng, Z. Qiao, X. Liu, H. Liu, J. Zhang and J. Ding. 2019. Fabrication of electrospun polymer nanofibers with diverse morphologies. *Molecules*, 24: 834. doi: 10.3390/molecules24050834.

Wu, J., Z. Qin, L. Qu, H. Zhang, F. Deng and M. Guo. 2019. Natural hydrogel in American lobster: A soft armor with high toughness and strength. *Acta Biomater.*, 88: 102–110. doi: 10.1016/j.actbio.2019.01.067.

Wu, T., X. Wang and K. Kito. 2015. Effects of pressures on the mechanical properties of corn straw bio-board. *Eng. Agric. Environ. Food.*, 8: doi: 10.1016/j.eaef.2015.07.003.

Xing, Y., M. Brewer, H. El-Gharabawy, G. Griffith and P. Jones. 2018. Growing and testing mycelium bricks as building insulation materials. *IOP Conference Series: Earth and Environmental Science*, 121: 022032. doi: 10.1088/1755-1315/121/2/022032.

Xu, G., L. Wang, J. Liu and J. Wu. 2013. FTIR and XPS analysis of the changes in bamboo chemical structure decayed by white-rot and brown-rot fungi. *Appl. Surf. Sci.*, 280. doi: 10.1016/j.apsusc.2013.05.065.

Yang, L., D. Park and Z. Qin. 2021. Material function of mycelium-based bio-composite: A review. *Front. Mater.*, 8: 1–17. doi: 10.3389/fmats.2021.737377.

Ziegler, A.R., S.G. Bajwa, G.A. Holt, G. McIntyre and D.S. Bajwa. 2016. Evaluation of physico-mechanical properties of mycelium reinforced green biocomposites made from cellulosic fibers. *Appl. Eng. Agric.*, 32: 931–938. doi: 10.13031/aea.32.11830.

11

Fungi-Derived Composite Nanopapers

Andreas Mautner

1. Introduction

Renewable natural materials were among the first resources next to stones early humans adopted in a variety of applications. For instance, wood or other plant or animal matter (e.g., natural fibers, wool, and leather) is provided for clothing, shelter, and basic tools. In addition, also the ignition and preservation of fire would have not been possible without these materials. Developments during ancient times brought about the use of new classes of materials, such as metals and the first synthetic materials, for instance, steel or bronze, but also innovations regarding natural fibers based upon natural macromolecules such as cellulose, e.g., in papyrus (Kabasci 2020). With the commencement of the industrial revolution, enormous progress sparked in this regard, which was tremendously reinforced during the 20th century. Up to the beginning of the past century, biomacromolecules extracted from natural resources were the primary raw materials for material and fuel production, however, they were majorly substituted by fossil/petroleum-based compounds (Pérez and Samain 2010). Moreover, material innovations safeguarded improved health yielding increased prosperity and extended life expectancies owing to facilitating higher hygiene standards and medical procedures impossible without them. For instance, developments in modern medicine would not have been possible without, typically synthetic, polymers. In general, new processing technologies and lightweight materials for the simple shaping of complex structures offered new pathways in the

Institute for Environmental Biotechnology, Department IFA, University of Natural Resources and Life Sciences Vienna, Konrad-Lorenz-Straße 20, 3430 Tulln an der Donau, Austria.
Email: andreas.mautner@boku.ac.at

development of devices supporting construction or mobility but also everyday life. Although these developments improved many aspects of life on earth in conjunction with industrial production at large, this progress came at the expense of environmental pollution. Water sources were negatively affected through industrial and communal efflux but also uncontrolled disposal of plastic waste in landfills contributes to environmental pollution. Thus, disposed of plastics eventually end up in the sea creating, e.g., the Great Pacific Garbage Patch and generally polluting vast areas of oceans and water resources. In this regard, two important goals, among the 17 UN sustainable development goals (UN 2020), need to be achieved in order to provide an intact environment for future generations: Remediation of water as well as treatment of waste water (#6 clean water and sanitation) and, preferably degradable, alternatives to synthetic polymer and composite materials based on renewable resources (#12 responsible consumption and production). One approach that has the potential to address both these goals is applications based on papers, in particular (composite) nanopapers because of their inherent properties at the nanoscale.

Nanopapers are structures composed of random networks of natural fibers at the nanoscale, commonly nanocellulose (NC), which provide particular properties yielding strong structures but also excellent filtration capabilities. Using NC in such macroscopic 2D sheets, i.e., nanopapers, has been proven to be of great potential (Nakagaito et al. 2005, Nogi and Yano 2008, Henriksson et al. 2011, Ansari et al. 2014, Mautner et al. 2017) and thus, nanocellulose has attracted extensive interest over the past two decades (Klemm et al. 2011), with a wide array of composite (Lee et al. 2014, Oksman et al. 2016) but also membrane applications (Metreveli et al. 2014, Mautner 2020). Since plants absorb carbon dioxide to nurture their growth, and thus vast amounts of CO_2 are stored, utilizing plant-based materials would also support actions mitigating the climate crisis (Crowther et al. 2015, Røyne et al. 2016). Nanofibrils can, however, also be derived from other natural polysaccharides, for instance, chitin. Thus, a further form of life should be considered when it comes to providing potential raw materials for the purpose of fabricating renewable nanopaper materials: fungi.

2. Chitin and Chitosan

2.1 Chitin

Chitin is based on a backbone of $(1\rightarrow4)$ N-acetyl-β-D-glucosamine and was first described by Braconnot during the early 1800s. Thus, it was discovered thirty years prior to cellulose (Braconnot 1811, Children 1824, Payen and Hebd 1838), but research into and applications of chitin have always slightly lagged behind cellulose, which was due to the ubiquitous and easy harvest of cellulose (Nawawi et al. 2019a). Eventually, most progress attained for cellulose was also adopted for chitin. In analogy to the function of cellulose in green plants, primarily chitin's chief biological function is to constitute the scaffold conferring mechanical support to the structure of the fungal cell wall or the exoskeleton of crustaceans and insects (Rinaudo 2007). Chitin can also be found in the shape of fibrils in fungal mycelia, which can be considered the "roots", i.e., the vegetative part, of fungi. The function of providing structural support and integrity is attained differently in the forms of life mentioned, based on their diverse physio-chemical properties.

In crystallographic terms, chitin naturally occurs in three allomorphs (α, β, and γ). α-Chitin is primarily found in "hard" structures, for example in fungi, arthropods, and crustaceans and is commonly considered the most abundant form of chitin, while β-chitin, which is existing in marine diatoms or the pen of squids, as well as γ-chitin, which is present in the stomach of squids and cocoon fibers of beetles), are prevalent in more flexible structures. In orthorhombic α-chitin, two molecules run antiparallel per unit cell while β-chitin is monoclinic, and γ-chitin, which still causes disagreement over its existence, appears to be just a variation of α-chitin (Blackwell 1969, Minke and Blackwell 1978, Jang et al. 2004, Rinaudo 2007, Kaya et al. 2017a, Seenuvasan et al. 2020). In α-chitin strong intersheet and intrasheet hydrogen bonds form a network as a consequence of the crystalline structure, while β-chitin exhibits weak intrasheet H-bonds, which facilitate higher reactivity in reactions to attach functional groups and higher affinity to solvents than α-chitin (Rudall and Kenchington 1973, Kurita et al. 1993, Kurita et al. 1994, Jang et al. 2004). β-Chitin as well as γ-chitin can be irreversibly transformed into α-chitin by acid treatment (Saito et al. 1997, Yui et al. 2007).

2.2 *Chitosan*

Chitin is biocompatible, biodegradable, highly available, and non-toxic as well as active against microbes and displays little immunogenicity. However, it is insoluble in common solvents and intractable, which are attractive qualities in diverse applications but limit the wide utilization of chitin (Pillai et al. 2009). Chitosan, which is a valuable deacetylated chitin derivative that is more soluble, already in a mildly acidic environment, circumvents this drawback thus having found widespread applications. Chitosan and chitin can be distinguished based on their solubility in different media (Pillai et al. 2009). While chitin exhibits excellent stability against solvents as well as acids and alkaline media, chitosan is soluble in acidic conditions. This allows for easy processing, for instance by processes based on dissolution and casting. The differences in solubility are due to the chemical constitution of these polysaccharides (Figure 1), which define their applications either for solution-casting techniques or papermaking from aqueous suspension, respectively. This makes them attractive for use in the respective method. Chitin is based on an N-acetylglucosamin backbone, while the monomer of chitosan is glucosamine, derived by deacetylation of N-acetylglucosamin. One hundred percent pure chitin or chitosan, respectively, are very rare and, hence, commonly the degree of acetylation (DA) or degree of deacetylation (DDA) are used to define the level of (de)acetylation. The DA is the number of acetyl groups per saccharide monomer, while the DDA is the complementary value, hence one minus DA. Even though not 100% accurate, the definition goes down to the solubility (see above), at a DA greater than 50% (DDA smaller than 50%) materials are usually referred to as chitin and at a DA of less than 50% to chitosan. Figure 1 shows examples of chitin exhibiting a DA of 75% and chitosan having a DA of 25%, respectively (Pillai et al. 2009).

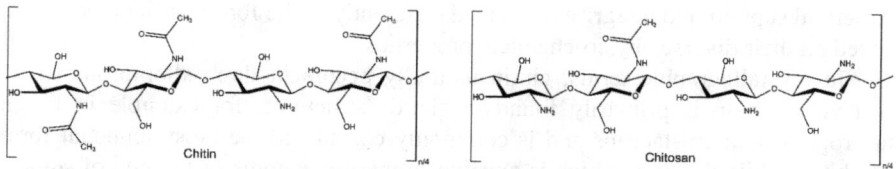

Figure 1: Structure of chitin (left) and chitosan (right) with degrees of acetylation of 0.75 and 0.25, respectively (Pillai et al. 2009).

Owing to its abundant free amine groups in addition to hydroxy groups (Ghaee et al. 2010), chitosan has an outstanding capability to adsorb ions (Vold et al. 2003, Sarode et al. 2019), which are based on the capacity to attract anions through electrostatic interactions and metal cations by chelation, respectively (Guibal 2004). Thus, chitosan can be efficiently used as adsorbent (Findon et al. 1993, Bilal et al. 2013).

3. Sources of Chitin

In nature, chitin is associated with different by-compounds depending on the form of life it is present in. This is similar to cellulose fibers that in plant matter are associated with lignin and hemi-celluloses. In shellfish, for instance, chitin is typically accompanied by minerals as well as sclerotized proteins (Nawawi et al. 2019a), whereas fungal chitin is associated with various polysaccharides like glucans, mannans, or chitosan dependent on the species (Muzzarelli 2011). So far most studies were based on chitin extracted from animal sources for the fabrication of films or (nano)papers (Muzzarelli and Pariser 1978, Kataoka and Ando 1979, Yusof et al. 2004, Ifuku et al. 2011, Kadokawa et al. 2011, Fan et al. 2012, Duan et al. 2013, Hassanzadeh et al. 2014, Ifuku et al. 2014, Mushi et al. 2014a, Mushi et al. 2014b, Jin et al. 2016, Kaya et al. 2017b, Casteleijn et al. 2018, Kaya et al. 2018, Kim et al. 2018). This was down to the high water content of fungi and the presence of further, mostly amorphous, polysaccharides, which diminish the yield of chitin, requiring added effort to extract pure chitin fibers. On the other hand, for the extraction of chitin from crustaceans, severe demineralization is obligatory which is not necessary for chitin derived from fungi.

Aside from the long-lasting preference for using chitin from animal sources, it was recognized quite early on that there is vast potential in fungal chitin (Allan et al. 1978). Chitin nanomaterials have been proposed as a potential reinforcing material in composites, however, huge inconsistency and insecurity in the availability of crustaceans, which are furthermore related to seasonal and regional fluctuations, intensified the need for considering chitin derived from fungi as a tangible alternative to chitin from animal sources (Nawawi et al. 2019a). Most importantly, fungal chitin natively exists in the shape of nanofibrils. These fibrils are prone to extraction from fungal biomass/cell walls with a very mild alkaline treatment and subsequent low-energy blending (Nawawi et al. 2019b, Jones et al. 2020a). Because of this easy isolation and extraction process of renewable nanofibrils of high availability, fungal chitin constitutes a feasible alternative to animal chitin (Figure 2) and also to

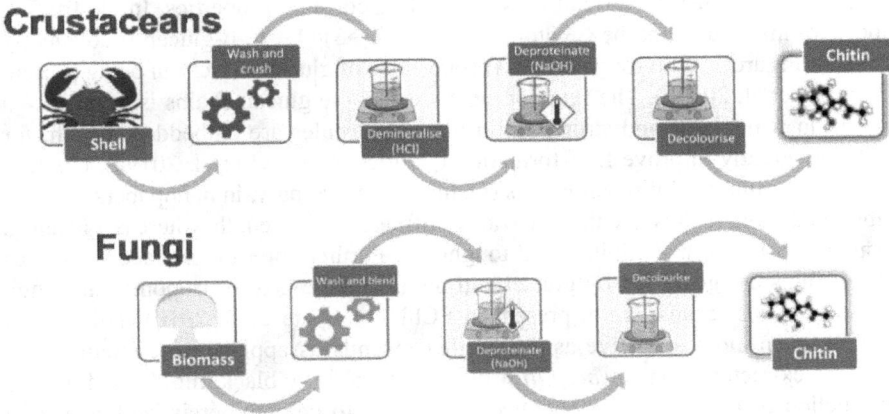

Figure 2: Extraction processes for chitin from crustacean- and fungi-derived raw materials, including mechanical (blending or crushing) and chemical (deproteination, demineralization, and decolorization) processes (Reprinted under CC BY 4.0 license from Jones et al. 2020a).

cellulose nanofibrils which are usually extracted from plant sources, e.g., paper pulp, necessitating high energy input during defibrillation (Jones et al. 2019a, Nawawi et al. 2019b). Fabricating nanopapers from natural nanofibrils is chiefly attractive for fungal chitin fibrils taking into account their nanoscale sizes which are already present in their native state. Thus, chitin nanofibrils but in particular nanopapers fabricated therefrom are an important approach for new materials based on chitin, with potential applications in composites materials or packaging (Jones et al. 2020b), as well as in procedures for the treatment of water (Janesch et al. 2020).

3.1 *Chitin from Fungal Resources*

In fungal chitin, mostly amorphous polysaccharides such as glucans are covalently bond to the backbone of the chitin macromolecule. This is the reason why chitin derived from fungal biomass was so far largely neglected as the presence of additional polysaccharides was commonly considered a disadvantage and glucan was mostly eliminated to obtain pure chitin. However, this apparent disadvantage was at last found to be an advantage for glucans do influence physico-chemical properties as well as the surface of chitin fibrils that are dissimilar to crustacean chitin (Nawawi et al. 2019a). Many publications studied the nature of covalent links between the chitin backbone and other polysaccharides like glucans, galactomannan, or manno-proteins, respectively. Thus, other than for crustacean-derived chitin, where merely residual fractions of other polysaccharides or proteins are present, fungal chitin contains significant amounts of the polysaccharide glucan. Their share is frequently higher than the chitin content itself, dependent on the species of the fungi (Sietsma and Wessels 1979, Surarit et al. 1988, Hartland et al. 1994, Kollár et al. 1995. Kollár et al. 1997, Fontaine et al. 2000, Heux et al. 2000, Cabib and Arroyo 2013, Stalhberger et al. 2014, Nawawi et al. 2019a, Janesch et al. 2020).

Frequently white button mushrooms (*Agaricus bisporus*, AB) served as model system for the study of fungi-derived chitin for their high availability from large-scale cultivation in industrial agriculture safeguarding constant properties. In the fruiting bodies of these fungi, at the C3 atom of chitin, $(1{\rightarrow}3)/(1{\rightarrow}6)$-β-glucan is covalently bonded (Figure 3) with roughly equal proportions of glucan and chitin being present (Nawawi et al. 2019b). The function of those flexible glucan chains is to provide a matrix in which stiff and strong chitin macromolecules are embedded, which, for instance, greatly improve film-formation qualities (Nawawi et al. 2019a), together constituting fungal chitin nanofibrils (FChNF). Accordingly, in nanopapers or films fabricated from FChNF, chitin provides stiffness and strength whereas glucan is responsible for both flexibility and toughness. Furthermore, the surface properties of the films are governed by glucan and the allergenic protein tropomyosin, often present in crustaceans, is not present in FChNF (Lopata et al. 2010). This makes FChNF even more attractive as this further extends its applicability. Exemplarily, FChNF extracted from *Aspergillus niger*, which is a black mold used in the production of penicillin, has obtained permission to be commercialized as a food supplement (Versali et al. 2009, EFSA 2010). Another advantage of FChNF over crustacean chitin and even nanocellulose is that FChNF are natively nano-sized and thus nanofibrils are easily isolated by mild extraction processes requiring much less energy for their extraction compared to the high amount of energy necessary for mechanical and chemical extraction processes utilized for extracting both crustacean-derived nanochitin fibrils or even nanocellulose (Khor 2001). This enables the use of renewable nanofibrils with higher sustainability potentially replacing synthetic polymer fibrils or even glass or carbon fibers.

In mushrooms, FChNF is present in the fruiting bodies, i.e., stalk and cap, and often extracted, for instance, from white AB serving as a classic system exhibiting appropriate stability in their composition and hence properties. AB is nutritious and also used as functional food supplement as a free radical scavenger and antioxidant (Lindequist et al. 2005, Guan et al. 2013, Lin et al. 2017). Chitin, glucan, and proteins are the main constituents of these fungi in their cell wall with soluble proteins as well

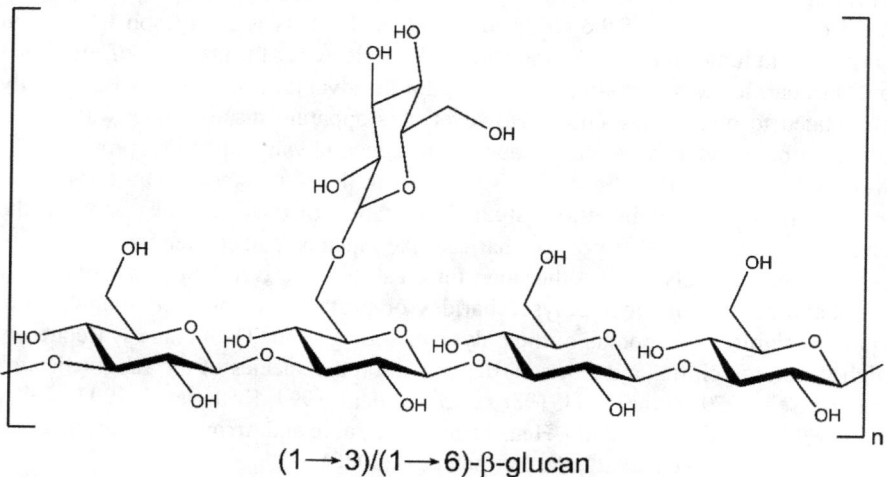

$(1{\longrightarrow}3)/(1{\longrightarrow}6)$-β-glucan

Figure 3: Structure of glucans in *A. bisporus*: $(1{\rightarrow}3)/(1{\rightarrow}6)$-β-glucan (Nawawi et al. 2019b).

as polysaccharides with no structural function also being present (Hammond 1979). There are also secondary compounds exhibiting cytostatic, anti-mutagenic, and genoprotective activity present, which is mostly caused by lectins and the enzyme tyrosinase that are potentially advantageous for human health (Yu et al. 1993, Shi et al. 2004, Lindequist et al. 2005). It has to be mentioned, though, that AB is an edible mushroom, whereby using food as material triggers ethical issues. Thus, albeit studies on FChNF based on edible mushrooms provided important insights into the utility of FChNF, alternatives would be most welcome which could, for instance, be non-edible mushrooms such as tree brackets or even the mycelium of fungi, i.e., the vegetative part of fungi. For such raw materials, ethical concerns regarding the use of food for material applications are much less pressing.

3.2 *Mycelium as Raw Material for Fungal Chitin*

Collectively, hyphae, which are chitinous micro-fibers that form the vegetative part of filamentous fungi, are called mycelium (Jones et al. 2019b). During the last decade, mycelium has attracted tremendous commercial and academic attention as a natural binder of biomass, for instance in packaging and textile materials, construction applications, or thermal and acoustic insulation, by fashioning environmentally innocuous and economic materials through the digestion and bonding of organic biomass (Holt et al. 2012, López Nava et al. 2016, Jones et al. 2017, Camere and Karana 2018, Jones et al. 2018, Karana et al. 2018, Appels et al. 2019). Exemplarily, shoe or textile manufacturers have demonstrated products based on leather-like materials derived from mycelium (Forbes-Magazine 2021, Jones et al. 2021).

In mycelium composites, a continuous matrix phase based on fibrous mycelium (Pelletier et al. 2013, Thakur and Siṅgha 2013) interfaces and thus combines with a dispersed phase of biomass that also constitutes the substrate for mycelium growth by which it is partially digested (Figure 4). Through this process the volume of the composite increases, in which organic biomass' function is being a filler and feedstock within and for a hyphal microfilament network (Holt et al. 2012, Pelletier et al. 2013, Thakur and Siṅgha 2013, Haneef et al. 2017, Islam et al. 2017, Jones et al. 2017). The natural biological growth allows for the fabrication of environmentally benign substitutes for synthetic materials such as planar polymer films and sheets and are manufactured by manufacturing processes demanding low amounts of energy (Haneef et al. 2017). Furthermore, 3-dimensional low-density objects of larger volume in the shape of foams or light-weight materials such as polymers can be facilitated (Holt et al. 2012, Pelletier et al. 2013, Travaglini et al. 2013, López Nava et al. 2016, Camere and Karana 2018, Karana et al. 2018). Key advantages of mycelium materials over conventional synthetic materials include biodegradability as well as low carbon footprint, cost, energy demand, and environmental impact (Arifin and Yusuf 2013, Haneef et al. 2017). On the other hand, very low density caused by a high porosity in mycelium materials often results in low mechanical properties that are commonly on the level of polymer foams. The reason for this is that so far usually agricultural waste and side stream materials, usually lignocellulosic fibers, were utilized as substrate for typical mycelium composites which in turn are composed of a composite

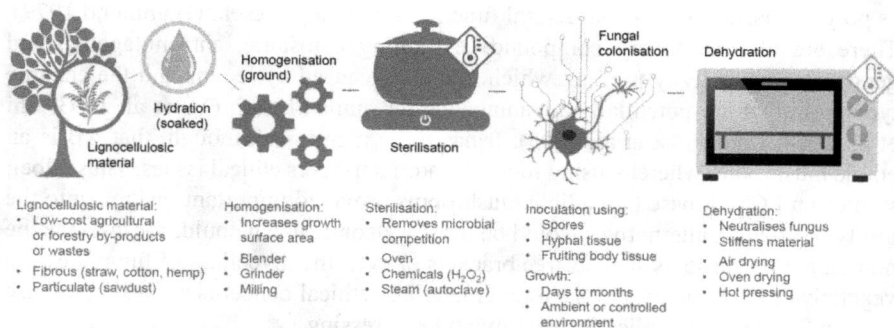

Figure 4: Scheme of the fabrication process of mycelium composites showing the main stages, purpose, and potential variations for each step of the production process (Reprinted under CC BY 4.0 license from Jones et al. 2020b).

constituted of fibrous fungal mycelium with partly digested biomass. Thus, ultimate tensile strengths of up to 10 MPa or regularly much less are achieved (Haneef et al. 2017, Appels et al. 2018). The low quality of side-stream substrates used in the production of mycelium materials is responsible for limited mechanical strength. The low strength agricultural waste-stream, side-stream, or by-product substrates of low strength themselves limit the mechanical performance of the composites and, in addition, are often connected to the hyphal filament matrix by few and thus overall weak bonds (Appels et al. 2019). The use of such low-quality biomass is attributed to the original idea behind utilizing mycelium composites which was the attempt of upgrading waste biomass (Jones et al. 2021). Mycelium composites also suffer from the low strength of the fibrous mycelium matrix itself caused by the presence of non-structural elements, such as cytoplasm, proteins, and lipids (Kavanagh 2005).

One option to improve the mechanical performance of mycelium-derived materials could be to exploit agricultural side streams commonly utilized merely as feedstock, substrate, and nutrient source for the growth of mycelium. This would eliminate these low-strength side-stream or by-products as fillers from the final composites. Instead, structural components of fungal biomass, i.e., chitin and other polysaccharides, could be extracted from the grown composite by eliminating non-structural compounds from the growth culture (Jones et al. 2019a). Such a process would permit the transformation of agricultural or industrial biomass side-streams into renewable polymers initiated by fungal growth. Natural polymer fibers, i.e., FChNF, are thus obtained from the hyphae cell walls within the mycelium biomass, which, similar to their appearance in fungal fruiting bodies, are natively present with dimensions at the nanoscale (Rinaudo 2007, Jones et al. 2017). The hydrogen bond network along the polysaccharide backbone of those fibers is responsible for rigidity and strength, with FChNF exhibiting a very high tensile strength of ~ 1.6 to 3.0 GPa (Webster and Weber 2007, Bamba et al. 2017). Alike FChNF from mushroom fruiting bodies, mycelium-derived chitin-nanofibrils are linked to flexible and branched β-glucan or other polysaccharide macromolecules constituting a tough and strong native nanocomposite architecture (Nawawi et al. 2019b). A further analogy to fungi fruiting body-derived chitin is that mycelium-sourced FChNF constitutes an abundant, cheap, and renewable alternative to seasonally as well as regionally limited and possibly allergenic crustacean chitin that is easily isolated (Di Mario et al. 2008, Ifuku 2014, Hassainia et al. 2018). On top of that, chitin derived from mycelium does not directly compete with food supply casting it more sustainable than mushroom-fruiting body-derived FChNF. Last but not least, the mycelium grows faster than the fruiting bodies.

4. Composite Nanopapers

Similar to ordinary office paper, which primarily is a random network of flexible cellulose fibers with diameters on the microscale, nanopapers consist of flexible fibrils, but on a different size scale, the nano-scale. Nanopapers were originally established for cellulose nanofibrils (Lee et al. 2012) as versatile network structures with utility in composite materials (Lee et al. 2014, Benítez and Walther 2017) or water treatment applications (Mautner 2020). Because of the high specific surface areas and abundance of hydroxy groups caused by the nano-scale of the fibrils forming the network structure, and accordingly many contact points and thus interactions, nanopapers exhibit excellent mechanical properties and membrane separation efficiency. Such developments initially targeting cellulose nanofibrils can be translated to another type of biomacromolecule nanofibril: fungal chitin-glucan (CG). Beneficial for the fabrication of nanofibrils in this regard is the native nanoscale dimension of FChNF which helps to minimize the energy demand required. Subsequently, also nanopapers fabricated from FChNF could be manufactured with lower amounts of energy than cellulose nanopapers. Furthermore, additional functional N-acetyl groups positioned at C2 of the macromolecular backbone are present (Nawawi et al. 2019b, Nawawi et al. 2020b). Constituting a native composite material, formed of chitin fibrils within a matrix of glucans, nanopapers based on FChNF combine the mechanical strength and stiffness of chitin with the flexibility and toughness of the glucan matrix phase, whereby FChNF were proposed beneficial over cellulose nanopapers.

4.1 Fungal Chitin-Glucan Nanopapers

Frequently, for the production of planar sheets, i.e., nanopapers, membranes, or films from chitinous materials, chitosan is utilized for its high solubility, whereas fabrication of such structures from chitin is much less attractive due to its limited solubility (Janesch et al. 2020). This is generally limiting the applicability of crustacean chitin, albeit, in specialized applications, e.g., organic solvent filtration, this low solubility constitutes an important advantage. For fungal chitin, the case is different. Here, film formation properties are proposed to be improved over animal-derived chitin owing to the presence of glucan covalently linked to the macromolecular chitin backbone. This matrix of glucan embedding chitin fibrils is anticipated to support the fabrication of strong and tough nanopaper networks (Nawawi et al. 2019b, Nawawi et al. 2020b). Furthermore, native fungal chitin-glucan fibrils exhibit diameters in the nanometer region thus constituting a renewable resource for nanofibrils and also manufacture of nanopapers from FChNF is possible with low energy demand. Easy and cheap access to nanopapers is important for instance in the domain of water treatment membranes. For cellulose nanopapers it was reported that the diameters of the fibrils shaping nanopaper size-exclusion membranes determine their pore size (Mautner et al. 2014). For FChNF this correlation is also anticipated, hence, pore sizes of nanopapers from fungal chitin were also proposed to be in the low nm range. The polar N-acetamide group coupled to the C2 of the polysaccharide backbone opens up a further possibility to apply FChNF nanopapers in membrane applications. This functional group exhibits attraction towards ions, e.g., metal ions, which facilitates adsorption of this type of pollutants. Thus, membranes exerting separation function in both the adsorption regime and size-exclusion based on FChNF would be possible. Moreover, deacetylation of N-acetyl groups yields glucosamine, which exercises even stronger interaction with charged moieties.

In order to study the utility of fungal raw materials for nanopaper fabrication, FChNF were extracted and characterized. Initially, FChNF were sourced from a model fungus: white button mushroom *A. bisporus*. For a natural material AB is highly available at appropriately constant composition since AB is cultivated in large-scale agriculture. In order to evaluate differences of FChNF from various parts of the fungi, whole fruiting bodies and stalk or cap, respectively, alone, were the constituents used to isolate FChNF from AB. For successful isolation of FChNF, it is key to avoid cleavage of glucan. Hence, a procedure mild enough to preserve the content and constitution of the native glucan fraction had to be developed. Opposed, for extracting chitin fibrils from crustaceans usually harsh protocols are employed using strong alkaline and acidic solutions as well as elevated temperatures. Applying these conditions to fungal raw material would result in hydrolyzing covalent bonds eventually cleaving glucan from the chitin backbone. Thus, for the extraction of FChNF, treatment of fungal biomass with hot water was performed in order to remove water-soluble constituents, e.g., low molecular weight (M_W) sugars or proteins, followed by a mild alkaline treatment to completely remove proteins and other byproducts (Nawawi et al. 2019b, 2020b). Elemental and polysaccharide analysis was utilized to investigate the chemical composition of FChNF. It was found that about 50 wt.-% of the AB extract was glucan while the other half was chitin. Nanoscale diameters of FChNF of about 15 nm were confirmed by electron microscopy, which happens to be thinner than chitin fibrils derived from crustaceans and also compared to common unmodified cellulose nanofibrils that are usually extracted from paper pulp.

After the successful extraction of FChNF, nanopapers were fabricated and surface properties were investigated. It was reported that the presence of glucan acting as an amorphous matrix for chitin fibrils improved and facilitated film formation (Figure 5 top). This circumvented a disadvantage of chitin fibrils extracted from crustaceans yielding nanopapers of poor mechanical performance. Furthermore, glucan as well as residues of hydrophobins also caused surfaces of FChNF nanopapers to be more hydrophobic. This was indicated by water contact angle values that were

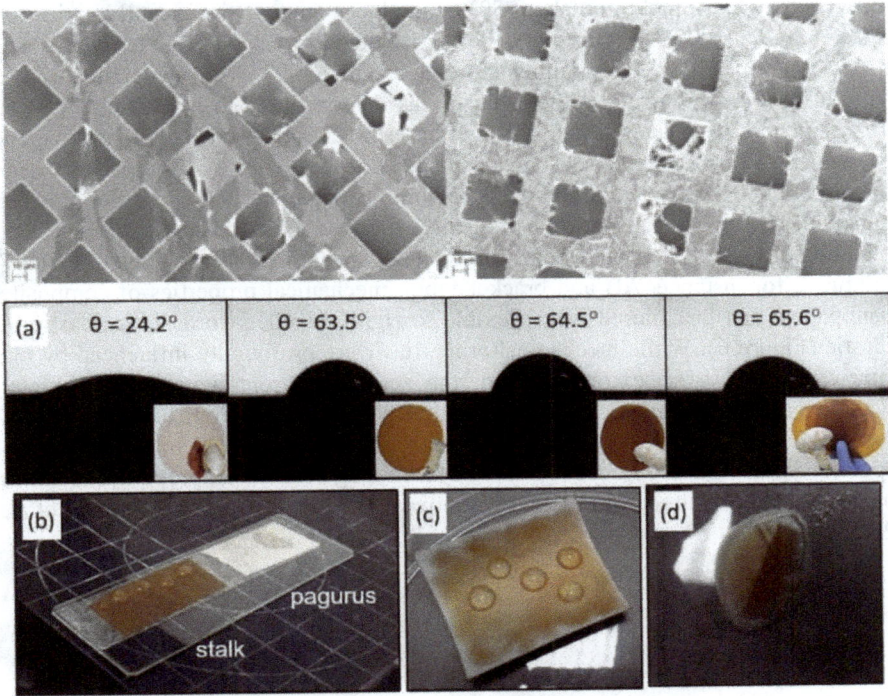

Figure 5: Top: Suspensions of chitin-glucan (left) and crustacean chitin (right) dried on a Cu grid indicating film formation properties due to the absence or presence, respectively, of glucan. Bottom: (a) Images of water droplets residing on various nanopaper surfaces and water contact angles θ of water droplets 1 min after deposition on (from left to right) *C. pagurus*, AB stalk, AB cap, and whole AB; pictures of water drops 1 h after deposition on (b) stalk and *C. pagurus*, (c) AB coated blotting paper, and (d) AB coated filter paper (Reprinted from Nawawi et al. 2020b, Copyright 2020 with permission from Elsevier).

higher for FChNF (θ = 65°) compared to chitin extracted from *Cancer pagurus* (θ = 24°) as shown in Figure 5 (bottom). Varying grammage and accordingly thickness of FChNF nanopapers had a decisive influence on the mechanical performance. An optimal nanopaper grammage of 80 grams per square meter was identified, which is similar to common office paper. Optimized FChNF nanopapers exhibited a tensile strength (200 MPa) that was a multiple of printing paper and on a level with nanopapers manufactured from unmodified nanocellulose (Mautner et al. 2018) that have to be extracted by harsh defibrillation, which is a process requiring a lot of energy (Nawawi et al. 2019b, 2020b). This demonstrated that the presence of glucan in FChNF derived from white button mushroom and subsequently in nanopapers produced thereof is responsible for mechanical and surface properties that vary tremendously from those of chitin derived from animal sources.

Both mechanical stiffness and strength were very high for FChNF nanopapers, which was attributed to the diameters of FChNF which are natively in the nanoscale, while it is also possible to tailor mechanical properties. Whereas in white button mushrooms glucan makes up about 50%, the glucan content significantly varies for different fungi species. For instance, FChNF isolated from *Daedaleopsis confragosa*, bracket fungi, FChNF contain much more glucan up to 99% (Nawawi et al., 2020a). This in turn reinforces the effect glucan was demonstrated to exhibit on FChNF, i.e., tremendously increased flexibility, with tree bracket-derived FChNF nanopapers even outperforming AB-derived nanopapers in terms of flexibility. This also opens up the possibility to tune the mechanical behavior of nanopapers by preparing composite nanopapers from various grades of FChNF derived from different species, i.e., blending of FChNF extracted from various fungi species. By formulation of FChNF mixtures, for instance, AB and bracket fungi, mechanical properties of composite nanopapers can be tailored from flexible to rigid, in other words from elastic to plastic (Figure 6). While mechanical properties were profoundly influenced by the species FChNF were extracted from, the surface hydrophilicity/hydrophobicity was, in general, the same for FChNF nanopapers prepared from brackets or AB, respectively, as well as composite nanopapers fabricated from blends of those two species. Applying this approach facilitates fashioning mechanical properties of chitin nanopapers as required but preserving the properties of the surface (Nawawi et al. 2020a).

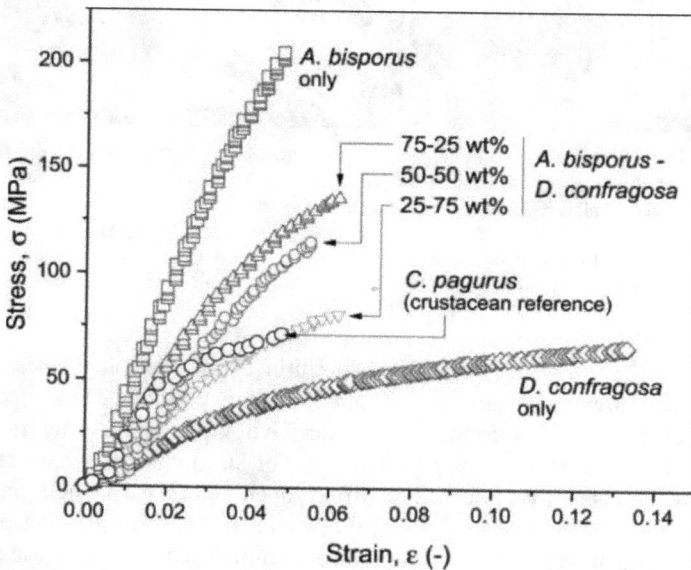

Figure 6: Stress-strain curves from tensile tests of FCHNF nanopapers derived from AB, *D. confragosa*, and blends containing various proportions thereof compared to chitin papers obtained from *C. pagurus* (Reprinted under CC BY 4.0 license from Nawawi et al. 2020a).

While high mechanical properties of random network structures in nanopapers similar to cellulose nanopapers are caused by native nanoscale dimensions of FChNF, such nanopapers are also anticipated to possess pores with diameters in the nm range as well (Mautner et al. 2014, 2015). Hence, also FChNF composite nanopapers derived from AB were proposed to be suitable for ultrafiltration processes utilizing the size exclusion mechanism. In addition, interactions of the N-acetyl group in FChNF with various pollutants would facilitate the adsorption of, e.g., (heavy) metal ions from water (Yousefi et al. 2021). AB-derived FChNF nanopapers were tested for their membrane characteristics and an inverted logarithmic correlation between grammage, which is proportional to nanopaper thickness, and permeance was found. These properties are comparable to cellulose nanopapers (Mautner et al. 2015). FChNF nanopapers exhibited pore sizes of about 10 nm, which is similar to their fibril diameter (15 nm). The pore size was determined by measuring the molecular weight cut-off (MWCO) for poly(ethylene glycol) which was 17 kDa (Figure 7). Moreover, the very high solvent stability of chitin can be utilized advantageously by not only being useful in water filtration but also in another emerging field: organic solvent nanofiltration (OSNF) (Marchetti et al. 2014, Mautner et al. 2014). Rejection of polystyrene macromolecules with an M_W of 45 kDa dissolved in tetrahydrofuran was achieved, which relates to a pore size of 12 nm. Differences in the pore size for various filtration media and macromolecules are caused by different swelling behavior in different media and differences in the coil structure of water-soluble polymers (e.g., poly(ethylene glycol)) and organic solvent soluble polymers (e.g., polystyrene). This indicated that FChNF composite nanopapers are capable of facilitating the filtration of nanoparticles in the tight UF range for both aqueous solution and organic solvents, which was further proven by the rejection of gold nanoparticles (Yousefi et al. 2021).

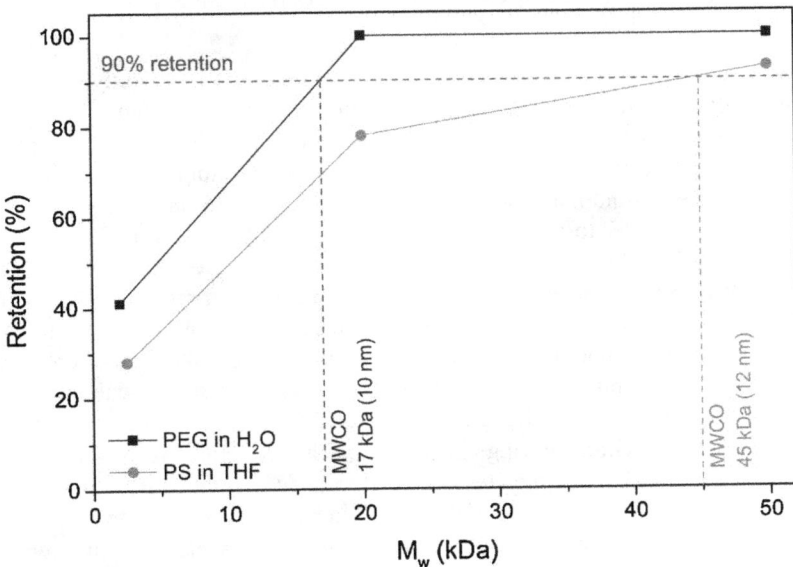

Figure 7: Rejection of macromolecules by FChNF nanopapers: MWCO for (a) poly(ethylene glycol) (PEG) dissolved in water and (b) polystyrene (PS) dissolved in tetrahydrofuran (THF). MWCO is defined as the M_W of a solute which is rejected 90% by a membrane (Reprinted from Yousefi et al. 2021, Copyright 2020 with permission from Elsevier).

FChNF nanopaper filters were shown to be able to reject nanoparticles by a size exclusion mechanism based on the pore size of their nanofibril network and are also capable of adsorbing (heavy) metal ions, exhibiting an adsorption capacity (q_E) for copper, which was utilized as a model compound for heavy metal ions, higher than 40 mg g^{-1}. Slightly lower q_E for FChNF compared to crustacean-derived chitin was explained by the presence of glucans. Glucans, making up about half of the FChNF, inherently reduce the content of chitin per unit mass, which in turn also reduces the amount of functional N-acetyl amino groups on the surface of FChNF, the primary functional groups taking part in adsorption. In order to improve the membrane and filter performance of FChNF nanopapers in filtration applications, i.e., permeance and q_E, FChNF were blended with cellulose microfibers to manufacture hierarchical composite papers. This approach was primarily directed at enhancing permeance by increasing both pore size and porosity but also to improve adsorption capacities by making functional groups of FChNF better available to adsorbates by spreading them along the wider network of cellulose microfibrils. Highly porous and thus permeable composite filters were obtained with an areic adsorption of 805 mg copper ions per m^2 membrane area, which is equal to a q_E of 81 mg g^{-1}. Increased q_E was caused by better availability of N-acetyl amino groups present at the surface of FChNF within a highly porous network of cellulose microfibrils as opposed to pure, dense FChNF nanopapers being shaped only of CG nanofibrils (Yousefi et al. 2021).

4.2 Chitosan-Glucan Composite Nanopapers by Deacetylation of Fungal Chitin

Cleavage of acetyl groups from CG, i.e., deacetylation, is commonly executed by strong alkaline media, e.g., caustic soda, yielding chitosan-glucan. Chitosan is a derivative of chitin in which N-acetyl groups are removed leaving behind primary amine groups. Because of the lone pair at the nitrogen atom of the amine group and reduced steric hindrance, adsorption of ions onto chitosan is easier facilitated than to chitin. In order to deacetylate chitin usually a hydrolysis protocol in strong alkaline media is followed, for instance in concentrated aqueous sodium hydroxide solution. Deacetylation of crustacean chitin takes place under rather harsh conditions at high temperatures and in a concentrated alkaline solution. During this process also removal of certain compounds that naturally exist in the exoskeleton of crustaceans is facilitated. More importantly, for there is no glucan present covalently bonded to chitin fibers, hydrolysis of glucan is no issue that needs to be considered for crustacean-derived chitin during treatment with concentrated NaOH solution. However, for FChNF, due to the presence of glucan, deacetylation needs to be carried out with care in order to not cleave glucans, which are responsible, for instance, for good film formation properties. Cleavage of glucan takes place when exposed to similar conditions used for deacetylation reactions, hence, the protocol to deacetylate fungal chitin needs to be carefully controlled in order to not remove glucan (Nawawi et al. 2019b).

Deacetylation of FChNF from AB was selected as the model system. After initial extraction of chitin-glucan from the fruiting bodies, deacetylation was performed using aqueous sodium hydroxide solutions of different concentrations (20, 40, and 60%) to yield chitosan fibrils connected to glucan (CSG) in the form of a gel (Janesch et al. 2020). The various CSG grades were submitted to analysis of the chemical composition. Unfortunately, that was not straightforward as three principal components (chitin, chitosan, and glucan) are present, and all of these compounds are polysaccharides of similar chemical composition. Thus, a variety of analytical techniques (carbohydrate and elemental analysis as well as Fourier-Transform infrared (FT-IR) and nuclear magnetic resonance (NMR) spectroscopy) had to be combined in order to determine the contents of these components. It was revealed that increasing both NaOH concentration and treatment time yielded extracts with higher overall contents of chitosan (~ 39%). This value had to be corrected for the content of glucan, which was as determined by carbohydrate analysis, in order to obtain a value of the degree of deacetylation (DDA) of chitin/chitosan macromolecules that was 58%. Already at an NaOH concentration of 20%, approximately one-third of glucan present in the original, non-deacetylated FChNF from AB was cleaved, however, the major content of glucan was preserved (Janesch et al. 2020). The partial removal of glucan yielded a distinct influence on the mechanical performance of nanopapers prepared from CSG. Tensile strength and Young's modulus were both decreased dependent on the NaOH concentration and duration of treatment, as determined by tensile tests. For nanopapers fabricated from FChNF and deacetylated with moderate concentrations of NaOH (20%), a strain to failure of more than 5%, similar to non-deacetylated FChNF nanopapers were found, but specimens exhibited significant yielding at stresses higher than 100 MPa, resulting in about half the tensile strength of CG nanopapers. At higher concentrations of NaOH (40%) both strength and strain were significantly reduced and yielding was initiated already at lower stresses. Apart from partial removal of glucan this was attributed to hydrolysis/chain cleavage of the polysaccharide backbone of all polysaccharides present, including chitin, taking place at elevated temperature in an alkaline medium.

One goal of deacetylating FChNF was to impact the surface properties by changing the functional groups attached to the surface of FChNF, however, the filtration performance was influenced, too. Reducing the number of acetyl groups attached to the surface of FChNF yielded the anticipated effect on the surface charge of the nanofbrils. The pH at which no net charge is present, i.e., the iso-electric point, increased and a more negative zeta (ζ)-potential plateau indicated more negative surface charge. These two effects hint at more amine groups, as opposed to N-acetyl amino groups, attached to the surface of FChNF and accordingly also to the surface of nanopapers (Figure 8). The membrane properties of FChNF nanopapers prepared from the different CSG grades were also altered, with permeances being somewhat lower than for CG nanopapers. This phenomenon was caused by the stronger interaction of more hydrophilic amine groups with water which yielded a chromatographic effect. The second important membrane property aside from permeance is the rejection of pollutants, either by virtue of size exclusion or adsorption. All grades of chitin/chitosan-glucan composite nanopapers exhibited pores with diameters of roughly 10 nm, which was confirmed by their ability to reject 10 nm Au nanoparticles. This result was comparable to chitin-glucan nanopapers and thus in alignment with the

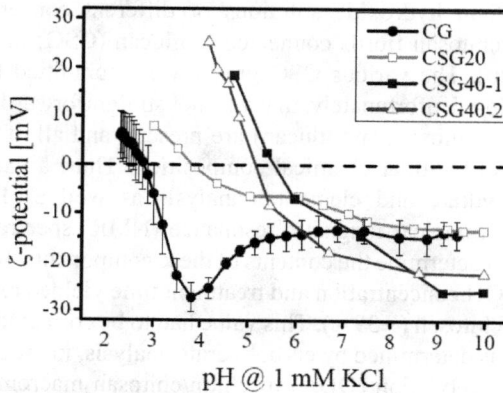

Figure 8: Surface charge as expressed by the ζ-potential = f(pH) for CG (black dots) nanopapers and partially deacetylated CSG nanopapers. CSG20-1: 20% NaOH, 1 h, open orange squares; CSG40-1: 40% NaOH, 1 h, full blue squares; CSG40-2: 40% NaOH, 2 h, open green triangles (Reprinted from Janesch et al. 2020, Copyright 2019 with permission from Elsevier).

hypothesis which proposes that the fibril diameter controls the pore size and thus the rejection performance of size exclusion filters. CSG nanofibrils were on average of the same diameters as CG nanofibrils, whereby also the pore size of networks formed by such nanofibrils was approximately in this size domain. Thus, the retention of pure C(S)G nanopapers towards nanoscale contaminants was substantial considering their fully renewable origin, however, the permeance of such nanopapers was still to be optimized.

An approach to tailor permeance properties while still conserving rejection properties was to create hierarchical composite filters of C(S)G with cellulose microfibrils, i.e., paper pulp fibers, similar to CG-cellulose composite filters that were designed to facilitate high permeance in conjunction with the appropriate rejection of pollutants. Astoundingly, adsorption capacities of up to 162 mg g^{-1} were reported (Janesch et al. 2020). This demonstrated that various types of acetylated/deacetylated FChNF nanopapers exhibiting high adsorption capacities could be fabricated by a papermaking process without the requirement for dissolving CSG in order to prepare films by casting methods. FChNF nanopapers thus fabricated exhibited size-exclusion properties capable of rejecting viruses, but also adsorbent nanopapers filters with high q$_E$ towards heavy metal ions could be fashioned.

4.3 *Chitin-Glucan Composite Nanopapers Derived from Mycelium*

Mushrooms, i.e., their fruiting bodies, are particularly suitable as a source for extracting fungal chitin-glucan that can eventually also be used as a precursor for the fabrication of chitosan-glucan nanofibrils. From these extracts manufacture of nanopapers exhibiting suitable properties for applications in structural composite and filtration applications is facilitated. Often white button mushrooms served as a model source for CG for it is already produced at large-scale cultivation facilities and hence their properties are reasonably reproducible for a natural product. While white button mushrooms are an almost ideal model system, it has to be emphasized that there are implications present when edible biomass is used as a material. Competition with food supply is inevitable when utilizing edible mushrooms, which is an ethical issue. Thus for improved sustainability of the raw material for FChNF, i.e., circumventing the use of fruiting bodies of edible mushrooms, culturing and using the vegetative part of mushrooms, that is mycelium, as raw material for fungal chitin, and subsequently, chitosan would be important in order to avoid competition with food supply (Jones et al. 2017, 2020b, 2021). Similar to the hyphae of mushroom fruiting bodies, mycelium primarily consists of chitin covalently bonded to polysaccharides such as glucan, forming FChNF. Importantly, chitin fibrils in mycelium are natively present as nanofibrils, too (Jones et al. 2017).

Usually, mycelium materials come in the form of composites joined together with the substrate they have been grown on and are applied in the form they were prepared. This allows for the fabrication of cheap renewable materials but these materials typically suffer from low mechanical properties. Low mechanical properties of such materials are caused by commonly utilizing low-grade biomass as a substrate that is usually of low mechanical integrity. Even though the ingrowth of mycelium binds the substrate still the originally weak constituents sacrifice better overall mechanical performance. To circumvent this drawback, higher-grade biomass could be utilized, which, on the other hand, would reduce the efficiency and sustainability of such materials, whose origins were down to the concept of upgrading low-grade biomass else considered a waste material. Another option to obtain mycelium-based materials of better performance is to grow mycelium on biocompatible, abundant, and cheap agricultural waste or side-streams but then to separate FChNF from the substrate-mycelium compound cultured and utilize these mycelium-based nanofibrils for the production of composite nanopapers. Initially, appropriate culture media and substrates had to be identified. Ideally, still following the concept of upcycling agricultural side-streams, those could be for instance blackstrap molasses, rice hulls, wheat straw, or sugarcane bagasse (Jones et al. 2019b). It was reported that blackstrap molasses were particularly suitable a choice for growing mycelium. Compared to high-quality growth media frequently used as nutrients in laboratory experiments for instance, malt extract, these substrates actually outperformed common growth media.

Upon a suitable culture medium had been identified, the second step in the process was screening species of fungi for high efficacy synthesizing FChNF (Jones et al. 2019a). It was found that not only the yield and rate of growth majorly differ between various species, but also different combinations of chitin and polysaccharides such as glucans or chitosan were generated. From the mycelium cultures grown on blackstrap molasses, structural polymer composite fibrils were extracted. This extraction of FChNF included hot-water and alkaline treatment. Extracted FChNF were then homogeneously dispersed in aqueous suspension and subsequently nanopapers fabricated by a vacuum filtration process of the dispersion and consolidation of the filter cake by hot-pressing (Figure 9).

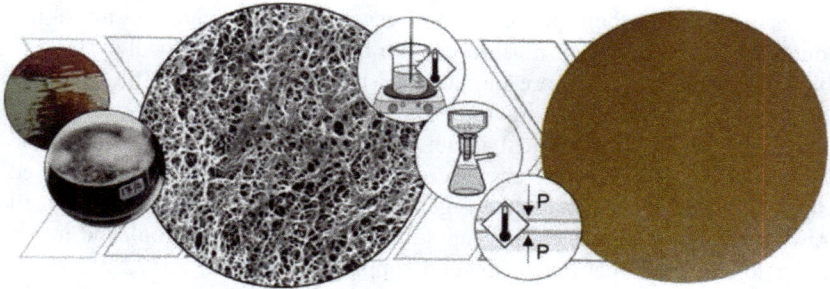

Figure 9: Process for the fabrication of composite nanopapers from mycelium cultures. The culture medium consisting of molasses is first inoculated with the respective species which then grows forming a thick hyphal network. The biomass-mycelium culture is then treated using sodium hydroxide, residues collected, vacuum-filtered, and finally hot-pressed (Reprinted from Jones et al. 2019a, Copyright 2019 with permission from the American Chemical Society).

Mycelium chitin-composite fibres were produced from *Trametes versicolor*, *Mucor genevensis*, and *Allomyces arbuscular*, with dry contents being on the same level with yields reported for both white button mushrooms fruiting bodies and chitin derived from animal sources (Jones et al. 2019a, Nawawi et al. 2019b, 2020b). Chemical, compositional, elemental, and carbohydrate analysis revealed a higher glucan-content in extracts from mycelium cultures compared to AB fruit bodies as shown by lower contents of glucosamine and nitrogen. Similar to fruiting body-derived FChNF, for which higher glucan contents result in lower mechanical stiffness and strength of nanopapers produced from such extracts, also for these mycelium-derived FChNF, higher glucan contents resulted in lower mechanical integrity than the outstanding quality of CG nanopapers derived from fruit bodies of white button mushroom. Not only high glucan contents were responsible for this result but also the presence of other saccharide fractions (predominantly galactose but also maltose or arabinose) which both inherently cause lower chitin contents. Furthermore, the biomineralization of calcium caused impurities that were detrimental to high mechanical strength and stiffness (Jones et al. 2019a). Nevertheless, even though with mycelium-derived nanopapers mechanical properties of nanopapers prepared from FChNF derived from fruiting bodies of AB could not be obtained, mechanical properties of crustacean-based chitin nanopapers were met or even outperformed (Nawawi et al. 2019b). Still, there is vast space for improvements by optimizing extraction and purification processes. Apart from the study of mechanical properties, the properties of the surface of the nanopapers need to be considered. Similar to nanopapers from fungal fruiting body-derived FChNF, the presence of glucan played a major role in surface properties. The surface properties of nanopapers fabricated from FChNF extracted from mycelium grown on suitable agricultural residues could be tailored from hydrophobic to hydrophilic (Figure 10). Thus, there is a wide spectrum of properties available for nanopapers fabricated from sustainable raw material (Jones et al. 2019a).

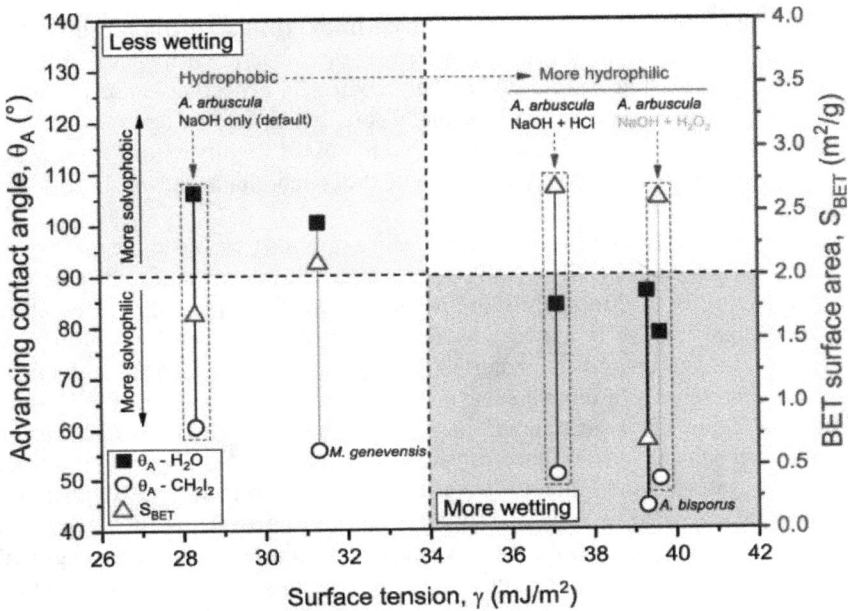

Figure 10: Advancing contact angles (θ_A, open circles (CH_2I_2) and blue full squares (H_2O)), specific surface area (S_{BET}, red hollow triangles, $m^2 \ g^{-1}$), and surface tension (mJ m^{-2}) for *M. genevensis* and *A. arbuscula* NaOH treated mycelium-derived nanopapers, *A. bisporus* fruiting body-derived nanopapers, and *A. arbuscula* NaOH and H_2O_2 or HCl, respectively, treated nanopapers. The transition between hydrophilic and hydrophobic properties following HCl or H_2O_2 treatment is marked by red dashed boxes (Reprinted from Jones et al. 2019a, Copyright 2019 with permission from the American Chemical Society).

5. Conclusion

Nano-sized fibers made from a variety of materials are a central class of materials that exhibit outstanding mechanical properties important for example in composite materials. Considering environmental problems associated with synthetic materials of non-renewable origin, by producing such nanofibers from sustainable raw materials by utilization of renewable resources ecologically benign solutions could be enabled. At present, nanofibers from renewable resources are commonly based on (nano)cellulose or crustacean-derived chitin fabricated by energy-intensive defibrillation of microfibers. Raw materials are for instance based on wood pulp, which consumes a high amount of energy and/or requires harsh chemical treatment. Chitin fibers extracted from fungi natively possess nanoscale dimensions, which reduces the energy demand for defibrillation and moreover provides additional surface functionality. Glucans are covalently bonded to fungal chitin fibers which facilitate the fabrication of composite nanopapers by an aqueous papermaking process that is scalable. Nanopaper networks possess mechanical properties that are prone to tuning by selecting the species utilized for the extraction of nanofibrils. Glucans and residues of hydrophobins are also responsible for more hydrophobic surfaces compared to papers or films derived from crustacean chitin. This allows for structural applications or filtration processes not feasible for highly hydrophilic chitin nanopapers. Chitosan-glucan can be prepared by deacetylation of fungal CG and exhibits greater attraction toward metal ions which allows for more efficient adsorption processes. This comes, however, at the cost of slightly reduced mechanical strength caused by reduced flexibility owing to the concomitant hydrolysis of glucans during the deacetylation reaction.

To improve sustainability and address issues caused by using edible mushrooms for material purposes, mycelium, the vegetative part of fungi, can also be employed as raw material for the isolation of chitin-glucan nanofibrils, avoiding competition with food supply which is the case when using fruiting bodies of mushrooms. At present, mycelium composite materials are typically comprised of agricultural residues which serve as a substrate that is bond by growing mycelium. The intention behind this approach is an upgrade of material side and waste streams, yet, the resulting mechanical performance of such composites is usually only moderate for the substrates provide merely low mechanical performance. Such composites can, nevertheless, also be used as raw materials for the isolation of fungal chitin nanofibrils. In such a process the substrate serves as a nutrient for mycelium growth. From FChNF extracted from mycelium cultures, nanopapers can be manufactured similar to CG isolated from mushroom fruiting bodies. Mechanical properties of those nanopapers are still lower compared to nanopapers from mushroom fruiting bodies because higher amounts of by-products are present. Optimization of extraction and purification processes could help to mitigate their influence and enable the development of composite nanopapers exhibiting appreciable mechanical properties from mycelium.

References

Allan, G.G., J.R. Fox and N. Kong. 1978. A critical evaluation of the potential sources of chitin and chitosan. pp. 64–78. *In*: Muzzarelli, R.A.A. and Pariser, E.R. (eds.). Proceedings of 1st International Conference on Chitin/Chitosan, MIT Sea Grant Program, 78–7.

Ansari, F., S. Galland, M. Johansson, C.J.G. Plummer and L.A. Berglund. 2014. Cellulose nanofiber network for moisture stable, strong and ductile biocomposites and increased epoxy curing rate. *Compos. Part A Appl. Sci.*, 63: 35–44.

Appels, F., J. Dijksterhuis, C.E. Lukasiewicz, K.M.B. Jansen, H.A.B. Wösten and P. Krijgsheld. 2018. Hydrophobin gene deletion and environmental growth conditions impact mechanical properties of mycelium by affecting the density of the material. *Sci. Rep.-UK*, 8: 4703.

Appels, F., S. Camere, M. Montalti, E. Karana, K.M. Jansen, J. Dijksterhuis, P. Krijgsheld and H.A. Wösten. 2019. Fabrication factors influencing mechanical, moisture-and water-related properties of mycelium-based composites. *Mater. Des.*, 161: 64–71.

Arifin, Y.H. and Y. Yusuf. 2013. Mycelium fibers as new resource for environmental sustainability. *Procedia Engineer.*, 53: 504–508.

Bamba, Y., Y. Ogawa, T. Saito, L.A. Berglund and A. Isogai. 2017. Estimating the strength of single chitin nanofibrils via sonication-induced fragmentation. *Biomacromolecules*, 18: 4405–4410.

Benítez, A.J. and A. Walther. 2017. Cellulose nanofibril nanopapers and bioinspired nanocomposites: A review to understand the mechanical property space. *J. Mater. Chem.* A, 5: 16003–16024.

Bilal, M., J.A. Shah, T. Ashfaq, S.M.H. Gardazi, A.A. Tahir, A. Pervez, H. Haroon and Q. Mahmood. 2013. Waste biomass adsorbents for copper removal from industrial wastewater—A review. *J. Hazard. Mater.*, 263 Part 2: 322–333.

Blackwell, J. 1969. Structure of β-chitin or parallel chain systems of poly-β-(1→4)-N-acetyl-D-glucosamine. *Biopolymers*, 7: 281–298.

Braconnot, H. 1811. Sur la nature des champignons. *Annales de Chimie et de Physique*, 79: 265–304.

Cabib, E., J. Arroyo and E. Karana. 2018. Fabricating materials from living organisms: An emerging design practice. *J. Clean. Prod.*, 186: 570–584.

Casteleijn, M.G., D. Richardson, P. Parkkila, N. Granqvist, A. Urtti and T. Viitala. 2018. Spin coated chitin films for biosensors and its analysis are dependent on chitin-surface interactions. *Colloids Surf., A*, 539: 261–272.

Children, J.G. 1824. On the nature of mushrooms. *Zool. J.*, 1: 101–111.

Crowther, T.W., H.B. Glick, K.R. Covey, C. Bettigole, D.S. Maynard, S.M. Thomas, J.R. Smith, G. Hintler, M.C. Duguid, G. Amatulli, M.N. Tuanmu, W. Jetz, C. Salas, C. Stam, D. Piotto, R. Tavani, S. Green, G. Bruce, S.J. Williams, S.K. Wiser, M.O. Huber, G.M. Hengeveld, G.J. Nabuurs, E. Tikhonova, P. Borchardt, C.F. Li, L.W. Powrie, M. Fischer, A. Hemp, J. Homeier, P. Cho, A.C. Vibrans, P.M. Umunay, S.L. Piao, C.W. Rowe, M.S. Ashton, P.R. Crane and M.A. Bradford. 2015. Mapping tree density at a global scale. *Nature*, 525: 201.

Di Mario, F., P. Rapana, U. Tomati and E. Galli. 2008. Chitin and chitosan from Basidiomycetes. *Int. J. Biol. Macromol.*, 43: 8–12.

Duan, B., C. Chang, B. Ding, J. Cai, M. Xu, S. Feng, J. Ren, X. Shi, Y. Du and L. Zhang. 2013. High strength films with gas-barrier fabricated from chitin solution dissolved at low temperature. *J. Mater. Chem. A*, 1: 1867–1874.

EFSA Panel on Dietetic Products, Nutrition & Allergies (NDA). (2010). Scientific Opinion on the safety of 'Chitin-glucan' as a Novel Food ingredient. *EFSA Journal*, 8: 1687.

Fan, Y., H. Fukuzumi, T. Saito and A. Isogai. 2012. Comparative characterization of aqueous dispersions and cast films of different chitin nanowhiskers/nanofibers. *Int. J. Biol. Macromol.*, 50: 69–76.

Findon, A., G. McKay and H.S. Blair. 1993. Transport studies for the sorption of copper ions by chitosan. *J. Environ. Sci. Health.*, Part A, 28: 173–185.

Fontaine, T., C. Simenel, G. Dubreucq, O. Adam, M. Delepierre, J. Lemoine, C.E. Vorgias, M. Diaquin and J.P. Latgé. 2000. Molecular organization of the alkali-insoluble fraction of *Aspergillus fumigatus* cell wall. *J. Biol. Chem.*, 275: 27594–27607.

Forbes Magazine. 2021. https://www.forbes.com/sites/timnewcomb/2021/04/22/creating-adidas-mushroom-based-stan-smith-mylo-sneakers/ Accessed 10.08.2022.

Ghaee, A., M. Shariaty-Niassar, J. Barzin and T. Matsuura. 2010. Effects of chitosan membrane morphology on copper ion adsorption. *Chem. Eng. J.*, 165: 46–55.

Guan, W., X. Fan and R. Yan. 2013. Effect of combination of ultraviolet light and hydrogen peroxide on inactivation of Escherichia coli O157:H7, native microbial loads, and quality of button mushrooms. *Food Control*, 34: 554–559.

Guibal, E. 2004. Interactions of metal ions with chitosan-based sorbents: A review. *Sep. Purif. Technol.*, 38: 43–74.

Hammond, J.B.W. 1979. Changes in composition of harvested mushrooms (*Agaricus bisporus*). *Phytochemistry*, 18: 415–418.

Haneef, M., L. Ceseracciu, C. Canale, I.S. Bayer, J.A. Heredia-Guerrero and A. Athanassiou. 2017. Advanced materials from fungal mycelium: Fabrication and tuning of physical properties. *Sci. Rep.-UK*, 7: 41292.

Hartland, R.P., C.A. Vermeulen, J.H. Sietsma, J.G.H. Wessels and F.M. Klis. 1994. The linkage of (1–3)-β-glucan to chitin during cell wall assembly in Saccharomyces cerevisiae. *Yeast*, 10: 1591–1599.

Hassainia, A., H. Satha and S. Boufi. 2018. Chitin from Agaricus bisporus: Extraction and characterization. *Int. J. Biol. Macromol.*, 117: 1334–1342.

Hassanzadeh, P., W. Sun, J.P. de Silva, J. Jin, K. Makhnejia, G.L.W. Cross and M. Rolandi. 2014. Mechanical properties of self-assembled chitin nanofiber networks. *J. Mater. Chem. B*, 2: 2461–2466.

Henriksson, M., L. Fogelström, L.A. Berglund, M. Johansson and A. Hult. 2011. Novel nanocomposite concept based on cross-linking of hyperbranched polymers in reactive cellulose nanopaper templates. *Compos. Sci. Technol.*, 71: 13–17.

Heux, L., J. Brugnerotto, J. Desbrières, M.F. Versali and M. Rinaudo. 2000. Solid State NMR for determination of degree of acetylation of chitin and chitosan. *Biomacromolecules*, 1: 746–751.

Holt, G., G. Mcintyre, D. Flagg, E. Bayer, J. Wanjura and M. Pelletier. 2012. Fungal mycelium and cotton plant materials in the manufacture of biodegradable molded packaging material: Evaluation study of select blends of cotton byproducts. *J. Biobased. Mater. Bio.*, 6: 431–439.

Ifuku, S. 2014. Chitin and chitosan nanofibers: Preparation and chemical modifications. *Molecules*, 19: 18367–18380.

Ifuku, S., A. Ikuta, H. Izawa, M. Morimoto and H. Saimoto. 2014. Control of mechanical properties of chitin nanofiber film using glycerol without losing its characteristics. *Carbohydr. Polym.*, 101: 714–717.

Ifuku, S., S. Morooka, A. Norio Nakagaito, M. Morimoto and H. Saimoto. 2011. Preparation and characterization of optically transparent chitin nanofiber/(meth)acrylic resin composites. *Green Chem.*, 13: 1708–1711.

Islam, M., G. Tudryn, R. Bucinell, L. Schadler and R. Picu. 2017. Morphology and mechanics of fungal mycelium. *Sci. Rep.-UK*, 7: 13070.

Janesch, J., M. Jones, M. Bacher, E. Kontturi, A. Bismarck and A. Mautner. 2020. Mushroom-derived chitosan-glucan nanopaper filters for the treatment of water. *React. Funct. Polym.*, 146: 104428.

Jang, M.-K., B.-G. Kong, Y.-I. Jeong, C.H. Lee and J.-W. Nah. 2004. Physicochemical characterization of α-chitin, β-chitin, and γ-chitin separated from natural resources. *J. Polym. Sci., Part A: Polym. Chem.*, 42: 3423–3432.

Jin, J., D. Lee, H.-G. Im, Y.C. Han, E.G. Jeong, M. Rolandi, K.C. Choi and B.-S. Bae. 2016. Chitin nanofiber transparent paper for flexible green electronics. *Adv. Mater.*, 28: 5169–5175.

Jones, M., T. Huynh, C. Dekiwadia, F. Daver and S. John. 2017. Mycelium composites: A review of engineering characteristics and growth kinetics. *J. Bionanosci.*, 11: 241–257.

Jones, M., T. Bhat, T. Huynh, E. Kandare, R. Yuen, C. Wang and S. John. 2018. Waste-derived low-cost mycelium composite construction materials with improved fire safety. *Fire Mater.*, 42: 816–825.

Jones, M., K. Weiland, M. Kujundzic, J. Theiner, H. Kählig, E. Kontturi, S. John, A. Bismarck and A. Mautner. 2019a. Waste-derived low-cost mycelium nanopapers with tunable mechanical and surface properties. *Biomacromolecules*, 20: 3513–3523.

Jones, M.P., A.C. Lawrie, T.T. Huynh, P.D. Morrison, A. Mautner, A. Bismarck and S. John. 2019b. Agricultural by-product suitability for the production of chitinous composites and nanofibers utilising Trametes versicolor and Polyporus brumalis mycelial growth. *Process Biochem.*, 80: 95–102.

Jones, M., A. Gandia, S. John and A. Bismarck. 2021. Leather-like material biofabrication using fungi. *Nat. Sustain.*, 4: 9–16.

Jones, M., M. Kujundzic, S. John and A. Bismarck. 2020a. Crab vs. mushroom: A review of crustacean and fungal chitin in wound treatment. *Mar. Drugs*, 18: 64.

Jones, M., A. Mautner, S. Luenco, A. Bismarck and S. John. 2020b. Engineered mycelium composite construction materials from fungal biorefineries: A critical review. *Mater. Des.*, 187: 108397.

Kabasci, S. 2020. Chapter 4—Biobased plastics. pp. 67–96. *In*: Letcher, T.M. (ed.). Plastic Waste and Recycling, Academic Press.

Kadokawa, J.-I., A. Takegawa, S. Mine and K. Prasad. 2011. Preparation of chitin nanowhiskers using an ionic liquid and their composite materials with poly(vinyl alcohol). *Carbohydr. Polym.*, 84: 1408–1412.

Karana, E., D. Blauwhoff, E. Hultink and S. Camere. 2018. When the material grows: A case study on designing (with) mycelium-based materials. *Int. J. Des.*, 12: 119–136.

Kataoka, S. and T. Ando. 1979. Regenerated Chitin film from the solution of trichloroacetic acid systems. *Kobunshi Ronbunshu*, 36: 175–181.

Kavanagh, K. 2005. Fungi: Biology and Applications. Hoboken, NJ: Wiley.

Kaya, M., M. Mujtaba, H. Ehrlich, A.M. Salaberria, T. Baran, C.T.R. Amemiya, L. Galli, L. Akyuz, I. Sargin and J. Labidi. 2017a. On chemistry of γ-chitin. *Carbohydr. Polym.*, 176: 177–186.

Kaya, M., I. Sargin, I. Sabeckis, D. Noreikaite, D. Erdonmez, A.M. Salaberria, J. Labidi, V. Baublys, and V. Tubelytė. 2017b. Biological, mechanical, optical and physicochemical properties of natural chitin films obtained from the dorsal pronotum and the wing of cockroach. *Carbohydr. Polym.*, 163: 162–169.

Kaya, M., A.M. Salaberria, M. Mujtaba, J. Labidi, T. Baran, P. Mulercikas and F. Duman. 2018. An inclusive physicochemical comparison of natural and synthetic chitin films. *Int. J. Biol. Macromol.*, 106: 1062–1070.

Khor, E. 2001. Chapter 1—The relevance of Chitin. pp. 1–8. *In*: Khor, E. (ed.). Chitin, Elsevier Science Ltd.

Kim, K., M. Ha, B. Choi, S.H. Joo, H.S. Kang, J.H. Park, B. Gu, C. Park, C. Park, J. Kim, S.K. Kwak, H. Ko, J. Jin and S.J. Kang. 2018. Biodegradable, electro-active chitin nanofiber films for flexible piezoelectric transducers. *Nano Energy*, 48: 275–283.

Klemm, D., F. Kramer, S. Moritz, T. Lindstroem, M. Ankerfors, D. Gray and A. Dorris. 2011. Nanocelluloses: A new family of nature-based materials. *Angew. Chem. Int. Ed.*, 50: 5438–5466.

Kollár, R., E. Petráková, G. Ashwell, P.W. Robbins and E. Cabib. 1995. Architecture of the yeast cell wall: The linkage between Chitin and β(1→3)-Glucan. *J. Biol. Chem.*, 270: 1170–1178.

Kollár, R., B.B. Reinhold, E. Petráková, H.J.C. Yeh, G. Ashwell, J. Drgonová, J.C. Kapteyn, F.M. Klis and E. Cabib. 1997. Architecture of the yeast cell wall: β(1→6)-Glucan interconnects mannoprotein, β(1→3)-Glucan, and chitin. *J. Biol. Chem.*, 272: 17762–17775.

Kurita, K., K. Tomita, S. Ishii, S.-I. Nishimura and K. Shimoda. 1993. β-chitin as a convenient starting material for acetolysis for efficient preparation of N-acetylchitooligosaccharides. *J. Polym. Sci., Part A: Polym. Chem.*, 31: 2393–2395.

Kurita, K., S. Ishii, K. Tomita, S.-I. Nishimura and K. Shimoda. 1994. Reactivity characteristics of squid β-chitin as compared with those of shrimp chitin: High potentials of squid chitin as a starting material for facile chemical modifications. *J. Polym. Sci., Part A: Polym. Chem.*, 32: 1027–1032.

Lee, K.-Y., Y. Aitomäki, L.A. Berglund, K. Oksman and A. Bismarck. 2014. On the use of nanocellulose as reinforcement in polymer matrix composites. *Compos. Sci. Technol.*, 105: 15–27.

Lee, K.-Y., T. Tammelin, K. Schulfter, H. Kiiskinen, J. Samela and A. Bismarck. 2012. High performance cellulose nanocomposites: comparing the reinforcing ability of bacterial cellulose and nanofibrillated cellulose. *ACS Appl. Mat. Inter.*, 4: 4078–4086.

Lin, Q., Lu, Y., J. Zhang, W. Liu, W. Guan and Z. Wang. 2017. Effects of high CO_2 in-package treatment on flavor, quality and antioxidant activity of button mushroom (*Agaricus bisporus*) during postharvest storage. *Postharvest Biol. Tec.*, 123: 112–118.

Lindequist, U., T.H.J. Niedermeyer and W.-D. Jülich. 2005. The pharmacological potential of mushrooms. *Evid.-Based Compl. Alt.*, 2: 906016.

Lopata, A.L., R.E. O'Hehir and S.B. Lehrer. 2010. Shellfish allergy. *Clin. Exp. Allergy*, 40: 850–858.

López Nava, J.A., J. Méndez González, X. Ruelas Chacón and J.A. Nájera Luna. 2016. Assessment of edible fungi and films bio-based material simulating expanded polystyrene. *Mater. Manuf. Processes*, 31: 1085–1090.

Marchetti, P., M.F. Jimenez Solomon, G. Szekely and A.G. Livingston. 2014. Molecular separation with organic solvent nanofiltration: A critical review. *Chem. Rev.*, 114: 10735–10806.

Mautner, A. 2020. Nanocellulose water treatment membranes and filters: A review. *Polym. Int.*, 69: 741–751.

Mautner, A., K.-Y. Lee, P. Lahtinen, M. Hakalahti, T. Tammelin, K. Li and A. Bismarck. 2014. Nanopapers for organic solvent nanofiltration. *Chem. Commun.*, 50: 5778–5781.

Mautner, A., K.-Y. Lee, T. Tammelin, A.P. Mathew, A.J. Nedoma, K. Li and A. Bismarck. 2015. Cellulose nanopapers as tight aqueous ultra-filtration membranes. *React. Funct. Polym.*, 86: 209–214.

Mautner, A., J. Lucenius, M. Österberg and A. Bismarck. 2017. Multi-layer nanopaper based composites. *Cellulose*, 24: 1759–1773.

Mautner, A., F. Mayer, M. Hervy, K.-Y. Lee and A. Bismarck. 2018. Better together: Synergy in nanocellulose blends. *Philos. Trans. Royal Soc.*, 376: 20170043.

Metreveli, G., L. Wågberg, E. Emmoth, S. Belák, M. Strømme and A. Mihranyan. 2014. A size-exclusion nanocellulose filter paper for virus removal. *Adv. Healthc. Mater.*, 3: 1546–1550.

Minke, R. and J. Blackwell. 1978. The structure of α-chitin. *J. Mol. Biol.*, 120: 167–181.

Mushi, N.E., N. Butchosa, Q. Zhou and L.A. Berglund. 2014a. Nanopaper membranes from chitin-protein composite nanofibers-structure and mechanical properties. *J. Appl. Polym. Sci.*, 131: 40121.

Mushi, N.E., S. Utsel and L.A. Berglund. 2014b. Nanostructured biocomposite films of high toughness based on native chitin nanofibers and chitosan. *Front. Chem.*, 2: 99.

Muzzarelli, R.A.A. 2011. Chitin nanostructures in living organisms. pp. 1–34. *In*: Gupta, N.S. (ed.). Chitin: Formation and Diagenesis, Springer Netherlands.

Muzzarelli, R.A.A. and E.R. Pariser. 1978. Proceedings of 1st International Conference on Chitin/ Chitosan. *1ˢᵗ International Conference on Chitin/Chitosan* (pp. 64–78). Cambridge, Massachusetts: MIT Sea Grant Program, 78–7.

Nakagaito, A.N., S. Iwamoto and H. Yano. 2005. Bacterial cellulose: The ultimate nano-scalar cellulose morphology for the production of high-strength composites. *Applied Physics A*, 80: 93–97.

Nawawi, W.M.F.W., M.P.E. Jones, Kontturi, A. Mautner and A. Bismarck. 2020a. Plastic to elastic: Fungi-derived composite nanopapers with tunable tensile properties. *Compos. Sci. Technol.*, 198: 108327.

Nawawi, W.M.F.W., K.-Y. Lee, E. Kontturi, A. Bismarck and A. Mautner. 2020b. Surface properties of chitin-glucan nanopapers from Agaricus bisporus. *Int. J. Biol. Macromol.*, 148: 677–687.

Nawawi, W.M.F.W., K.-Y. Lee, E. Kontturi, R.J. Murphy and A. Bismarck. 2019b. Chitin nanopaper from mushroom extract: natural composite of nanofibers and glucan from a single biobased source. *ACS Sustain. Chem. Eng.*, 7: 6492–6496.

Nawawi, W., M. Jones, R.J. Murphy, K.-Y. Lee, E. Kontturi and A. Bismarck, A. 2019a. Nanomaterials derived from fungal sources—is it the new hype? *Biomacromolecules*, 21: 30–55.

Nogi, M. and H. Yano. 2008. Transparent nanocomposites based on cellulose produced by bacteria offer potential innovation in the electronics device industry. *Adv. Mater.*, 20: 1849–1852.

Oksman, K., Y. Aitomäki, A.P. Mathew, G. Siqueira, Q. Zhou, S. Butylina, S. Tanpichai, X. Zhou and S. Hooshmand. 2016. Review of the recent developments in cellulose nanocomposite processing. *Composites Part A*, 83: 2–18.

Payen, A. and C.R. Hebd. 1838. Mémoire sur la composition du tissu propre des plantes et du ligneux. *Comptes rendus de l'Académie des Sciences*, 7: 1052–1056.

Pelletier, M.G., G.A. Holt, J.D. Wanjura, E. Bayer and G. McIntyre. 2013. An evaluation study of mycelium based acoustic absorbers grown on agricultural by-product substrates. *Ind. Crop. Prod.*, 51: 480–485.

Pérez, S. and D. Samain. 2010. Structure and engineering of celluloses.pp. 25–116. *In*: Horton, D. (ed.). Adv. Carbohydr. Chem. Biochem., Academic Press.

Pillai, C.K.S., W. Paul and C.P. Sharma. 2009. Chitin and chitosan polymers: Chemistry, solubility and fiber formation. *Prog. Polym. Sci.*, 34: 641–678.

Rinaudo, M. 2007. Chitin and chitosan—Properties and applications. *Prog. Polym. Sci.*, 31: 603–632.

Røyne, F., D. Peñaloza, G. Sandin, J. Berlin and M. Svanström. 2016. Climate impact assessment in life cycle assessments of forest products: Implications of method choice for results and decision-making. *J. Clean. Prod.*, 116: 90–99.

Rudall, K.M. and W. Kenchington. 1973. The chitin system. *Biolog. Rev.*, 48: 597–633.

Saito, Y., J.L. Putaux, T. Okano, F. Gaill and H. Chanzy. 1997. Structural aspects of the swelling of β chitin in HCl and its conversion into α chitin. *Macromolecules*, 30: 3867–3873.

Sarode, S., P. Upadhyay, M.A. Khosa, T. Mak, A. Shakir, S. Song and A. Ullah. 2019. Overview of wastewater treatment methods with special focus on biopolymer chitin-chitosan. *Int. J. Biol. Macromol.*, 121: 1086–1100.

Seenuvasan, M., G. Sarojini and M. Dineshkumar. 2020. Chapter 6—Recovery of chitosan from natural biotic waste. pp. 115–133. *In*: Varjani, S., Pandey, A., Gnansounou, E., Khanal, S.K. and Raveendran, S. (eds.). Current Developments in Biotechnology and Bioengineering, Elsevier.

Shi, Y.-l., A.E. James, I.F.F. Benzie and J.A. Buswell. 2004. Genoprotective activity of edible and medicinal mushroom components. *Int. J. Med. Mushrooms*, 6: 1–14.

Sietsma, J.H. and J.H.G. Wessels. 1979. Evidence for covalent linkages between chitin and β-Glucan in a fungal wall. *Microbiology*, 114: 99–108.

Stalhberger, T., C. Simenel, C. Clavaud, V.G.H. Eijsink, R. Jourdain, M. Delepierre, J.-P. Latgé, L. Breton and T. Fontaine. 2014. Chemical organization of the cell wall polysaccharide core of malassezia restricta. *J. Biol. Chem.*, 289: 12647–12656.

Surarit, R., P.K. Gopal and M.G. Shepherd. 1988. Evidence for a glycosidic linkage between chitin and glucan in the cell wall of Candida albicans. *J. Gen. Microbiol.*, 134: 1723–1730.

Thakur, V.K. and A.S. Singha. 2013. *Biomass-based Biocomposites*. Shropshire, U.K.: Smithers Rapra Technology Ltd.

Travaglini, S., J. Noble, P.G. Ross and C.K.H. Dharan. 2013. Mycology Matrix Composites. *28ᵗʰ ASC Technical Conference*. State College, PA.

UN Sustainable Development Goals 2020. https://sustainabledevelopment.un.org/ Accessed 10.08.2022.

Versali, M.-F., F. Clerisse, J.-M. Bruyere and S. Gautier. 2009. Cell wall derivatives from biomass and preparation thereof. US.

Vold, I.M.N., K.M. Vårum, E. Guibal and O. Smidsrød. 2003. Binding of ions to chitosan-selectivity studies. *Carbohydr. Polym.*, 54: 471–477.

Webster, J. and R. Weber. 2007. *Introduction to Fungi*. Cambridge: Cambridge University Press.

Yousefi, N., M. Jones, A. Bismarck and A. Mautner. 2021. Fungal chitin-glucan nanopapers with heavy metal adsorption properties for ultrafiltration of organic solvents and water. *Carbohydr. Polym.*, 253: 117273.

Yu, L., D.G. Fernig, J.A. Smith, J.D. Milton and J.M. Rhodes. 1993. Reversible inhibition of proliferation of epithelial cell lines by *Agaricus bisporus* (Edible Mushroom) lectin. *Cancer Res.*, 53: 4627–4632.

Yui, T., N. Taki, J. Sugiyama and S. Hayashi. 2007. Exhaustive crystal structure search and crystal modeling of β-chitin. *Int. J. Biol. Macromol.*, 40: 336–344.

Yusof, N.L.B.M., L.Y. Lim and E. Khor. 2004. Flexible chitin films: structural studies. *Carbohydr. Res.*, 339: 2701–2711.

Nutraceuticals

12

Nutraceutical Profile of Four Wild Edible Mushrooms of the Western Ghats

Venugopalan Ravikrishnan and *Kandikere R Sridhar**

1. Introduction

Mushrooms represent one of the major non-timber minor forest products useful in human nutrition, medicine and plant growth promotion. Compared to plants and animals, mushrooms are relatively one of the major untapped resources having innumerable benefits for mankind. The knowledge of the use of wild mushrooms as a secondary source of food and medicine is well-known worldwide (Boa 2004). The history of the use of mushrooms in human medicine dates back to 100 AD, in China. In the Indian Subcontinent, ethnic practices pertaining to edible and medicinal mushrooms were known as early as 1700–1100 BC (Wasson 1971). Purkayastha and Chandra (1985) have documented as many as 283 species of wild mushrooms which are consumed by tribal communities in India. Mushrooms are emerging as a potential alternative inexpensive nutrient source to cater to the needs of the teeming population.

Based on the edibility, mushrooms could be classified into two major groups: (i) Ascomycetes (Discomycetes or cup fungi) (e.g., truffles); (ii) Basidiomycetes (Phragmobasidiomycetes: Auriculariales, Septabasidiales and Tremellales) (e.g., gilled/poroid fungi) (Alexander 2013). Among the non-conventional food commodities, wild edible mushrooms are highly relished oriental foodstuffs

Department of Biosciences, Mangalore University, Mangalagangotri, Karnataka, India.
* Corresponding author: kandikere@gmail.com

worldwide owing to their texture, aroma and nutritional versatility (Manzi et al. 2001). Mushrooms possess adequate dietary components especially protein, trace elements, unsaturated fatty acids, fiber, carbohydrates, vitamins and pigments. Although initially mushrooms were considered to be valuable only for nutritional propose (Dupont et al. 2017). However, recent developments changed the perception of the use of mushrooms as a source of nutraceuticals, medicine and cosmetics (Boa 2004, Hyde et al. 2010, Thawthong et al. 2014, Hapuarachchi et al. 2018).

In view of the increasing understanding of the relevance of mushrooms in human welfare, number of nutritionally and medicinally important mushrooms are being domesticated and many of which are under commercial cultivation. Some of the most common mushrooms under cultivation worldwide include *Agaricus bisporus*, *A. bitorquis* (Button mushrooms), *Auricularia auricula* (Jew's ear mushroom), *Coprinus comatus* (Shaggy Mane), *Ganoderma lucidum* (Reishi), *Grifola frondosa* (Maitake), *Hericium erinaceus*, *H. coralloides* (Lion's head), *Lentinula edodes* (Shiitake), *Pholiota microspora* (Nameko), *Flammulina velitupes* (Enoki mushroom), *Pleurotus* spp. (Oyster mushroom), *Tremella fuciformis* (White jelly) and *Volvariella volvacea* (Paddy straw mushroom), etc. (Marshall and Nair 2009). Owing to increased cultivation, in the last 30 years, the global production of mushrooms has hiked up to 30-folds (Royse et al. 2017). One of the major advantages of mushroom cultivation is the bioconversion of inexpensive lingo-cellulosic agricultural wastes into utility products.

The most common mushrooms cultivated worldwide include *Agaricus* spp. (button mushrooms), *Auricularia* (ear mushroom), *Coprinus* (Shaggy Mane), *Ganoderma* (Reishi), *Grifola frondosa* (Maitake), *Hericium* (lion's head), Shiitake (*Lentinula edodes*), Nameko (*Pholiota*), Oyster mushroom (*Pleurotus* spp.), *Tremella* (white jelly) and *Volvariella volvacea* (straw mushroom) (Marshall and Nair 2009). Several wild mushrooms are regularly hunted for consumption purposes in Northern India, which include *Coprinus comatus*, *Lactarius deliciosus*, *Lycoperdon perlatum*, *Macrolepiota procera*, *Morchella* spp., *Pleurotus* spp., *Podaxis pistillaris*, *Russula virescens*, *R. cyanoxantha*, *Tuber melanosporum* and *Termitomyces* spp., etc. (Atri et al. 2016). In the Western Ghats, exclusively based on traditional knowledge, 51 species of wild mushrooms are reported to be consumed and the most favored mushrooms in this regard are *Astraeus hygrometricus*, *Clitocybe infundibuliformis*, *Fistulina hepatica*, *Lentinus sajor-caju*, *Pleurotus* spp. and *Scleroderma citrinum* and *Termitomyces* spp. (Karun and Sridhar 2017).

Mushrooms serve as a potential food source owing to their versatility to provide nutrients deficient in plant food sources and they are not as expensive as animal foods. Wild mushrooms are of special interest as nutraceuticals in developing countries due to their capability to combat protein-energy malnutrition (PEM) and their therapeutic significance. They are also known to fulfill nutritional requirements to the stipulated standards of FAO-WHO. In view of the above the nutraceutical constituents of four wild edible mushrooms (*Boletus edulis*, *Hygrocybe alwisii*, *Lentinus squarrosulus* and *Termitomyces schimperi*) collected from the foothills of the Western Ghats have been evaluated and presented in the present manuscript.

2. Nutraceutical Analysis

2.1 Mushrooms and Processing

Young sporophores of *Boletus edulis*, *Hygrocybe alwisii*, *Lentinus squarrosulus* and *Termitomyces schimperi* were collected from three locations of the foothills of the Western Ghats region (as replicates). They were identified based on standard manuals (Jordan 2004, Phillips 2006, Mohanan 2011, Buczacki 2012). After determining the moisture content, they were dried separately on aluminum spreadsheets in an oven ($58 \pm 2°C$) until attaining constant mass and part of the dried mushrooms were powdered and individually preserved in air-tight containers for use in experiments for nutraceutical evaluation.

2.2 Proximal and Mineral Analysis

Gravimetric analysis was adapted and carried out to assess the moisture content of replicate samples of fresh mushrooms and the dried flours of individual mushrooms so prepared (AOAC, 2006). The micro-Kjeldahl method as given by Humphries (1956) was used to find out the crude protein content ($N \times 6.25$) (Humphries 1956). Total lipids, crude fiber and ash contents were assessed gravimetrically based on AOAC (2006) methodology. The carbohydrate content was determined using the phenol-sulphuric acid method given by DuBois et al. (1956). The energy value of the mushroom flour was calculated by the formula suggested by Manzi et al. (2004).

The quantification of the macro-minerals (Na, K, Ca and Mg) and micro-minerals (Fe, Cu, Zn and Se) in the individual mushroom flours were carried out using Atomic Absorption Spectroscopy (AAS) following AOAC (2006) methodology. The amount of phosphorous as orthophosphate in the individual samples was assessed by the vanado molybdophosphoric acid method (APHA 1995). Depending upon the proportion of Na, K, Ca and P present in the evaluated mushroom samples the ratios of Na/K and Ca/P were also calculated.

2.3 Amino Acid Analysis

The amino acid analysis was carried out using High-Performance Liquid Chromatography (HPLC) as per the methodology given by Fierabracci et al. (1991). The total essential amino acids (TEAA) to total amino acids (TAA) ratio was also calculated. The score of essential amino acids (EAAS) was determined by dividing the content of individual EAA by the EAA as per the available standard reference pattern (FAO-WHO 1991). The protein efficiency ratios (PER) were calculated using the evaluated proportion of amino acid in the mushroom flours employing the formula given by Alsmeyer et al. (1974).

2.4 Fatty Acid Analysis

To determine the amount of specific fatty acids in the individual mushroom flour, the Soxhlet method was used (Xiao et al. 2012). Fatty acid methyl esters (FAME) were determined through the esterification of the fatty acids using gas chromatography as per the protocol given by Nareshkumar (2007). Based upon the data generated fatty acid ratios were also calculated [TUFA/TSFA, TPUFA/TMUFA, $C_{14:0}+C_{15:0}+(C_{16:0}/C_{18:0})$, $C_{18:1}/C_{18:2}$ and ω-6/ω-3].

2.5 Data Analysis

The contents of moisture, crude protein, total lipid, crude fiber, ash and carbohydrate in the four mushroom samples were statistically analyzed and compared using One-way ANOVA (Sigma Plot # 11, Systat Software Inc., USA).

3. Proximal Qualities

3.1 Moisture

The moisture content of the mushrooms studied varied between 72.8% in *L. squarrosulus* to 93.5% in *H. alwisii* (Table 1). This component in mushrooms seems to depend on the moisture present in their substrates. Those occurring on leaf litter or soil (e.g., *B. edulis*, *H. alwisii* and *T. schimperi*) possess more moisture compared to wood-preferring mushrooms (*L. squarrosulus*). Depending on the moisture content, mushrooms have a varied shelf life and those with high moisture are prone to early deterioration. *Lentinus* spp. has better shelf life compared to other mushrooms primarily because of less moisture. Hence, harvested mushrooms need to be processed as soon as possible (e.g., refrigeration, drying and cooking) to avoid deterioration in quality.

In the assessment of nutritional components and dry matter of foodstuff is important to express proximal qualities and minerals on a dry weight basis. Usually, freshly collected mushrooms are reported to possess dry matter ranging from 5–15%

Table 1: The proximate composition of four wild mushrooms of the Western Ghats on a dry weight basis (n = 3 ± SD) ([a]Cheftel et al. 1985, USDA 1999, Longvah 2000[b], Ensminger et al. 1990, Matz 1991, USDA 1999, Belderok 2000).

	Boletus edulis	*Hygrocybe alwisii*	*Lentinus squarrosulus*	*Termitomyces schimperi*	*Soyabean*[a]	*Wheat*[b]
Moisture (%)	83.58 ± 0.46	93.47 ± 0.05	72.75 ± 0.20	90.52 ± 1.84	8.4–8.5	8–18
Crude protein (g/100 g)	22.60 ± 0.32	24.99 ± 0.00	12.50 ± 0.00	33.06 ± 1.90	36.5–40.0	10–16
Total lipids (g/100 g)	5.97 ± 0.03	2.80 ± 0.40	6.32 ± 0.05	2.40 ± 0.00	17.5–20.0	1.5–2.0
Crude fiber (g/100 g)	8.60 ± 0.40	7.60 ± 0.80	32.00 ± 1.50	11.17 ± 1.25	9.31	2.0–2.7
Ash (g/100 g)	1.38 ± 0.04	0.28 ± 0.02	1.11 ± 0.15	1.02 ± 0.05	3.2–4.9	1.2–3.0
Carbohydrates (g/100 g)	61.4 ± 0.05	64.3 ± 1.20	48.1 ± 0.20	52.4 ± 0.65	30.2–35.0	65.4–78.0
Calorific value (kJ/100 g)	1650 ± 7	1622 ± 6	1264 ± 2	1541 ± 21	1739–1810	1377–1431

(Beluhan and Ranogajec 2011). However, the dry matter range was as high as 16.5% and 27% in *B. edulis* and *L. squarrosulus*, respectively. Crisan and Sands (1978) depicted documented that the moisture content in mushrooms depends on the time of harvest, period of maturation and conditions of the environment (e.g., temperature and humidity). This indicates the necessity to exercise special attention to the processing of mushrooms like *Termitomyces schimperi* and *Hygrocybe alwisii* which possess high moisture content (90.5% and 93.5%, respectively). Such high moisture content seems to be responsible for promoting high microbial growth, enzyme activity and rate of perishability (Johnsy et al. 2011, Okoro and Achuba 2012).

3.2 *Crude Protein*

Edible mushrooms are highly valued owing to their high-quality proteins (Johnsy et al. 2011). Among the presently evaluated mushrooms, *L. squarrosulus* possesses the least amount of protein content (12.5%) in comparison to all other mushrooms while *T. schimperi* possesses the highest amount (33.1%) (Table 1). *Lentinus squarrosulus* contains double the amount of crude protein as compared to the quantity of crude protein reported in this species from the southwest region (Ghate and Sridhar 2019). The presently evaluated amount of crude protein in *T. schimperi* is on the higher side than reported in *T. globules* and *T. schimperi* from the Western Ghats and *T. clypeatus* from the southwest region (Sudheep and Sridhar 2014, Karun et al. 2018, Ghate and Sridhar 2019). This variation is primarily because the protein value of edible mushrooms is reported to be influenced by a number of factors especially the type of mushroom, location of harvest, time of harvest, nature of substratum, developmental stage, sampled part, nitrogen level and size of the pileus (Colak et al. 2009, Beluhan and Ranogajec 2011, Kakon et al. 2012).

Except for *L. squarrosulus*, protein content in all other evaluated mushrooms is comparable with the protein content in edible legumes (22.6–33.1%). However, c it is on the lower side in comparison to soybean but much more in comparison to wheat. Except for *L. squarrosulus*, the protein content in other mushrooms is reported to be comparable or even higher than vegetables (Kakon et al. 2012). Sanmee et al. (2003) and Kumar et al. (2013), reported protein content in edible mushrooms to vary from 28.9% to 39.1%. In comparison except for in *T. schimpheri* (33.06%), in all other presently evaluated mushrooms the amount of crude proteins is largely on the lower side ranging from 12.50 % in *L. squarrosulus* to 24.99% in *H. alwisii*. Edible mushrooms are highly valued owing to their high-quality proteins (Johnsy et al. 2011). Although mushrooms are a rich source of crude proteins, but they are reported to stand below animal meat in this regard, however, the overall proportion of crude proteins in them is reported to be comparable to several foodstuffs including milk (Kakon et al. 2012). The protein value of edible mushrooms is influenced by a number of factors especially the type of mushroom, location of harvest, time of harvest, nature of substratum, developmental stage, sampled part, nitrogen level and size of pileus (Colak et al. 2009, Beluhan and Ranogajec 2011, Kakon et al. 2012).

3.3 Total Lipids

The total lipids content in the presently evaluated mushrooms ranged from 2.4% in *T. schimperi* to 6.3% in *L. squarrosulus* (Table 1). The amount of total lipids evaluated presently in *L. squarrosulus* is on the lower side in comparison to the total lipids reported in the samples of this species from the southwest region by Ghate and Sridhar (2019), so also *T. schimperi* compared to *T. globulus* found in the Western Ghats (Sudheep and Sridhar 2014).

When compared with edible oil seeds (e.g., soybean), edible mushrooms possess a low amount of lipids and a high amount of crude proteins. Thus mushrooms are highly suitable for human nutrition as a low-calorie diet (Beluhan and Ranogajec 2011). The presently evaluated mushrooms evaluated possess a higher amount of total lipids in comparison to wheat, but not as high as soybean.

3.4 Crude Fiber

Mushrooms are known to possess insoluble as well as soluble fibers (e.g., β-glucans and chitosans). The insoluble fibers are known to aid in digestion, while the soluble fibers are reported to help in combating cardiovascular diseases by lowering cholesterol levels (Aletor and Aladetimi 1995, Sanmee et al. 2003, Manzi et al. 2004). In addition, to being rich in dietary fiber, edible mushrooms are also known for anti-tumorigenic and hypocholesterolemic qualities (Chihara 1993). The presently evaluated mushrooms also contain a good amount of crude fibers which ranged from 7.6% in *H. alwisii* to 32% in *L. squarrosulus* (Table 1). In the case of *L. squarrosulus* its value is comparable to the material of this species collected from the southwest region of the Western Ghats (Ghate and Sridhar 2019). In the case of *T. schimperi* the crude fibre content is more than documented by Sudheep and Sridhar (2014) again in the material from the Western Ghats.

Amongst the presently evaluated mushrooms *L. squarrosulus* and *T. schimperi* possess higher crude fiber content in comparison to soybean and wheat (Table 1). Mushrooms are known for insoluble as well as soluble fibers (e.g., β-glucans and chitosans). The insoluble fibers aid in digestion, while the soluble fibers have the capacity to combat cardiovascular diseases by lowering cholesterol levels (Aletor and Aladetimi 1995, Sanmee et al. 2003, Manzi et al. 2004). In addition, being rich in dietary fiber edible mushrooms are also known for anti-tumorigenic and hypocholesterolemic qualities (Chihara 1993).

3.5 Ash

The ash contents of the samples were fairly low, which ranged from 0.3% in *H. alwisii* to 1.4% in *B. edulis* (Table 1) and is on the lower side in comparison to soybean and wheat. Such a low ash content in mushrooms is indicative of the presence of a low amount of mineral content in them.

3.6 Carbohydrates

The edible mushrooms are highly valued for their carbohydrate content, which is reported to range from 35% to 70% (Manzi et al. 1999, Cheung 2010, Johnsy et al. 2011) and as a potential source of carbohydrates, these are comparable to cereals particularly wheat, in possessing glucans, monosaccharides, disaccharides, sugar alcohol, glycogen and chitin as part of their composition (Ensminger et al. 1990, Matz 1991, Kurztman 1997, USDA 1999, Belderok 2000). In the presently evaluated mushrooms the carbohydrate content ranged from 48.1% in *L. squarrosulus* to 64.3% in *H. alwisii* (Table 1). In *L. squarrosulus* its value is comparable to the sample of this species collected from southwest region (Ghate and Sridhar 2019). However, its content in *T. schimperi* is on the higher side than *T. globulus* from the Western Ghats as reported by Sudheep and Sridhar (2014), *T. clypeatus* as well as *T. umkowaan* in the southwest region as reported by Karun et al. (2018) and Ghate and Sridhar (2019). When compared with soybean the carbohydrate content in the presently evaluated mushrooms is on the higher side. Carbohydrate content in mushrooms studied is higher than in soybean. The edible mushrooms are highly valued owing to their carbohydrate content, which ranges from 35% to 70% (Manzi et al. 1999, Cheung 2010, Johnsy et al. 2011). Similar to cereals (e.g., wheat), mushrooms are a potential source of carbohydrates (glucans, monosaccharides, disaccharides, sugar alcohol, glycogen and chitin) (Ensminger et al. 1990, Matz 1991, Kurztman 1997, USDA 1999, Belderok 2000).

3.7 Calorific Value

The calorific value in the presently evaluated mushrooms ranged from 1264 kJ/100 g in *L. squarrosulus* to 1650 kJ/100 g in *B. edulis* (Table 1). In comparison Ghate and Sridhar (2019) reported higher calorific value in the material of *L. squarrosulus*, *T. clypeatus* and *T. umkowaan* collected from southwest region (Karun et al. 2018, Ghate and Sridhar 2019), while in *T. schimperi* it is reported to be comparable to *T. globulus* collected from Western Ghats. However, the energy value in the presently evaluated mushrooms is comparable to wheat, while it is on the lower side when compared to soybean (Table 1).

3.8 Comparison Among Mushrooms

In the comparison of the moisture content of mushrooms by one-way ANOVA with the Holm-Sidak method a significant difference ($p < 0.001$) was noticed. Similarly, the crude protein, crude lipid, crude fiber and carbohydrate contents in the four evaluated mushrooms significantly varied from one another in their respective proportion ($p < 0.001$). Except for *T. schimperi* vs. *L. squarrosulus* ash content also differed significantly among the evaluated mushrooms ($p < 0.001$).

4. Mineral Profile

Among the nine minerals evaluated, copper and selenium contents were present in lesser amounts and in comparison, phosphorous content was present in maximum amounts (Table 2). While working on *L. squarrosulus* collected from the southwest region of the Western Ghats, Ghate and Sridhar (2019) reported a lower amount of minerals in comparison to the presently evaluated sample of this species. Sodium, potassium and calcium contents in the presently evaluated *T. schimperi* were found to be on the lower side, while phosphorus content is on the higher side in comparison to their values reported in *T. globulus* from the Western Ghats by Sudheep and Sridhar (2014). As for phosphorus is concerned, this component was found to be more in *T. schimperi* in comparison to its value in *T. umkowaan* as reported by Karun et al. (2018) collected from the southwest region of the Western Ghats.

The mineral content of *L. squarrosulus* was lower than the same species in the southwest region (Ghate and Sridhar 2019). Sodium, potassium and calcium contents of *T. Schimperi* are lower, while phosphorus content is higher than *T. globulus* from the Western Ghats (Sudheep and Sridhar 2014). The phosphorus component of *T. schimperi* is more than *T. umkowaan* in the southwest region (Karun et al. 2018).

Table 2: Mineral composition (mg/100 g) of four wild mushrooms of the Western Ghats on a dry weight basis (n = 3 ± SD) ([a]USDA, 1999, [b]NRC-NAS, 1989, [c]Pennington and Young, 1990).

	Boletus edulis	*Hygrocybe alwisii*	*Lentinus squarrosulus*	*Termitomyces schimperi*	Soyabean[a]	NRC-NAS pattern[b]; Pennington and Young[c]
Sodium	0.66 ± 0.07	0.84 ± 0.01	0.91 ± 0.02	1.22 ± 0.00	2	120–200[b]
Potassium	17.52 ± 0.68	14.32 ± 0.70	18.64 ± 0.00	14.88 ± 0.18	1797	500–700[b]
Calcium	6.07 ± 0.06	0.02 ± 0.00	0.05 ± 0.01	7.45 ± 0.30	277	600[b]
Phosphorus	252.66 ± 0.37	281.15 ± 1.14	40.92 ± 0.84	320.44 ± 0.27	704	500[b]
Magnesium	6.58 ± 0.94	0.99 ± 0.08	1.50 ± 0.20	6.25 ± 0.05	280	60[b]
Iron	7.78 ± 0.05	1.81 ± 0.39	0.29 ± 0.05	2.17 ± 0.05	15.1	10[b]
Copper	0.02 ± 0.00	0.04 ± 0.01	0.01 ± 0.00	0.32 ± 0.03	1.66	0.6–0.7[b]
Zinc	0.13 ± 0.00	0.04 ± 0.01	0.01 ± 0.00	0.11 ± 0.02	4.89	5.0[b]
Selenium	0.11 ± 0.01	0.09 ± 0.03	0.08 ± 0.02	0.15 ± 0.00		0.05–0.2[c]
Na/K ratio	0.04	0.06	0.05	0.08	0.001	0.24–0.29
Ca/P ratio	0.02	< 0.01	< 0.01	0.02	0.39	1.2

Although mushrooms contain a sufficient amount of both macro and micronutrients, however, despite these mushrooms are not considered superior candidates for minerals as they do not meet the NRC-NAS (1989) standards in this regard. Out of the presently documented minerals, the amount of phosphorus was found to be the highest compared to other minerals in the evaluated mushrooms. Selenium, another important mineral was found within the recommended range for human consumption (Pennington and Young 1990). In all the evaluated mushrooms the ratio of Na/K is < 1, which is almost comparable to the ratio of Na/K in soybean and NRC-NAS standards (USDA 1999, NRC-NAS 1989), which is reported to have direct medicinal relevance particularly in combating the blood pressure (Yusuf et al. 2007). Unlike legumes, mushrooms are reported to possess a low amount of sodium and potassium content (Bahl 1998). All such foodstuffs containing a low amount of sodium and high amount of potassium content (Na/K ratio, < 0.6) are reported to be suitable for diet formulation (Nieman et al. 1992). As for the Ca/P ratio in the evaluated mushrooms is concerned (< 1), the values obtained are quite low and do not conform to the NRC-NAS (1989) standards which can be visualized from the comparative values given in Table 2. The Ca/P ratio of > 1 is reported to help in controlling the drain of calcium in urine which has a role in giving strength to the bones (Shills and Young 1988).

5. Amino Acids and Protein Efficiency Ratio

5.1 Amino Acids

Mushrooms contain a variable amount of essential and non-essential amino acids. The relative value of the detected amino acids in different mushrooms is given in Table 3. The quantity of most of the essential amino acids (EAA) and EAA score of *L. squarrosulus* is comparable with the values obtained by Ghate and Sridhar (2019) in the sample of this species collected from the southwest region of Western Ghats. As for *T. schimperi* is concerned the percentage of cystine, methionine, leucine, tyrosine, phenylalanine, lysine and histidine contents in this mushroom is higher in comparison to their values evaluated by Sudheep and Sridhar (2014) in *T. globulus* from the Western Ghats. Also, the proportion of methionine and tyrosine in *T. schimperi* collected in the southwest region is on the higher side than *T. clypeatus* (Ghate and Sridhar 2019). The amount of cystine, methionine and tyrosine in *T. schimperi* is also higher in comparison to the evaluated values of these amino acids in *T. umkowaan* from the southwest region of Western Ghats (Karun et al. 2018).

In the table above (Table 3) the proportion of indispensable and dispensable amino acids in the four wild edible mushrooms has been compared with other food items including soybean, rice, wheat and whole eggs. Although some dispensable amino acids of mushrooms are low in concentration, several of them are similar to or higher than soybean, cereals and whole egg. Amongst the indispensable amino acids, except for tryptophan, which was present below detectable level and a few more exceptions, EAA meets the FAO-WHO (1991) stipulated standards, which have been reflected in the EAA score (Table 4). With exception of cystine and tryptophan, the

Table 3: The comparison of the amount of different amino acids (mg/100 g protein) in four wild mushrooms of the Western Ghats with other food items (n = 3 ± SD; BDL, below detectable level) ([a]Bau et al. 1994, [b]Livsmedelsverk 1988, [c]Ensminger et al. 1990, Pomeramz 1998, USDA 1999 Lookhart and Bean 2000, Posner, 2000, [d]FAO, 1970, [e]FAO-WHO 1991, [f]Cystine + methionine, [g]Tyrosine + phenylalanine, [h]Ratio of total essential amino acid/total amino acid ratio).

	Boletus edulis	Hygrocybe alwisii	Lentinus squarrosulus	Termitomyces schimperi	Soybean[a]	Rice[b]	Wheat[c]	Whole egg[d]	FAO-WHO[e]
Dispensable amino acids									
Glutamic acid	4.80 ± 0.05	12.48 ± 0.06	5.20 ± 0.10	13.67 ± 0.30	17	15.2	35.5–36.9	13.0	
Aspartic acid	5.48 ± 0.02	6.90 ± 0.01	4.75 ± 0.04	6.36 ± 0.11	11	8.8	3.7–4.2	9.6	
Serine	7.53 ± 0.03	8.88 ± 0.03	4.91 ± 0.02	8.09 ± 0.05	5.7	5.4	3.7–4.8	7.6	
Proline	5.44 ± 0.06	5.85 ± 0.04	7.41 ± 0.05	5.73 ± 0.01	4.9	4.3	11.4–11.7	4.2	
Alanine	5.96 ± 0.02	7.68 ± 0.01	6.96 ± 0.04	7.11 ± 0.05	4.2	5.8	2.8–3.0	5.9	
Glycine	6.60 ± 0.09	8.61 ± 0.02	6.88 ± 0.00	8.67 ± 0.21	4.0	4.5	3.2–3.5	3.3	
Arginine	3.89 ± 0.04	4.92 ± 0.10	5.43 ± 0.03	5.78 ± 0.15	7.1	7.7	3.1–3.8	6.1	
Indispensable amino acids									
Threonine	3.69 ± 0.01	4.79 ± 0.04	4.01 ± 0.00	3.94 ± 0.06	3.8	3.2	2.2–3.0	5.1	3.4
Valine	6.08 ± 0.04	5.71 ± 0.00	5.77 ± 0.06	5.37 ± 0.08	4.6	6.6	3.7–4.5	6.9	3.5
Cystine	0.09 ± 0.01	0.15 ± 0.05	0.08 ± 0.07	0.64 ± 0.00	1.7	1.2	1.6–2.6	5.9	2.5[f]
Methionine	3.23 ± 0.02	1.27 ± 0.01	3.60 ± 0.05	1.73 ± 0.00	1.2	2.6	0.9–1.5	3.4	
Isoleucine	3.88 ± 0.10	4.49 ± 0.07	4.33 ± 0.06	4.38 ± 0.50	4.6	4.3	3.4–4.1	6.3	2.8
Leucine	4.91 ± 0.04	6.58 ± 0.01	5.68 ± 0.06	5.95 ± 0.09	7.7	8.2	6.5–7.2	8.8	6.6
Tyrosine	11.39 ± 0.00	2.93 ± 0.07	3.20 ± 0.03	4.17 ± 0.41	3.4	3.7	1.8–3.2	4.2	6.3[g]
Phenylalanine	2.69 ± 0.00	3.98 ± 0.09	3.78 ± 0.01	3.28 ± 0.05	4.8	5.1	4.5–4.9	5.7	
Tryptophan	BDL	BDL	BDL	BDL	1.2	1.3	0.7–1.0	1.7	1.1
Lysine	12.97 ± 0.03	7.05 ± 0.07	13.07 ± 0.08	6.38 ± 0.05	6.1	3.7	1.8–2.4	7.0	5.8
Histidine	2.55 ± 0.01	2.73 ± 0.00	3.10 ± 0.01	2.45 ± 0.07	2.5	2.4	1.9–2.6	2.4	1.9
TEAA/TAA ratio[b]	0.44	0.39	0.49	0.36	0.44	0.45	0.31–0.35	0.54	

Table 4: The essential amino acid score (EAAS) as well as the protein efficiency ratio (PER) of four wild mushrooms of the Western Ghats.

	Boletus edulis	*Hygrocybe alwisii*	*Lentinus squarrosulus*	*Termitomyces schimperi*
EAA score				
Threonine	1.09	1.41	1.18	1.16
Valine	1.74	1.63	1.65	1.53
Cystine + Methionine	1.33	0.57	1.47	0.95
Isoleucine	1.38	1.60	1.55	1.56
Leucine	0.74	1.00	0.86	0.90
Tyrosine + Phenylalanine	2.23	1.10	1.11	1.18
Lysine	2.23	1.22	2.25	1.10
Histidine	1.34	1.44	1.63	1.29
PER				
PER1	1.30	2.04	1.56	1.76
PER2	0.57	2.21	1.77	1.80
PER3	−6.80	1.68	1.81	0.16
Mean PER	–	1.98	1.72	1.24
Quality of protein	Poor	Moderate	Moderate	Poor

amount of rest of the EAA in the evaluated mushrooms is at par with or higher than their values in soybean, rice, wheat and whole egg. The total EAA/total amino acid (TAA) ratio in *T. schimperi* (0.36) is on the lower side as compared to *L. squarrosulus* (0.49). However, these values are either comparable to or higher than the soybean, rice and wheat (0.31–0.45). An increase in TEAA/TAA ratio demonstrates the good composition of EAA in these mushrooms.

As is apparent from the data in Table 4, mushrooms contain almost all essential amino acids required for human nourishment. This is in conformity with the observations of Kakon et al. (2012). Wild edible mushrooms serve as a potential source of nutrition especially proteins, carbohydrates and EAA and thus can play an important role in tackling protein-energy malnutrition (Alam et al. 2008). Interestingly, all mushrooms studied have lysine content almost similar to or higher than soybean, cereals (rice and wheat) and the whole egg including FAO-WHO (1991) standards. Such high lysine-containing mushrooms are reported as suitable foodstuff for vegetarians who are deprived of animal proteins (Verma et al. 1987). The presence of a sufficient quantity of glutamic acid in the assessed mushrooms adds to their palatability since this amino acid is reported to serve as a natural source of flavor similar to the common food additive like monosodium glutamate (Yamaguchi 1971).

5.2 *Protein Efficiency Ratio*

The quality of protein is reported to depend on the Protein Efficiency Ratio (PER) status (Friedman 1996). Based on PER status, foodstuffs have been grouped into three categories: < 1.5, poor quality; 1.5–2, moderate quality; > 2, high quality. Accordingly, the PER_1 and PER_2 were highest in *H. alwisii* (high quality) followed by *T. schimperi* (moderate quality), while the PER_3 was highest in *L. squarrosulus* (moderate quality) (Table 4). Based on the mean values of PER status, *H. alwisii* qualifies for the highest quality parameters, followed by *L. squarrosulus* and *T. schimperi*, which fall under moderate protein quality grouping, while in comparison it is poor in *B. edulis* and *T. schimperi*. The PER status of *L. squarrosulus* is also lower in comparison to the value reported for this species collected from the scrub jungles in the southwest region by Ghate and Sridhar (2019).

6. Fatty Acids Profile

The fatty acid profile of the evaluated mushrooms showed interesting results, particularly with respect to the presence of both saturated and unsaturated fatty acids including the favourable fatty acid ratios (Table 5). Palmitic acid was detected in the highest proportion in the evaluated mushrooms followed by stearic acid. The percentage of saturated fatty acids in all four evaluated mushrooms was found to be comparable to soybean and wheat. As for the presence of unsaturated fatty acids in these mushrooms is concerned, these are composed of a blend of ω-9, ω-6 and ω-3 fatty acids. The linoleic acid was found to be present in the highest amount followed by the oleic acid, while the ecosadienoic acid was present in the least quantity. In all the evaluated mushrooms ω-6 linoleic acid was documented. *B. edulis* was found to contain three ω-6 and two ω-3 fatty acids, while *L. squarrosulus* has one each of ω-6 and ω-3 fatty acids. In comparison, *T. schimperi* consists of only one ω-6 fatty acid, while ω-3 fatty acids were altogether absent in this mushroom. The oleic and linoleic acid contents were not as high as in soybean and wheat (Table 5).

Except for *H. alwisii* and *T. schimperi*, in all other evaluated mushrooms, the amount of unsaturated fatty acids (USFA) was found to be more in comparison to the amount of saturated fatty acids (SFA). The overall value of USFA was highest in *B. edulis* followed by *H. alwisii*. Among the five fatty acid ratios calculated (Table 5), there is an increase in the first two ratios (TUFA/TSFA and TPUFA/TMUFA) and a decrease in the last three ratios $[(C_{14:0}+C_{15:0} + (C_{16:0}/C_{18:0}),$ $C_{18:1}/C_{18:2}$ and ω-6/ω-3)]. This is reported to be important from a nutrition point of view (Supriya et al. 2012). The TUFA/TSFA ratio was found to be highest in *B. edulis* followed by *L. squarrolulus*, while the TPUFA/TMUFA ratio was highest in *L. squarrolulus* followed by *H. alwisii*. The TPUFA/TMUFA ratios of the studied mushrooms is on the lower side in comparison to the value of the ratio of TPUFA/ TMUFA reported by Pastor-Cavada et al. (2009) in 30 taxa of *Vicia*, which is a wild legume (1.2–6 vs. 1.3–7.5). Foodstuffs having $C_{18:0}$ fatty acid as one of the components are known to be less hypercholesterolemic in comparison to those having fatty acid $C_{16:0}$ instead of $C_{18:0}$ (Baer et al., 2003), hence lower ratios of $C_{16:0}/C_{18:0}$ and $C_{18:1}/C_{18:2}$ are believed to be more suitable for human diet (Pastor-Cavada et al. 2009). The ratios $C_{14:0}+C_{15:0} + (C_{16:0}/C_{18:0})$ (in *H. alwisii*), $C_{18:1}/C_{18:2}$ (in *L. squarrosulus*) and ω-6/ω-3 (in *H. alwisii* and *T. schimperi*) were found to be least depicting their versatility as for fatty acid composition is concerned. An

Table 5: The fatty acid methyl esters of four wild mushrooms of the Western Ghats (g/100 g total lipids) (n = 3 ± SD; BDL, below detectable level; Tr, Trace) ([a]Wahnon et al. 1988, Cho 1989, [b]Davis et al. 1980, Pomeranz 1998).

	Boletus edulis	*Hygrocybe alwisii*	*Lentinus squarrosulus*	*Termitomyces schimperi*	Soybean[a]	Wheat[b]
Saturated fatty acids						
Capric acid (C10:0)	BDL	BDL	BDL	0.03 ± 0.00		
Lauric acid (C12:0)	0.16 ± 0.05	1.14 ± 0.00	0.59 ± 0.09	0.37 ± 0.00	Tr-4.5	
Myristic acid (C14:0)	0.45 ± 0.03	0.85 ± 0.00	1.02 ± 0.01	3.15 ± 0.12	Tr-4.5	
Pentadecanoic acid (C15:0)	BDL	BDL	0.58 ± 0.00	1.35 ± 0.10		
Palmitic acid (C16:0)	16.12 ± 0.01	32.10 ± 0.62	10.73 ± 0.20	53.73 ± 0.94	11–11.6	11–32
Heptadecanoic acid (C17:0)	0.27 ± 0.00	0.20 ± 0.00	0.47 ± 0.03	1.46 ± 0.00		
Stearic acid (C18:0)	5.48 ± 0.03	14.01 ± 0.29	3.23 ± 0.20	0.18 ± 0.00	2.5–4.1	0–4.6
Arachidic acid (20:0)	BDL	BDL	BDL	1.75±0.06	Tr	
Heneicosanoic acid (C21:0)	BDL	BDL	BDL	BDL		
Behenic acid (C22:0)	0.49 ± 0.02	1.90 ± 0.00	0.37 ± 0.02	1.85 ± 0.30		
Tricosanoic acid (C23:0)	BDL	BDL	BDL	BDL		
Lignoceric acid (C24:0)	BDL	5.46 ± 0.04	BDL	4.57 ± 0.72		
Unsaturated fatty acids						
Myristoleic acid (C14:1) (ω-5)	BDL	BDL	BDL	BDL		
Palmitoleic acid (C16:1) (ω-7)	0.57 ± 0.10	BDL	0.85 ± 0.40	BDL	Tr	
Palmitelaidic acid (16:1) (ω-7)	BDL	0.36 ± 0.00	BDL	0.94 ± 0.00		
Oleic acid (C18:1) (ω-9)	14.30 ± 0.10	11.59 ± 0.41	5.22 ± 0.20	11.62 ± 0.83	21.1–22	11–29
Linoleic acid (C 18:2) (ω-6)	25.86 ± 0.31	32.25 ± 1.06	19.41 ± 0.07	13.39 ± 0.10	52.4–54	44–74
Linolelaidic acid (18:2) (ω-6)	1.42 ± 0.00	BDL	BDL	BDL		

Table 5 contd. ...

...Table 5 contd.

	Boletus edulis	Hygrocybe alwisii	Lentinus squarrosulus	Termitomyces schimperi	Soybean[a]	Wheat[b]
Linolenic acid (18:3) (ω-3)	1.17 ± 0.02	BDL	4.20 ± 0.30	BDL	7.1–7.5	0.7–4.4
Eicosadienoic acid (20:2) (ω-6)	0.13 ± 0.00	BDL	BDL	BDL		
Eicosapentaenoic acid (20:5) (ω-3)	1.18 ± 0.10	BDL	BDL	BDL		
Docosahexaenoic acid (22:6) (ω-3)	BDL	BDL	BDL	BDL		
Total and ratio						
Total saturated fatty acids (TSFA)	22.97	56.02	16.99	69.38		
Total unsaturated fatty acids (TUFA)	44.63	43.84	29.68	25.01		
TUFA/TSFA	1.94	0.78	1.75	0.36		
TPUFA/TMUFA	2.00	2.78	3.89	1.15		
$C_{14:0}+C_{15:0}+(C_{16:0}/C_{18:0})$	3.39	3.14	4.92	303		
$C_{18:1}/C_{18:2}$	0.52	0.36	0.27	0.87		
ω-6/ω-3	11.66	–	4.62	–		

increase in the ratio of ω-6/ω-3 has been reported to result in an increased risk for cardiovascular disease (Weber 1989, Gebauer et al. 2006). In the mushrooms studied presently, this ratio ranges from 4.6 in *L. squarrolulus* to 11.7 in *B. edulis*, which is comparatively lower than the ratio of ω-6/ω-3 reported in 30 taxa of the wild legume *Vicia* (2.6–16.2) by Pastor-Cavada et al. (2009). Unlike *L. squarrolulus* in our study, *L. sajor-caju* also possesses many ω-3 as well as ω-6 fatty acids (Lata and Atri 2021). Many benefits are reported to be associated with the foodstuffs having a low ω-6 to ω-3 ratio. For example, a ratio value of 2.5 is known to reduce the cell proliferation in the rectum of patients suffering from colorectal cancer; a ratio value of 2–3 is known to combat the inflammation process in patients suffering from rheumatoid arthritis; a ratio value of 5 is reported to be advantageous to the patients of asthma (Simopoulos 2002). However, in the recommendation of FAO-WHO (1998), the ratio of ω-6/ω-3 has been fixed between 5 and 10. The reported ratio of ω-6/ω-3 in the ancient (Paleolithic) diet is as low as 0.79, however, in the modern diet around the world it stands between 1.00 and 50 thereby forecasting the danger of susceptibility of the consumer population to several diseases (Simopoulos 2004). In spite of their low total lipid content in them, several wild edible mushrooms contain the required proportion of necessary fatty acids for the human diet.

7. Conclusion and Outlook

Among the four wild edible mushrooms evaluated, dry matter yield was the highest in *L. squarrosulus* owing to low moisture content as it grows on the woody substrate. The crude protein content is comparable with legumes, while low in total lipids and rich in crude fiber. Ash content is low, which is a direct reflection of the presence of a low quantity of minerals. Carbohydrates and calorific values are comparable to cereals and legumes. All the mushrooms possess low minerals with an appropriate Na/K ratio (< 1). Except for tryptophan, all essential amino acids are present in an adequate proportion which is either similar to or higher than legumes, cereals and whole eggs which meets the FAO-WHO (1991) standards. The evaluated mushrooms also contain many ω-6 and ω-3 fatty acids with ratios favourable for human consumption. The nutritional composition of *L. squarrosulus* and *T. schimperi* showed variation with their counterparts or related species occurring in the Western Ghats and west coast region.

Amanita hemibapha although edible, local people in the foothill of the Western Ghats do not consume it due to its bright yellow-red mosaic appearance. Wild mushrooms serve as an alternative to plant and animal-derived foods and provide a livelihood to the tribals or local dwellers. Their nutraceutical value provides nourishment as well as combat several lifestyle-acquired diseases. As these mushrooms emerge in huge quantities during the southwest monsoon season, appropriate methods of collection, transport, processing and preservation are necessary. In addition, habitat conservation and strategies to enhance the production of wild mushrooms are equally important. Changes in nutraceutical qualities in preserved as well as cooked samples need to be evaluated so as to generate additional knowledge for processing the mushrooms appropriately without major losses in quality.

Acknowledgements

The first author (VR) gratefully acknowledges the facilities provided by the Department of Biosciences, Mangalore University. Thanks are due to the research guide Late Dr. M. Rajashekhar for guidance and the Board of Research in Nuclear Sciences (BRNS), Bhabha Atomic Research Centre, Mumbai, India for funding this research.

References

Alam, N., R. Amin, A. Khan, I. Ara, M.J. Shim, M.W. Lee and T.S. Lee. 2008. Nutritional analysis of cultivated mushrooms in Bangladesh—*Pleurotus ostreatus, Pleurotus sajor-caju, Pleurotus florida* and *Calocybe indica*. *Mycobiology*, 36: 228–232.

Aletor, V.A. and O.O. Aladetimi. 1995. Compositional studies on edible tropical species of mushrooms. *Food Chem.*, 54: 265–268.

Alexander, H. 2013. Morphology and classification. pp. 3–33. *In*: Chang, S.T. and Hayes, W.A. (eds.). The Biology and Cultivation of Edible Mushrooms. Academic Press, London.

Alsmeyer, R.H., A.E. Cunningham and M.L. Happich. 1974. Equations predict PER from amino acid analysis. *Food Technol.*, 28: 34–38.

AOAC. 2006. Official methods of analysis of AOAC international, 18th Edition. Association of Official Analytical Chemists, Arlington, VA, USA.

APHA. 1995. Standard methods for the examination of water and wastewater, 19th Edition. American Public Health Association Publications, Washington DC, USA.

Atri, N.S., B. Kumari, S. Kumar, R.C. Upadhyay, Lata and A. Gulati. 2016. Nutritional profile of wild edible mushrooms of North India. pp. 372–395. *In*: Deshmukh, S.K., Misra, J.K., Tewari, J.P. and Popp, T. (eds.). Fungi - Applications and Management Strategies. CRC Press, Boca Raton.

Baer, D.J., J.T. Judd, P.M. Kris-Etherton, G. Zhao and E.A. Emken. 2003. Stearic acid absorption and its metabolizable energy value are minimally lower than other fatty acids in healthy men fed mixed diets. *J. Nutr.*, 133: 4129–4134.

Bahl, N. 1998. Handbook on Mushroom. 2nd edition, pp. 21–23, IBH Publication Co. Ltd., Oxford.

Bau, H.M., C.F. Vallaume, F. Evard, B. Quemener, J.P. Nicolas and L. Mejean. 1994. Effect of solid state fermentation using *Rhizopus oligosporus* sp. T-3 on elimination of antinutritional substances and modification of biochemical constituents of defatted rape seed meal. *J. Sci. Food Agric.*, 65: 315–322.

Belderok, B. 2000. Developments in bread making processes. *Pl. Foods Hum. Nutr.*, 55: 1–86.

Beluhan, S. and A. Ranogajec. 2011. Chemical composition and non-volatile components of Croatian wild edible mushrooms. *Food Chem.*, 124: 1076–1082.

Boa, E.R. 2004. Wild Edible Fungi a Global Overview of their Use and Importance to People. Food and Agriculture Organization of the United Nations, Rome.

Buczacki, S. 2012. Collins Fungi Guide. Harper-Collins Publishers, London.

Cheftel, J.C., J.L. Cuq and D. Lorient. 1985. Proteinesalimentaires, Tec. and Doc. Lavoisier, Paris.

Cheung, P.C.K. 2010. The nutritional and health benefits of mushrooms. *Nutr. Bull.*, 35: 292–299.

Chihara, G. 1993. Medicinal aspects of lentian isolated from *Lentinus edodes* (Berk), pp. 261–266, Chinese University Press, Hong Kong.

Cho, B.H.S. 1989. Soybean oil: Its nutritional value and physical role related to polyunsaturated fatty acid metabolism. Am. Soybean Assoc. Tech. Bull. # 4HN6.

Colak, A., O. Faiz and E. Sesli. 2009. Nutritional composition of some wild edible mushrooms. *Turk. J. Biochem.*, 34: 25–31.

Crisan, E.V. and A. Sands. 1978. Edible mushrooms: Nutritional value. pp. 137–165. *In*: Chang, S.T. and Hayes, W.A. (eds.). The Biology and Cultivation of Edible Mushrooms. Academic Press, New York.

Davis, K.R., N. Litteneker, D. LeTourneau and J. McGinnis. 1980. Evaluation of the nutrient composition of wheat - 1: Lipid constituents. *Cereal Chem.*, 57: 178–184.

Dubois, M., K.A. Gilles, J.K. Hamilton, P.A. Rebers and F. Smith. 1956. Colorimetric method for determination of sugars and related substances. *Anal. Chem.*, 28: 350–356.

Dupont, J., S. Dequin, T. Giraud, F.L. Tacton, S. Marsit, J. Ropars, F. Richad and M.A. Selosse. 2017. Fungi as a source of food. *In*: The Fungal Kingdom. Microbiology Spectrum 5: FUNK-0030-2016.

Ensminger, N.E., J.E. Oldfield and W.W. Heinenann. 1990. Feeds and Nutrition. Ensminger Publishing Company, USA.

FAO. 1970. Amino acid content of foods and biological data on proteins. Food and Agricultural Organization of United Nations, Rome.

FAO-WHO. 1998. Preparation and use of food-based dietary guidelines. Report of Joint FAO-WHO Consultation, Technical Report Series # 880: FAO, Geneva, Rome.

FAO-WHO. 1991. Protein Quality Evaluation. Reports of a Joint FAO-WHO Expert Consultation, Food and Nutrition Paper # 51, pp. 1–66, Food and Agriculture Organization of the United Nations, FAO, Rome.

Fierabracci, V., P. Masiello, M. Novelli and E. Bergamini. 1991. Application of amino acid analysis by high performance liquid chromatography with phenyl isothiocyanate derivatization to the rapid determination of free amino acids in biological samples. *J.Chromatogr. B Biomed. Sci. Appl.*, 570: 285–291.

Friedman, M. 1996. Nutritional value of proteins from different food sources—A review. *J. Agric. Food Chem.*, 44: 6–29.

Gebauer, S.K., T.L. Psota, W.S. Harris and P.M. Kris-Etherton. 2006. n-3 Fatty acid dietary recommendations and food sources to achieve essentiality and cardiovascular benefits. *Am. J. Clin. Nutr.*, 83: 1526S–1535S.

Ghate, S.D. and K.R. Sridhar. 2019. Nutritional attributes of two wild mushrooms of southwestern India. pp. 105–120. *In*: Sridhar, K.R. and Deshmukh, S.K. (eds.). Advances in Macrofungi: Diversity, Ecology and Biotechnology. CRC Press, Boca Raton.

Hapuarachchi, K.K., W.A. Elkhateeb, S.C. Karunarathna, C.R. Cheng, A.R. Bandara, P. Kakumyan, K.D. Hyde, G.M. Daba and T.C. Wen. 2018. Current status of global *Ganoderma* cultivation, products, industry and market. *Mycosphere*, 9: 1025–1052.

Humphries, E.C. 1956. Mineral composition and ash analysis. pp. 468–502. *In*: Peach, K. and Tracey, M.V. (eds.). Modern Methods of Plant Analysis, Volume #1. Springer, Berlin.

Hyde, K.D., A.H. Bahkali and M.A. Moslem. 2010. Fungi—an unusual source for cosmetics. *Fungal Divers.*, 43: 1–9.

Johnsy, G., D. Sargunam, M.G. Dinesh and V. Kaviyarasan. 2011. Nutritive value of edible wild mushrooms collected from the Western Ghats of Kanyakumari district. *Bot. Res. Int.*, 4: 69–74.

Jordan, M. 2004. The Encyclopaedia of fungi of Britain and Europe. Francis Lincoln Publishers Ltd., London.

Kakon, A.J., M.B.K. Choudhury and S. Saha. 2012. Mushroom is an ideal food supplement. *Journal of Dhaka National Medical College & Hospital*, 18: 58–62.

Karun, N.C. and K.R. Sridhar. 2017. Edible wild mushrooms in the Western Ghats: Data on the ethnic knowledge. *Data in Brief*, 14: 320–328.

Karun, N.C., K.R. Sridhar and C.N. Ambarish. 2018. Nutritional potential of *Auricularia auricula-judae* and *Termitomyces umkowaan*—The wild edible mushrooms of Southwestern India. pp. 281–301. *In*: Gupta, V.K., Treichel, H., Shapaval, V., De Oliveira, L. and Tuohy, M.G. (eds.). Microbial Functional Foods and Nutraceuticals, John Wiley & Sons Ltd., NJ USA.

Kumar, R., A. Tapwal, S. Pandey, R.K. Borah, D. Borah and J. Borgohain. 2013. Macro-fungal diversity and nutrient content of some edible mushrooms of Nagaland, India. *Bioscience*, 5: 1–7.

Kurztman, R.H. 1997. Nutrition from mushrooms, understanding and reconciling available data. *Mycoscience*, 38: 247–253.

Lata and N.S. Atri. 2021. Fatty acid profile of an indigenous strain of *Lentinus sajor-caju* (Basidiomycota). *Ukr. Bot. J.*, 78: 327–334.

Livsmedelsverk, S. 1988. Energiochnäringsämnen. The Swedish Food Administration, Stockholm, Sweden.

Longvah, T. 2000. Nutritive value of northeast Indian plant foods. *Nutr. News*, 21: 1–6.

Lookhart, G. and S. Bean. 2000. Cereal proteins: composition of the major fractions and methods for identification. pp. 363–383. *In*: Kulp, K. and Ponte Jr, J.G. (eds.). Hand book of Cereal Science and Technology. Marcel Dekker Inc., New York.

Manzi, P., A. Aguzzi and L. Pizzoferrato. 2001. Nutritional value of mushrooms widely consumed in Italy. *Food Chem.*, 73: 321–325.

Manzi, P., L. Gambelli, S. Marconi, V. Vivanti and L. Pizzoferrato. 1999. Nutrients in edible mushrooms: An inter-species comparative study. *Food Chem.*, 65: 477–482.

Manzi, P.S., A.A. Marconi and L. Pizzoferrato. 2004. Commercial mushroom nutritional quality and effect of cooking. *Food Chem.*, 84: 201–206.

Marshall, E. and N.G. Nair. 2009. Make Money by Growing Mushrooms. Infrastructure and Agro-Industries Division, Food and Agriculture Organization of the United Nations, Rome.

Matz, S.A. 1991. The Chemistry and Technology of Cereals as Food and Feed. Van Nostrand Reinhold, New York.

Mohanan, C. 2011. Macrofungi of Kerala. Handbook # 27, Kerala Forest Research Institute, Peechi, India.

Nareshkumar, S. 2007. Capillary gas chromatography method for fatty acid analysis of coconut oil. *J. Pl. Crops*, 35: 23–27.

Nieman, D.C., D.E. Butterworth and C.N. Nieman. 1992. Nutrition. Wm C Brown Publishers, Dubuque.

NRC-NAS. 1989. Recommended dietary allowances, National Academy Press, Washington DC.

Okoro, I.O. and F.I. Achuba. 2012. Proximate and mineral analysis of some wild edible mushrooms. *Afr. J.Biotechnol.*, 11: 7720–7724.

Pastor-Cavada, E., R. Juan, J.E. Pastor, M. Alaiz and J. Vioque. 2009. Fatty acid distribution in the seed flour of wild *Vicia* species from Southern Spain. *J. Am. Oil Chem. Soc.*, 86: 977–983.

Pennington, J.A.T. and B. Young. 1990. Iron, zinc, copper, manganese, selenium and iodine in foods from the United States total diet study. *J. Food Comp. Anal.*, 3: 1661–1684.

Phillips, R. 2006. Mushrooms. Pan Macmillan, London.

Pomeranz, Y. 1998. Chemical composition of kernel structures. pp. 97–158. *In*: Pomeranz, Y. (ed.). Wheat Chemistry and Technology. American of Cereal Chemists Inc., Minnesotta.

Posner, E.S. 2000. Wheat. pp. 1–30. *In*: Kulp, K. and Ponte Jr, J.G. (ed.). Hand book of Cereal Science and Technology. Marcel Dekker, New York.

Purkayastha, R.P. and A. Chandra. 1985. Manual of Indian edible mushrooms. Today and Tomorrow's Printers and Publishers, New Delhi.

Royse, D.J., J. Baars and Q. Tan. 2017. Current overview of mushroom production in the world. pp. 5–13. *In*: Zied, D.C. and Pardo-Giminez, A. (eds.). Edible and Medicinal Mushrooms: Technology and Applications. John Wiley & Sons Ltd., Hoboken.

Sanmee, R.B., D.P. Lumyong, K. Izumori and S. Lumyong. 2003. Nutritive value of popular wild edible mushrooms from northern Thailand. *Food Chem.*, 84: 527–532.

Shills, M.E.G. and V.R. Young. 1988. Modern nutrition in health and disease. pp. 276–282. *In*: Neiman, D.C., Buthepodorth, D.E. and Nieman, C.N. (eds.). Nutrition. WmC Brown, Dubugue, USA.

Simopoulos, A.P. 2002. The importance of the ratio of omega-6/omega-3 essential fatty acids. *Biomed. Pharmacother.*, 56: 365–379.

Simopoulos, A.P. 2004. Omega-6/omega-3 essential fatty acid ratio and chronic diseases. *Food Rev. Int.*, 20: 77–90.

Sudheep, N.M. and K.R. Sridhar. 2014. Nutritional composition of two wild mushrooms consumed by the tribals of the Western Ghats of India. *Mycology*, 5: 64–72.

Supriya, P., K.R. Sridhar, S. Nareshkumar and S. Ganesh. 2012. Impact of electron beam irradiation on fatty acid profile of *Canavalia* seeds. *Food Bioproc. Technol.*, 5: 1049–1060.

Thawthong, A., S.C. Karunarathna, N. Thongklang, E. Chukeatirote, P. Kakumyan, S. Chamyuang, L.M. Rizal, P.E. Mortimer, J. Xu, P. Callac and K.D. Hyde. 2014. Discovering and domesticating wild tropical cultivatable mushrooms. *Chiang Mai J. Sci.*, 41: 731–764.

USDA. 1999. Nutrient Data Base for Standard Reference Release # 13, Food Group 20: Cereal Grains and Pasta, *U.S.* Department of Agriculture, Agricultural Research Service, USA, Agriculture Handbook # 8–20.

Verma, R.N., G.B. Singh and K.S. Bilgrami. 1987. Fleshy fungal flora of NEH India-I. Manipur and Meghalaya. *Ind. Mush. Sci.*, 2: 414–421.

Wahnon, R., S. Mokady and U. Cogan. 1988. Fatty acid composition in soybean oil. Proceedings of 19th World Congress, International Society for Fat Research, Tokyo.

Wasson, R.G. 1971. Soma: Divine Mushroom of Immortality, Ethnomycology Studies # 1, Harvest special. Harcourt Brace Jovanovich, New York.

Weber, P.C. 1989. Are we what we eat? Fatty acids in nutrition and in cell membranes: cell functions and disorders induced by dietary conditions. *In*: Fish fats and your health. Svanoy Foundation, Norway Report # 4: 9–18.

Xiao, L., S.A. Mjøs and B.O. Haugsgjerd. 2012. Efficiencies of three common lipid extraction methods evaluated by calculating mass balances of the fatty acids. *J. Food Comp. Anal.*, 25: 198–207.

Yamaguchi, S., T. Yoshikawa, S. Ikeda and T. Ninomiya. 1971. Measurement of the relative taste intensity of some a-amino acids and 50-nucleotides. *J. Food Sci.*, 36: 846–849.

Yusuf, A.A. B.M. Mofio and A.B. Ahmed. 2007. Proximate and mineral composition of *Tamarindus indicas* Linn 1753 seeds. *Sci. World J.*, 2: 1–4.

13

Medicinal Mushrooms as Nutritional Management for Post COVID-19 Syndromes

Nehal M El-Deeb,[1,]* *Hesham Ali El Enshasy,*[1,2]
Aya M Fouad[3] *and Ahmed M Kenawy*[1,]*

1. Introduction

Mushrooms are a huge category of fungi that exists on earth since millions of years ago. They evolved and diversified into three categories (edible, poisonous, and magical). Mushrooms have been always closely associated with human activities and they were used by ancient Egyptian Pharos for medicinal uses and are the main cornerstone in ancient traditional Chinese medicine. Nowadays, mushrooms represent an important category of novel food that play crucial roles in healthy food composition. The value of mushrooms as nutritional agents relays on their properties, including low fat, low calories, and low sodium contents. In addition, the presence of a high amount of dietary fiber, vitamins, proteins, and different important mineral elements (Cu, Mn, Se, and Zn) makes mushrooms an excellent nutritional agent. Several metabolites produced from mushrooms can offer plenty of health

[1] City of Scientific Research and Technological Applications (SRTA-City), New Borg El-Arab City, Alexandria, Egypt.
[2] Institute of Bioproduct Development (IBD), Universiti Teknologi Malaysia (UTM), Skudai, Johor Bahru, Malaysia.
[3] Faculty of Pharmacy, Damanhour University, Egypt.
* Corresponding authors: nehalmohammed83@gmail.com; aatta75@yahoo.com

benefits. For Example, β-glucan, proteoglycan, ergosterol, calcaelin, ganoderic acid, hispolon, flavonoids, lectins, laccase, lentinan, phenolics, nucleosides, nucleotides, and triterpenoids can act as antioxidant, antiaging, anticancer, antimicrobial, anti-inflammatory, anti-obesity, and immunomodulatory agents.

In this respect, these biological molecules can effectively reduce many health problems associated with heart, gut, hypertension, and metabolic disorders and improve the immune system. As a health promoter, edible mushrooms are now more widespread than before, including this much attention given to the research activities performed to explore the ingredients of a diverse group of edible mushroom species. These metabolites are needed to supply the body with its needs via staple functional food. In addition, due to the aforementioned health value of mushrooms, there is an increasing need for testing such a powerful source of immune system boosters against newly emerging diseases such as COVID-19. More and more attention should be paid to the bio-supplementation of food products with mushroom nano-byproducts and their bioactive compounds. Also, the therapeutic properties and the health benefits of medicinal and edible mushrooms should be reassessed for a new healthy lifestyle.

2. Mushrooms Biodiversity

Mushrooms usually exist in tropical regions. They have attractive colors and shapes and are closely related to human modern life activities. Mushrooms have a huge morphological diversity among all groups of fungi. They have spread into most niches and play diverse roles in the ecosystem (Rahi and Malik 2016). Despite their importance, their evolution is poorly known due to the lack of inclusive molecular phylogenetic studies. Taxonomically, mushrooms represent basidiomycotina and ascomycotina phyla (Naranjo-Ortiz and Gabaldón 2019). In an attempt to provide molecular evidence using data from gene family analysis and genomics, 5,284 species have been studied to generate a comprehensive phylogenetic tree (Figure 1) to shape morphological evolution in mushrooms (Varga et al. 2019). Despite the nutritional and medicinal importance of mushrooms, their classification is still in its initial stages.

2.1 Types of Mushrooms

There is a discrepancy about the real number of mushrooms in the pieces of literature and the number of mushroom species is estimated to be 150,000–160,000 on earth. Out of them, only 10% of the named ones are known ($\approx$ 15,000 species) (Wasser 2014). Over the years, knowledge about the safety of mushroom observations have, been accumulated and was used as a means to classify mushrooms (Li et al. 2021). Mushrooms can be divided into three different categories, according to their nutritional values and their effects after eating them, edible, poisonous, and magic mushrooms. The first category, edible mushrooms, combines mushrooms that can be eaten without causing harm to the body or its physiology. This group represents the majority of all known mushrooms; however, most of them are too

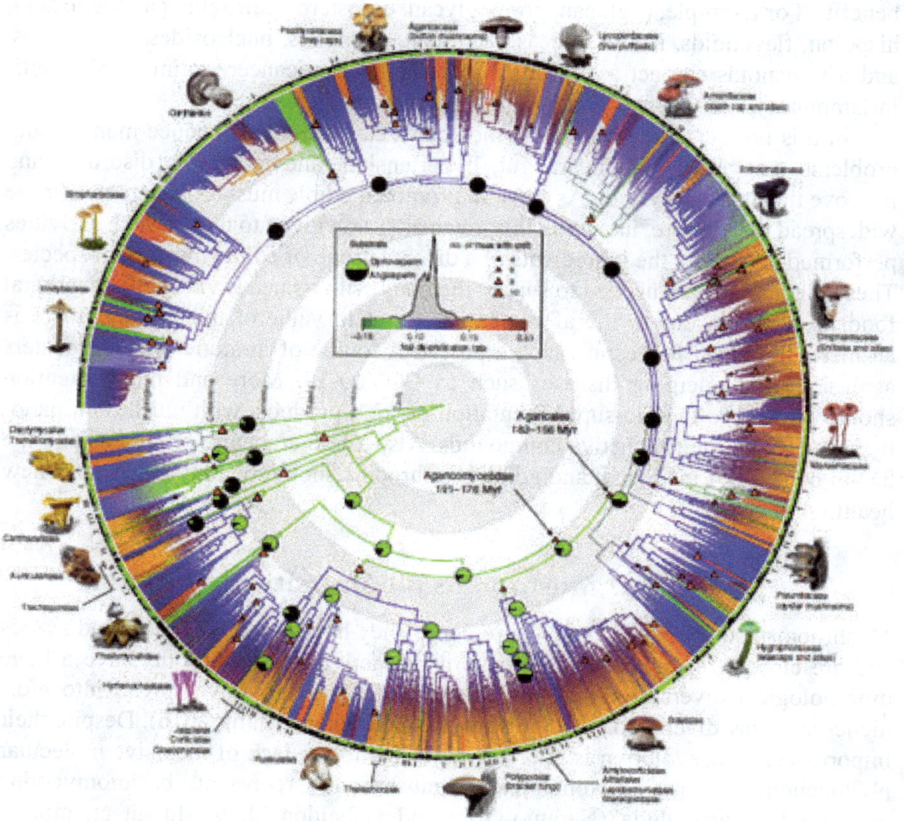

Figure 1: Phylogenetic relationship among most mushrooms (Varga et al. 2019).

slimy, tough, or bitter to be used as food. Many mushrooms are only edible in certain stages of their life cycle. The second category is poisonous mushrooms which are known to have effects ranging from mild gastric/intestinal discomfort up to death. They are dangerous mushrooms that can be very similar to edible ones; hence, it is of great importance to correctly identify wild mushrooms before using them in a meal. The third category is magic mushrooms, which is the group of the least common types among all known mushrooms. Mushrooms belonging to this category have hallucinogenic properties due to the occurrence of psilocybin, the psychedelic compound in hallucinogenic mushrooms. Psilocybin has mind-altering effects that are highly affected by the mental state of the person (Cavanna et al. 2022). These wild mushrooms grow in many areas but are listed as a schedule 1 drug in the United States of America.

3. Bioactive Ingredients in Mushrooms

Edible mushrooms are consumed for their high nutritional importance, medicinal properties, and their sweetened taste. Many edible-mushrooms species that exist in the wild are usually cultivated for consumption (Li et al. 2021). Only a few mushrooms (20 mushroom species) are cultivated for production, about 35 mushroom species are cultivated for commercial purposes, and around 20 of them are cultivated for the industrial level, but the remaining species are uncultivable (Sánchez 2004). Edible mushrooms contain several bioactive components including vitamins, polysaccharides, proteins, polyunsaturated fatty acids, phenolics, amino acids, dietary fibers, and minerals (Zeb and Lee 2021). These components were proven to have several health benefits and medicinal properties that enable mushrooms to be used as anticancer, antidiabetic, antiaging, antiviral, antimicrobial, antioxidant, immune-stimulatory, cholesterol-lowering, and anti-inflammatory agents (Alves et al. 2012, Chen et al. 2014, Lo and Wasser 2011, Roda et al. 2020). In spite of all the above-mentioned good features, a few mushroom species were studied and several species are waiting. Nowadays, mushrooms and their above-mentioned bioactive components are drawing the attention of medical experts and pharmacologists.

3.1 Lectins

Lectins are carbohydrate-binding proteins, which are among the most important medicinal components of mushrooms. They are the most significant mushroom constituents and major metabolites that dictate a mushroom's bioactivities (Singh et al. 2010). They are produced by different edible mushrooms, including *Grifola frondosa*, *Tricholoma mongolicum*, *Agaricus bisporus*, *Boletus satanus*, *Agaricus campestris*, *Flammulina velutipes*, *Ganoderma lucidum*, and *Volvariella volvacea*. They were reported to have active properties for the immune system, increase insulin secretion, and have anticancer effects (Hassan et al. 2015). For example, lectins isolated from *Clitocybe nebularis* showed immune-stimulatory effects on the dendritic cells (Švajger et al. 2011).

3.2 Glucans

Glucans are distinctive constituents in mushrooms that showed anticancer, immune-stimulatory, anti-inflammatory, and antioxidant activities (Cerletti et al. 2021). They exist in different edible mushrooms like *Auricularia auricular* (Jelly ears), *Pleurotus ostreatus* (Oyster), *Ganoderma lucidum* (Reishi), and *Lentinula edodes* (Shiitake). Beta-glucan, produced by *Pleurotus pulmonarius* is an effective anti-inflammatory that confers anti-nociceptive properties and inhibits cytokines. Glucans separated from *Pleurotus pulmonarius* ceased colon cancer development via cell proliferation regulation, suppressing inflammation, and induction pf apoptosis (Caz et al. 2015, Zhu et al. 2015).

3.3 Phenolics

Phenolic extracts from mushrooms were investigated in many studies and showed efficacy in solving a wide range of health problems. Different phenolic acids (cinnamic, p-hydroxybenzoic, salicylic, syringic, gallic, p-coumaric, ferulic, chlorogenic, and caffeic acids in addition to flavonoid could be found in all types of mushrooms. p-coumaric, caffeic, and gallic acids are the main groups of phenolics and they play essential roles in the biological activities of mushrooms (Abdelshafy et al. 2021). Phenolic compounds and polyphenols provide mushrooms with high antioxidant activity. Also, they showed anticancer effects against both human kidney and ovarian cancer cell lines (Gąsecka et al. 2018). For example, flavonoids such as naringenin, myricetin, quercetin, rutin, hesperetin, and morin are polyphenols that exhibited antiproliferative effects (Tatipamula and Kukavica 2021).

3.4 Terpenoids

Terpenoids monoterpenoids, diterpenoids, and sesquiterpenoids were proven to possess antimicrobial, antineurodegenerative, antimalarial, anticholinesterase, anticancer, antioxidant, and anti-inflammatory properties (Dasgupta and Acharya 2022). Phytosterols (ergosterol peroxide and ergosterols) are the main sterol group in edible mushrooms that act as a precursor of vitamin D2. In mushrooms, they showed more effective anti-inflammatory effects than the nonsteroidal drug indomethacin (Zhang et al. 2022). High concentrations of ergosterols might exist in several mushroom species such as *Lentinus edodes*, *Agaricus bisporus*, *Grifola frondose*, *Agaricus bisporus*, and *Pleurotus ostreatus* (El-Ramady et al. 2022).

4. Medicinal Mushrooms

Mushrooms are known to own several medicinal properties based on the environmental conditions and growth stage. These properties were used in folk medicine across the world for a long time to treat human diseases. The early herbs collectors were keen to know the medicinal value of mushrooms more than their suitability as a food source. The roles of mushrooms in medicine were described in many scientific reports. Out of the 150,000 mushroom species around the world, about 700 are recorded to harbor more than 130 medicinal properties. These types of mushrooms contain essential metabolites. For example, alkaloids, polysaccharides, polyphenols, terpenoids, sesquiterpenes, and metal-chelating agents can be used for the treatment of different human diseases. Recently, medicinal mushrooms are attracting more attention and are paving the way for new research avenues to discover how these types of mushrooms treat human diseases through different mechanisms by reducing oxidative stress, regulating neurotrophins synthesis, modulation the activity of acetylcholine esterase, and protecting neurons (El-Ramady et al. 2022, Lo and Wasser 2011, Wasser 2014).

4.1 Applications of Medicinal Mushrooms

Medicinal mushrooms could be used in different medicinal applications based on the target disease according to (Panda and Luyten 2021) as follow:

4.1.1 Anticancer Agent

This group combines mushrooms their extracts are able to treat cancer, including colorectal, renal, oral, leukaemia, and cervical, breast, lung, liver, and prostate cancer. Methanol and ethanol extracts, polysaccharides, and ganoderic acid are bioactive constituents that could be used in cancer treatment. Medicinal mushrooms that contain these metabolites are including *Agaricus brasiliensis*, *Boletus edulis*, *Ganoderma lucidum*, *P. pulmonarius*, *Pleurotus ostreatus*, *Tremella mesenterica*, *Ophiocordyceps sobolifera*, and *Sanghuangprous vaninii* (El-Ramady et al. 2022, Panda and Luyten 2021).

4.1.2 Antidiabetic Agent

This group includes different glucose-lowering types of mushrooms, for example, *Phellinus linteus* and *Grifola frondose* produce bioactive polysaccharides (Lo and Wasser 2011).

4.1.3 Cardiovascular Diseases Treatment

Agaricus bisporus along with *Tricholoma matsutakei* are among the species belonging to this group. The compounds that may treat cardiovascular diseases could be attributed to the crude extract of peptides found in them (Klupp et al. 2015, Panda and Luyten 2021).

4.1.4 Immune-function Agent

The Immune-function group combines several species such as *Trametes pubescens* and *Pleurotus ostreatus*. Lectin extracted from this group could improve immune system efficacy (Panda and Luyten 2021, Singh et al. 2010, Wasser 2014).

4.1.5 Rheumatoid Arthritis Therapeutic Agent

Some mushroom species are located within this group such as *Psilocybin* spp. and *Grifola frondose*. The extracted polysaccharides or Psilocin may help in relieving the symptoms of rheumatoid arthritis (Cavanna et al. 2022).

4.1.6 Antiviral Agent

Major viral diseases that affect humans such as influenza, hepatitis B (HBV) and hepatitis C viruses (HCV), human immunodeficiency virus (HIV), herpes simplex virus (HSV) (Abdelshafy et al. 2021, Dasgupta and Acharya 2022, El-Ramady et al. 2022). In mushrooms, there are many compounds with an antiviral activity that can be isolated and utilized as antiviral agents from different mushroom species (Panda and Luyten 2021). As anti-HIV, from *Lentinus edodes*, different components were extracted including polycarboxylated water-solubilized lignin. Meanwhile, the ubiquitin-like protein was extracted from *Pleurotus ostreatus* (Abdelshafy et al. 2021). Many triterpenoids were used from *Ganoderma lucidum* (Klupp et al. 2015), while *Ganoderma colossus* produces lanostane triterpenes. In addition, different mushrooms showed anti-HSV activity due to the presence of bioactive compounds sulfated polysaccharide from *Agaricus brasiliensis*, and proteoglycan from *Ganoderma lucidum* (Chun et al. 2021). For example, bioactive compounds that showed antiviral activity against the H5N1 influenza virus such as hispidin were obtained from *Phellinus baumii*, phenolic extracts of *Inonotus hispidus*, and sesquiterpenoid from *Phellinus igniarius* (Quang et al. 2006). In addition, lectins and polysaccharides extracted from *Pleurotus ostreatus* and *Antrodia camphorate*, respectively were used in the treatment of HBV (Dudekula et al. 2020). While, mycelia and solid culture extracts of *Lentinula edodes* were used for treating HCV (Matsuhisa et al. 2015). Many other antiviral mushrooms were reported to act against other viruses including polysaccharides from *Auricularia auricular* against Newcastle virus, sterols from *Hypsizygus marmoreus* against Epstein–Barr viral infection and Murill extract from *Agaricus blazei* for the foot-and-mouth virus. Recently, scientific pieces of evidence were found and reported about the biological activity of some mushrooms, components against COVID-19, including polysaccharides, ergosterol, and colosso-lactones (Abdelshafy et al. 2021, Wasser 2014).

5. Management of Post-COVID-19 Syndromes

Since 2019, the globe is facing a pandemic known as the severe acute respiratory syndrome coronavirus 2 (SARS-CoV-2) or known as the "COVID-19 pandemic". Nowadays, patients, after their recovery, may face many symptoms that may last over four weeks to months post-infection. Sometimes they may disappear or return in such conditions known as post-COVID-19 conditions (Paudyal et al. 2022).

People with post-COVID-19 symptoms return experience reported the following symptoms:

5.1 General Symptoms

- Fatigue or tiredness that interrupts daily activities
- Symptoms that may get worse, especially after mental and/or physical effort
- Fever

5.2 Respiratory and Heart Symptoms

- Difficulty in breathing with or without breath shortness
- Chest pain
- Cough
- Fast heart beating (heart palpitations)

5.3 Neurological Symptoms

- Difficulty to think or concentrate (brain fog)
- Headache
- Sleep problems
- Dizziness when you stand up

- Pins-and-needles feelings
- Change in taste or smell ability
- Depression
- Anxiety

5.4 Digestive Symptoms

- Diarrhea
- Stomach pain

5.5 Other Symptoms

- Joint or muscle pain
- Rash
- Changes in menstrual cycles

Scientists and healthcare professionals deal with the overall health status recovery from post-COVID-19 symptoms. There's currently no universal strategy that was approved by the health care professionals to diagnose and treat post-COVID-19 conditions to seek the main cause of long COVID-19 symptoms remains unknown (DiSogra 2020). Multiple mechanisms might be responsible for prolonged symptoms in an individual. Scientists argue that long-lasting symptoms of post-COVID-19 infection might be attributed to a prolonged inflammatory process (Paudyal et al. 2022). Healthcare providers try to manage symptoms as they are different from one patient to another and they alter in duration and severity. In the early-stage treatment, recommended for COVID-19, various complementary and alternative medicines (CAMs) approaches were established to be used in clinical practices such as the implementation of zinc and vitamin C in the treatment plan, as they were proven to play an important role in maintaining the functionality of the immune system (DiSogra 2020, Souza et al. 2020). This integration may also introduce synergistic effects to the key determinant steps in the immune system response.

6. Nutraceuticals and Dietary Supplements for Long COVID Management

Nutrients can play an important role in optimizing immunity to help lower the potential risk that may cause severe progress of COVID-19 infection. In October 2021 the British Association for Nutrition and Lifestyle Medicine suggested three nutritional therapy strategies to help long haulers (DiSogra 2020).

The first plan was to decrease the hyper-inflammatory events caused by long COVID-19 through plant-based diets, which were shown to be effective in reducing inflammatory responses. This plan included different varieties of fruits, vegetables, seeds, nuts, and legumes. Plant-based foods are the source of dietary fibers, an unusual component in western diets (Chavda et al. 2022). Lack of these fibers is usually associated with hyper-inflammatory conditions such as metabolic disorders and autoimmune diseases. In addition, the human body nature is not fit to digest dietary fibers, instead, the fibers are digested by gut microbes. Gut microbiomes use dietary fiber as a source of food and produce by-products including short-chain fatty acids and other compounds. These by-products help to regulate T-regulatory cells and prevent hyperactive immune responses (Minkenberg 2010). Cognition problems and memory losses due to long-haulers are generally described as "brain fog" and could be correlated to viral gut dysbiosis. Omega-3s can maintain the normal cognition and memory functions of the brain. Omega-3s (DHA and EPA) are abundant in oily fish, including sardines and salmon. The plant could be a good source of a plethora of phytochemicals that act as natural antioxidants. Quercetin, a flavonoid with the proven benefits of anti-inflammation, is abundantly found in a variety of vegetables (Chavda et al. 2022, DiSogra 2020). Plant-based diet regime does not require the elimination of meat from daily diets. Other animal protein sources could be utilized to nourish our body with its needs of essential amino acids, iron, omega-3, and vitamin B12 (Storz 2021).

The second strategy aims to the mindfulness of healthy diet intake, balanced diet boosts the immune system. Choosing healthy foods is not enough for an individual to eat well as consume the right amount of food. Nutraceuticals mean "nutrients found in food that is beneficial to an individual's health". These nutrients could be obtained from diverse sources of diets or as dietary supplements. The quantity of nutraceuticals in natural food products is inadequate to offer health benefits and hence food supplements are needed to fulfil the human body's requirements (Dubey et al. 2022).

The third strategy aims to support symptoms by the usage of vitamins and minerals, including vitamin D, vitamin C, and Zinc which were proven to fortify our immune system. Vitamin D, a highly important component for bone health and the immune system, has the ability to regulate the anti-inflammatory pathways by controlling the well-reported cytokine storm in many COVID-19 patients. Pro-Vitamin D (Calcitriol) plays an important role in protecting the lung from severe damage via regulating ACE2 gene expression in lung tissues in COVID-19 (Paudyal et al. 2022).

Vitamin C with wide antioxidant and anti-inflammatory properties was encouraged for intravenous injection experiments that showed great results. It is evident that vitamin C causes a speed up in natural killer (NK) cell proliferation without affecting their functionality. Moreover, vitamin C decreases reactive oxygen species (ROS) generation, responsible for the activation of inflammasome and NLRP3. This activation affects the maturation and the release of diverse cytokines (IL1beta and IL-18) which play a key role in the inflammatory systemic syndrome that characterizes sepsis. ICAM-1 expression was found to be blocked by Vitamin C which simultaneously blocks the activation of NFKappa-B, which is involved in the different processes (neoplastic, inflammatory, and apoptotic processes) by inhibiting TNF-α (Abdelshafy et al. 2021, Dubey et al. 2022, Wasser 2014).

Zinc is a vital element immune system efficacy. It showed diverse roles in the development of active immune system during innate and adaptive immune responses and guarantee its maintenance. Also, it is responsible for the integrity of the epithelial barriers. Zinc was found to be able to control the development of T cells, modulate their activity, and reduce cytokine storming. Exposure to such cytokine storm may lead to systemic immune response damage and result in acute respiratory distress syndrome (ARDS) followed by multiple organ failure. Zinc insufficiency reduces NK cells and cytolytic T cells activities, which play important roles in destroying viruses, bacteria, and tumor cells. Zinc has shown a direct antiviral activity by reducing RNA virus's replication, in such fashion that makes Zn an essential element for immune response post viral infection (Dubey et al. 2022, Mahmud et al. 2022). On the other hand, probiotics have been traditionally used as nutritional supplements. New scientific research supports the antiviral and overall immune system-strengthening health benefits of probiotics. They are living microorganisms that can adhere to gut epithelial cells and inhibit pathogenic germs from adhering to the gut walls. Probiotics, once attached to gut epithelial cells, release chemicals antibacterial substances such as short-chain fatty acids, antibacterial peptides, and bacteriocins, which inhibit pathogen growth (Al-Madboly et al. 2020, Panda and Luyten 2021). Probiotics also regulate intestinal permeability and help in maintaining healthy gut through the reduction of inflammation events and supporting the epithelial barrier due to increase the secretion of mucin (highly glycosylated material) into the intestinal in order to form a mucus film that prevents the accessibility of pathogenic bacteria to gut wall. Probiotics may improve the function of the intestinal epithelium barrier by increasing the tight junction protein expression (Akour 2020, Bottari et al. 2021). It is noteworthy that the depletion of essential gut commensals such as *Lactobacillus* and *Bifidobacterium* was documented in some COVID-19 patients (Kurian et al. 2021). Therefore, restoring homeostasis of gut microbiota is important for patients.

7. Nano-Nutraceuticals and Long COVID-19

Various nanotechnology-based delivery methods were used for the successful delivery of natural medicinal products. Several methods including metal NPs, polymeric NPs, liposomes, magnetic NPs, solid lipid NPs, inorganic NPs, nanocapsules, quantum dots, nanospheres, dendrimers, nanoemulsions, polymeric micelles have been employed (Dubey et al. 2022, Mahmud et al. 2022).

Curcumin, an active ingredient in turmeric with antiviral properties, is poorly absorbed from the gastrointestinal tract when a conservative amount in the form of capsules and/or tablets is applied due to its lipophilic properties. It was found that nano-micelles dissolve the compound with an efficiency reached 100%. In such a way, Sina-curcumin soft gels were found to be completely soluble in the acidic environment of the stomach, releasing micro micelles with stability that reached 6 hours. Moreover, Glycyrrhizic acid nanoparticles (GANPs) target loci that show significant inflammatory responses. Moreover, GANPs showed anti-inflammatory properties as well as antiviral activities that reduced organ damage in sick mice and gave them a survival advantage (Dubey et al. 2022, Zhao et al. 2021).

8. Medicinal Mushrooms for Post-COVID-19 Syndrome Management

Aside from being immune modulators and anti-inflammatory products, medicinal mushrooms have the power of healing and organ support (areas of action or tropisms). Also, they are a natural source of several vitamins (B, C, D), and essential minerals, and are capable of changing the structure of our gut microbiota (prebiotics). The ability of medicinal mushrooms to enrich gut microbiota after COVID-19 could be of particular importance, while studies still examine the relationship between gut microbiota and the SARS-CoV-2 virus. In addition, the deleterious effects of COVID-19, such as those associated with cardiometabolic and other hallmark disorders; demonstrate dysregulation of the homeostatic function within the renin-angiotensin-aldosterone system (RAAS) (Abdullah et al. 2012, Abubakar et al. 2021, Bohn et al. 2020, Lichtenberger and Vijayan 2021). Vascular function was found to be maintained by RAAS. The integral membrane protein ACE2 exists in many body tissues, including endothelium, liver, lungs, heart, and kidney tissues. ACE2 regulation disturbance affects RAAS intensely, revealing symptoms, involving oxidative-stress and hyper inflammation. Medicinal mushrooms were studied for their ACE inhibitory, anti-inflammatory, antioxidant, and antiplatelet activities (Mohamad Ansor et al. 2013, Roncero-Ramos and Delgado-Andrade 2017) (Figure 2).

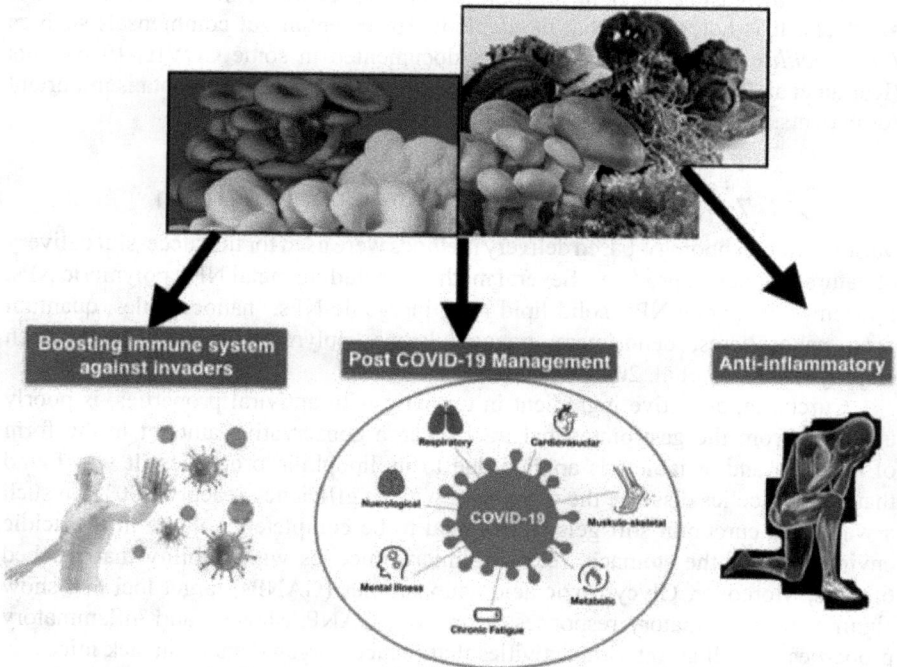

Figure 2: Medicinal mushrooms for post-COVID-19 syndrome management.

In medicinal mushrooms, active compounds, e.g., sterols, triterpenes, phenolic compounds, and polysaccharides have shown metabolic-modulating abilities (Dicks and Ellinger 2020). These include lowering activities for blood pressure, cholesterol, triglyceride, and glycaemia. The ACE inhibitory action of mushroom extracts was investigated using hot water and alcohol (Mohamad Ansor et al. 2013). Hot water extract of *Pleurotus* spp and *G. lucidum* showed a strong ACE inhibitory activity due to the presence of a plethora of phenolic compounds and mushrooms antioxidant capacity. However, variations were found due to the extraction method and the used species (Mohamad Ansor et al. 2013). Several peptides known to be ACE inhibitors were obtained from the *In vitro* digestion of *P. ostreatus* (Gargano et al. 2017).

A randomized, double-blind prevention trial was carried out in the Democratic Republic of the Congo to test the herbal mixture (Tomeka®), which contains several food-based nutrients (soy, which is regarded for its potent ACE2 inhibition) and *A. bisporus*. The study investigated the interacting effect of the tested mixture on different COVID-19 markers, including angiotensin-II and angiotensin (Kalač 2013, Lichtenberger and Vijayan 2021, Mansueto et al. 2020). Nutritional components may indirectly fortify ACE inhibition via the interception of viral entry or throughout the regulation and enhancement of biomarkers linked to other systems (Lichtenberger and Vijayan 2021). For example, high levels of sodium ions can cause impairment of the endothelial vasculature and increase the risk of hypertension, but symptoms might be improved when high levels of potassium ion were applied. Therefore, mushrooms with contain high potassium content and contain low sodium should be a good source of ACE inhibitors (Mansueto et al. 2020, Mingyi et al. 2019).

Dexamethasone was found to be the only treatment that resulted in a survival advantage to COVID-19 patients. However, the results still unclear for long term benefits and if secondary infections might occur. Natural polysaccharides, particularly mushroom polysaccharides, become frequent dietary constituents after attracting the attention of functional food experts (Zhang et al. 2020).

Mushroom polysaccharides can enhance the immunological function of many organs and cells, and increase immunity (Orhan and Senol Deniz 2020). The antiviral and anti-inflammatory properties of mushrooms, besides the presence of specific polysaccharides, mushroom have become excellent candidates for new antiviral therapeutics. Consequently, natural compounds, which are extracted from mushrooms with potential antiviral activities that could act as COVID-19 therapies, seem to be an attractive source of new drugs (Slomski 2021).

Some interesting mushrooms (*Cordyceps sinensis, Inonotus obliquus, Agaricus blazei Murill, Agaricus subrufescens, Grifola frondosa, Ganoderma lucidum, Hericium erinaceus, Lentinula edodes, Pleurotus ostreatus, Trametes versicolor*, and *Poria cocos*) also showed pleiotropic effects in reducing cytokines production that could provide a multimodal approach to COVID-19 management. As a result, diet supplementation with nutraceuticals-based products based on ethno-mycological mushroom expertise is a promising step in preventing and treating the present epidemic (Schieffer et al. 2021).

9. Management of COVID-19 Associated Cardio-metabolic Disorders

Active metabolites of many medicinal mushrooms have shown biological activity towards reducing the risk of cardio-metabolic disease associated with COVID-19. One-third of 40–60 years old COVID-19 patients were identified as being suffering from comorbidities (CVD and hypertension) (Kalač 2013). Metabolic disorders such as obesity, diabetes, and hyper-lipidaemia are involved in disease severity (Kała et al. 2020, Kalač 2013). Furthermore, meta-analyses results indicated that there was an increasing need for intensive care for older patients with cardiovascular morbidities and increased death rate among them (Kalač 2013).

Lovastatin was extracted from the mycelium and fruiting bodies of mushrooms. This compound is belonging to the statin group (Atlı et al. 2019). It inhibits HMG-CoA reductase, rate limiting enzyme in cholesterol biosynthesis. Therefore, medicinal mushrooms showing lovastatin's mechanisms could be a promising solution for high cholesterol. This capability was observed in the extract of *Pleurotus* spp. fruiting bodies. Also, other mushrooms showed this ability such as *H. erinaceus* and *A. bisporus* (Aramabašić Jovanović et al. 2021, Atlı et al. 2019, Gargano et al. 2017). The effects of many medicinal mushrooms were investigated employing different cardio-metabolic parameters. (Gargano et al. 2017) carried out a systematic review for eight clinical studies included patients with and diabetes and control individuals without type-2 diabetes utilizing fresh, cooked, or dry powder of *P. ostreatus mushroom*. Promising results were obtained from these studies, however, the poor methodology and insufficient reporting caused high inconsistency. Nevertheless, good effects were resulted in the glycaemic control, including reduction in fasting glucose level and/or 2 h postprandial, improved lipids profile, a reduction in antioxidant capacity, a reduction in blood pressure, and a reduction in food intake without weight change (Gargano et al. 2017). The total polysaccharides of *P. ostreatus* were added to high-fat diets that were applied for 4 weeks to streptozotocin-induced type-2 diabetic rats (Jeong et al. 2010). Reduction in blood glucose levels was found, improved insulin resistance, and liver glycogen content increased through the activation of Glycogen Synthase Kinase-2 (GSK-2) phosphorylation and GLUT4 translocation in muscle tissue. *A. bisporus* demonstrated increased HDL and reductions in both triglycerides and LDL in high-fat-fed rats, additionally, glucose levels were reduced in the induced type-2 diabetic rats upon treatment (Zhang et al. 2017). The anti-diabetic and antioxidant protective effects of *H. erinaceus* polysaccharides on the pancreas, kidney, and liver of STZ-induced rats were also reported (Li et al. 2016).

10. Effect of Mushroom on Altered Gut Microbiota after Long COVID-19

It is clear that the structure and activity of intestinal microbiota could impose health outcomes. For example, changes in gut microbiota may cause long-lasting inflammatory diseases, including thrombosis, atherosclerosis, asthma, and diabetes that partially due to speculated oxidative-stress (Hess et al. 2018). On the other hand, symbiosis decreases the risks of many metabolic diseases such as cardiovascular disorders, lowers postprandial glucose levels, improves laxation, and increases acceptance for weight management (Ma et al. 2021). These factors are all determinants associated with COVID-19 infection severity, which are associated with lifestyle; hence, they are applicable and modifiable in order to lower the associated risk. Luckily, the diversity and functions of gut microbiota are influenced by the digestible and non-digestible fibers and prebiotic properties of mushrooms. (Chang et al. 2015) compiled collected studies from different models of disease, which demonstrated a clear health improvement from edible mushroom polysaccharides (EMPs). The relation between beneficial host microbiota and EMPs suggests regulatory effects (Chang et al. 2015). EMPs benefits indicated were metabolism-related enhancements including, reduction in total cholesterol and lipid markers, reducing insulin resistance and inflammation, improvement signaling pathways, anti-obesity activities, morphology of intestine and gut mucosa integrity, and increasing pro-inflammatory cytokines inhibition (Chang et al. 2015).

Short-chain fatty acids provide important benefits to the body, including immune stimulation, colonic pH modulation, nutrient supply to the colonic epithelium, and reduced oxidative stress (Hess et al. 2018). However, the proportions of these organic acids differ based on the EMPs provided via different mushrooms (Chang et al. 2015). *G. lucidum* showed the most substantial microbiota-derived health benefits. *G. lucidum* polysaccharides S3 showed clear anti-obesity and anti-diabetic effects (Chen et al. 2020, Ganesan and X, 2018). Also, they restored gut microbiota of type-2 diabetes-induced rats to a normal level (Chen et al. 2020), in addition, *G. lucidum* mycelium polysaccharide positively affected the relative abundance of some beneficial bacteria such as *Roseburia*, *Lactobacillus*, and *Lachnospiracea* in mice induced with diethyldithiocarbamate (Hess et al. 2018). Health-promoting activities were observed when an increase in Roseburia was found after supplementing them with *G. lucidum* polysaccharide S3 in their diets.

11. Conclusion and Outlook

For centuries, mushrooms have been considered not only as healthy food and supplements but also as one of the main components of traditional medicines in many cultures. This is based on the rich content of many bioactive compounds and health claims in the treatment of different diseases. Among known mushroom ingredients, molecules of different molecular weights from a wide range of chemical classes such as polysaccharides, terpenoids, phenolics, and others proved to exhibit immunomodulating activities. With the increasing awareness about the importance of immunomodulating food ingredients in human health to fight against new emerging diseases such as COVID-19, mushrooms become one of the very attractive nutrients. Therefore, we believe that using mushrooms in the treatment of post-COVID-19 syndromes (PCS) will increase the demand for mushroom consumption. Therefore, more research is currently needed to understand underlying the mechanism of mushrooms' bioactive ingredients in the treatment of these new PCS diseases. The role of the individual bioactive molecules in pure form or in combination with other ingredients needs to be further studied in depth to develop a treatment protocol for PCS.

References

Abdelshafy, A.M., Z. Luo, T. Belwal, Z. Ban and L. Li. 2021. A comprehensive review on preservation of shiitake mushroom (Lentinus edodes): Techniques, Research Advances and Influence on Quality Traits. *Food Rev. Int.*, 2021: 1–34.

Abdullah, N., S.M. Ismail, N. Aminudin, A.S. Shuib and B.F. Lau. 2012. Evaluation of selected culinary-medicinal mushrooms for antioxidant and ace inhibitory activities. *Evid. Based Compl. Altern. Med.*, 2012: 464238. https://doi.org/10.1155/2012/464238.

Abubakar, M.B., D. Usman, G. El-Saber-Batiha, N. Cruz-Martins, I. Malami, K.G. Ibrahim, B. Abubakar, M.B. Bello, A. Muhammad, S.H. Gan, A.I. Dabai, M. Alblihed, A. Ghosh, R.H. Badr, D. Thangadurai and M.U. Imam. 2021. Natural products modulating angiotensin converting enzyme 2 (ace2) as potential COVID-19 therapies. *Front. Pharmacol.*, 12: 629935.

Akour, A. 2020. Probiotics and COVID-19: Is there any link? *Lett. in Appl. Microbiol.*, 71: 229–234.

Al-Madboly, L.A., N.M. El-Deeb, A. Kabbash, M.A. Nael, A.M. Kenawy and A.E. Ragab. 2020. Purification, characterization, identification, and anticancer activity of a circular Bacteriocin from *Enterococcus thailandicus. Front. Bioeng. Biotechnol.*, 8: 450.

Alves, M.J., I.C. Ferreira, J. Dias, V. Teixeira, A. Martins and M. Pintado. 2012. A review on antimicrobial activity of mushroom (Basidiomycetes) extracts and isolated compounds. *Planta medica*, 78: 1707–1718.

Aramabašić Jovanović, J., M. Mihailović, A. Uskoković, N. Grdović, S. Dinić and M. Vidaković. 2021. The effects of major mushroom bioactive compounds on mechanisms that control blood glucose level. *J. of Fungi*, 7: 58.

Atlı, B., M. Yamaç, Z. Yıldız and M. Şölener. 2019. Solid state fermentation optimization of *Pleurotus ostreatus* for lovastatin production. *Pharmac. Chem. J.*, 53: 858–864.

Bohn, M.K., A. Hall, L. Sepiashvili, B. Jung, S. Steele and K. Adeli. 2020. Pathophysiology of COVID-19: Mechanisms underlying disease severity and progression. *Physiol.*, 35: 288–301.

Bottari, B., V. Castellone and E. Neviani. 2021. Probiotics and Covid-19. *Inter. J. Food Sci. Nutr.*, 72: 293–299.

Cavanna, F., S. Muller, L.A. de la Fuente, F. Zamberlan, M. Palmucci, L. Janeckova, M. Kuchar, C. Pallavicini and E. Tagliazucchi. 2022. Microdosing with psilocybin mushrooms: A double-blind placebo-controlled study. *Transl. Psych.*, 12: 1–11.

Caz, V., A. Gil-Ramírez, C. Largo, M. Tabernero, M. Santamaría, R. Martin-Hernandez, F.R. Marín, G. Reglero and C. Soler-Rivas. 2015. Modulation of cholesterol-related gene expression by dietary fiber fractions from edible mushrooms. *J. Agric. and Food Chem.*, 63: 7371–7380.

Cerletti, C., S. Esposito and L. Iacoviello. 2021. Edible mushrooms and beta-glucans: Impact on human health. *Nutr.*, 13: 2195.

Chang, C.J., C.S. Lin, C.C. Lu, J. Martel, Y.F. Ko, D.M. Ojcius, S.F. Tseng, T.R. Wu, Y.Y.M. Chen and J.D. Young. 2015. *Ganoderma lucidum* reduces obesity in mice by modulating the composition of the gut microbiota. *Nat. Comm.*, 6: 1–19.

Chavda, V.P., A.B. Patel, D. Vihol, D.D. Vaghasiya, K.M.S.B. Ahmed, K.U. Trivedi and D.J. Dave. 2022. Herbal remedies, nutraceuticals, and dietary supplements for COVID-19 management: An update. *Clin. Complem. Med. Pharmacol.*, 2: 100021.

Chen, M., D. Xiao, W. Liu, Y. Song, B. Zou, L. Li, P. Li, Y. Cai, D. Liu and Q. Liao. 2020. Intake of *Ganoderma lucidum* polysaccharides reverses the disturbed gut microbiota and metabolism in type 2 diabetic rats. *Intern. J. Biol. Macromol.*, 155: 890–902.

Chen, S.N., C.S. Chang, M.H. Hung, S. Chen, W. Wang, C.J. Tai and C.L. Lu. 2014. The effect of mushroom beta-glucans from solid culture of *Ganoderma lucidum* on inhibition of the primary tumor metastasis. *Evid. Based Compl. Alter. Med.*, 2014: 252171.

Chun, S., J. Gopal and M. Muthu. 2021. Antioxidant activity of mushroom extracts/polysaccharides-their antiviral properties and plausible Anti COVID-19 properties. *Antioxdant*, 10: 1899.

Dasgupta, A. and K. Acharya. 2022. Bioactive terpenoids from mushrooms. pp. 145–154. *In*: Gupta, V. (ed.). New and Future Developments in Microbial Biotechnology and Bioengineering, 1st Edition. Elsevier.

Dicks, L. and S. Ellinger. 2020. Effect of the intake of oyster mushrooms (*pleurotus ostreatus*) on cardiometabolic parameters—A systematic review of clinical trials. *Nutrition*, 12: 1134.

DiSogra, R.M. 2020. COVID-19 "Long-Haulers": The emergence of auditory/vestibular problems after medical intervention. *Can. Audiol.*, 7: 180–188.

Dubey, A.K., S.K. Chaudhry, H.B. Singh, V.K. Gupta and A. Kaushik. 2022. Perspectives on nano-nutraceuticals to manage pre and post COVID-19 infections. *Biotechnol. Rep.*, 33: e00712.

Dudekula, U.T., K. Doriya and S.K. Devarai. 2020. A critical review on submerged production of mushroom and their bioactive metabolites. *3 Biotech.*, 10: 1–12.

El-Ramady, H., N. Abdalla, K. Badgar, X. Llanaj, G. Törős, P. Hajdú, Y. Eid and J. Prokisch. 2022. Edible mushrooms for sustainable and healthy human food: nutritional and medicinal attributes. *Sustainability*, 14: 4941.

Ganesan, K. and B. Xu. 2018. Anti-obesity effects of medicinal and edible mushrooms. *Molecules*, 23: 2880.

Gargano, M.L., L.J.L.D. van Griensven, O.S. Isikhuemhen, U. Lindequist, G. Venturella, S.P. Wasser and G.I. Zervakis. 2017. Medicinal mushrooms: Valuable biological resources of high exploitation potential. *Plant Biosys.*, 151: 548–565.

Gąsecka, M., M. Siwulski and M. Mleczek. 2018. Evaluation of bioactive compounds content and antioxidant properties of soil-growing and wood-growing edible mushrooms. *J. Food Proc. and Preserv.*, 42: e13386.

Hassan, M.A.A., R. Rouf, E. Tiralongo, T.W. May and J. Tiralongo. 2015. Mushroom lectins: Specificity, structure and bioactivity relevant to human disease. *Intern. J. Molec. Sci.*, 16: 7802–7838.

Hess, J., Q. Wang, T. Gould and J. Slavin. 2018. Impact of Agaricus bisporus mushroom consumption on gut health markers in healthy adults. *Nutrition*, 10: 1402.

Jeong, S.C., Y.T. Jeong, B.K. Yang, R. Islam, S.R. Koyyalamudi, G. Pang, K.Y. Cho and C.H. Song. 2010. White button mushroom (*Agaricus bisporus*) lowers blood glucose and cholesterol levels in diabetic and hypercholesterolemic rats. *Nutr. Res.*, 30: 49–56.

Kała, K., A. Kryczyk-Poprawa, A. Rzewińska and B. Muszyńska. 2020. Fruiting bodies of selected edible mushrooms as a potential source of lovastatin. *Eur. Food Res. Technol.*, 246: 713–722.

Kalač, P. 2013. A review of chemical composition and nutritional value of wild-growing and cultivated mushrooms. *J. Sci. Food Agric.*, 93: 209–218.

Klupp, N.L., D. Chang, F. Hawke, H. Kiat, H. Cao, S.J. Grant and A. Bensoussan. 2015. *Ganoderma lucidum* mushroom for the treatment of cardiovascular risk factors. *Cochrane Database Syst. Rev.*, 17: CD007259.

Kurian, S.J., M.K. Unnikrishnan, S.S. Miraj, D. Bagchi, M. Banerjee, B.S. Reddy, G.S. Rodrigues, M.K. Manu, K. Saravu, C. Mukhopadhyay and M. Rao. 2021. Probiotics in prevention and treatment of COVID-19: Current perspective and future prospects., *Arch. Med. Res.*, 52: 582–594.

Li, H., Y. Tian, N. Menolli Jr, L. Ye, S.C. Karunarathna, J. Perez-Moreno, M.M. Rahman, M.H. Rashid, P. Phengsintham and L. Rizal. 2021. Reviewing the world's edible mushroom species: A new evidence-based classification system. *Comp. Rev. Food Sci. Food Safety*, 20: 1982–2014.

Li, K., C. Zhuo, C. Teng, S. Yu, X. Wang, Y. Hu, G. Ren, M. Yu and J. Qu. 2016. Effects of *Ganoderma lucidum* polysaccharides on chronic pancreatitis and intestinal microbiota in mice. *Inter. J. Biol. Macromolec.*, 93: 904–912.

Lichtenberger, L.M. and K.V. Vijayan. 2021. Is COVID-19-Induced platelet activation a cause of concern for patients with cancer? *Cancer Res.*, 81: 1209–1211.

Lo, H.C. and S.P. Wasser. 2011. Medicinal mushrooms for glycemic control in diabetes mellitus: History, current status, future perspectives, and unsolved problems. *Int. J. Med. Mushrooms*, 13: 401–26.

Ma, G., H. Du, Q. Hu, W. Yang, F. Pei and H. Xiao. 2021. Health benefits of edible mushroom polysaccharides and associated gut microbiota regulation. *Crit. Rev. Food Sci. and Nutr.*, 62: 6646–6663.

Mahmud, N., M.I. Anik, M.K. Hossain, M.I. Khan, S. Uddin, M. Ashrafuzzaman and M.M. Rahaman. 2022. Advances in nanomaterial-based platforms to combat covid-19: Diagnostics, preventions, therapeutics, and vaccine developments. *ACS Appl. Bio Materials*, 5: 2431–2460.

Mansueto, G., M. Niola and C. Napoli. 2020. Can COVID 2019 induce a specific cardiovascular damage or it exacerbates pre-existing cardiovascular diseases? *Pathol. Res. and Pract.*, 216: 153086.

Matsuhisa, K., S. Yamane, T. Okamoto, A. Watari, M. Kondoh, Y. Matsuura and K. Yagi. 2015. Anti-HCV effect of Lentinula edodes mycelia solid culture extracts and low-molecular-weight lignin. *Biochem. and Biophys. Res. Commun.*, 462: 52–57.

Mingyi, Y., T. Belwal, H.P. Devkota, L. Li and Z. Luo. 2019. Trends of utilizing mushroom polysaccharides (MPs) as potent nutraceutical components in food and medicine: A comprehensive review. *Trends in Food Sci. Technol.*, 92: 94–110.

Minkenberg, R. 2010. Result disclosure on clinical trials gov—first experiences and challenges. *Pharmac. Prog.*, 3: 51–56.

Mohamad Ansor, N., N. Abdullah and N. Aminudin. 2013. Anti-angiotensin converting enzyme (ACE) proteins from mycelia of *Ganoderma lucidum* (Curtis) P. Karst. *BMC Compl. Altern. Med.*, 13: 256.

Naranjo-Ortiz, M.A. and T. Gabaldón. 2019. Fungal evolution: Diversity, taxonomy and phylogeny of the fungi. *Biol. Rev.*, 94: 2101–2137.

Orhan, I.E. and F.S. Senol Deniz. 2020. Natural products as potential leads against coronaviruses: Could they be encouraging structural models against SARS-CoV-2? *Natural Prod. Bioprosp.*, 10: 171–186.

Panda, S.K. and W. Luyten. 2021. Medicinal mushrooms: Clinical perspective and challenges. *Drug Discov. Today*, 27: 636–651.

Paudyal, V., S. Sun, R. Hussain, M.H. Abutaleb and E.W. Hedima. 2022. Complementary and alternative medicines use in COVID-19: A global perspective on practice, policy and research. *Res. Social Adm. Pharm.*, 18: 2524–2528.

Quang, D.N., T. Hashimoto and Y. Asakawa. 2006. Inedible mushrooms: A good source of biologically active substances. *Chem. Rec.*, 6: 79–99.

Rahi, D.K. and D. Malik. 2016. Diversity of mushrooms and their metabolites of nutraceutical and therapeutic significance. *J. Mycol.*, 2016: 1–8.

Roda, E., F. De Luca, C. Di Iorio, D. Ratto, S. Siciliani, B. Ferrari, F. Cobelli, G. Borsci, E.C. Priori and S. Chinosi. 2020. Novel medicinal mushroom blend as a promising supplement in integrative oncology: A multi-tiered study using 4t1 triple-negative mouse breast cancer model. *Int. J. of Molec. Sci.*, 21: 3479.

Roncero-Ramos, I. and C. Delgado-Andrade. 2017. The beneficial role of edible mushrooms in human health. *Curr. Opin. in Food Sci.*, 14: 122–128.

Sánchez, C. 2004. Modern aspects of mushroom culture technology. *Appl. Microbiol. Biotechnol.*, 64: 756–762.

Schieffer, E., B. Schieffer and D. Hilfiker-Kleiner. 2021. Cardiovascular diseases and COVID-19: Pathophysiology, complications and treatment. *Herz.*, 46: 107–114.

Singh, R.S., R. Bhari and H.P. Kaur. 2010. Mushroom lectins: Current status and future perspectives. *Crit. Rev. Biotechnol.*, 30: 99–126.

Slomski, A. 2021. Trials test mushrooms and herbs as anti–COVID-19 agents. *JAMA*, 326: 1997–1999.

Souza, A.C.R., A.R. Vasconcelos, P.S. Prado and C.P.M. Pereira. 2020. Zinc, vitamin D and vitamin C: Perspectives for COVID-19 with a focus on physical tissue barrier integrity. *Front. in Nutr.*, 7: 606398.

Storz, M.A. 2021. Lifestyle adjustments in long-COVID management: Potential benefits of plant-based diets. *Curr. Nutr. Rep.*, 10: 352–363.

Švajger, U., J. Pohleven, J. Kos, B. Štrukelj and M. Jeras. 2011. CNL, a ricin B-like lectin from mushroom *Clitocybe nebularis*, induces maturation and activation of dendritic cells via the toll-like receptor 4 pathway. *Immunol.*, 134: 409–418.

Tatipamula, V.B. and B. Kukavica. 2021. Phenolic compounds as antidiabetic, anti-inflammatory, and anticancer agents and improvement of their bioavailability by liposomes. *Cell Biochem. Func.*, 39: 926–944.

Varga, T., K. Krizsán, C. Földi, B. Dima, M. Sánchez-García, S. Sánchez-Ramírez, G.J. Szöllősi, J.P. Szarkándi, V. Papp, L. Albert, W. Andreopoulos, C. Angelini, V. Antonín, K.W. Barry, N.L. Bougher,

P. Buchanan, B. Buyck, V. Bense, P. Catcheside and L.G. Nagy. 2019. Megaphylogeny resolves global patterns of mushroom evolution. *Nat. Ecol. and Evol.*, 3: 668–678.

Wasser, S. 2014. Medicinal mushroom science: Current perspectives, advances, evidences, and challenges. *Biomed. J.*, 37: 345–56.

Zeb, M. and C.H. Lee. 2021. Medicinal properties and bioactive compounds from wild mushrooms native to North America. *Molecules*, 26: 251.

Zhang, C., J. Li, C. Hu, J. Wang, J. Zhang, Z. Ren, X. Song and L. Jia. 2017. Antihyperglycaemic and organic protective effects on pancreas, liver and kidney by polysaccharides from *Hericium erinaceus* SG-02 in streptozotocin-induced diabetic mice. *Sci. Rep.*, 7: 1–13.

Zhang, R., Y. Han, D.J. McClements, D. Xu and S. Chen. 2022. Production, characterization, delivery, and cholesterol-lowering mechanism of phytosterols: A Review. *J. Agric. Food Chem.*, 70: 2483–2494.

Zhang, X., Y. Tan, Y. Ling, G. Lu, F. Liu, Z. Yi, X. Jia, M. Wu, B. Shi and S. Xu. 2020. Viral and host factors related to the clinical outcome of COVID-19. *Nature*, 583: 437–440.

Zhao, Z., Y. Xiao, L. Xu, Y. Liu, G. Jiang, W. Wang, B. Li, T. Zhu, Q. Tan, L. Tang, H. Zhou, X. Huang and H. Shan. 2021. Glycyrrhizic acid nanoparticles as antiviral and anti-inflammatory agents for COVID-19 treatment. *ACS Appl. Mat. Interf.*, 13: 20995–21006.

Zhu, F., B. Du, Z. Bian and B. Xu. 2015. Beta-glucans from edible and medicinal mushrooms: Characteristics, physicochemical and biological activities. *J. Food Comp. Anal.*, 41: 165–173.

14

Health Benefits of Mushrooms on Obesity and Chronic Diseases

Hassiba Benbaibeche,[1] Mohamed El Fadel Ousmaal,[1] Amira Rebah,[1] Yasmine Fatima Akchiche,[1] Hesham Ali El Enshasy,[2,3] Daniel Joe Dailin,[2] Dalia Sukmawati[4] and Ali Zineddine Boumehira[1,5,]*

1. Introduction

Obesity has increased worldwide and reached pandemic proportions (Meldrum et al. 2017). Obesity is associated with non-communicable chronic diseases (NCCD), including type 2 diabetes, hypertension, dyslipidemia, chronic kidney disease, cardiovascular disease, nonalcoholic fatty liver disease and certain types of cancer (Swinburn et al. 2011). Several factors are involved in the pathogenesis of obesity, including eating behavior, genetic predisposition, sedentary lifestyle, environment, food cravings and gut dysbiosis (Benbaibeche et al. 2015, Singer-Englar et al. 2019). The NCCD constitute a health burden, their treatment adds a load on healthcare systems. In fact, healthcare use increased among obese people leveling up medical

[1] University of Algiers, Faculty of Sciences, Algiers, Algeria.
[2] Institute of Bioproduct Development (IBD), Universiti Teknologi Malaysia (UTM), Johor, Malaysia.
[3] City of Scientific Research and Technology Applications, New Burg Al Arab, Alexandria, Egypt.
[4] Faculty of Mathermatics and Natural Science, Universitas Negeri Jakarta, Jakarta Timur 13220, Indonesia.
[5] Ecole Nationale Supérieure Agronomique - ENSA, El Harrach, Algiers, Algeria.
* Corresponding author: ali.boumehira@edu.ensa.dz

expenditures. Finkelstein et al. (2009) showed that obesity was associated with increases in per capita inpatient expenditures by 45.5%, outpatient and physician office expenditures by 26.9%, and an 80.4% increase in prescription medication expenditures in comparison with normal weight expenditures. Mushrooms are regarded as health-promoting foods, some studies suggest the impacts of mushrooms on cognition, weight management, oral health, and cancer risk. Mushrooms may support healthy immune and inflammatory responses through interaction with the gut microbiota, enhancing the development of adaptive immunity, and improving immune cell functionality (Feeney et al. 2014).

2. Mushrooms and Obesity

2.1 *Obesity*

Nowadays, obesity has reached the status of a global epidemic, affecting both developing and developed countries (Grotto et al. 2019). The World Health Organization (WHO) emphasizes that this disease is increasing very rapidly; in 2016, more than 1.9 billion adults were overweight, of which 650 million were obese, and in 2019, 38 million children were overweight or obese (Dubey et al. 2019). It is predicted that 1.12 billion people will be obese worldwide by 2030 (Ganesan and Xu, 2018). Thereby, WHO considers obesity to be one of the most serious and neglected public health issues threatening the world today (Grotto et al. 2019).

Obesity is considered as a complex metabolic disease characterized by an excessive accumulation of fat in adipose tissue and corresponds to the translation of an imbalance between energy intake and expenditure (Ganesan and Xu 2018, Mustafa et al. 2022).

Most obese people are predisposed to health problems, such as diabetes, cardiovascular disease, cancer, sleep apnea, osteoarthritis; longevity and quality of life are also affected (Martel et al. 2017, Grotto et al. 2019, Mustafa et al. 2022). Obesity is not an acute disease that develops instantly. Its appearance is gradual and occurs during childhood and/or adolescence following the complex interactions of several factors including heredity, food, lifestyle, societal determinants, environment and infectious agents (Ganesan and Xu 2018, Dubey et al. 2019, Mustafa et al. 2022). Being overweight is mainly caused by excessive consumption of high-calorie foods and lack of physical activity. Moreover, the increasingly sedentary nature of daily tasks, labor-saving technology, modified modes of transport and urbanization strongly promote obesity (Grotto et al. 2019, Mustafa et al. 2022).

Currently, various investigations have revealed that nutrigenomics and gut microbiota are the main determinants of obesity. Age and gender also influence the development of obesity. In Spain, men suffer from obesity more than women. In addition, the highest obesity rate is observed in the 50–60 age group and among those who had a pregnancy at an advanced age (Mustafa et al. 2022). Education level is also associated with being overweight. Epidemiological studies on obesity have revealed an inverse relationship between the prevalence of obesity and the level of education (Mustafa et al. 2022). Genetics influence more than 70% of the establishment of obesity. However, the exact mechanism responsible for this development is not well known. According to Neel, owner of the "thrifty genes" theory; the genes that predispose to obesity are present in populations that suffer from hunger. These individuals develop a so-called "obesogenic" environment and can become extremely obese (Mustafa et al. 2022). Obesity is classified into two broad categories: monogenic, inherited according to a Mendelian pattern, which is generally rare and polygenic or common, which is the result of a hundred polymorphisms each having an effect (Loos and Yeo 2022).

2.2 Mushrooms as Potential Anti-Obesity Agents

2.2.1 Bioactive Compounds

Due to their rich and diverse composition, mushrooms are recognized as one of the most important dietary supplements (Ganesan and Xu 2018). Therefore, they can be used in the treatment of lifestyle-related diseases such as obesity.

Mushrooms contain various bioactive compounds, which have favourable effects on the human body and protect it from diseases (Table 1).

2.2.2 Impact of Bioactive Compounds on Obesity

Studies have shown that the dietary fibers of mushrooms can play a role in the treatment and prevention of obesity, and this by their binding to lipids which thus inhibits their absorption and lowers the level of cholesterol in the blood. These indigestible carbohydrates are mainly represented by chitins, β-glucans and mannans. After fermentation in the large intestine, they release less energy than what is released by glucose (Mustafa et al. 2022).

Pleurotus ostreatus, often called an oyster mushroom, is one of the most consumed mushrooms in the world after the button mushroom *Agaricus bisporus*. It contains twice as much β-glucan as *A. bisporus* (Mustafa et al. 2022).

The dietary fibers of mushrooms can also bind to bile acids and thus increase their secretion in the faeces. Since bile acids are produced from cholesterol, the action of fibers can lead to a reduction of this lipid in the blood (Martel et al. 2017, Ganesan and Xu 2018, Dubey et al. 2019).

Ergosterol peroxide, a compound of sterol nature present in mushrooms, has many positive effects on anti-obesogene in particular, the decrease in the accumulation of fatty acids in 3T3-L1 cells, the inhibition of the upward expression of mRNA involved in the regulation of sterols in the body and the increase of AMPK expression. In addition, ergosterol peroxide allows inhibition of the expression of unsaturated fat synthase, unsaturated fat translocase and acetyl-coenzyme A carboxylase involved in the synthesis and transport of unsaturated fatty acids to a long chain. Mushrooms have a very remarkable appetite suppressant effect, indeed once consumed, their chewing increases the secretion of gastric acid and saliva, which leads to increased gastric distension and gives a feeling of satiety. In addition, the fibers present in this type of food increase in volume in the stomach and lead to satiety (Mustafa et al. 2022). It is recognized that dysbiosis at the level of the intestinal flora has an impact on the installation of obesity and that the polysaccharides of mushrooms participate

Table 1. Anti-obesity effect of some mushrooms.

Mushroom	Anti-obesity effect	Reference
Agaricus bisporus	β-(1,4)-D-glucan reduces the amount of fat and the expression of the peroxisome proliferator-activated receptor (PPAR) gene, triglyceride transfer protein (MTP) gene, liver fatty acid binding protein (L-FABP) gene and intestinal fatty acid-binding protein (I-FABP) gene.	Li et al. 2019
	Phytosterols extracted reduce LDL cholesterol and plasma cholesterol.	Lin et al. 2009
Flammulina velutipes	Acetone fraction inhibits intracellular lipid accumulation and metabolic syndrome.	Nobushi et al. 2013
Ganoderma lucidum	Dietary supplement of *G. lucidum* suppresses expression of adipogenic transcription factors PPAR-γ, sterol regulatory binding element protein-1c (SREBBP-1c) and CCAAT/ enhancer binding protein-α (C/EBP-α), as well as enzymes and proteins associated with lipid synthesis, transport, and storage: fatty acid synthase (FAS), acyl-CoA synthetase-1 (ACS1, fatty binding protein-4 (FABP4, fatty acid transport protein-1 (FATP1), and perilipin.	Thyagarajan-Sahu et al. 2011
Grifola frondosa	*G. frondosa* regulates the ceramide levels and restored lipid metabolism via the suppression of Toll-like receptor 4/nuclear factor kappa-B signaling, which is involved in inflammation and IR.	Jiang et al. 2021
Hericium erinaceus	*H. erinaceus* powder decreases the amounts of fat tissue, plasma levels of total cholesterol, and leptin.	Hiraki et al. 2017
	Ethanol extracts decreases body weight gain, fat weight, and serum and hepatic triacylglycerol levels. It acts as an agonist of PPARα.	Hiwatashi et al. 2010
Inonotus obliquus	*Obliquus* extract can control the body weight and attenuate the lipid accumulation accompanied with the decreased levels of free fatty acids, TG, and cholesterol in the serum, liver, and adipose tissue through inhibiting the expressions of several genes involved in lipid synthesis.	Wu et al. 2015
Pleurotus citrinopileatus	*P. citrinopileatus* water extract (PWE) reduces the weight gain, fat accumulation, and food intake. PWE decreases the serum TG, cholesterol and LDL.	Sheng et al. 2019
Tremella fuciformis	Polysaccharide as an anti-obesity prebiotic, polysaccharide prevented the variation of 3T3-L1 adipocytes by decreasing the expression of mRNA.	Jeong et al. 2008

in the restoration of this imbalance, therefore it would be beneficial to consume them (Martel et al. 2017, Ma et al. 2021).

3. Mushrooms and Hypertension

3.1 Hypertension

Hypertension is a serious public health concern that affects people all over the world. Its high prevalence imposes a significant burden on both developing and developed countries (Bromfield and Muntner 2013, Mohammed Nawi et al. 2021). Almost one-third of the adult global population (more than one billion), suffers from hypertension, with two-thirds living in low- and middle-income countries (LMICs). This incidence increases with age, reaching 60% of the population over the age of 60. According to a 2015 poll, one in every four women and one in every five men had hypertension (Mohammed Nawi et al. 2021, Mamdouh et al. 2022). Meanwhile, current predictions show that the number of hypertensive patients will rise by 15–20% to approximately 1.5 billion by 2025, and would certainly rise even more by 2040 (Iqbal and Jamal 2022, Gentile et al. 2022). Furthermore, elevated blood pressure is one of the main risk factors for premature death, cardiovascular diseases (CVDs), chronic renal disease, and dementia worldwide. It causes more than seven million premature deaths globally each year, accounting for 12.8% of all fatalities, and is more deadly than diabetes (3.4%) and obesity (4.8%) combined (Brunström et al. 2021, Mohammed Nawi et al. 2021, Gentile et al. 2022). In fact, hypertension (HTN) is still one of the five major causes of death across the world, with cardiovascular illnesses contributing to more than 40% of all deaths (Amponsem-Boateng et al. 2019). Hypertension is a common chronic health problem, characterized by persistently high blood pressure. A person is considered to be hypertensive if his or her systolic blood pressure and/or diastolic blood pressure, measured on two distinct days, is ≥140 mmHg and ≥ 90 mmHg, respectively (Mohammed Nawi et al. 2021, Iqbal and Jamal 2022). Whereas for a teenager over the age of 13, hypertension is diagnosed from a systolic blood pressure of 130 mmHg and/or diastolic blood pressure of 80 mmHg. Though, to confirm this diagnosis, blood pressure measures should be done during separate clinical appointments (Mohammed Nawi et al. 2021). The International Society of Hypertension (ISH) divides blood pressure (BP) into four categories: normal (< 130/85 mmHg), high-normal (130–139/85–89 mmHg), grade 1 hypertension (140–159/90–99 mmHg), and grade 2 hypertension (160/100 mmHg). Additionally, prehypertension is seen as a novel category of hypertension, associated with a higher future risk of developing both hypertension and cardiovascular diseases in the future. In adults, it was characterized as systolic blood pressure of 120 mmHg-139 mmHg and/or diastolic blood pressure of 80–89 mmHg, and in adolescents, blood pressure (BP) of 120/80 mmHg (Nugroho et al. 2022, Malik et al. 2022). Although the definition and classification of hypertension have changed over time, there is general agreement that persistent arterial pressure of 140/90 mmHg or more is considered excessive and should be treated (Iqbal and Jamal 2022). It is essential to measure blood pressure regularly, as it may not show signs or symptoms. Moreover, most people with hypertension are unaware of the problem, which can lead to more dangerous and potentially deadly complications. Therefore, it is inappropriate to see high blood pressure as a mild disease, especially given its reputation as a "silent killer" (WHO 2021, Wijaya et al. 2022). Hence, detecting hypertension early helps to avoid serious repercussions later in life (Kumawat et al. 2021).

3.1.1 Risk Factors for Hypertension

Understanding the causes of high blood pressure is essential for preventing it. As a multifactorial condition, hypertension is influenced by a variety of constitutive, hereditary, and environmental factors (Salem-Sokhn et al. 2021). According to the World Health Organization (WHO), the risk factors for developing hypertension are divided into nonmodifiable and modifiable risk factors (WHO 2021). The first category includes a family history of hypertension (Tozo et al. 2022), advanced age (Shukuri et al. 2019, Mamdouh et al. 2022), and the presence of additional comorbidities such as diabetes and chronic renal disease (Hsu and Tain 2021, Mohammed Nawi et al. 2021). The second category consists of diet, physical inactivity, smoking, alcohol use, and being overweight or obese (WHO 2021). In fact, several studies have demonstrated that a high risk of hypertension was linked to excessive salt consumption (Grillo et al. 2019, Hunter et al. 2022), a diet high in saturated and trans fats (Widiyanto et al. 2021), a poor intake of fruits and vegetable legumes (Rosário et al. 2018, Shukuri Tewelde and Shaweno 2019, Widiyanto et al. 2021), physical inactivity (Díaz-Martínez et al. 2018, Gamage and Seneviratne 2021, Malik et al. 2022), consumption of tobacco and alcohol (Mishra et al. 2022, Singh et al. 2022), as well as obesity/overweight (Tang et al. 2022, Modjadji et al. 2022).

3.1.2 *How to Deal with Hypertension*

Since hypertension is a chronic disorder, it needs not just pharmacological but also non-pharmacological therapy (Wijaya et al. 2022). Guidelines all agree that high blood pressure should be treated with a systolic blood pressure (SBP) aim of less than 140 mm Hg in patients without comorbidities and under 130 mmHg in those with hypertension and known cardiovascular disease (Brunström and Carlberg 2018, Campbell et al. 2022). However, less than 14% of persons with hypertension have their blood pressure regulated at systolic/diastolic BP < 140/90 mm Hg, despite the fact that safe, well-tolerated, and affordable treatments are available (Al-Makki et al. 2022). In addition, even though pharmacological treatments using antihypertensive drugs have good benefits, they can generate side effects in patients such as weariness, sexual dysfunction, cardiac arrhythmias, renal failure, fetal anomalies and allergic responses, especially when taken for an extended period of time. Furthermore, when additional illnesses are present along with hypertension the body's response to anti-hypertensive medications may be impacted and the body's metabolic functions may suffer, possibly worsening hypertension itself (Yahaya Rahman and Abdullah 2014). Therefore, the use of lifestyle modifications for the prevention and adjuvant treatment of hypertension is being backed by more and more data (Valenzuela et al. 2021). The effectiveness of several non-pharmacological interventions in lowering blood pressure and preventing hypertension has been proven (Widiyanto et al. 2021, Valenzuela et al. 2021, Wijaya et al. 2022). The most efficient lifestyle interventions include losing weight, consuming less sodium and more potassium, exercising more, abstaining from alcohol, and adopting a diet that emphasized fruits and vegetables, such as the Diet Approaches to Stop Hypertension (DASH), which combines a number of elements that are helpful in regulating blood pressure (Widiyanto et al. 2021, Valenzuela et al. 2021, WHO 2021, Wijaya et al. 2022). In short, well-managed hypertension and a healthy lifestyle improve the quality of life and lowers the risk of complications (Mohammed Nawi et al. 2021).

3.2 Mushrooms as a Potential Hypertension Therapy

The quest for safe, efficient, and new therapeutic agents from a natural source is necessary due to the conflict between the benefits and adverse effects of synthetic antihypertensive medications (Yahaya Rahman and Abdullah 2014). As previously said, improving your diet and leading an active lifestyle are two natural strategies to reduce your blood pressure (Elgendy et al. 2022). In this regard, dietary mushrooms are one of the emerging approaches in the non-pharmacological treatment of hypertension. Mushrooms are edible fungi that offer a plethora of medicinal and nutritional bioactive compounds, promoting their use in maintaining global public health (Chaturvedi et al. 2018, Uffelman et al. 2022). For centuries, they have served as both food and medicine, making them an integral part of the human diet. In reality, they are a naturally occurring source of various classes of bioactive substances referred to as secondary metabolites (phenols, terpenoids, polysaccharides, oligopeptides, sterols, etc.). They include a variety of enzymes with different qualities (hypocholesterolemic, hypoglycemic, hypotensive, anti-inflammatory, antioxidant, etc.), as well as being high in protein, fiber, vitamins (B, C, D, E, H, and K), minerals (iron, potassium, manganese, selenium, and zinc), and carbohydrates (β-glucanes) and low in salt and fat. Furthermore, they possess ergothioneine and glutathione, two key sulfur-containing antioxidants known to play an important role in avoiding chronic illness and early mortality, as well as supporting healthy aging. All of this turns them into food that has substantial therapeutic advantages for human health. Actually, they are recognized as a functional food for the prevention and treatment of various disorders, including hypertension. The term "mushroom nutraceuticals" is frequently used to describe therapeutic mushrooms, however, both edible and medicinal mushrooms contain biological antihypertensive substances (Chaturvedi et al. 2018, Ba et al. 2021). The mushroom's low salt concentration and high potassium content (182–395 mg/100 g) justify its use in an antihypertensive diet, as potassium-rich foods help lower blood pressure. Indeed, several studies have been conducted to investigate the antihypertensive effects of mushrooms, which have piqued the interest of several scientists worldwide. Thus, various mushrooms have been shown to reduce blood pressure, including *Agaricus bisporus* (J.E. Lange) Imbach, *Armillariella mellea* (Vahl) P. Kumm., *Ganoderma lucidum* (Curtis) P. Karst., *G. adspersum* (Schulz.) Donk, *Pleurotus cornucopiae* (Paulet) Rolland, *P. eryngii* (DC.) Quél., *P. ostreatus* (Jacq.) P., *P. pulmonarius* (Fr.), *Volvariella volvacea* (Bull.) Singer and *Inonotus obliquus* (Chaga). The latter (Chaga) has already been used in folk medicine. It treats high blood pressure in a variety of ways: exerting anti-oxidant effects on the molecular mechanism involved in the oxidative state and vascular functions as well as having anti-inflammatory properties, lowering total cholesterol, bad cholesterol (LDL), and triglycerides and increasing the production of nitric oxide (NO), which is essential for vascular health. (Badalyan et al. 2021). Nevertheless, it has been stated that most of the hypotensive properties of mushrooms are due to their ability to inhibit the angiotensin-converting enzyme (ACE), which controls the activity of the renin-angiotensin-aldosterone system (RAAS), one of the fundamental blood pressure regulators. However, in contrast to synthetic ACE inhibitors such as captopril, enalapril, and lisinopril, mushrooms have no side effects. Furthermore, a *Tricholoma giganteum* anti-hypertensive peptide has been discovered to be more effective than the widely used anti-hypertensive drugs captopril. Other mushrooms like *Pleurotus species, Grifola frondosa, Pholiota adipose, Hericium erinaceus, Hypsizigus marmoreus, Agaricus bisporus, Flammulina velutipes,* and *G. lucidum* have also been shown to contain ACE inhibitory anti-hypertensive peptides (Yahaya et al. 2014). Otherwise, research on spontaneously hypertensive rats showed that the auto-digested reishi (*Ganoderma lingzhi*) extract could be a great source of hypotensive peptides that may be used to treat hypertension or be included in functional foods (Tran et al. 2014). A further study found that a mushroom diet might be used as a preventative measure to lower the risk of pregnancy-induced hypertension and to manage neonatal birth weight while lowering co-morbidities including gestational weight gain, diabetes, etc. (Sun and Niu 2020). In addition, consuming mushrooms has been found to be significantly inversely associated with the risk of death from all causes (Ba et al. 2021). Therefore, mushrooms may be described as an unrivalled source of healthy food and medications, with a bright future as an alternative medicine branch, including the treatment of hypertension (Chaturvedi et al. 2018).

4. Mushrooms and Cardiovascular Disease

Cardiovascular diseases (CVDs) are a group of disorders of the heart and blood vessels which are the first leading cause of death in developed countries (Fitzmaurice et al. 2017). CVDs are complex diseases associated with multiple factors such as high triglycerides and low-density lipoprotein cholesterol (LDLc) levels, LDLc oxidation, increased platelet aggregation and hypertension (Lemieux et al. 2001, Seo et al. 2015).

The therapeutic contribution of mushrooms essentially lies in the reduction of the factors mentioned below. This therapeutic power of mushrooms is reflected in the reduction of hypercholesterolemia (Shimada et al. 2003, Nabubuya et al. 2010), atherogenesis (Kim et al. 2017), inflammation (Martin 2010), platelet aggregation (Lu et al. 2014) and oxidative stress (Ames et al. 2018). Given the complexity of CVDs, the particular therapeutic capacity of mushrooms could be due to the diversity of bioactive molecules that mushrooms contain. Polysaccharides, glycosides, alkaloids, volatile oils, terpenoids, tocopherols, phenolics, flavonoids, lectins, enzymes and organic acids are among the bioactive substances present in fruiting bodies, cultured mycelium, and cultured broth (Patel and Goyal 2012, Chang et al. 2012). The most important polysaccharide in modern medicine is β-glucan, which is the best-known and most adaptable metabolite with a wide range of biological activities (Kumar et al. 2021).

Hypercholesterolemia is a key risk factor for cardiovascular disorders, and several mushroom species have been shown to decrease lipid levels in both animals and people. In cardiovascular disorders associated with hypercholesterolemia, blood cholesterol levels must be controlled to both prevention and therapy (Kumar et al. 2021). Atherosclerosis is the most common cause of cardiovascular disease, in which fat, cholesterol, calcium, and other chemicals form plaques in the arteries walls. Atherosclerosis is increasingly recognized as a chronic inflammatory disease as well as a hyperlipidemia disorder. Kim et al. suggested that consumption of *Portobello* mushroom and *Shiitake* mushroom may be effective in preventing high-fat–induced atherosclerosis in LDLr −/− mice through modulation of lipid metabolism and immune function (Kim et al. 2017). This anti-atherogenic activities could be due to the presence, in these mushrooms, of both soluble and insoluble fiber, as well as bioactive antioxidant compounds including ergothioneine, which has been demonstrated to have potential antiatherogenic effect in cell culture systems and preliminary animal investigations (Smith et al. 2020). In addition, it has been shown that *Termitomyces microcarpus* mushrooms intake reduces the prevalence of disorders associated with high blood lipids, and it has also been reported that the mushrooms' high fiber content can lower total serum cholesterol, LDLc, and triglycerides (Nabubuya et al. 2010). These hypocholesterimiant effects of mushrooms are partly related to the presence of fungal sterols (ergosterol and derivatives) and polysaccharides (β-glucans) which have shown in vitro their inhibitory activities on the absorption and biosynthesis of cholesterol (Gil-Ramirez et al. 2014). Otherwise, inhibition of S-adenosyl-L-homocysteine hydrolase (involved in the hepatic phospholipid metabolism) could be a potential target for eritadenine ($2(R),3(R)$-dihydroxy-4-(9-adenyl) butanoic acid) from *Lentinula edodes* (Shiitake mushrooms) which leads to a decrease of cholesterol levels (Shimada et al. 2003).

The mushroom's anti-inflammatory properties can be of great interest in the treatment of CVDs. Indeed, many aqueous, methanolic, ethanolic, and ethyl acetate mushroom extracts have been demonstrated to reduce the generation of inflammatory mediators by down-regulating their gene expression (Elsayed et al. 2014). Pre-incubation of human aortic endothelial cells with ergothioneine suppressed IL-1–stimulated production of adhesion molecules and reduced monocyte adherence, two well-known markers of inflammation (Martin 2010).

In addition, antioxidants may play a substantial role in the prevention of chronic diseases due to the crucial function of oxidative stress in the development of many chronic diseases such as cardiovascular disease. The inherent antioxidant qualities of mushroom components ergothioneine and glutathione may be one of the molecular mechanisms behind the link between mushroom consumption and a lower risk of all-cause mortality (Ames et al. 2018). Based on its antioxidant power, ergothioneine was consistently found as the primary metabolite linked with a lower risk of cardiometabolic disease and mortality in a recent unbiased plasma metabolomics investigation (Smith et al. 2020).

Several studies have shown that some mushroom species do have antiplatelet aggregation properties and therefore, antiatherogenic activity. Lu et al. (2014) found that the crude extract of *Antrodia camphorata* was able to inhibit platelet aggregation by reducing collagen in a dose-dependent manner.

Caused by thrombus formation inside blood vessels disrupting blood flow, thrombosis is a major complication of cardiovascular disease. The standard treatment for thrombosis is the use of intravenous insertion of thrombolytic agents. Many fibrin(ogen)olytic enzymes have been identified and described from various mushroom species in the last decade. A fibrinolytic enzyme isolated from *Cordyceps militaris* was recently investigated (Liu et al. 2015a). This enzyme was able to induce fibrin and fibrinogen degradation, implying that it functions similarly to plasmin (Amirullah et al. 2018).

5. Anti-Diabetic Potential of Mushrooms

Diabetes mellitus is a metabolic disorder characterized by hyperglycemia resulting from impairment of insulin secretion, defective insulin action or both (Punthakee et al. 2018). It is one of the biggest epidemics in human history of the present century (Zimmet 2017). Data from the International Diabetes Federation on the prevalence of diabetes show that more than half a billion people had diabetes. Estimates suggest that by 2045 the number of diabetic patients is expected to increase to 783 million (Sun et al. 2022). The two main types of diabetes are type 2 diabetes mellitus (*T2DM*) and type 1 diabetes mellitus (T1DM). T2DM may range from predominant insulin resistance with relative insulin deficiency to a predominantly insulin secretory defect with insulin resistance (IR). T1DM results from autoimmune destruction of the pancreatic β-cell leading to absolute insulin deficiency (Punthakee et al. 2018).

Insulin acts mainly through phosphorylation of the protein kinase B (PKB, Akt) and phosphoinositide 3-kinase (PI3K). Akt activates glucose metabolism via the activation of phosphofructokinase and deactivation of glycogen synthase kinase 3 (GSK-3) by an increase in glucose utilization and reduction of gluconeogenesis in the liver and muscle (Huang et al. 2018). Akt phosphorylation activates glucose transporter 4 (GLUT4) which increases glucose uptake in muscles leading to reduce blood glucose levels (BGL) (Saini et al. 2010, Wondmkun et al. 2020). In T2DM, the PI3K/Akt pathway is damaged in various tissues as a result of IR.

Several animal studies shed light on the anti-diabetic effects of mushroom's secondary metabolites, such as polysaccharides and terpenoids which reduce IR by lowering blood glucose, inhibit enzymes involved in the hydrolyze of oligosaccharides and promote proliferation in ß-cells.

Polysaccharides from *Hericium erinaceus* (Cai et al. 2020), *Grifola frondosa* (Chen et al. 2019), and *Inonotus obliquus* (Wang et al. 2017) reduce fasting blood glucose (FBG) and IR, via activation of PI3K/Akt signaling pathways and improvement of the glucose transportation by activating GLUT4. *Agaricus bisporus* (Balakrishnan et al. 2018), *Agaricus blazei* Murrill (Oh et al. 2010) and *Cordyceps sinensis* (Li et al. 2006) polysaccharides lowers BGL. *Ganoderma lucidum* extract, reduced BGL by suppressing the liver gene expression of phosphoenolpyruvate carboxykinase (PEPCK) which is involved in the gluconeogenesis pathway (Seto et al. 2009).

The anti-diabetic properties of terpenoids were associated with the inhibition of enzymes involved in glucose metabolism, such as α-glucosidase and α-amylase which hydrolyze oligosaccharides to monosaccharaides and increase BGL (Nazaruk et al. 2015). Mushroom terpenoids from *Ganoderma lucidum* (Fatmawati et al. 2011), *Inonotus obliquus* (Ying et al. 2014), and *Tremella fuciformis* (Ma et al. 2021), have an *α-glucosidase* inhibitory effect. Moreover, polysaccharides from

Inonotus obliquus (Wang et al. 2018) and *Silphium perfoliatum* inhibit α-amylase and α-glucosidase activity (Guo et al. 2020).

Furthermore, polysaccharides can reduce injury to β cells and improve their functionality in diabetic animals; polysaccharides from *Ganoderma atrum* (Zhu et al. 2013), *Grifola frondosa* (Lei et al. 2012), *Agaricus bisporus*, and *Hericium erinaceus* (Zhang et al. 2017) maintain β cells growth.

Up to here, a limited number of studies examined anti-diabetic effects of mushrooms on human subjects using crude mushroom extract. Effect of the culinary *Pleurotus ostreatus* and *P. cystidiosus* mushrooms (consumed as freeze-dried powders at a dose of 50 mg/kg/body weight) in type 2 diabetic patients showed a reduction in postprandial serum glucose levels and increased serum insulin levels (Jayasuriya et al. 2015). Dietary consumption of *Agaricus bisporus* reduces diabetes risk factors and can occur over time in adults predisposed to type 2 diabetes (Calvo et al. 2016). The mushroom *Agaricus blazei Murill* in combination with metformin and gliclazide improved IR among treated subjects with T2DM (Hsu et al. 2007). Supplementation by ganopoly, a polysaccharide fraction extracted from *Ganoderma lucidum* in patients with T2DM for 12 weeks decreased a glycosylated hemoglobin (HbA1c), FBG, postprandial glucose and increase insulin levels (Gao et al. 2004).

6. Hepatoprotective Effects of Mushrooms

The liver is an active organ that plays a central role in metabolism and detoxification. The major metabolic functions of the liver are glucose and lipid metabolism and detoxification of toxic metabolites, drugs and xenobiotics (Chiang 2014). Liver damage involves oxidative stress and lipid peroxidation that increases hepatic enzymes alanine transaminase (ALT) and aspartate transaminase (AST) (Li et al. 2015, Lee et al. 2008) in the blood. Many liver diseases impair liver metabolic functions and progress to nonalcoholic fatty liver disease (NAFLD), hepatic fibrosis, and cirrhosis (Chiang 2014).

Some studies were conducted with purified or semi-purified molecules extracted from mushrooms to show their hepatoprotectives effects. *Ganoderma lucidum* is one of the most studied mushrooms. *G. lucidum* extracts, mainly polysaccharides or triterpenoids show protective activities against liver damage; in the CCL4 and D-galactosamine liver-injured mice, administration of total triterpenoids extracts inhibited the increase of serum ALT and liver triglyceride levels, antagonized the decrease of the antioxidant enzymes; superoxide dismutase (SOD) and glutathione peroxidase (GSH-Px), and inhibited the increase of malondialdehyde (MDA) a degradation product of lipid peroxides (Shi et al. 2008, He et al. 2008). Ganoderenic acid is one active compound of the extract, with a potent inhibitory effect on β-glucuronidase, an indicator of hepatic damage (Kim et al. 1999). Moreover, treatment with G. *lucidum* polysaccharides for 12 weeks reduced hepatitis B e antigen and hepatitis B viruses (HBV) DNA in patients with HBV infection (Wasser 2005).

Polysaccharides from the *Pleurotus ostreatus* and the crude mycelial polysaccharide from *Pholiota Dinghuensis* decreased ALT and AST levels in blood enhanced the activities of SOD and GSH, and reduced the formation of MDA against CCL4-induced acute liver injury in rat and mice respectively (Zhu et al. 2019, Gan et al. 2012). Polysaccharides enhanced the antioxidant effect to attenuate apoptosis. The administration of heteropolysaccharide isolated from *Pleurotus eryngii* to mice induced reduction of liver lipid peroxidation and the elevation of the hepatic antioxidant system. The polysaccharide has the potential to mitigate IR, oxidative stress, and liver dysfunction (Ren et al. 2014).

Anthraquinol is a ubiquinone derivative isolated from the basidioma and the mycelium of *Antrodia cinnamomea*. In human hepatoma cell lines (HepG2 cells), pretreatment with anthraquinol inhibits ethanol-induced AST, ALT, reactive oxygen species (ROS), nitric oxide (NO), MDA productions and GSH depletion. Western blot and RT-PCR analysis showed that antroquinonol enhanced Nrf-2 activation and its downstream antioxidant gene HO-1 via the MAPK pathway. This mechanism was then confirmed in vivo in an acute ethanol-intoxicated mouse model: elevation of serum ALT and AST levels, hepatocellular lipid peroxidation and GSH depletion were prevented by an ethanolic *A. cinnamomea* extract in a dose-dependent manner (Kumar et al. 2011). A recent study evaluates the effects of two polysaccharide-peptides (PSI and PSII) extracted from *Pleurotus citrinopileatus* in a hepatoma cell model (HepG2 cells) of nonalcoholic fatty liver disease (NAFLD). PSI and PSII increased the survival rates of injured cells, elevated the intracellular activity of SOD, and decreased intracellular triglyceride content and ALT and AST release (Huang et al. 2022).

Other studies using the whole mushroom with non-purified or semi-purified extract showed that aqueous extract of *Panus giganteus* (Wong et al. 2012), *Antrodia camphorate* and *Armillariella tabescens* (Lu et al. 2007) and phenolic compounds of *Macrocybe gigantea* (Acharya et al. 2012), *Pleurotus cornucopiae* (El Bohi et al. 2009), *Lentinula edodes* (Itoh et al. 2010) reduce the level of classical markers of hepatic damage in mice or rat.

7. Mushrooms in Neurodegenerative Diseases

Neurodegeneration is a progressive loss of neurons and their processes with a corresponding progressive impairment in neuronal function (Jack et al. 2015). Alzheimer's disease (AD), Parkinson's disease (PD), and Amyotrophic lateral sclerosis are three of the major neurodegenerative diseases (NDs) (Checkoway et al. 2011). Most NDs share common molecular mechanisms, including aggregation of misfolded proteins, oxidative damage, mitochondrial dysfunction, neurotrophic impairment, and neuroinflammatory responses (Jellinger et al. 2010).

AD is the most common NDs, characterized by a progressive decline of cognition. In the brain, the pathological hallmarks of AD is the deposition of amyloid beta (Aβ) plaques and the formation of neurofibrillary tangles (NFTs) composed of hyperphosphorylated tau protein, which causes synaptic degeneration and neuronal death (Ittner and Götz 2011).

Hericium erinaceus is among the most characterized medicinal mushrooms and with a focus on the nervous system (Friedman 2015). The nutraceutical properties of edible mushrooms help to prevent AD progression (Phan et al. 2017).

In the AD mouse model, terpenoids from the mycelia of *Hericium erinaceus* (HE) prevent AD-related pathology, a 30-day administration of erinacine A and S diminished cerebral Aβ plaque and glial cell activation and enhanced hippocampal neurogenesis (Tzeng et al. 2018). Erinacine C suppresses the overproduction of NO, IL-6, TNF-α, and iNOS in LPS-activated BV2 microglia. The anti-neuroinflammatory properties of Erinacine C occur via inhibition of NF-κB expression (Wang et al. 2019). HE fruiting the body also alleviates the Aβ mediated damage in PC12 cells (Liu et al. 2015b). The aqueous extract of *Ganoderma lucidum* attenuated Aβ-induced synaptotoxicity by preserving the synaptic density protein, synaptophysin in rat cortical neurons via JNK and P39 kinase pathway (Lai et al. 2008).

Nerve Growth Factor (NGF) is a neurotrophic factor that maintains and organizes neurons functionally. Methanol extracts from HE enhance the NGF mRNA expression in human astrocytoma cells, via the activation of the JNK pathway (Mori et al. 2008). The HE aqueous extract prevents AD progression by enhancing the acetylcholine and choline acetyltransferase concentrations in both the serum and the hypothalamus in the AD mouse model (Zhang et al. 2015). The polysaccharide extract of *Coprinus comatus* and *Coprinellus truncorum* and the extract of *Phellinus pini* inhibits the acetylcholinesterase (Pejin et al. 2019, Im et al. 2016). *Pleurotus ostreatus* polysaccharide extract enhances cognitive function, increased antioxidant enzymes like superoxide dismutase (SOD) and glutathione peroxidase, and inhibited the formation of Aβ and tau protein aggregates in a rat model (Zhang et al. 2015). *Coriolus versicolor* also increased antioxidant activity (SOD and catalase) and inhibited pro-inflammatory cytokines (IL-1β, IL-6 and TNF-α), which improve spatial memory in a mouse model of AD (Fang et al. 2015).

PD is the second most common NDs, resulting from nigrostriatal dopaminergic neurodegeneration which gives rise to motor disturbances including resting tremor, limb rigidity, and postural impairments (Braak et al. 2003). Nonmotor symptoms include mood disorder, depression, and cognitive impairment (Goldman 2014).

Oral administration of *Hericium erinaceus* mycelia in rat model of PD led to improvement in oxidative stress and dopaminergic lesions in the striatum and substantia nigra (Wong et al. 2011). *Agaricus blazei* extract shows neuroprotective activity in the rotenone-induced PD model (Venkateshgobi et al. 2018). Methanol extract of fruiting bodies from *Ganoderma lucidum* demonstrated neuroinflammatory suppression in primary rat microglia activated by LPS and neuron-glia co-culture model. The extract of *G. lucidum* reduces NO, TNF- α, IL-1 β, and superoxide release and downregulates TNF-α and IL-1β mRNA expression (Zhang et al. 2011).

Multiple sclerosis (MS) is an autoimmune disease of the central nervous system, where the myelin sheath-coated neurons are destructed. The disease symptoms include muscle spasms, fatigue, impaired memory and vision, and paralysis (Warner 2009).

The extract of *H. erinaceus* demonstrated regulatory action on rate and duration of myelination *in vitro*. The process of the myelin sheath formation in the presence of extract *H. erinaceus* proceeded faster and was completed by 26 days *in vitro* (in controls, it was finished in 31 days) (Kolotushkina et al. 2003).

In the mice model of MS, a water-ethanol extract of *Phellinus igniarius* (Piwep) may have a high therapeutic potential for ameliorating MS progression. The Piwep suppresses demyelination and infiltration of encephalitogenic immune cells including CD4+ T cells, CD8+ T cells, and macrophages in the spinal cord. Piwep also reduced the expression of vascular cell adhesion molecule-1 (VCAM-1) in the spinal cord and inhibited the proliferation of lymphocytes and secretion of interferon-γ in the lymph node (Li et al. 2014).

8. Mushrooms and Cancer

Cancer is a major global threat and is still one of the leading causes of mortality in 30 developed countries (Fitzmaurice et al. 2017). To fight against cancer, several classic therapeutic approaches have been used like treatments include surgery, chemotherapy, radiation, hormone-targeted therapies and immunotherapies. These therapeutic approaches may have a substantial impact on the quality of life during or after cancer treatment due to various side effects such as long-term damage to the heart, lungs, kidneys, or reproductive organs (Urruticoechea et al. 2010). In addition to their anti-tumor effects, medicinal mushrooms can reduce the side effects associated with the use of conventional treatments. They are frequently used in East Asia to cure a variety of diseases, particularly in complementary cancer treatment (Jeitler et al. 2020).

Several active compounds in a single mushroom species have the ability to influence several cancer-related pathways in a synergistic manner (Blagodatski et al. 2018, Joseph et al. 2018). This has led to increased interest in the use of medicinal mushrooms in Western countries, especially after the publication of a significant number of pre-clinical studies which showed that they have anti-cancer and regenerative characteristics, little is known about their therapeutic utility (Jeitler et al. 2020).

The most studied species of medicinal mushrooms were *Lentinula edodes* (Islam et al. 2021), *Aspergillus oryza* (Konishi et al. 2021), and *Antrodia cinnamomea* (Tsai et al. 2016). *Coriolus versicolor* (Chay et al. 2017), *Agaricus blazei* Murrill (Tangen et al. 2015), *Agaricus sylvaticus* (Costa Fortes et al. 2010), *Ganoderma lucidum* (Zhao et al. 2012). In these studies, the whole mushroom, its extract or one of its compounds was administered orally. Compounds present in mushrooms prevent cancer proliferation by an anti-mitotic activity, a mitotic kinase inhibitor activity, topoisomerase inhibitor activity, and by inhibition of angiogenesis causing apoptosis of cancer cells (Kumar et al. 2021). Indeed, heptelidic acid produced by *Aspergillus oryza* inhibited tumor growth by causing apoptosis in pancreatic cancer cells via p38 signaling activation (Konishi et al. 2021). Furthermore, heptelidic acid had an anti-tumor impact by inhibiting cancer cell proliferation in extra-intestinal organs by transferring bacteria-derived anti-tumor molecules, which is a novel anti-tumor mechanism (Konishi et al. 2021). In addition, polysaccharide krestin and polysaccharide peptide are bioactive extracts of *Coriolus versicolor*. Ohwada et al. observed that patients with stage II and III colorectal cancer who had conventional treatment with 3 grams of polysaccharide krestin per day had a higher percentage of

5-year disease-free survival and a lower relative risk of regional metastase (Ohwada et al. 2006). A meta-analysis of three trials with 1,094 colorectal cancer patients found that those who took polysaccharide krestin had significantly better overall survival and disease-free survival than those who did not (Sakamoto et al. 2006). In a meta-analysis of 8,009 gastric cancer patients from eight randomized controlled trials, polysaccharide krestin users had a better survival rate (Oba et al. 2007).

The benefits of consuming mushrooms are not limited to their antitumor activity but also to improving the quality of life by reducing the side effects linked to conventional therapy and by restoring hematological parameters (Ahn et al. 2004). Traditional cancer treatment and the disease itself are known to cause decreased quality of life and fatigue (Bottomley 2002). Studies have shown that various elements of quality of life improved after mushroom intake. Furthermore, in other research, emotional and mental well-being, as well as mood ratings (especially anxiety and despair) also improved (Ahn et al. 2004). In addition, several mushroom species are suspected to have an antifatigue effect (Geng et al. 2017). Thirty-seven patients with advanced adenocarcinomas of the breast, lungs, stomach, liver, and colon region who were undergoing chemotherapy were enrolled and randomly assigned to *Antrodia cinnamomea* or placebo for a 30-day supplemental treatment (Tsai et al. 2016). Except for sleep, which was greatly better in the verum group, there were no significant changes between the groups in the assessments. Using data from 48 breast cancer patients who were on endocrine therapy and had cancer-related fatigue. When compared to a placebo, a 4-week therapy with *Ganoderma lucidum* spore powder enhanced overall quality of life (Zhao et al. 2012).

In an interesting study, 100 patients with various gynaecological malignancies (cervical, endometrial, and ovarian) received chemotherapy with *Agaricus blazeii murill Kyowa* or a placebo (Ahn et al. 2004). It was found that, unlike the placebo group, patients under *Agaricus blazeii murill Kyowa* improved their emotions (anxiety, depression, mental stability), as well as their physical strength. In addition, sleep disturbances impact 30–50% of cancer patients, increasing the risk of depression, exhaustion, increased pain, and lower survival chances (Otte et al. 2015). Indeed, some research found a link between medicinal mushroom use and improved sleep quality (Tsai et al. 2016, Chay et al. 2017).

Side effects reduction by conventional therapy is also a target for mushrooms. Indeed, when patients with multiple myeloma were treated with *Agaricus blazei murill*, their immunological condition improved dramatically in terms of white blood cells and immunoglobins, resulting in fewer infections (Tangen et al. 2015). Medicinal mushrooms may contribute to the treatment of bone marrow suppression caused by chemotherapy and radiotherapy (Hofer and Pospisil 2011). In vitro, their major ingredient, ß-glucans, has haematological effects and aids bone marrow regeneration (Sorimachi et al. 2001).

Otherwise, nutritional supplementation with *Agaricus sylvaticus* had positive haematological and glycaemic benefits in cancer patients. Indeed, it was noticed that 3 or 6 months of nutritional supplementation with *Agaricus sylvaticus* was able to reduce fasting plasma glucose, lipid and liver parameters with normalization of blood pressure (Fortes et al. 2009).

9. Conclusion

Mushrooms and their metabolites have several beneficial effects in the prevention and treatment of several chronic diseases. They have been applied for centuries in different cultures by traditional medicine practitioners. Mushrooms have been usually applied as whole, or unfractionated extracts and containing molecules. However, some recent products used semi-purified extracts to claim a health benefit based on the determined bioactive compounds. However, further work is needed for a better understanding of the molecular interaction of mushroom bioactive molecules and human pathophysiology. On the other hand, using mushrooms as natural bioresources in the reduction, control, and treatment of chronic diseases has also a socio-economic impact as it will help mushroom farmers to develop a high-value product with a higher sales price.

References

Acharya, K., S. Chatterjee, G. Biswas, A. Chatterjee and G.K. Saha. 2012. Hepatoprotective effect of a wild edible mushroom on carbon tetrachloride-induce hepatoprotective in mice. *Int. J. Pharm. Pharm. Sci.*, 4: 285–288.

Ahn, W.S., D.J. Kim, G.T. Chae, J.M. Lee, S.M. Bae, J.I. Sin, Y.W. Kim, S.E. Namkoong and I.P. Lee. 2004. Natural killer cell activity and quality of life were improved by consumption of a mushroom extract, *Agaricus blazei* Murill Kyowa, in gynecological cancer patients undergoing chemotherapy. *Int. J. Gynecol. Cancer*, 14: 589–594.

Al-Makki, A., D. DiPette, P.K. Whelton, M.H. Murad, R.A. Mustafa, S. Acharya, H. Mamoun Beheiry, B. Champagne, K. Connell, M.T. Cooney, N. Ezeigwe, T.A. Gaziano, A. Gidio, P. Lopez-Jaramillo, U.I. Khan, V. Kumarapeli, A.E. Moran, M.M. Silwimba, B. Rayner, A. Sukonthasan, J. Yu, N. Saraffzadegan, K.S. Reddy and T. Khan. 2022. Hypertension pharmacological treatment in adults: A world health organization guideline executive summary. *Hypertension*, 79(1): 293–301.

Ames, B.N. 2018. Prolonging healthy aging: longevity vitamins and proteins. *Proc. Natl. Acad. Sci.*, 115(43): 10836–10844.

Amirullah, N.A., N.Z. Abidin and N. Abdullah. 2018. The potential applications of mushrooms against some facets of atherosclerosis: A review. *Food Res Int.*, 105: 517–536.

Amponsem-Boateng, C., W. Zhang, T.B. Oppong, G. Opolot and E.K.D. Kyere. 2019. A cross-sectional study of risk factors and hypertension among adolescent Senior High School students. *Diabetes Metab. Syndr. Obes.*, 12: 1173.

Ba, D.M., X. Gao, J. Muscat, L. Al-Shaar, V. Chinchilli, X. Zhang, P. Ssentongo, R.B. Beelman and J.P. Richie. 2021. Association of mushroom consumption with all-cause and cause-specific mortality among American adults: Prospective cohort study findings from NHANES III. *Nutr. J.*, 20(1): 1–11.

Badalyan, S.M., A. Barkhudaryan and S. Rapior. 2021. The cardioprotective properties of Agaricomycetes mushrooms growing in the territory of Armenia. *Int. J. Med. Mushrooms*, 23(5): 21–31.

Balakrishnan, P. and C.T. Loganayagi. 2018. Antihyperglycemic activity of *Agaricus bisporus* mushroom extracts on alloxan induced diabetic rats. *Int. J. Pharma. Res. Health Sci.*, 6: 2475–2479.

Benbaibeche, H., E.M. Haffaf, G. Kacimi, B. Oudjit, N.A. Khan and E.A. Koceïr. 2015. Implication of corticotropic hormone axis in eating behaviour pattern in obese and type 2 diabetic participants. *Br. J. Nutr.*, 113: 1237–1243.

Blagodatski, A., M. Yatsunskaya, V. Mikhailova, V. Tiasto, A. Kagansky and V.L. Katanaev. 2018. Medicinal mushrooms as an attractive new source of natural compounds for future cancer therapy. *Oncotarget*, 9: 29259–29274.

Bottomley, A. 2002. The cancer patient and quality of life. *Oncologist*, 7: 120–125.

Braak, H., K. Del Tredici, U. Rüb, R.A. De Vos, E.N.J. Steur and E. Braak. 2003. Staging of brain pathology related to sporadic Parkinson's disease. *Neurobiol. Aging*, 24(2): 197–211.

Bromfield, S. and P. Muntner. 2013. High blood pressure: The leading global burden of disease risk factor and the need for worldwide prevention programs. *Curr. Hypertens. Rep.*, 15(3): 134–136.

Brunström, M. and B. Carlberg. 2018. Association of blood pressure lowering with mortality and cardiovascular disease across blood pressure levels: A systematic review and meta-analysis. *JAMA Intern. Med.*, 178(1): 28–36.

Cai, W.D., Z.C. Ding, Y.Y. Wang, Y. Yang, H.N. Zhang and J.K. Yan. 2020. Hypoglycemic benefit and potential mechanism of a polysaccharide from *Hericium erinaceus* in streptozotoxin-induced diabetic rats. *Process Biochem.*, 88: 180–188.

Calvo, M.S., A. Mehrotra, R.B. Beelman, G. Nadkarni, L. Wang, W. Cai and J. Uribarri. 2016. A retrospective study in adults with metabolic syndrome: Diabetic risk factor response to daily consumption of *Agaricus bisporus* (white button mushrooms). *Plant Foods Hum. Nutr.*, 71(3): 245–251.

Campbell, N.R., M.P. Burnens, P.K. Whelton, S.Y. Angell, M.G. Jaffe, J. Cohn and P. Ordunez. 2022. 2021 World Health Organization guideline on pharmacological treatment of hypertension: Policy implications for the region of the Americas. *Lancet Reg. Health Am.*, 9: 100219.

Chang, S.T. and S.P. Wasser. 2012. The role of culinary-medicinal mushrooms on human welfare with a pyramid model for human health. *Int. J. Med. Mushrooms*, 14(2): 95–134.

Chaturvedi, V.K., S. Agarwal, K.K. Gupta, P.W. Ramteke and M.P. Singh. 2018. Medicinal mushroom: Boon for therapeutic applications. *3 Biotech.*, 8(8): 1–20.

Chay, W.Y., C.K. Tham, H.C. Toh, H.Y. Lim, C.K. Tan, C. Lim and S.P. Choo. 2017. *Coriolus versicolor* (Yunzhi) use as therapy in advanced hepatocellular carcinoma patients with poor liver function or who are unfit for standard therapy. *J. Altern. Compl. Med.*, 23(8): 648–652.

Checkoway, H., J.I. Lundin and S.N. Kelada. 2011. Neurodegenerative diseases. *IARC Sci. Publ.*, (163): 407–419.

Chen, Y., D. Liu, D. Wang, S. Lai, R. Zhong, Y. Liu, C. Yang, B. Liu, M.R. Sarker and C. Zhao. 2019. Hypoglycemic activity and gut microbiota regulation of a novel polysaccharide from *Grifola frondosa* in type 2 diabetic mice. *Food and Chem. Toxicol.*, 126: 295–302.

Chiang, J. 2014. Liver physiology: Metabolism and detoxification. pp. 1770–1782. *In*: McManus, L.M. and Mitchell, R.N. (eds.). Pathobiology of Human Disease, Elsevier, San Diego.

Costa Fortes, R., V. Lacorte Recova, A. Lima Melo and M.R. Carvalho Garbi Novaes. 2010. Life quality of postsurgical patients with colorectal cancer after supplemented diet with *Agaricus sylvaticus* fungus. *Nutr. Hosp.*, 25: 586–596.

Diaz-Martinez, X., F. Petermann, A.M. Leiva, A. Garrido-Mendez, C. Salas-Bravo, M.A. Martínez and C. Celis-Morales. 2018. Association of physical inactivity with obesity, diabetes, hypertension and metabolic syndrome in the Chilean population. *Rev. Med. Chil.*, 146(5): 585–595.

Dubey, S.K., V.K. Chaturvedi, D. Mishra, A. Bajpeyee, A. Tiwari and M.P. Singh. 2019. Role of edible mushroom as a potent therapeutics for the diabetes and obesity. *3 Biotech.*, 9(12): 450.

El Bohi, K.M., Y. Hashimoto, K. Muzandu, Y. Ikenaka, Z.S. Ibrahim, A. Kazusaka and M. Ishizuka. 2009. Protective effect of *Pleurotus cornucopiae* mushroom extract on carbon tetrachloride-induced hepatotoxicity. *Jpn. J. Vet. Res.*, 57(2): 109–118.

Elgendy, M.F., A.E.A.E. Dawah, M.A. Elawady, S.Y.E. Zidan and M.A. Elmahdy. 2022. Lifestyle modification and its effect on the control of hypertension. *Egypt. J. Hosp. Med.*, 89(1): 4811–4816.

Elsayed, E.A., H. El Enshasy, M.A. Wadaan and R. Aziz. 2014. Mushrooms: A potential natural source of anti-inflammatory compounds for medical applications. *Mediators Inflamm.*, 2014: 805841.

Fang, X., Y. Jiang, H. Ji, L. Zhao, W. Xiao, Z. Wang and G. Ding. 2015. The synergistic beneficial effects of Ginkgo flavonoid and *Coriolus versicolor* polysaccharide for memory improvements in a mouse model of dementia. *Evid. Based Complement Alternat. Med.*, (9): 128394.

Fatmawati, S., K. Shimizu and R. Kondo. 2011. Ganoderol B: A potent α-glucosidase inhibitor isolated from the fruiting body of *Ganoderma lucidum*. *Phytomedicine*, 18(12): 1053–1055.

Feeney, M.J., J. Dwyer, C.M. Hasler-Lewis, J.A. Milner, M. Noakes, S. Rowe, M. Wach, R.B. Beelman, J. Caldwell and M.T. Cantorna. 2014. Mushrooms and health summit proceedings. *J. Nutr.*, 144: 1128S–1136S.

Finkelstein, E.A., J.G. Trogdon, J.W. Cohen and W. Dietz. 2009. Annual medical spending attributable to obesity: Payer-and service-specific estimates: Amid calls for health reform, real cost savings are more likely to be achieved through reducing obesity and related risk factors. *Health Aff.*, 28: w822–w831.

Fitzmaurice, C., C. Allen, R.M. Barber, L. Barregard, Z.A. Bhutta, H. Brenner, D.J. Dicker, O. Chimed-Orchir, R. Dandona and L. Dandona. 2017. Global, regional, and national cancer incidence, mortality, years of life lost, years lived with disability, and disability-adjusted life-years for 32 cancer groups, 1990 to 2015: A systematic analysis for the global burden of disease study. *JAMA Oncol.*, 3: 524–548.

Fortes, R.C., M.R.C.G. Novaes, V.L. Recova and A.L. Melo. 2009. Immunological, hematological, and glycemia effects of dietary supplementation with *Agaricus sylvaticus* on patients' colorectal cancer. *Exp. Biol. Med.*, 234: 53–62.

Friedman, M. 2015. Chemistry, nutrition, and health-promoting properties of *Hericium erinaceus* (Lion's Mane) mushroom fruiting bodies and mycelia and their bioactive compounds. *J. Agric. Food Chem.*, 63(32): 7108–7123.

Gamage, A.U. and R.D.A. Seneviratne. 2021. Physical inactivity, and its association with hypertension among employees in the district of Colombo. *BMC Public. Health.*, 21(1): 1–11.

Gan, D., L. Ma, C. Jiang, M. Wang and X. Zeng. 2012. Medium optimization and potential hepatoprotective effect of mycelial polysaccharides from *Pholiota dinghuensis* Bi against carbon tetrachloride-induced acute liver injury in mice. *Food Chem. Toxicol.*, 50: 2681–2688.

Ganesan, K. and B. Xu. 2018. Anti-obesity effects of medicinal and edible mushrooms. *Molecules*, 23(11): 2880. doi: 10.3390/molecules23112880.

Gao, Y., J. Lan, X. Dai, J. Ye and S. Zhou. 2004. A phase I/II study of Ling Zhi mushroom *Ganoderma lucidum* (W. Curt.: Fr.) Lloyd (Aphyllophoromycetideae) extract in patients with type II diabetes mellitus. *Int. J. Med. Mushrooms*, 6: 8–40.

Geng, P., K.C. Siu, Z. Wang and J.Y. Wu. 2017. Antifatigue functions and mechanisms of edible and medicinal mushrooms. *BioMed. Res. Int.*, 2017: 9648496.

Gentile, G., K. Mckinney and G. Reboldi. 2022. Tight blood pressure control in chronic kidney disease. *J. Cardiovasc. Dev. Dis.*, 9(5): 139.

Gil-Ramírez, A., A. Ruiz-Rodríguez, F.R. Marín, G. Reglero and C. Soler-Rivas. 2014. Effect of ergosterol-enriched extracts obtained from *Agaricus bisporus* on cholesterol absorption using an in vitro digestion model. *J. Funct. Foods*, 11: 589–597.

Goldman, J.G. and R. Postuma. 2014. Premotor and non-motor features of Parkinson's disease. *Curr. Opin. Neurol.*, 27(4): 434–441.

Grillo, A., L. Salvi, P. Coruzzi, P. Salvi and G. Parati. 2019. Sodium intake and hypertension. *Nutrients*, 11(9): 1970. doi: 10.3390/nu11091970.

Grotto, D., I.F. Camargo, K. Kodaira, L.G. Mazzei, J. Castro, R.A.L. Vieira, C.D.C. Bergamaschi and L.C. Lopes. 2019. Effect of mushrooms on obesity in animal models: Study protocol for a systematic review and meta-analysis. *Syst. Rev.*, 8: 1–5.

Guo, Y., H. Shang, J. Zhao, H. Zhang and S. Chen. 2020. Enzyme-assisted extraction of a cup plant (*Silphium perfoliatum* L.) Polysaccharide and its antioxidant and hypoglycemic activities. *Process Biochem.*, 92: 17–28.

He, H., J.P. He, Y.J. Sui, S.Q. Zhou and J. Wang. 2008. The hepatoprotective effects of *Ganoderma lucidum* peptides against carbon tetrachloride-induced liver injury in mice. *J. Food. Biochem.*, 32: 628–641.

Hiraki, E., S. Furuta, R. Kuwahara, N. Takemoto, T. Nagata, T. Akasaka, B. Shirouchi, M. Sato, K. Ohnuki and Shimizu, K. 2017. Anti-obesity activity of Yamabushitake (Hericium erinaceus) powder in ovariectomized mice, and its potentially active compounds. *J. Nat. Med.*, 71: 482–491.

Hiwatashi, K., Y. Kosaka, N. Suzuki, K. Hata, T. Mukaiyama, K. Sakamoto, H. Shirakawa and M. Komai. 2010. Yamabushitake mushroom (Hericium erinaceus) improved lipid metabolism in mice fed a high-fat diet. *Biosci. Biotechnol. Biochem.*, 74: 1447–1451.

Hofer, M. and M. Pospíšil. 2011. Modulation of animal and human hematopoiesis by β-glucans: A review. *Molecules*, 16: 7969–7979.

Hsu, C.-H., Y.-L. Liao, S.-C. Lin, K.-C. Hwang and P. Chou. 2007. The mushroom *Agaricus blazei* Murill in combination with metformin and gliclazide improves insulin resistance in type 2 diabetes: A randomized, double-blinded, and placebo-controlled clinical trial. *J. Altern. Complement. Med.*, 13: 97–102.

Hsu, C.-N. and Y.-L. Tain. 2021. Targeting the renin–angiotensin–aldosterone system to prevent hypertension and kidney disease of developmental origins. *Int. J. Mol. Sci.*, 22(5): 2298. doi: 10.3390/ijms22052298.

Huang, X., G. Liu, J. Guo and Z. Su. 2018. The PI3K/AKT pathway in obesity and type 2 diabetes. *Int. J. Biol. Sci.*, 14: 1483.

Huang, Y., Y. Gao, X. Pi, S. Zhao and W. Liu. 2022. *In vitro* hepatoprotective and human gut microbiota modulation of polysaccharide-peptides in *Pleurotus citrinopileatus*. *Front. Cell. Infect. Microbiol.*, 553. doi.org/10.3389/fcimb.2022.892049.

Hunter, R.W., N. Dhaun and M.A. Bailey. 2022. The impact of excessive salt intake on human health. *Nat. Rev. Nephrol.*, 18: 321–335.

Im, K.H., T.K. Nguyen, J.K. Kim, J. Choi and T.S. Lee. 2016. Evaluation of anticholinesterase and inflammation inhibitory activity of medicinal mushroom *Phellinus pini* (Basidiomycetes) fruiting bodies. *Int. J. Med. Mushrooms*, 18: 1011–1022.

Iqbal, A.M. and S.F. Jamal. 2022. Essential hypertension. In Stat Pearls, Stat StatPearls [Internet]. Treasure Island (FL): StatPearls Publishing; 2022 Jan-. Available from: https://www.ncbi.nlm.nih.gov/books/NBK539859/.

Islam, S., T. Kitagawa, B. Baron, K. Kuhara, H. Nagayasu, M. Kobayashi, I. Chiba and Y. Kuramitsu. 2021. A standardized extract of cultured *Lentinula edodes* mycelia downregulates cortactin in gemcitabine-resistant pancreatic cancer cells. *Oncol. Lett.*, 22: 1–8.

Itoh, A., K. Isoda, M. Kondoh, M. Kawase, M. Kobayashi, M. Tamesada and K. Yagi. 2009. Hepatoprotective effect of syringic acid and vanillic acid on concanavalin a-induced liver injury. *Biol. Pharm. Bull.*, 32: 1215–1219.

Ittner, L.M. and J. Götz. 2011. Amyloid-β and tau—A toxic pas de deux in Alzheimer's disease. *Nat. Rev. Neurosci.*, 12: 67–72.

Jack Jr, C.R., H.J. Wiste, S.D. Weigand, D.S. Knopman, M.M. Mielke, P. Vemuri, V. Lowe, M.L. Senjem, J.L. Gunter and D. Reyes. 2015. Different definitions of neurodegeneration produce similar amyloid/neurodegeneration biomarker group findings. *Brain*, 138: 3747–3759.

Jayasuriya, W.B.N., C.A. Wanigatunge, G.H. Fernando, D.T.U. Abeytunga and T.S. Suresh. 2015. Hypoglycaemic activity of culinary *Pleurotus ostreatus* and *P. cystidiosus* mushrooms in healthy volunteers and type 2 diabetic patients on diet control and the possible mechanisms of action. *Phytother. Res.*, 29: 303–309.

Jeitler, M., A. Michalsen, D. Frings, M. Hübner, M. Fischer, D.A. Koppold-Liebscher, V. Murthy and C.S. Kessler. 2020. Significance of medicinal mushrooms in integrative oncology: A narrative review. *Front. Pharmacol.*, 11: 580656.

Jellinger, K.A. 2010. Basic mechanisms of neurodegeneration: a critical update. *J. Cell. Mol. Med.*, 14: 457–487.

Jeong, H.J., S.J. Yoon and Y.R. Pyun. 2008. Polysaccharides from edible mushroom Hinmogi (Tremella fuciformis) inhibit differentiation of 3T3-L1 adipocytes by reducing mRNA expression of 3T3-L1 by reducing mRNA expression of PPARγ, C/EBPα, and Leptin. *Food. Sci. Biotechnol.*, 17: 267–273.

Jiang, X., J. Hao, Z. Liu, X. Ma, Y. Feng, L. Teng, Y. Li and D. Wang. 2021. Anti-obesity effects of *Grifola frondosa* through the modulation of lipid metabolism via ceramide in mice fed a high-fat diet. *Food. Funct.*, 12: 6725–6739.

Joseph, T.P., W. Chanda, A.A. Padhiar, S. Batool, S. LiQun, M. Zhong and M. Huang. 2018. A preclinical evaluation of the antitumor activities of edible and medicinal mushrooms: A molecular insight. *Integr. Cancer. Ther.*, 17: 200–209.

Kim, D.-H., S.-B. Shim, N.-J. Kim and I.-S. Jang. 1999. β-Glucuronidase-inhibitory activity and hepatoprotective effect of *Ganoderma lucidum*. *Biol. Pharm. Bull.*, 22: 162–164.

Kim, S., M. Thomas and M. Meydani. 2017. Anti-atherogenic potential of edible mushrooms. *FASEB. J.*, 31: 973.973–973.973.

Kolotushkina, E., M. Moldavan, K.Y. Voronin and G. Skibo. 2003. The influence of *Hericium erinaceus* extract on myelination process *in vitro*. *Fiziol. Zh.*, 49: 38–45.

Konishi, H., S. Isozaki, S. Kashima, K. Moriichi, S. Ichikawa, K. Yamamoto, C. Yamamura, K. Ando, N. Ueno and H. Akutsu. 2021. Probiotic *Aspergillus oryzae* produces anti-tumor mediator and exerts anti-tumor effects in pancreatic cancer through the p38 MAPK signaling pathway. *Sci. Rep.*, 11: 1–12.

Kumar, K.S., F.-H. Chu, H.-W. Hsieh, J.-W. Liao, W.-H. Li, J.C.-C. Lin, J.-F. Shaw and S.-Y. Wang. 2011. Antroquinonol from ethanolic extract of mycelium of *Antrodia cinnamomea* protects hepatic cells from ethanol-induced oxidative stress through Nrf-2 activation. *J. Ethnopharmacol.*, 136: 168–177.

Kumar, K., R. Mehra, R.P. Guiné, M.J. Lima, N. Kumar, R. Kaushik, N. Ahmed, A.N. Yadav and H. Kumar. 2021. Edible Mushrooms: A comprehensive review on bioactive compounds with health benefits and processing aspects. *Foods*, 10: 2996. doi: 10.3390/foods10122996.

Kumawat, D.S., S. Roy Bhowmik and R. Sethi. 2021. Early identification of hypertensive patients in apparently asymptomatic cohort and risk stratification as per new AHA guidelines (2017) among non-teaching employee of KGMU, Lucknow. *IOSR-JNHS*, 10(5): 48–53.

Lai, C.S.W., M.S. Yu, W.H. Yuen, K.F. So, S.Y. Zee and R.C.C. Chang. 2008. Antagonizing β-amyloid peptide neurotoxicity of the anti-aging fungus *Ganoderma lucidum*. *Brain Res.*, 1190: 215–224.

Lee, H.S., H.H. Kim and S.K. Ku. 2008. Hepatoprotective effects of *Artemisiae capillaris* Herba and *Picrorrhiza rhizoma* combinations on carbon tetrachloride–induced subacute liver damage in rats. *Nutr. Res.*, 28: 270–277.

Lei, H., S. Guo, J. Han, Q. Wang, X. Zhang and W. Wu. 2012. Hypoglycemic and hypolipidemic activities of MT-α-glucan and its effect on immune function of diabetic mice. *Carbohydr. Polym.*, 89: 245–250.

Lemieux, I., B. Lamarche, C. Couillard, A. Pascot, B. Cantin, J. Bergeron, G.R. Dagenais and J.-P. Després. 2001. Total cholesterol/HDL cholesterol ratio vs LDL cholesterol/HDL cholesterol ratio as indices of ischemic heart disease risk in men: the Quebec Cardiovascular Study. *Arch. Inter. Med.*, 161: 2685–2692.

Li, L., G. Wu, B.Y. Choi, B.G. Jang, J.H. Kim, G.H. Sung, J.Y. Cho, S.W. Suh and H.J. Park. 2014. A mushroom extract Piwep from *Phellinus igniarius* ameliorates experimental autoimmune encephalomyelitis by inhibiting immune cell infiltration in the spinal cord. *Biomed. Res. Int.*, 2014: 218274. doi: 10.1155/2014/218274.

Li, S., H.Y. Tan, N. Wang, Z.J. Zhang, L. Lao, C.W. Wong and Y. Feng. 2015. The role of oxidative stress and antioxidants in liver diseases. *Int. J. Mol. Sci.*, 16(11): 26087–26124.

Li, S., G. Zhang, Q. Zeng, Z. Huang, Y. Wang, T. Dong and K.W.K. Tsim. 2006. Hypoglycemic activity of polysaccharide, with antioxidation, isolated from cultured Cordyceps mycelia. *Phytomedicine*, 13: 428–433.

Li, X., Y. Xue, L. Pang, B. Len, Z. Lin, J. Huang, Z. ShangGuan and Y. Pan. 2019. Agaricus bisporus-derived β-glucan prevents obesity through PPAR γ downregulation and autophagy induction in zebrafish fed by chicken egg yolk. *Int. J. Biol. Macromol.*, 125: 820–828.

Lin, X., L. Ma, S.B. Racette, C.L. Anderson Spearie and R.E. Ostlund. 2009. Phytosterol glycosides reduce cholesterol absorption in humans. *Am. J. Physiol. Gastrointest. Liver Physiol.*, 296: G931–G935.

Liu, X., N.-K. Kopparapu, X. Shi, Y. Deng, X. Zheng and J. Wu. 2015a. Purification and biochemical characterization of a novel fibrinolytic enzyme from culture supernatant of *Cordyceps militaris*. *J. Agric. Food. Chem.*, 63: 2215–2224.

Liu, Z., Q. Wang, J. Cui, L. Wang, L. Xiong, W. Wang and C. Mao. 2015b. Systemic screening of strains of the lion's mane medicinal mushroom *Hericium erinaceus* (Higher Basidiomycetes) and its protective effects on Aβ-Triggered neurotoxicity in PC12 cells. *Int. J. Med. Mushrooms*, 17(3): 219–229.

Loos, R.J. and G.S. Yeo. 2022. The genetics of obesity: From discovery to biology. *Nat. Rev. Genet.*, 23: 120–133.

Lu, W.J., S.C. Lin, C.C. Lan, T.Y. Lee, C.H. Hsia, Y.K. Huang and J.R. Sheu. 2014. Effect of *Antrodia camphorata* on inflammatory arterial thrombosis-mediated platelet activation: The pivotal role of protein kinase C. *Sci. World J.*, 2014: 745802.

Lu, Z.-M., W.-Y. Tao, X.-L. Zou, H.-Z. Fu and Z.-H. Ao. 2007. Protective effects of mycelia of *Antrodia camphorata* and *Armillariella tabescens* in submerged culture against ethanol-induced hepatic toxicity in rats. *J. Ethnopharmacol.*, 110: 160–164.

Ma, X., M. Yang, Y. He, C. Zhai and C. Li. 2021. A review on the production, structure, bioactivities and applications of *Tremella polysaccharides*. *Int. J. Immunopathol. Pharmacol.*, 35: 20587384211000541.

Malik, K.S., K.A. Adoubi, J. Kouame, M. Coulibaly, M.-L. Tiade, S. Oga, M. Ake, O. Ake and L. Kouadio. 2022. Prevalence and risks factors of prehypertension in africa: A systematic review. *Ann. Glob. Health.*, 88(1): 13. doi: 10.5334/aogh.2769.

Mamdouh, H., W.K. Alnakhi, H.Y. Hussain, G.M. Ibrahim, A. Hussein, I. Mahmoud, F. Alawadi, M. Hassanein, M. Abdullatif and K. AlAbady. 2022. Prevalence and associated risk factors of hypertension and pre-hypertension among the adult population: Findings from the Dubai Household Survey, 2019. *BMC Cardiovasc. Disord.*, 22: 1–9.

Martel, J., D.M. Ojcius, C.-J. Chang, C.-S. Lin, C.-C. Lu, Y.-F. Ko, S.-F. Tseng, H.-C. Lai and J.D. Young. 2017. Anti-obesogenic and antidiabetic effects of plants and mushrooms. *Nat. Rev. Endocrinol.*, 13: 149–160.

Martin, K.R. 2010. The bioactive agent ergothioneine, a key component of dietary mushrooms, inhibits monocyte binding to endothelial cells characteristic of early cardiovascular disease. *J. Med. Food*, 13: 1340–1346.

Meldrum, D.R., M.A. Morris and J.C. Gambone. 2017. Obesity pandemic: Causes, consequences, and solutions—but do we have the will? *Fertil. Steril.*, 107: 833–839.

Mishra, V.K., S. Srivastava, T. Muhammad and P. Murthy. 2022. Relationship between tobacco use, alcohol consumption and non-communicable diseases among women in India: evidence from National Family Health Survey-2015–16. *BMC Public Health*, 22: 1–12.

Modjadji, P., M. Bokaba, K.E. Mokwena, T.S. Mudau, K.D. Monyeki and P.M. Mphekgwana. 2022. Obesity as a risk factor for hypertension and diabetes among truck drivers in a logistics company, South Africa. *Appl. Sci.*, 12(3): 1685. doi 10.3390/app12031685.

Mohammed Nawi, A., Z. Mohammad, K. Jetly, M.A. Abd Razak, N.S. Ramli, W.A.H. Wan Ibadullah and N. Ahmad. 2021. The prevalence and risk factors of hypertension among the urban population in southeast Asian countries: A systematic review and meta-analysis. *Int. J. Hypertens*, 2021: 6657003.

Mori, K., Y. Obara, M. Hirota, Y. Azumi, S. Kinugasa, S. Inatomi and N. Nakahata. 2008. Nerve growth factor-inducing activity of *Hericium erinaceus* in 1321N1 human astrocytoma cells. *Biol. Pharm. Bull.*, 31: 1727–1732.

Mustafa, F., H. Chopra, A.A. Baig, S.K. Avula, S. Kumari, T.K. Mohanta, M. Saravanan, A.K. Mishra, N. Sharma and Y.K. Mohanta. 2022. Edible mushrooms as novel myco-therapeutics: Effects on lipid level, obesity, and BMI. *J. Fungi.*, 8: 211.

Nabubuya, A., J. Muyonga and J. Kabasa. 2010. Nutritional and hypocholesterolemic properties of *Termitomyces microcarpus* mushrooms. *Afr. J. Food. Agric. Nutr. Dev.*, 10(3): doi 10.4314/ajfand. v10i3.54081.

Nazaruk, J. and M. Borzym-Kluczyk. 2015. The role of triterpenes in the management of diabetes mellitus and its complications. *Phytochem. Rev.*, 14: 675–690.

Nobushi, Y., Y. Hamada and K. Yasukawa. 2013. Inhibitory effects of the edible mushroom Flammulina velutipes on lipid accumulation in 3T3-L1 cells. *J. Pharm. Nutr. Sci.*, 3: 222–227.

Nugroho, P., H. Andrew, K. Kohar, C.A. Noor and A.L. Sutranto. 2022. Comparison between the world health organization (WHO) and international society of hypertension (ISH) guidelines for hypertension. *Ann. Med.*, 54: 837–845.

Oba, K., S. Teramukai, M. Kobayashi, T. Matsui, Y. Kodera and J. Sakamoto. 2007. Efficacy of adjuvant immunochemotherapy with polysaccharide K for patients with curative resections of gastric cancer. *Cancer. Immunol. Immunother.*, 56: 905–911.

Oh, T.W., Y.A. Kim, W.J. Jang, J.I. Byeon, C.H. Ryu, J.O. Kim and Y.L. Ha. 2010. Semipurified fractions from the submerged-culture broth of *Agaricus blazei* Murill reduce blood glucose levels in streptozotocin-induced diabetic rats. *J. Agric. Food. Chem.*, 58: 4113–4119.

Ohwada, S., T. Ogawa, F. Makita, Y. Tanahashi, T. Ohya, N. Tomizawa, Y. Satoh, I. Kobayashi, M. Izumi and I. Takeyoshi. 2006. Beneficial effects of protein-bound polysaccharide K plus tegafur/uracil in patients with stage II or III colorectal cancer: Analysis of immunological parameters. *Oncol. Rep.*, 15: 861–868.

Otte, J.L., J.S. Carpenter, S. Manchanda, K.L. Rand, T.C. Skaar, M. Weaver, Y. Chernyak, X. Zhong, C. Igega and C. Landis. 2015. Systematic review of sleep disorders in cancer patients: Can the prevalence of sleep disorders be ascertained? *Cancer. Med.*, 4: 183–200.

Patel, S. and Goyal, A. 2012. Recent developments in mushrooms as anti-cancer therapeutics: A review. *3 Biotech.*, 2: 1–15.

Pejin, B., K. Tešanović, D. Jakovljević, S. Kaišarević, F. Šibul, M. Rašeta and M. Karaman. 2019. The polysaccharide extracts from the fungi *Coprinus comatus* and *Coprinellus truncorum* do exhibit AChE inhibitory activity. *Nat. Prod. Res.*, 33(5): 750–754.

Phan, C.W., P. David and V. Sabaratnam. 2017. Edible and medicinal mushrooms: Emerging brain food for the mitigation of neurodegenerative diseases. *J. Med. Food.*, 20(1): 1–10.

Punthakee, Z., R. Goldenberg and P. Katz. 2018. Definition, classification and diagnosis of diabetes, prediabetes and metabolic syndrome. *Can. J. Diabetes*, 42: S10–S15.

Ren, D., Y. Zhao, Y. Nie, X. Lu, Y. Sun and X. Yang. 2014. Chemical composition of *Pleurotus eryngii* polysaccharides and their inhibitory effects on high-fructose diet-induced insulin resistance and oxidative stress in mice. *Food Funct.*, 5(10): 2609–2620.

Saini, V. 2010. Molecular mechanisms of insulin resistance in type 2 diabetes mellitus. *World J. Diabetes*, 1(3): 68–75.

Sakamoto, J., S. Morita, K. Oba, T. Matsui, M. Kobayashi, H. Nakazato, Y. Ohashi and Meta-Analysis Group of the Japanese Society for Cancer of the Colon Rectum. 2006. Efficacy of adjuvant immunochemotherapy with polysaccharide K for patients with curatively resected colorectal cancer: a meta-analysis of centrally randomized controlled clinical trials. *Cancer Immunol. Immunother.*, 55(4): 404–411.

Salem-Sokhn, E., A. Salami, M. Fawaz, A.H. Eid and S. El Shamieh. 2021. *Helicobacter Pylori* interacts with serum vitamin D to influence hypertension. *Curr. Aging Sci.*, 14(1): 26–31.

Seo, H.S. and M.H. Choi. 2015. Cholesterol homeostasis in cardiovascular disease and recent advances in measuring cholesterol signatures. *J. Steroid Biochem. Mol. Biol.*, 153: 72–79.

Seto, S.W., T.Y. Lam, H.L. Tam, A.L. Au, S.W. Chan, J.H. Wu, P.H. Yu, G.P. Leung, S.M. Ngai, J.H. Yeung, P.S. Leung, S.M. Lee and Y.W. Kwan. 2009. Novel hypoglycemic effects of *Ganoderma lucidum* water-extract in obese/diabetic (+db/+db) mice. *Phytomedicine*, 16(5): 426–436.

Sheng, Y., C. Zhao, S. Zheng, X. Mei, K. Huang, G. Wang and X, He. 2019. Anti-obesity and hypolipidemic effect of water extract from Pleurotus citrinopileatus in C57 BL/6J mice. *Food Sci Nutr.*, 7: 1295–1301.

Shi, Y., J. Sun, H. He, H. Guo and S. Zhang. 2008. Hepatoprotective effects of *Ganoderma lucidum* peptides against D-galactosamine-induced liver injury in mice. *J. Ethnopharmacol.*, 117(3): 415–419.

Shimada, Y., T. Morita and K. Sugiyama. 2003. Eritadenine-induced alterations of plasma lipoprotein lipid concentrations and phosphatidylcholine molecular species profile in rats fed cholesterol-free and cholesterol-enriched diets. *Biosci. Biotechnol. Biochem.*, 67(5): 996–1006.

Shukuri, A., T. Tewelde and T. Shaweno. 2019. Prevalence of old age hypertension and associated factors among older adults in rural Ethiopia. *Integr. Blood Press. Control*, 12: 23–31.

Singer-Englar, T., G. Barlow and R. Mathur. 2019. Obesity, diabetes, and the gut microbiome: An updated review. *Expert. Rev. Gastroenterol. Hepatol.*, 13: 3–15.

Singh, P.K., R. Dubey, L. Singh, N. Singh, C. Kumar, S. Kashyap, S.V. Subramanian and S. Singh. 2022. Mixed effect of alcohol, smoking, and smokeless tobacco use on hypertension among adult population in India: A nationally representative cross-sectional study. *Int. J. Environ. Res. Public Health*, 19(6): 3239. doi: 10.3390/ijerph19063239.

Smith, E., F. Ottosson, S. Hellstrand, U. Ericson, M. Orho-Melander, C. Fernandez and O. Melander. 2020. Ergothioneine is associated with reduced mortality and decreased risk of cardiovascular disease. *Heart*, 106(9): 691–697.

Sorimachi, K., K. Akimoto, Y. Ikehara, K. Inafuku, A. Okubo and S. Yamazaki. 2001. Secretion of TNF-alpha, IL-8 and nitric oxide by macrophages activated with *Agaricus blazei* Murill fractions *in vitro*. *Cell Struct. Funct.*, 26(2): 103–108.

Sun, H., P. Saeedi, S. Karuranga, K. Pinkepank, K. Ogurtsova, B.B. Duncan and D.J. Magliano. 2022. IDF Diabetes Atlas: Global, regional and country-level diabetes prevalence estimates for 2021 and projections for 2045. *J. Diabetes Res.*, 183: 109119.

Sun, L. and Z. Niu. 2020. A mushroom diet reduced the risk of pregnancy-induced hypertension and macrosomia: a randomized clinical trial. *Food Nutr. Res.*, 64: 10.29219/fnr.v64.4451.

Swinburn, B.A., G. Sacks, K.D. Hall, K. McPherson, D.T. Finegood, M.L. Moodie and S.L. Gortmaker. 2011. The global obesity pandemic: shaped by global drivers and local environments. *Lancet*, 378: 804–814.

Tang, N., J. Ma, R. Tao, Z. Chen, Y. Yang, Q. He, Y. Lv, Z. Lan and J. Zhou. 2022. The effects of the interaction between BMI and dyslipidemia on hypertension in adults. *Sci. Rep.*, 12(1): 927. doi: 10.1038/s41598-022-04968-8.

Tangen, J.M., A. Tierens, J. Caers, M. Binsfeld, O.K. Olstad, A.M. Trøseid, J. Wang, G.E. Tjønnfjord and G. Hetland. 2015. Immunomodulatory effects of the *Agaricus blazei* Murrill-based mushroom extract AndoSan in patients with multiple myeloma undergoing high dose chemotherapy and autologous stem cell transplantation: a randomized, double blinded clinical study. *BioMed. Res. Int.*, 2015: 718539.

Thyagarajan-Sahu, A., B. Lane and D. Sliva. 2011. ReishiMax, mushroom based dietary supplement, inhibits adipocyte differentiation, stimulates glucose uptake and activates AMPK. *BMC Complement. Altern. Med.*, 11: 1–14.

Tozo, T.A., B.O. Pereira, F.J.D. Menezes Junior, C.M. Montenegro, C. M.M. Moreira and N. Leite. 2022. Family history of hypertension: Impact on blood pressure, anthropometric measurements and physical activity level in schoolchildren. *Int. J. Cardiovasc. Sci.*, 35: 382–390.

Tran, H.B., A. Yamamoto, S. Matsumoto, H. Ito, K. Igami, T. Miyazaki, R. Kondo and K. Shimizu. 2014. Hypotensive effects and angiotensin-converting enzyme inhibitory peptides of Reishi (*Ganoderma lingzhi*) auto-digested extract. *Molecules*, 19(9): 13473–13485.

Tsai, M.Y., Y.C. Hung, Y.H. Chen, Y.H. Chen, Y.C. Huang, C.W. Kao, Y.L. Su, H.H. Chiu and K.M. Rau. 2016. A preliminary randomised controlled study of short-term *Antrodia cinnamomea* treatment combined with chemotherapy for patients with advanced cancer. *BMC complement. Med. Ther.*, 16(1): 322.

Tzeng, T.T., C.C. Chen, C.C. Chen, H.J. Tsay, L.Y. Lee, W.P. Chen and Y.J. Shiao. 2018. The cyanthin diterpenoid and sesterterpene constituents of *Hericium erinaceus* mycelium ameliorate Alzheimer's disease-related pathologies in APP/PS1 transgenic mice. *Int. J. Mol. Sci.*, 19(2): 598.

Uffelman, C., Y. Wang, E. Davis, N. Chan and W. Campbell. 2022. Effects of mushroom consumption on cardiometabolic disease risk factors: A Systematic review of randomized controlled trials. *Curr. Dev. Nutr.*, 6(Supplement_1): 51. doi: 10.1093/cdn/nzac047.051.

Urruticoechea, A., R. Alemany, J. Balart, A. Villanueva, F. Vinals and G. Capella. 2010. Recent advances in cancer therapy: An overview. *Current Pharmaceut. Design*, 16(1): 3–10.

Valenzuela, P.L., P. Carrera-Bastos, B.G. Gálvez, G. Ruiz-Hurtado, J.M. Ordovas, L.M. Ruilope and A. Lucia. 2021. Lifestyle interventions for the prevention and treatment of hypertension. Nature reviews. *Cardiology*, 18(4): 251–275.

Venkateshgobi, V., S. Rajasankar, W.M.S. Johnson, K. Prabu and M. Ramkumar. 2018. Neuroprotective effect of *Agaricus blazei* extract against rotenone-induced motor and nonmotor symptoms in experimental model of parkinson's disease. *Int. J. Nutr. Pharmacol. Neurol. Dis.*, 8(2): 59–65.

Wang, C., X. Gao, R.K. Santhanam, Z. Chen, Y. Chen, L. Xu, C. Wang, N. Ferri and H. Chen. 2018. Effects of polysaccharides from *Inonotus obliquus* and its chromium (III) complex on advanced glycation end-products formation, α-amylase, α-glucosidase activity and H_2O_2-induced oxidative damage in hepatic L02 cells. *Food Chem. Toxicol.*, 116(Pt B): 335–345.

Wang, J., C. Wang, S. Li, W. Li, G. Yuan, Y. Pan and H. Chen. 2017. Anti-diabetic effects of Inonotus obliquus polysaccharides in streptozotocin-induced type 2 diabetic mice and potential mechanism via PI3K-Akt signal pathway. *Biomed. Pharmacother.*, 95: 1669–1677.

Wang, L.Y., C.S. Huang, Y.H. Chen, C.C. Chen, C.C. Chen and C.H. Chuang. 2019. Anti-inflammatory effect of erinacine C on NO production through down-regulation of NF-κB and activation of Nrf2-mediated HO-1 in BV2 microglial cells treated with LPS. *Molecules*, 24(18): 3317.

Wasser, S.P. 2005. Reishi or ling zhi (*Ganoderma lucidum*). pp. 603–622. *In:* Coates, P.M., Blackman, M.R., Cragg, G.M., Levine, M., Moss, J. and White, J.D. (eds.). Encyclopaedia of Dietary Supplements. CRC Press.

WHO - World Health Organisation. 2021. "Hypertension". August 25, 2021. https://www.who.int/news-room/fact-sheets/detail/hypertension.

Widiyanto, A., S.I. Putri, A.S. Fajriah and J.T. Atmojo. 2021. Prevention of hypertension at home. *J. Qualit. Public Health*, 4(2): 301–308.

Wijaya, I.K., M.Y. Tahir, M.T. Talib, H. Tasa and S. Mulyani. 2022. The effect of brisk walking on blood pressure in hypertension patients: A literature review. *KnE Life Sci./5th International Conference in Nursing (IVCN)*, 327–334.

Wondmkun, Y.T. 2020. Obesity, insulin resistance, and type 2 diabetes: Associations and therapeutic implications. *Diabetes Metabol. Syndr. Obes.*, 13: 3611–3616.

Wong, K.H., M. Naidu, P. David, M.A. Abdulla, N. Abdullah, U.R. Kuppusamy and V. Sabaratnam. 2011. Peripheral nerve regeneration following crush injury to rat peroneal nerve by aqueous extract of medicinal mushroom *Hericium erinaceus* (Bull.: Fr) Pers. (Aphyllophoromycetideae). *Evid. Based Compl. Altern. Med.*, 2011: 580752.

Wong, W.L., M.A. Abdulla, K.H. Chua, U.R. Kuppusamy, Y.S. Tan and V. Sabaratnam. 2012. Hepatoprotective effects of *Panus giganteus* (Berk.) corner against thioacetamide-(TAA-) induced liver injury in rats. *Evid. Based Compl. Alter. Med.*, 2012: 170303.

Wu, T., Q. Shu, K. Yang, X. Xie, X. Wang, Y. Wang, A. Guo, N. Yuan, B. Zhao and B. Chi. 2015. Ameliorating effects of Inonotus obliquus on high fat diet-induced obese rats. *Acta. Biochim. Biophys. Sin.*, 47: 755–757.

Yahaya, N.F.M., M.A. Rahman and N. Abdullah. 2014. Therapeutic potential of mushrooms in preventing and ameliorating hypertension. *Trends. Food. Sci. Technol.*, 39: 104–115.

Ying, Y.M., L.Y. Zhang, X. Zhang, H.B. Bai, D.E. Liang, L.F. Ma and Z.J. Zhan. 2014. Terpenoids with alpha-glucosidase inhibitory activity from the submerged culture of *Inonotus obliquus*. *Phytochemistry*, 108: 171–176.

Zhang, C., J. Li, C. Hu, J. Wang, J. Zhang, Z. Ren and L. Jia. 2017. Antihyperglycaemic and organic protective effects on pancreas, liver and kidney by polysaccharides from *Hericium erinaceus* SG-02 in streptozotocin-induced diabetic mice. *Sci. Rep.*, 7(1): 1–13.

Zhang, H., L. Wu, E. Pchitskaya, O. Zakharova, T. Saito, T. Saido and I. Bezprozvanny. 2015. Neuronal store-operated calcium entry and mushroom spine loss in amyloid precursor protein knock-in mouse model of Alzheimer's disease. *J. Neurosci.*, 35: 13275–13286.

Zhang, R., S. Xu, Y. Cai, M. Zhou, X. Zuo and P. Chan. 2011. *Ganoderma lucidum* protects dopaminergic neuron degeneration through inhibition of microglial activation. *Evid. Based Compl. Altern. Med.*, 2011: 156810.

Zhao, H., Q. Zhang, L. Zhao, X. Huang, J. Wang and X. Kang. 2012. Spore powder of *Ganoderma lucidum* improves cancer-related fatigue in breast cancer patients undergoing endocrine therapy: A pilot clinical trial. *Evid. Based Complemnt. Altern. Med.*, 2012: 809614.

Zhu, B., Y. Li, T. Hu and Y. Zhang. 2019. The hepatoprotective effect of polysaccharides from *Pleurotus ostreatus* on carbon tetrachloride-induced acute liver injury rats. *Int. J. Biol. Macromol.*, 131: 1–9.

Zhu, K., S. Nie, C. Li, S. Lin, M. Xing, W. Li, D. Gong and M. Xie. 2013. A newly identified polysaccharide from *Ganoderma atrum* attenuates hyperglycemia and hyperlipidemia. *Int. J. Biol. Macromolec.*, 57: 142–150.

Zimmet, P.Z. 2017. Diabetes and its drivers: the largest epidemic in human history? Clin. *Diabetes and Endocrinol.*, 3: 1. doi.org/10.1186/s40842-016-0039-3.

15

Functional Role of Macrofungi as Prebiotics and Health Perspectives

Amira M Galal Darwish,[1,] Adel Aksoy,[2]*
Heba FFM Idriss,[3] Maizatulakmal Yahayu[4]
and Hesham Ali El Enshasy[4,5,6]

1. Introduction

Mushrooms are basidiomycetes that are large enough to be seen with a characteristic fruiting body. Different varieties of mushrooms show different colors, surfaces, and bioactivities. In a cold moist environment, they usually grow on soil surfaces and leaves or on their food sources like decaying wood. Mushrooms are well-known as beneficial food with low-calorie sources of fiber, vitamins, minerals and amino acids. They also contain a variety of active compounds that may have nutraceutical and medicinal effects (Firenzuoli et al. 2008). At least 2000 species of macrofungi

[1] Department of Food Technology, Arid Lands Cultivation Research Institute (ALCRI), City of Scientific Research and Technological Applications (SRTA-City), 21934 Alexandria, Egypt.
[2] Eskil Vocation of High School, Laboratory Veterinary Science, Aksaray University, Aksaray, Turkey.
[3] Department of Soil, Plant and Food Sciences, Bari Aldo Moro University, Bari, Italy.
[4] Institute of Bioproduct Development (IBD), Universiti Teknologi Malaysia (UTM), Johor Bahru, Johor, Malaysia.
[5] School of Chemical and Energy Engineering, Faculty of Engineering, Universiti Teknologi Malaysia (UTM), Johor Bahru, Johor, Malaysia.
[6] City of Scientific Research and Technology Application (SRTA-City), New Burg Al Arab, Alexandria, Egypt.
Emails: adilaksoy@aksaray.edu.tr.; h.idriss1@studenti.uniba.it; maizatul@ibd.utm.my; henshasy@ibd.utm.my
* Corresponding author: amiragdarwish@yahoo.com/adarwish@srtacity.sci.eg

are known as edible species including genera *Agaricus, Auricularia, Cordiceps, Flammulina, Grifola Hericium, Lactarius, Lentinus Pisolithus, Pleurotus, Russula* and *Tramella* (Assemie and Abaya 2022).

Around the world, polysaccharide-rich fungi have been used to maintain healthy bowel function, blood sugar and lipid levels, and improve the immune system, inflammation, and cancers. Due to their non-toxic and environmentally friendly qualities, mushrooms with their nutraceutical components are employed as prebiotics in food processing, aquaculture, poultry and pharmaceutical industries. Some parts of the mushrooms including mycelium, stem, dried powder extract and polysaccharides have also been used as growth enhancers, immunostimulants, improving enzymes activity in the digestive system, antimicrobials and enhancing health conditions (Mohan et al. 2022).

This chapter focused on the benefits of macrofungi and their derivatives as prebiotics including medicinal mushrooms, which may prevent/treat various disorders due to different types of polysaccharides/ derivatives that can be considered as potential dietary supplements or nutraceuticals. Future opportunities are based on revealing novel bioactive compounds, considering safety in the edible mushroom species that exert functional and pharmacological effects due to functional metabolites.

2. History of Mushroom Diet as "Food of the Gods"

For decades, people have eaten mushrooms as one of the key strategies for prehistoric communities to survive as a rich source of food and medicine. Mushrooms were revered in ancient Egypt as immortality plants and were given by God Osiris to human beings. Due to their unique flavor, mushrooms were declared to be a delicacy solely fit for Egyptian royalty. Commoners were not only barred from eating them, but also from touching them (Buller 1914, Kotowski 2019). A research study in Max Planck Institute for Evolutionary Anthropology in Leipzig, showed that mushrooms were consumed by human as early as the Upper Palaeolithic Period and was first discovered from the examination of dental calculus from the teeth found in the Lower Magdalenian burial of a woman namely "Red Lady", where the ochre covered remains were found in 2010 at El Miron cave in Cantabria (located in Northern Spain) (Power et al. 2015). The discovery of fungi spores that produce mushrooms (boletes and agarics) points to the intentional consumption of fungus, especially given that both of these groups contain a variety of edible and therapeutic mushrooms. Due to their desirable sensory, alluring culinary qualities, and medicinal attributes. The Chalcolithic Tyrolean Iceman "Ötzi" carried several types of fungi (Kotowski 2019).

Since ancient times, mushrooms are well-known as the "Gods' food" and normally presented during festive seasons, meanwhile, the Vikings and Greeks considered that consuming mushrooms will give extra strength and passion for the preparation of war. Native American people also use the mushrooms in age-old rituals to cross the body and mental barrier. Being traditionally used by ancient people in the treatment of various diseases, *Ganoderma lucidum* is also known as Lingzhi (in China), Reishi (in Japan) and Mannentake (in Korea). It is also acknowledged for promoting health and longevity and is considered a combination of spiritual energy and a source of immortality. The Japanese also recognized this mushroom as a "10,000-year", while the Chinese have treasured this mushroom as an "elixir of life" (Power et al. 2015, El-Sheikha 2022).

3. Prebiotic Concept and Criteria

Prebiotics were first introduced as a "non-digestible food ingredient that beneficially affects the host by selectively stimulating the growth and/or activity of one or a limited number of bacteria in the colon, and thus improves host health" in 1995 by Glenn Gibson and Marcel Roberfroid (Gibson and Roberfroid 1995). This concept has defined that only selected carbohydrate groups are considered as prebiotics such as lactulose, galacto-oligosaccharide (GOS) as well as a short and long chain of β-fructans (fructo-oligosaccharides FOS and inulin). Other than that, a dietary prebiotic was also pre-defined as "a selectively fermented ingredients that resulted in specific changes in the constituents and/ or activity of the gastrointestinal microbiota, thus conferring benefit(s) upon host health" (Gibson et al. 2010) in the 6th Meeting of the International Scientific Association of Probiotics and Prebiotics (ISAPP 2008).

The classification of a compound as prebiotic is considered based on the following criteria; (i) resistance ability to the acidity of the stomach, unable to be hydrolyzed by mammalian enzymes, digestibility in the gastrointestinal tract; (ii) ability to be fermented by intestinal microbiota; (iii) ability to enhance the growth and activity of bacteria in the intestinal tract (Davani-Davari et al. 2019). The definition originated from the word "selectivity", which means the ability or potency of a prebiotic to stimulate a specific gut microbiota. Figure 1 Illustrates briefly the

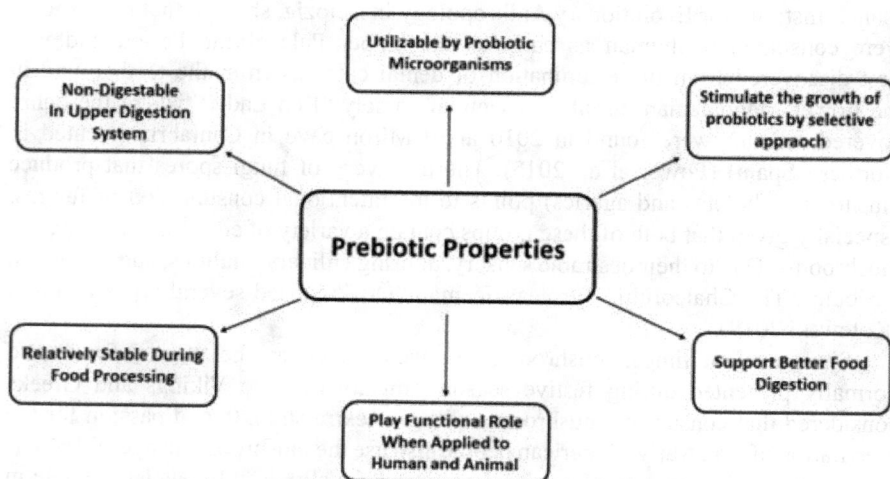

Figure 1: General requirements for potential prebiotic ingredient.

main requirements for potential prebiotics (Ramberg et al. 2010, Markowiak and Ślizewska 2017).

4. Mushrooms as Potential Prebiotics

Mushrooms contain a wide number of bioactive compounds from different classes including polysaccharides, glycoproteins, proteins, lipids, and secondary metabolites. These bioactive compounds can play a significant functional role in human and animal health. In addition, some of these compounds can increase the growth of probiotics and play a functional role as prebiotics.

4.1 Macrofungal Polysaccharides

The most active components of mushrooms that exhibit observable positive functional features such as antioxidant, immune-stimulatory, lipid-lowering, and anti-tumor action are polysaccharides including; glucose, mannose, galactose, fructose, arabinose, glucuronic acid, and β-D-glucan (Solano-Aguilar et al. 2018). However, their structural components play a crucial role in the biological actions of polysaccharides in mushrooms. The structural characterization of polysaccharides is mainly depending on the molecular weight, the arrangement of glycosidic bonds, monosaccharide identification, and the presence of functional groups. For instance, the primary chain of lentinan (from *Lentinula edodes*) is based on its main chain consisting of β-(1→3)-D-glucan (Meng et al. 2016). Meanwhile, *Grifola frondosa* polysaccharides (namely glucose, galactose, fructose, xylose, mannose and arabinose) exhibit anti-tumor potentials by stimulating macrophages and T cells. *Ganoderma lucidum* polysaccharide (GLP) chelates chromium ions, which had hypoglycemic effects. Further analyses of the function-structure relationship are useful enough to enhance further the application of mushroom polysaccharides in the hypoglycemia (Liu et al. 2022).

In Japan (1977), it was the first time to use Polysaccharopeptide Krestin (PSK) was for the treatment of gastric cancer patients as an adjuvant, while it was delayed in China till (1987). Proteoglycans (PSP) contain glucans that consist of the main chain and branches of β1–3 and β1–6. Both PSP and PSK are proteoglycans and possessed 100 kDa molecular weight but only differ from one another in the presence of glucose, mannose, fructose, xylose and galactose sugar which was previously extracted from *Trametes versicolor* mycelial culture (Bains et al. 2021).

4.2 Glycopeptides

Despite having a shorter chain of amino acids, glycopeptides share structural similarities with glycoproteins. Several naturally occurring anti-cancer mushroom polysaccharides are bonded to glycopeptides, proteoglucans, or glycoprotein remainders. The immune-stimulatory activity of therapeutic mushrooms could occur through different conjugations of polysaccharide–protein complexes or glycol conjugates. A higher level of complexity can be reached by the conjugation of polypeptides and proteins with polysaccharides. The glycopeptides isolated from *Marasmius androsaceus* showed an analgesic effect (Meng et al. 2016, Yang et al. 2019), *Pleurotus* sp. glycoprotein exhibited anti-cancer properties (Mohan et al. 2022). Table 1 shows the most potent functional compounds of mushrooms with their potential functional therapeutic roles. The polysaccharides mentioned are considered as non-toxic with no significant side effects when consumed (Singdevsachan et al. 2016).

Table 1: Potent functional compounds of mushrooms and their potential role. Modified from (Bhakta and Kumar 2013, Singdevsachan et al. 2016, Liu et al. 2022).

Functional compound	Mushroom name	Potential role
(1-3)-,(1-6) branchedglucan, a heteroglycan 9 (Fr.II) consisting of D-mannose, D-glucose, and D-galactose, a (1-6)-a-glucan and a (1-3)-, (1-6)-β-D-glucan	*Pleurotus florida*	Immunoenhancing, stimulates macrophages, splenocytes and thymocytes
- Polysaccharide fractions (IA-a and IA-b). IA-aisa1,4-α-glucan exhibiting 1,6 branching. This branching structure suggests that IA-a has a glycogen-like structure. The IA-b component appears to be an Arabinoxylan-like polysaccharide, mainly consisting of arabinose and xylose. - Lentinan [main chain consisting of β-(1→3)-D-glucan]	*Lentinula edodes*	Potently stimulated cytokine production, stimulate phagocytosis Immunomodulatory effects
Branched (1-3)-β-D-glucan	*Dictyophora indusiata*	Anti-inflammatory
Glucan	*Lentinus squarrosulus*	immunomodulatory
Grifloan (glucose, mannose, galactose, arabinose, xylose, and fructose)	*Grifola frondosa*	Antitumor activity by activating macrophages and T cells
Glucan-protein complex β-linked glucan	*Geastrum saccatum*	Anticancer
(1→3),(1→6)-D-polysaccharide, (1→3),(1→6)-D-polysaccharide	*Pleurotus ostreatus*	Increase gastrointestinal motility
β-(1→3)-D-glucan	*Sparassis crispa*	Lipid peroxidation inhibition
α-(1→3)-linked D-glucan	*Pleurotus eryngii*	Antiproliferative
β-(1→3)-D-glucan	*Termitomyces eurhizus*	Anti-aging effects
β-D-glucan	*Termitomyces microcapus*	Hepatoprotective activity
αβ-(1→4),(1→6)-glucan	*Calocybe indica*	Downregulate lipogenesis genes
(1→3)-linked glucose with branches at *O*-6	*Boletus erythropus*	Antimicrobial activity

5. Edible Macrofungi Species as Rich Source of Prebiotic

Abundant types of edible mushrooms have been explored for their potential bioactivities, but most of the species remain to be unexplored with a few available in the market. The majority of macrofungi species as edible and medicinal belong to *Pleurotus* spp. (*Pleurotus citrinopileatus*, *P. djamor*, *P. ostreatus*, *P. pulmonarius* and *P. abalonus*), *Lentinus* spp. (*Lentinus dicholamellatus* and *L. squarrosulus*), *Agaricus* spp. (*Agaricus bisporus*, *A. bitorquis* and *A. campestris*), *Boletus* spp. (*Boletus badius* and *B. edulis*), *Terfezia* spp. (*Terfezia arenaria*, *T. boudieri*, *T. claveryi* and *T. leptoderma*), *Armillaria* spp. (*Armillaria mellea* and *A. matsutake*), *Auricularia* spp. (*Auricularia auricula-judae*, *A. polytricha* and *A. auricular*), *Morchella* spp. (*Morchella hortensis* and *M. esculenta*), *Hericium* spp. (*Hericium erinaceus* and *H. coralloides*), *Grifola* spp. (*Grifola frondosa* and *G. gargal*), *Ganoderma lucidum*, *Termitomyces fuliginosus*, *Lycoperdon utriforme*, *Phlebopus marginatus*, *Coriolus versicolor*, *Cantharellus cibarius*, *Lactarius deliciosus*, *Tuber melanosporum*, *Ustilago maydis*, *Flammulina uelutipes*, *Pholiota nameko*, *Tricholoma matsutake*, *Coprinus fimetarius*, *Cordyceps sinensis*, *Lentinula edodes*, *Voluariella uoluacea*, *Stropharia rugoso-annulata*, *Tremella fuciformis*, *Calocybe indica*, *Delastria rosea*, *Tirmania pinoyi* and *Schizophyllum commune* (Læssøe and Hansen 2007, Karun and Sridhar 2017, Sawangwan et al. 2018). Figures 2 and 3 show the most common types of cultivable and wild mushrooms.

5.1 Pleurotus

Many mushroom species are either cultivated or found as wild mushrooms in various ecosystems. The genus *Pleurotus* was identified in 1775, and the oyster mushroom was added to it by the German mycologist Paul Kummer in 1871. The majority of Europe, including the UK, Ireland, and other countries, are home to *Pleurotus ostreatus*. It is found in several areas of North America and many regions of Asia, including Japan (El-Ramady et al. 2022).

Lentinus species is type of mushroom which is widely cultivated around the world, while the subgenus such as *Lentinus crinitus* is natively grown all over Americas and Colombia. In addition to high fiber content (57.18%), *Lentinus crinitus* contains relatively high protein (14.42%) comparing to other macrofungi; namely *Ganoderma lucidum* (8.59%), *Lentinus brunneofloccosus* (8.13%) and *Pleurotus ostreatus* (9.56%) (Dávila et al. 2020).

5.2 Agaricus

Agaricus sp. such as *Agaricus blazei* Murrill (ABM) was first cultivated in Japan due to its claimed health benefits and extensively used in oriental countries as a functional food and natural remedy, especially for cancer prevention and treatment (Firenzuoli et al. 2008). Umami taste peptides like Gly-Leu-Pro-Asp and Gly-His-Gly-Asp isolated from the mushroom *Agaricus bisporus* are reported to act as

Figure 2: Some edible mushroom species. (1) *Auricularia auricula-judae* (Wood Ear or Jelly Ear fungus), (2) *Lentinus edodes* (Shiitake mushrooms), (3) *Pleurotus citrinopileatus* (Golden Oyster Mushrooms), (4) *Pleurotus djamor* (pink oyster), (5) *Pleurotus ostreatus* (oyster mushroom), (6) *Pleurotus pulmonarius*, (7) *Termitomyces fuliginosus*, (8) *Lycoperdon utriforme* (Puffball mushroom), (9) *Lentinus squarrosulus*, (10) *Phlebopus marginatus*, (11) *Ganoderma lucidum* (Reishi mushroom), (12) *Agaricus bisporus*.

key molecules for kokumi taste. Kokumi's distinctive flavor is best characterized by mouthfeel, complexity, and continuity. When added to a chicken broth that is otherwise tasteless, kokumi taste compounds can improve the umami, sweet, and salty flavors (Das et al. 2021).

5.3 Ganoderma

Ganoderma sp. appears in trees, decaying woods or logs and remain in tropical and moderate weather countries which is mainly applicable in Traditional Chinese Medicine (TCM) due to the chemical constituents present. Polysaccharides possessed

Figure 3: Wild edible mushrooms are grown on wood/monocot stub. (a) *Amylosporus campbellii*, (b) *Coprinus disseminatoides*, (c) *Filoboletus manipularis*, (d) *Gyrodontium sacchari*, (e) *Hericium cirrhatum*, (f) *Lentinus squarrosulus*, (g) *Pleurotus cornucopiae*, (i, h) *Pleurotus eöus*, (j) *Polyporus arcularius*, (k) *Royoporus spathulatus* (Karun and Sridhar 2017).

an ultimate therapeutic potential in *G. lucidum* compared to other biopolymers namely proteins and nucleic acid including (1→3), (1→6)-a/β-glucans, glycoproteins, and water-soluble heteropolysaccharides. Extensive work on the bioactivity of these polysaccharides has been conducted to evaluate the anti-hypoglycemic effect, antitumor or anticancer, anti-fatigue, immunomodulatory properties, antioxidant, anti-inflammatory, anti-hypolipidemic and anti-decrepitude (longevity) properties. *G. lucidum* polysaccharides and glycoproteins potential role are concluded in Figure 4 (Liu et al. 2014, Li et al. 2021).

5.4 *Cordyceps*

Cordyceps sp. has been used as traditional Chinese medicine as well which named "soft gold" due to its high market value in China. Polysaccharides compounds in *C. sinensis* various showed pharmacological functions in *in vivo* and *in vitro* models respectively to improve growth and immunological functions (Mohan et al. 2022). *Cordyceps militaris* were also found to exhibit nutritional and therapeutic properties due to the presence of bioactive compounds namely cordycepic acid, cordycepin, fibrinolytic enzyme, ergosterol, xanthophylls (cordyxanthin-I, crodyxanthin-II, cordyxanthin-III and cordyxanthin-IV), adenosine, superoxide dismutase and organic selenium (Omak and Yilmaz-Ersan 2022).

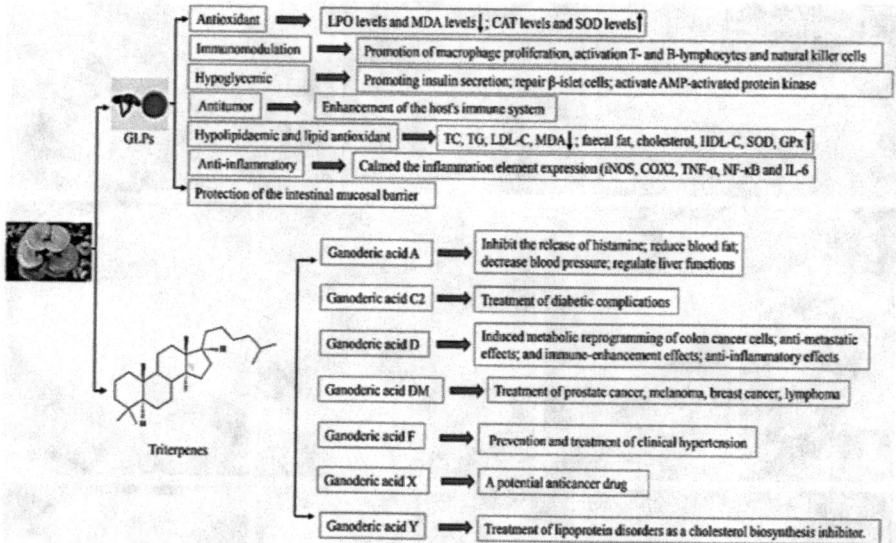

Figure 4: *Ganoderma lucidum* polysaccharides and glycoproteins potential role (Li et al. 2021).

5.5 *Schizophyllum*

The edible *Schizophyllum commune* (Split gill mushroom) is widely scattered around the world which happen to be in grey-whitish colour with the fan-shaped. The homopolysaccharides comprises of a linear chain of β-D-(1-3)-gluco-pyranosyl groups and β-d-(1-6)-glucopyranosyl groups produced pharmaceutical as immunomodulatory, antioxidant and anticancer, its polysaccharides (α- and β-1, 3-glucans) exhibited antioxidant and hypoglycemic potential and various biological functions agents (Ediriweera et al. 2015, Kumar et al. 2017).

5.6 *Flammulina*

A velvet shank mushroom namely *Flammulina velutipes* is also one of the common edible macrofungi which are widely cultivated species in Japan and Asian countries. *Flammulina velutipes* is an excellent source of protein, vitamins, minerals, unsaturated fatty, acidstriterpenes, polysaccharides and sterols. Its non-toxic polysaccharides CHFVP-1 (24.44 kDa) and CHFVP-2 (1,497 kDa) have anti-inflammatory, antioxidant, anti-aging, immune-modulatory and promote intestinal motility (Mahfuz et al. 2019, Liang et al. 2022).

5.7 *Grifola*

Grifola frondosa which is also known as the Maitake mushroom is a well-known edible mushroom native to China and Japan, respectively. In Chinese, it is known as the grey tree flower (hui-shu-hua) and this mushroom is grown in the northeastern part of Japan and eastern North America, Europe, Asia, the United States and

Canada. Other names for this mushroom include the king of mushrooms, sheep's head, hen-of-the-woods and also cloud mushroom. The β-glucan extract has been linked to a number of biological actions, including antibacterial, anticancer, antioxidant, and immunomodulatory properties. Polysaccharides from *Grifola frondosa* and *Grifola gargal* in daily nutrition showed improvement in feed consumption and enhance the performance of growth. Their D-fraction, a β-glucan complex is important in commercially complementary medicine and health products. In addition, other polysaccharides fractions were obtained from *G. frondosa* like X-fraction, MD-fraction, MZ-fraction, Grifolan and MT-α-glucan. These fractions gave several bioactivities properties such as an antitumor, antivirus, anti-diabetic, anti-inflammatory, immunomodulation, decreasing the glucose and fatty acid levels in blood-modified gut microbiota which support the antitumor effects of polysaccharides (Wu et al. 2021, Mustafa et al. 2022).

5.8 Hericium

Hericium erinaceum (Monkey head or Lion's Mane mushroom), is another edible mushroom with medicinal macrofungus spread across Europe, America and Asian countries and traditionally being consumed as a health tonic and TCM for the treatment or prevention of chronic gastritis, gastric ulcers, and other diseases related to digestive-tract in Asian countries. It consists of bioactive constituents namely lectins, lipids, proteins, erinacol, hericenone, erinacine and terpenoids with high nutritional and medicinal values to improve the immune system. Its polysaccharides L-glycero-α-D-manno-heptose (HEP) are the main bioactive compounds, composed of 3-O-methylrhamnose, L-fucose, D-galactose and D-glucose (Yang et al. 2018).

The HEP is a branched-polysaccharides which containing $(1\rightarrow6)$-linked-α-D-galactopyranosyl backbone, where the fructose branch linked to the O-2 and a fraction of glucose and 3-O-methyl rhamnose remain. Its health-promoting quality as antioxidant, anti-cancer, immunomodulatory, antibacterial, and hypolipidemic actions, are what attracted the attention. In the gastrointestinal tract of the digestive system, the fermentative bacteria in the colon and digestive enzymes in both the stomach and small intestine are involved in the molecular weight alteration, chemical constituents, conformation and structure, and thus facilitating intestinal absorption and bioactivity of its polysaccharides. Therefore, numerous studies have used simulated digestion of the gastrointestinal tract to assess the stability, bioactivity, and bioavailability of polysaccharides (Yang et al. 2018).

5.9 Calocybe

Calocybe indica (Milky white mushroom) has been used in many disciplines such as; pharmacology, antibiotics, and the food industry to create functional, nutraceutical foods and dietary supplements (Shashikant et al. 2022). *Calocybe* polysaccharide was reported for immunomodulating and cytotoxic potentials by increased antioxidant enzyme levels catalase (CAT) and superoxide dismutase (SOD) and total carotenoid levels (Das et al. 2021). The demand for this mushroom is elevating due to its nutraceutical value. The antibacterial activities of against different pathogenic and opportunistic pathogenic bacterial species were documented (Datta et al. 2020).

5.10 *Coriolus*

The *Coriolus* sp. (Turkey tail mushroom) is belong to the family of Basidiomycetes and is a polypore mushroom that is cultivated in a variety of hues. The polysaccharide peptides (PSPs) and polysaccharide krestin (PSK) are commonly used in China and Japan for a number of health-stimulating properties, including antioxidant, antibacterial, antitumor, immunomodulatory, and anticancer (Mohan et al. 2022).

5.11 *Truffles*

Mushrooms rose to the rank of a luxury commodity reserved for royalty. Affluent Romans relished truffles (*Tuber* spp. *P. Micheli* ex F.H. Wigg.). Pacific Northwest (PNW) forests harbor over 350 truffle species in 55 genera. In order to uncover a truffle, collectors must remove some of the forest duffs, as truffles are harder to be found than mushrooms. The main source of food for small mammals is truffles (Liu et al. 2018). Culinary edible truffles are predominantly used for flavoring, their aroma differs by species and they must be smelled to be appreciated. They are described as "fungal," "musky," "garlicky," "cheesy," "earthy," and "fruity". As their perfume is rapidly lost during cooking, truffles are frequently added just before the food is served and truffle-flavored goods are only lightly cooked. They provide more fragrance than flavor to a dish.

Any recipe that contains fat can benefit from adding truffles since some of its fragrant chemicals are fat-soluble. To add taste to pasta meals, soups, omelettes, dips, spreads, and ice cream, truffle flavor is infused into olive oil, butter, eggs, meat, cheese, and cream. Truffles take a long time to mature and only provide strong aromas when they are fully ripe. Due to their short maturation period and short shelf life, they are swiftly picked, marketed, and consumed. They need to be well-cleansed with a brush before being used as they grow underground, which could reduce their shelf life. Truffles are stored in a variety of ways, such as in waxy-paper bags or nestled in non-cooked rice, to allow them to "breathe" (Trappe et al. 2009). Some truffle species are exhibited in Figure 5.

6. Mushroom Diet for Weight Control

The number of obesity case increases exceeding threefold between 1980 to 2014. In 2014 itself, more than 600 million people were considered as obese and it is predicted that in 2030, around 1.12 billion people will experience the obesity. Either in the well-developed countries or in lower-middle-income nations; people tend to die because of overweight and obesity problem. Lack of exercise especially during the pandemic lockdown (2019–2021), and the consumption of high calories foods

Figure 5: Different types of truffle species. (a) *Tuber aestivum*, stereothecia, (b) *Balsamia polysperma*, ptychothecia, (c) *Fischerula macrospora*, solid ptychothecium, (d) *Choiromyces venosus*, solid ptychothecium, (e) *Tuber rufum*, stereothecia, (f) *Hydnotrya tulasnei*, ptychothecia.

are the main causes of obesity. Starchy foods are one of the main energy sources of the human daily diet intake despite of their role in human nutrition in current years where the excessive consumption of processed starch has shown to be associated with diverse health problem including obesity. The dietary mushrooms that have functional and medicinal features are highly associated with obesity and diabetes (Ganesan and Xu 2018, Zhao et al. 2022).

Mushrooms have antioxidant characteristics that support cells' antioxidant defense mechanisms and lower the risk of hypertension and dyslipidemia associated with obesity. Their regular use can aid in adjusting metabolic abnormalities, including obesity. The oyster mushroom, *P. ostreatus*, has bioactive compounds, may colonize a wide variety of lignocellulosic substrates, and grows in natural deposits including β-glucans as nutritional fibers, which aid as prebiotic in cardiometabolic health that have gained (Mustafa et al. 2022).

Ganoderma lucidum decreases overweight by modulating the microbiota as prebiotic to controlling obesity. Consuming mushrooms may help regulate microbiota, hence reducing insulin resistance associated with diabetes and obesity and preserving glucose homeostasis. In obese diabetic rats, polysaccharides from *Pleurotus tuberregium* mushrooms showed antihyperglycemic and antihyperlipidemic potential as well as a reduction in oxidative stress (Ganesan and Xu 2018).

Numerous pieces of research confirmed that mushrooms have anti-obesity properties. In clinical research, red meat was replaced in the obese subjects' usual diets with mushrooms. After evaluation for one year, less energy intake, a lower body mass index (BMI), lower body weight, small waist circumference, and lower both diastolic and systolic blood pressure were observed. The *G. gargal* anti-obesogenic effect after administration in mice was observed for 42 days and has reduced the glucose in blood, triglyceride and adipose tissue, as well as *G. lucidum* which reduced the body fats (Mustafa et al. 2022).

Studies showed that *Agaricus bisporus* supplementation has to improve colon fermentation profiles and is represented by the high level of short-chain fatty acids (SCFAs) production. It also helps in reducing the cholesterol level, serum glucose, triglyceride concentrations and liver enzyme activities (aspartate aminotransferase and alanine aminotransferase). The findings have suggested that *A. bisporus* exhibited health-promoting properties and are partially linked to the gastrointestinal microbiome and metabolic roles (Asad et al. 2020).

Basidiomycetes-X (BDM-X) has been described for a number of biological functions, and consumed traditionally by ancient people as a tea or medication to prevent or treat cancer. BDM-X (available in the market) is a material with low calories, high in dietary fibre like β-glucan which is well-known in treating obesity related disorder such as cardiovascular diseases, diabetic, and lipotoxicity of liver. BDM-X contains high antioxidant than *A. blazei* Murill and contributes to glucolipotoxicity because of the excessive reactive oxygen species and oxidative stress-activated cellular signals which are the mediators of insulin resistance and pancreatic ß-cell dysfunction (Mst et al. 2020).

7. Role of Mushroom in Gut Health Modulation

One of the elements that affect the microbiome taxonomic and functional composition is diet. Western-style, high-energy foods reduce microbial diversity and cause inflammation and autoimmune, which can result in metabolic disorders. Fiber-rich foods enhance the diversity of bacteria and their functional components, which aid in the control of lipid and glucose balance. A few days of consistent ingestion of a certain diet can change the phylogenetic makeup and functional potential of the gut microbiome. In particular, it has been demonstrated that consuming fungal fibers, which bacteria digest to produce SCFAs, improves the integrity of the gut wall, immunology, and metabolic balance (Asad et al. 2020).

The microbiota can be activated by several factors including probiotics (microorganisms that stimulate the microbiota), prebiotics (food molecules rich in oligosaccharides or polysaccharides), and synbiotics (a combination of probiotics and prebiotics). Studies concentrated on the health advantages connected to the prebiotics' synergistic selective effect on gut microbiota. Prebiotics has the ability to control the gut flora, which is crucial for controlling non-alcoholic fatty liver disease (NAFLD). The gut microbiota is abundant in protein-metabolizing enzymes that work through the manufacture of vitamin K and a number of vitamin B components, as well as microbial proteinases and peptidases in conjunction with human proteinases. Together with the innate and adaptive immune systems, the gut microbiota helps to modulate the gut immune system (Jayachandran et al. 2017).

There are several polysaccharides contained in mushroom prebiotics namely hemicellulose, chitin, α and β-glucans, mannans, xylans and galactans. *A. bisporus* (J. Lange) Imbach which belongs to Agaricaceae family (agaricomycetes) contains a high amount of protein, minerals, vitamins and dietary fibre, and thus contributes to several bioactivities such as antioxidant, anti-inflammatory, hypoglycemic, and hypo-cholesterolemic effects (Asad et al. 2020). *Grifola frondosa* polysaccharides help in increasing the lactic acid-producing bacteria (*Lactobacillus, Lactococcus, Streptococcus*) and SCFA-producing bacteria (*Allobaculum, Bifidobacterium, Ruminococcus*) (Mustafa et al. 2022).

Hericium erinaceus is rich in ketones, sinapine, glycoprotein, and alkaloids. Its polysaccharides which are the main active constituents in *H. erinaceus*, exhibited numerous pharmacological and pharmacokinetic properties, namely anti-cancer or anti-tumor, immunomodulatory activity, antioxidant, hepatoprotective activity, anti-hypolipidemic, anti-hypoglycemic, neuroprotective and neuroregenerative, anti-fatigue, and anti-aging activities, as well as protection of the gastrointestinal mucosa with their positive influences on the colonic health by altering the moisture contents, pH values, production of cecum and colonic contents by SCFA, and feces. Furthermore, *H. erinaceus* polysaccharides as prebiotics enhanced the colonization of beneficial bacteria, such as *Bifidobacteria, Faecalibacteria, Blautia, Butyricicoccus*, and *Lactobacilli* sp., which in turn has altered the abundance of *Verrucomicrobia, Firmicutes, Bacteroidetes* and *Proteobacteria* to normal levels (Li et al. 2021, Tian et al. 2022).

Mohan et al. (2022) have revealed the mechanism of mushroom polysaccharides as prebiotics to enhance the body weight, improve the digestive system, antioxidant enzymes, stress-tolerance, immune system response and immune-related gene expression levels of aquaculture animals by inhibiting the entry of microbes and decreasing the cortisol levels in host organisms.

Ganoderma sp. polysaccharides can influence the dominance and function of gut microbiota and confer kinds of benefits on their host by various metabolites such as SCFAs and branched short-chain fatty acids (BCFAs). The bioactive polysaccharide of *G. atrum* and *G. lucidum* revealed to simulate the human intestinal microbial ecosystem via microbial metabolites including acetic, propionic, butyric, and valeric acids, total SCFAs production reducing pH (Yang et al. 2022).

8. The Application of Macrofungi in Animal Nutrition

Based on the broad spectrum of functionality, the mushroom is mainly considered as animal feed in farming. The Basidiomycetes taxonomy is predominant amongst the species and rapid development of pharmacology has pushed mycotherapy to the margin among the applied treatments as natural origin effective drugs, especially with the pathogens' increased drug resistance of to antibiotics. The most popular applied species in this field are; *Agaricus* sp., *Lentinula edodes*, *Cantharellus cibarius*, and *Pleurotus* sp. mushroom, which evidenced their aid in enhancing poultry health and performance. They were reported to improve the ability to enhance immunity parameters in birds, antimicrobial, and antioxidant potentials, positively affect gut morphology and contribute as potential growth promoters (Bederska-Łojewska et al. 2017). Lentinan, a polysaccharide isolated from shiitake mushroom has exhibited antitumor activity in mice via various studied mechanisms such as preventing chemical and viral oncogenesis, suppressing cancer metastasis and recurrence in animal models and succeeding to aid in extending the overall survival rate of cancer patients, especially those having gastric and colorectal carcinomas (Singdevsachan et al. 2016).

Worldwide shortage in feed supplies encourages the use of agro-industry as feedstuffs in animal nutrition as prebiotics rich in protein, vitamins, minerals and antioxidants. Additionally, many alternative pharmaceutical products have been introduced as feed supplements in veterinary medicine from mushroom sources (Hassan et al. 2020). These bioactive substances have been efficient enough in refining and improving the efficacy of feedstuff, especially for heat-stressed or weak livestock (Yagi et al. 2019). Subsequently, after the edible parts are already collected, the residue namely spent mushroom substrate (SMS) contains 75–85% of unemployed nutrients. As one of the token sustainable environmental solutions; spent mushroom substrates (SMS) are studied for feeding dairy cattle instead of disposing of 5 million tons annually. SMS from *A. cornea*, *L. edodes*, *P. citrinopileatus*, *P. eryngii* and *P. ostreatus*, respectively which were grown on various lignocellulosic waste materials offered a promising added value by conversion from a low-cost, to nutritionally sufficient feed (Li et al. 2020).

9. Macrofungi Role in Sustainability: Perspectives and Challenges

Recently, the high demand towards macrofungi is expanded due to the popularity of the pleasant taste and specific flavor. The unique nutritional value is what makes macrofungi an important commodity especially due to high quality proteins (Niego et al. 2021). Mushrooms comprise of moisture (85–95%), carbohydrates (35–70%), protein (15–34.7%), cholesterol-free fat (10%), minerals (6–10.9%), and nucleic acids (3–8%). It also contains a large number of vitamins such as thiamine 1.4–2.2 mg (%), riboflavin 6.7–9.0 mg (%), nacin 60.6–73.3 mg (%), biotin, ascorbic acid 92–144 mg (%), pantothenic acid 21.1–33.3 mg (%), and folic acid 1.2–1.4 mg/100 g in dry weight basis. Mushrooms contain minerals such as potassium, iron, copper, zinc, manganese, essential amino acids, and unsaturated fatty acids. These unlimited nutritional benefits are accompanied by low calorie and sodium values. Being recognized as a food source a long time ago, mushrooms were also found capable of healing properties in TCM. Consequently, they can be considered as important dietary supplements (Valverde et al. 2015). For example, *Calocybe indica* is a rich source of vitamins, minerals, proteins, and amino acids, and since they are low in fat content, they make an ideal diet for heart patients. They are a good source of bioactive polysaccharides, such as β-glucans and polyphenols (flavonoids, alkaloids, and triterpenoids) and contain high antioxidant compounds and thus help in treating and preventing disorders of carcinogenesis, ageing, and physical injury, and infection, obesity, neurodegenerative diseases, and cardiovascular diseases (Shashikant et al. 2022).

The study of biotechnology provides opportunities to identify appropriate remedies for the mushroom-based accumulation of agro-waste in the environment. Mushrooms (*Pleurotus ostreatus* L.) green biotechnology includes the myco-remediation of polluted soil and water as well as bio-fermentation. The circular economy approach could be effectively achieved by using oyster mushrooms (*Pleurotus ostreatus* L.), as its culture medium is regarded as a crucial source for creating biofertilizers, animal feed, bioenergy, and bio-remediators. The production of bioethanol and enzymes such as lignin peroxidase, laccase, and manganese peroxidase can be achieved by applying mushrooms. Utilizing oyster mushrooms for the reduction, reuse, and recycling of agro-industrial wastes from the production of plant-based foods, animal-based foods, and non-food industries (El-Ramady et al. 2022).

Aiming to achieve Sustainable Development Goals (SDGs) in this field; some challenges should be considered in the near future research work that may pave the way towards enhanced health and environment:

- Mushrooms can be a good resource of protein based on the determination of protein content in several species of mushrooms, with the mechanism involved including the availability, quality and growth progress should be extensively studied in the near future since they can lower carbon footprint and provide an affordable source of high-quality protein.

- Application of the biological indicators and innovative bioassays for evaluation of toxicology of edible mushroom species together with the active metabolites' constituents for further studies.

- The expansion in fungal water-soluble polysaccharides treatments is required to provide stable compounds suitable for various applications. However, the limited information available concerning the hardness properties, stability condition in water and polysaccharides efficacy may encourage more research efforts on this point.

- For further identification of the functional role of macrofungi prebiotic role in growth stimulation and gut microbiome alteration; further clinical patterns need to be studied for evidence prove for their efficiency.

- The effects of medicinal mushrooms should be focused on in future studies.

- The use of various chemicals in solvent extraction, freeze-drying and processing of mushroom polysaccharides might contribute to side effects on the environment. The introduction to green extraction approaches will involve minimum usage of volatile organic chemicals and provide high yield and better quality than the conventional methods that need to be expanded in future research.

- Expanding large-scale field trials for consumer awareness, acceptability, cost-effectiveness, bioavailability, sustainability, and the utilization of mushrooms, adverse effects and concerns of edible species need to be the next stage of focus.

10. Conclusion

Mushrooms symbolize a growing attraction towards commercialization as food supplement/additives or drugs, thus, further expansions of these applications, this chapter discussed the benefits of macrofungi and their derivatives as prebiotics including medicinal mushrooms, which may prevent/treat various disorders due to different types of polysaccharides/derivatives that can be considered as potential dietary supplements or nutraceuticals. Future opportunities are based on revealing novel bioactive compounds, considering safety, in the edible mushroom species that exert functional and pharmacological effects due to functional metabolites, such as lectins, polyphenolics, terpenoids, ergosterols, and volatile organic compounds. Yet, research findings demonstrated that the mushroom polysaccharides can provide multiple health effects including; enhanced innate immune responses (with different identified or unidentified mechanisms), weight control, modulating the beneficial gut microbiota, which leads to protection against pathogen susceptibility. Consequently, further investigation is required to identify innovative sustainable technologies to isolate and purify these polysaccharides and their derivatives. On the other hand, the use of advanced novel ingredients to overcome future sustainable needs and challenges is in parallel with investigating the combined effects of probiotic organisms with the prebiotic bioactive molecules that clarify their synergistic effects. The safety analysis is crucial for food additives studied, considering regulation for implementation on the industrial scale, taking into account standardized labelling with high traceability to compromise the increased consumers' awareness.

References

Asad, F., H. Anwar, H.M.Yassine, M.I. Ullah, Aziz-Ul-rahman, Z. Kamran and M.U. Sohail. 2020. White button mushroom, *Agaricus bisporus* (Agaricomycetes), and a probiotics mixture supplementation correct dyslipidemia without influencing the colon microbiome profile in hypercholesterolemic rats. *Int. J. Med. Mushrooms*, 22: 235–244.

Assemie, A. and G. Abaya. 2022. The effect of edible mushroom on health and their biochemistry. *Int. J. Microbiol.*, 2022: 8744788.

Bains, A., P. Chawla, S. Kaur, A. Najda, M. Fogarasi and S. Fogarasi. 2021. Bioactives from mushroom: Health attributes and food industry applications. *Materials*, 14.

Bederska-Łojewska, D., S. Świątkiewicz and B. Muszyńska. 2017. The use of Basidiomycota mushrooms in poultry nutrition—A review. *Anim. Feed Sci. Technol.*, 230: 59–69.

Bhakta, M. and P. Kumar. 2013. Mushroom polysaccharides as a potential prebiotics. *Int. J. Heal. Sci. Res.*, 3: 77–84.

Buller, A.H.R. 1914. The fungus lore of the Greeks and Romans. *Trans. Br. Mycol. Soc.*, 5: 21–66.

Das, A.K., P.K. Nanda, P. Dandapat, S. Bandyopadhyay, P. Gullón, G.K. Sivaraman, D.J. McClements, B. Gullón and J.M. Lorenzo. 2021. Edible mushrooms as functional ingredients for development of healthier and more sustainable muscle foods: A flexitarian approach. *Molecules*, 26: 2463.

Datta, S., J. Dubey, S. Gupta, A. Paul, P. Gupta and A.K. Mitra. 2020. Tropical milky white mushroom, *Calocybe indica* (Agaricomycetes): An effective antimicrobial agent working in synergism with standard antibiotics. *Int. J. Med. Mushrooms*, 22: 335–346.

Davani-Davari, D., M. Negahdaripour, I. Karimzadeh, M. Seifan, M. Mohkam, S.J. Masoumi, A. Berenjian and Y. Ghasemi. 2019. Prebiotics: Definition, types, sources, mechanisms, and clinical applications. *Foods*, 8: 1–27.

Dávila, G.L.R., A.W. Murillo, F.C.J. Zambrano, M.H. Suárez and A.J.J. Méndez. 2020. Evaluation of nutritional values of wild mushrooms and spent substrate of *Lentinus crinitus* (L.) Fr. *Heliyon*, 6: 0–4.

Ediriweera, S.S., R.L.C. Wijesundera, C.M. Nanayakkara and O.V.D.S.J. Weerasena. 2015. Comparative study of growth and yield of edible mushrooms, *Schizophyllum commune* Fr., *Auricularia polytricha* (Mont.) Sacc. and *Lentinus squarrosulus* Mont. on lignocellulosic substrates. *Mycosphere*, 6: 760–765.

El-Ramady, H., N. Abdalla, Z. Fawzy, K. Badgar, X. Llanaj, G. Törős, P. Hajdú, Y. Eid and J. Prokisch. 2022. Green biotechnology of Oyster Mushroom (*Pleurotus ostreatus* L.): A sustainable strategy for myco-remediation and bio-fermentation. *Sustainability*, 14(8): 3667.

El-Sheikha, A.F. 2022. Nutritional profile and health benefits of *Ganoderma lucidum* "Lingzhi, Reishi, or Mannentake" as functional foods: Current scenario and future perspectives. *Foods*, 11: 1030.

Firenzuoli, F., L. Gori and G. Lombardo. 2008. The medicinal mushroom Agaricus blazei murrill: Review of literature and pharmaco-toxicological problems– *Evidence-based Complement. Altern. Med.*, 5: 3–15.

Ganesan, K. and B. Xu. 2018. Anti-obesity effects of medicinal and edible mushrooms. *Molecules*, 23: 2880.

Gibson, G. and M. Roberfroid.1995. Dietary modulation of the human colonic microbiota: Introducing the concept of prebiotics. *J. Nutr.*, 125: 1401–1412.

Gibson, G.R., K.P. Scott, R.A. Rastall, K.M. Tuohy, A. Hotchkiss, A. Dubert-Ferrandon, M. Gareau, E.F. Murphy, D. Saulnier, G. Loh, S. Macfarlane, N. Delzenne, Y. Ringel, G. Kozianowski, R. Dickmann, I. Lenoir-Wijnkoop, C.Walker and R. Buddington. 2010. Dietary prebiotics: current status and new definition. *Food Sci. Technol. Bull. Funct. Foods*, 7: 1–19.

Hassan, R.A., M.E. Shafi, K.M. Attia and M.H. Assar. 2020. Influence of Oyster mushroom waste on growth performance, immunity and intestinal morphology compared with antibiotics in broiler chickens. *Front. Vet. Sci.*, 7: 1–12.

Jayachandran, M., J. Xiao and B. Xu. 2017. A critical review on health promoting benefits of edible mushrooms through gut microbiota. *Int. J. Mol. Sci.*, 18.

Karun, N.C. and K.R. Sridhar. 2017. Edible wild mushrooms of the Western Ghats: Data on the ethnic knowledge. *Data Br.*, 14: 320–328.

Kotowski, M.A. 2019. History of mushroom consumption and its impact on traditional view on mycobiota—An example from Poland. *Microb. Biosyst.*, 4: 1–13.

Kumar Chandrawanshi, N., D. Kumar Tandia and S. Jadhav. 2017. Nutraceutical properties evaluation of *Schizophyllum Commune. Indian J. Sci. Res.*, 13: 57–62.

Læssøe, T. and K. Hansen. 2007. Truffle trouble: What happened to the Tuberales? *Mycol. Res.*, 111: 1075–1099.

Li, M., L. Yu, J. Zhao, H. Zhang, W. Chen, Q. Zhai and F. Tian. 2021. Role of dietary edible mushrooms in the modulation of gut microbiota. *J. Funct. Foods*, 83: 104538.

Li, T.H., P.F. Che, C.R. Zhang, B. Zhang, A. Ali and L.S. Zang. 2020. Recycling of spent mushroom substrate: Utilization as feed material for the larvae of the yellow mealworm *Tenebrio molitor* (coleoptera: Tenebrionidae). *PLoS One*, 15: 1–12.

Liang, Q., Q. Zhao, X. Hao, J. Wang, C. Ma, X. Xi and W. Kang. 2022. The effect of *Flammulina velutipes* polysaccharide on immunization analyzed by intestinal flora and proteomics. *Front. Nutr.*, 9: 1–17.

Liu, Q., J. Wu, P. Wang, Y. Lu and X. Ban. 2022. Neutral polysaccharides from *Hohenbuehelia serotina* with hypoglycemic effects in a Type 2 diabetic mouse model. *Front. Pharmacol.*, 13: 1–10.

Liu, W., L. Chen, Y. Cai, Q. Zhang and Y. Bian. 2018. Opposite polarity monospore genome de novo sequencing and comparative analysis reveal the possible heterothallic life cycle of *Morchella importuna*. *Int. J. Mol. Sci.*, 19: 1–22.

Liu, Y., J. Zhang, Q. Tang, Y. Yang, Q. Guo, Q. Wang, D. Wu and S.W. Cui. 2014. Physicochemical characterization of a high molecular weight bioactive β-d-glucan from the fruiting bodies of *Ganoderma lucidum*. *Carbohydr. Polym.*, 101: 968–974.

Mahfuz, S., T. He, S. Liu, D. Wu, S. Long and X. Piao. 2019. Dietary inclusion of mushroom (*Flammulina velutipes*) stem waste on growth performance, antibody response, immune status, and serum cholesterol in broiler chickens. *Animals*, 9: 1–11.

Markowiak, P. and K. Ślizewska. 2017. Effects of probiotics, prebiotics, and synbiotics on human health. *Nutrients*, 9: 1–30.

Meng, X., H. Liang and L. Luo. 2016. Antitumor polysaccharides from mushrooms: A review on the structural characteristics, antitumor mechanisms and immunomodulating activities. *Carbohydr. Res.*, 424: 30–41.

Mohan, K., D. Karthick Rajan, T. Muralisankar, A. Ramu Ganesan, K. Marimuthu and P. Sathishkumar. 2022. The potential role of medicinal mushrooms as prebiotics in aquaculture: A review. *Rev. Aquac.*, 1300–1332.

Mst, A., S. Sato and T. Konishi. 2020. Obesity preventive function of novel edible mushroom, Basidiomycetes-X (Echigoshirayukidake). Manipulations of insulin resistance and lipid metabolism. *J. Tradit. Complement. Med.*, 10: 245–251.

Mustafa, F., H. Chopra, A.A. Baig, S.K. Avula, S. Kumari, T.K. Mohanta, M. Saravanan, A.K. Mishra, N. Sharma and Y.K. Mohanta. 2022. Edible mushrooms as novel myco-therapeutics: Effects on lipid level, obesity, and BMI. *J. Fungi*, 8: 1–21.

Niego, A.G., S. Rapior, N. Thongklang, O. Raspé, W. Jaidee, S. Lumyong and K.D. Hyde. 2021. Macrofungi as a nutraceutical source: promising bioactive compounds and market value. *J. Fungi*, 7.

Omak, G. and L. Yilmaz-Ersan. 2022. Effect of *Cordyceps militaris* on formation of short-chain fatty acids as postbiotic metabolites. *Prep. Biochem. Biotechnol.*, 22: 1–9. doi: 10.1080/10826068.2022.2033992.

Power, R.C., D.C. Salazar-García, L.G. Straus, M.R. González Morales and A.G. Henry. 2015. Microremains from El Mirón Cave human dental calculus suggest a mixed plant-animal subsistence economy during the Magdalenian in Northern Iberia. *J. Archaeol. Sci.*, 60: 39–46.

Ramberg, J.E., E.D. Nelson and R.A. Sinnott. 2010. Immunomodulatory dietary polysaccharides: A systematic review of the literature. *Nutr. J.*, 9: 1–22.

Sawangwan, T., W. Wansanit, L. Pattani and C. Noysang. 2018. Study of prebiotic properties from edible mushroom extraction. *Agric. Nat. Resour.*, 52: 519–524.

Shashikant, M., A. Bains, P. Chawla and M. Fogarasi. 2022. The current status, bioactivity, food, and pharmaceutical approaches of *Calocybe indica*: A Review. *Antioxidants*, 11: 1145.

Singdevsachan, S.K., P. Auroshree, J. Mishra, B. Baliyarsingh, K. Tayung and H. Thatoi. 2016. Mushroom polysaccharides as potential prebiotics with their antitumor and immunomodulating properties: A review. *Bioact. Carbohydrates Diet. Fibre*, 7: 1–14.

Solano-Aguilar, G.I., S. Jang, S. Lakshman, R. Gupta, E. Beshah, M. Sikaroodi, B. Vinyard, A. Molokin, P.M. Gillevet and J.F. Urban. 2018. The effect of dietary mushroom agaricus bisporus on intestinal microbiota composition and host immunological function. *Nutrients*, 10: 1–16.

Tian, B., Y. Geng, T. Xu, X. Zou, R. Mao, X. Pi, W. Wu, L. Huang, K. Yang, X. Zeng and P. Sun. 2022. Digestive characteristics of *Hericium erinaceus* polysaccharides and their positive effects on fecal microbiota of male and female volunteers during *in vitro* fermentation. *Front. Nutr.*, 9: 1–17.

Trappe, J.M., R. Molina, D.L. Luoma, E. Cázares, D. Pilz, J.E. Smith, A. Michael, S.L. Miller and M.J. Trappe. 2009. Diversity, ecology, and conservation of Truffle fungi in forests of the Pacific Northwest. USDA-General Tech. Rep. PNW-GTR-772, Vol. PNW GTR 77.

Valverde, M.E., T. Hernández-Pérez and O. Paredes-López. 2015. Edible mushrooms: Improving human health and promoting quality life. *Int. J. Microbiol.*, 2015: 376387.

Wu, J.Y., K.C. Siu and P. Geng. 2021. Bioactive ingredients and medicinal values of *Grifola frondosa* (Maitake). *Foods*, 10(1): 95.

Yagi, F., Y. Minami, M. Yamada, K. Kuroda and M. Yamauchi. 2019. Development of animal feeding additives from mushroom waste media of shochu lees. *Int. J. Recycl. Org. Waste Agric.*, 8: 215–220.

Yang, L., X. Kang, W. Dong, L. Wang, S. Liu, X. Zhong and D. Liu. 2022. Prebiotic properties of *Ganoderma lucidum* polysaccharides with special enrichment of *Bacteroides ovatus* and *B. uniformis* in vitro. *J. Funct. Foods*, 92: 105069.

Yang, X., D. Xu, Q. Gao and Y. Gao. 2019. Isolation and characterization of glycopeptides with analgesic activities from *Marasmius androsaceus*. *Am. J. Plant Sci.*, 10: 1196–1205.

Yang, Y., C. Zhao, M. Diao, S. Zhong, M. Sun, B. Sun, H. Ye and T. Zhang. 2018. The prebiotic activity of simulated gastric and intestinal digesta of polysaccharides from the *Hericium erinaceus*. *Molecules*, 23: 1–14.

Zhao, H., L. Wang, M. Brennan and C. Brennan. 2022. How does the addition of mushrooms and their dietary fibre affect starchy foods. *J. Future Foods*, 2: 18–24.

Interaction with Fauna

16

Cordyceps militaris and its Applications

Katarzyna Kała, Karol Jędrejko,
Katarzyna Sułkowska-Ziaja and *Bożena Muszyńska**

1. Introduction

The *Cordyceps* genus has up to 750 identified species distributed worldwide (mainly reported from South Asia, Europe and North America). Representatives of this family are parasites of plants and insects. *Cordyceps militaris* (L.) Link (phylum Ascomycota; class Sordariomycetes; order Hypocreales; family *Cordycipitaceae*) is the second most widespread species in the genus *Cordyceps* and simultaneously the second most studied species of this genus (Kirk 2010). In English-language literature, it is known as the Scarlet Caterpillar Club. In China, it is popularly referred to as "Dong Chong Xia Cao" and is the second most commercialized species in China, Japan, and Korea.

Carl Linnaeus described this species in 1753 and designated its scientific name *Clavaria militaris*. After 80 years (1833), German mycologist Johann Heinrich Friedrich Link renamed it *Cordyceps militaris*, which is a generally accepted scientific name. Synonyms for *C. militaris* include *Clavaria militaris*, *Sphaeria militaris*, *Torrubia militaris*, *Clavaria granulosa*, *Corynesphaera militaris*, *Hypoxylon militare* and *Xylaria militaris* (Kirk 2010).

Cordyceps militaris is a parasite that attacks the larvae and pupae of nocturnal butterflies that resides in soil and litter (Figure 1A).

Jagiellonian University Medical College, Faculty of Pharmacy, Department of Pharmaceutical Botany, 9 Medyczna Str., 30–688 Kraków, Poland.
* Corresponding author: muchon@poczta.fm

Figure 1: *Cordyceps militaris* on a pupa (A) and cultivated (B) *C. militaris* in Kraków, Poland.

These larvae are hidden in the soil, decaying wood, or moss. After leaving the environment, they infect night butterfly caterpillars. The mycelium of the parasite grows at the expense of the host's internal organs and causes the host to mummify, forming a type of spore organ called a pseudosclerotium; from this, reproductive organs then grow (Kobayasi 1941, 1982, Kobayasi and Shimizu 1982). The fruiting bodies emerge from the ground, from the host pupae. Its most common hosts in the wild are the larvae (caterpillars) and pupae of various insect species belonging to the order Lepidoptera. In addition, shared hosts from other orders include *Ips sexdentatus*, *Lachnosterna quercina* and *Tenebrio molitor* (order Coleoptera), *Cimbex similis* (order Hymenoptera) and *Tipula paludosa* (order Diptera). Morphological diversity and adaptation of *C. militaris* to a wide range of host insects likely contribute to its occurrence in different geographic regions as well as ecological zones around the world (Sung and Spatafora 2004). They appear from summer to late autumn, growing singly or in groups. They are cylindrical in shape and slightly flattened. Stroma, or compact mycelium layer, is composed of tightly intertwined filamentous hyphae 20–60 mm high and 3–10 mm wide. It is divided into an infertile stem and a fertile head. The head is cylindrical and dark orange-yellow, sometimes red. Its surface is covered with pores. The stem is clearly demarcated by a different color and surface from the fertile part but is of similar width. It is clearly lighter, shiny, and light orange to ochre in color. It is more or less cylindrical. The stem remains in the soil as deep as the host larva/pupa was. The interior part of the fruiting body is white to pale orange. Filiform spores over 200 μm long, are divided into numerous secondary spores, cylindrical-spindle-shaped, smooth, hyaline 3.5–6 × 1–1.5 μm arranged one behind the other in a chain (Davis et al. 2012).

C. militaris is found mainly in deciduous and mixed forests, or at their edges, on insect larvae and pupae residing in the soil. We can find it in the lowlands and higher places, from June to December. It has a wide distribution range in North America, Asia and Europe (Breitenbach and Kränzlin 1981, Wilga 2010) (Figure 1B).

So far, *C. militaris*, is not authorized in Europe Union (EU) as an ingredient in foods. This is an unauthorised novel food that cannot be used for human consumption without prior approval from the European Commission (EC) and European Food Safety Authority (EFSA). Only *Cordyceps sinensis* is an authorized ingredient of foods in the EU. In the European Union, *C. militaris* is only found in products that are dietary supplements, while in Asia it has been a recognized food source (https://www.nutraveris.com/en/; https://webgate.ec.europa.eu/rasff-window/screen/search). In 2009, the Ministry of Health of China considered *C. militaris* as a new food resource, which can be commercially produced and used in food industries (Fenglin et al. 2020).

2. Nutraceutical Potential

It is important that both the fruit bodies as well as the mycelium of *C. militaris* can be used as nutraceuticals. The species *C. militaris* turned out to be a good source of protein as well as essential amino acids, meeting the nutritional requirements (FAO and WHO) for high nutritional quality food materials. It turns out that these proteins can serve as products that improve the nutritional properties of food, which is an alternative to plant proteins. This is particularly important as it is estimated that the requirement for high-quality protein will increase over the next twenty years. Importantly, in this species, the protein was determined in the amount of 29% (w/w), which is a higher amount compared to other edible mushrooms (*Lentinula edodes* 9.87–28.78%; *Lentinus tuberregium* 25%, truffle 20–27%, *Tricholoma matsutake* 20.3%) (Yu et al. 2021).

The total content of amino acids in fruit bodies of *C. militaris* species is 57.4 mg/g dry weight basis. Not only protein amino acids, but also non-protein amino acids were determined (Chan et al. 2015). GABA (γ-aminobutyric acid) is a non-protein amino acid synthesized in humans from glutamic acid with the enzyme glutamate decarboxylase (GAD) and vitamin B_6 (acting as a reaction cofactor). GABA is an inhibitory neurotransmitter in the central nervous system. It affects the hippocampus, hypothalamus, cerebellum, striatum, and spinal cord. The GABA regulates memory, sleep, learning and emotional processes such as anxiety as well as stress. It also shows anticonvulsant and myorelaxant activities (Boonstra et al. 2015). In the fruit bodies of *C. militaris*, the amount of GABA was found to be 756 μg/g of dry mass (Chen et al. 2012, Cohen et al. 2014). Some scientific research showed the low bioavailability of GABA after oral administration in humans. Differences were observed in the results of the pharmacokinetic and pharmacodynamic parameters. Other obstacles related to GABA include a short biological half-life and difficulty penetrating the blood-brain barrier. In medicine, only GABA derivatives have been used thus far in drugs (Hepsomali et al. 2020). Nevertheless, some scientific articles have confirmed the efficiency of oral GABA supplementation in humans. Oral administration of GABA at the dose of 100 mg/day improved sleep parameters, which is important due to the content in edible mushrooms, especially *C. militaris* (Li et al. 2015, Yamatsu et al. 2016).

Important proteinaceous substances present in *C. militaris* are also lectins, fungal immunomodulatory proteins, antifungal proteins, and other enzymes (Yu et al. 2021). Lectins are protein-sugar combinations (spare proteins) produced by mushrooms and plants. It has been shown that, compared to plant material, the role of lectins in mushrooms is more complicated. They are involved in processes at the cellular and molecular levels. In species of mushrooms such as *C. militaris*, lectins have importance depending on the growth phase of the mushrooms or the environment in which they grow. Lectins can also protect mushrooms against environmental toxins, such as bacteria, viruses and insecticides. Lectins induced interest due to their immunomodulatory activity, directly related to the stimulation of the maturation of cells of the immune system, and anti-tumor activity (Singh et al. 2014 Varrot et al. 2013). Lectins, detected in fruit bodies of *C. militaris* have been shown to possess mitogenic activity. Lectins bind to sugar residues on the surface of cells to initiate the agglutination process (Hassan et al. 2015, Singh et al. 2010). The anti-tumor effect is also possible due to the isolation of a new immunomodulatory CMIP protein that showed antimetastasis activity on a mouse model (4T1) tumor cell (Yu et al. 2021).

Another very important group of biologically active compounds is carbohydrates, including mushroom polysaccharides (Chen et al. 2013, Cohen et al. 2014). Moreover, an important factor affecting the content and chemical structure of polysaccharides is the source of *C. militaris*, including cultivation methods. Extraction methods, time and solvents used are also crucial, and directly affect the biological activity of this group of compounds (Lee et al. 2020, Zhang et al. 2019). The chemical structure is determined by the type of monosaccharides involve in the formation of the polysaccharides, their spatial configuration, linear sequence, location of glycosidic bonds, and degree of branching of the chain (Zhang et al. 2019). Mushroom polysaccharides, in addition to anticancer activity, exhibit antihyperglycemic and antihyperlipidemic effects. The *In vitro* and *in vivo* experiments demonstrated that the polysaccharides present in *C. militaris* exhibit also antioxidant, immunostimulatory, anti-inflammatory and hepatoprotective potentials (Lee et al. 2020, Zhang et al. 2019).

Two types of polysaccharides, intracellular and extracellular (IPCM, EPCM) with hypolipidemic activity, were isolated. IPCM appeared to be an α-pyran polysaccharide (average molecular weight: 32.5 kDa). It consists mainly of glucose (10.5%), mannose (51.94%) and galactose (37.25%). EPCM, on the other hand, is an α-pyran polysaccharide (average molecular weight: 20 kDa) that consists mainly of glucose (18.33%), mannose (44.51%), and galactose (35.38%) (Huang et al. 2018). In an *in vivo* study, EPCM treatment showed a potential effect on improving the serum lipid profile of mice, as seen by reducing triglycerides (TG) by 45.45%, serum total cholesterol (TC) by 20.05% and low-density lipoprotein cholesterol (LDL-C) by 52.63%, while IPCM treatment significantly reduced TG levels by 47.93%, TC by 20.74% and LDL-C by 38.25%. In addition, EPCM has been shown to alleviate hyperlipidemia. The mechanism of action is thought to be associated with an increase in serum lipoprotein lipase (LPL) expression and a decrease in hepatic 3-hydroxy-3-methylglutaryl-CoA reductase (HMGR) expression, whereas IPCM increased serum LPL expression. It has been proven that polysaccharides from *C. militaris* can be studied not only as immunostimulants but also as functional food or natural medicines to prevent hyperlipidemia (Huang et al. 2018).

Immunomodulatory activity, in turn, was tested by, among other things, induction of immune response *in vitro* after stimulating the activity of macrophages to generate interferon (IFN-γ), NO, reactive oxygen species (ROS), IL-1β, tumor necrosis factor (TNF-α), lymphocytes (T and B), and natural killer (NK) cells.

Polysaccharides also showed activity to increase phagocytosis by macrophages (Zhang et al. 2019). The activity of inhibiting the growth of tumor cells and inducing their apoptosis was also important (Zhang et al. 2019). It has been shown that total extracts obtained using water or ethanol (50%) from *C. militaris* tend to promote type 1 immunity, while total extracts obtained using ethanol (70–80%) promote type 2 immunity. The autoimmune diseases, e.g., rheumatoid arthritis related to type 1 immunity, and other diseases, e.g., asthma, atopic dermatitis, and allergic rhinitis are associated with type 2 immunity (Lee et al. 2020).

Other important polysaccharides isolated from *C. militaris* include CPS-1 composed of rhamnose, xylose, D-mannose, D-glucose, and D-galactose, which may be responsible for reducing inflammation as well as humoral immunity. However, no significant changes in the phagocytic function of innate immunity and cellular immunity were observed with this polysaccharide. An acid polysaccharide (APS) has been shown to modulate not only the expression of induced nitric oxide synthase (iNOS) and inflammatory cytokines (IL-1β, IL-6, TNF-α and IFN-γ), which may enhance antiviral activity, but also affect the expression of anti-inflammatory cytokines (IL-10), which may avoid excessive macrophage activation. A reduction in deaths due to the influenza A virus has been confirmed in *in vivo* studies. CMP-W1 and CMP-S1, on the other hand, are heteropolysaccharides that can affect the induction of splenocyte proliferation in a mouse model and have synergistic effects on concanavalin A-induced T cells as well as LPS-induced B cells (Lee et al. 2020).

One of the main carbohydrates present in *C. militaris* is D-mannitol. It belongs to polyols and is one of the very important products of metabolism. It is commonly referred to as cordycepinic acid and is a carbohydrate reserve and control factor of mushroom metabolic pathways and is a transporter of other compounds in osmoregulation. It is effective in renal failure, diuretic and antitussive in humans (Cohen et al. 2014). Glucose, mannose, and galactose are the most vital monosaccharides that form saccharide polymers. Their content varies depending on whether *C. militaris* is extracted from mycelium from *in vitro* cultures or in fruiting body form. In further scientific studies on the polysaccharides of *C. militaris*, other monosaccharides were detected (rhamnose, arabinose, and xylose) (Zhang et al. 2019).

A very important group of compounds with a health-promoting potential are fatty acids. The fatty acid profile indicates almost 70% of unsaturated fatty acids (Ashraf et al. 2020). In the species, *C. militaris*, four free sterols (campesterol, ergosterol, cholesterol, and β-sitosterol) have been determined. Also marked 10 free fatty acids, including lauric acid, stearic acid, docosanoic acid, myristic acid, palmitoleic acid, palmitic acid, pentadecanoic acid, lignoceric acid, linoleic acid, and oleic acid (Yang et al. 2009). One of the unsaturated fatty acids determined with the highest content in this species was linoleic acid (Hur 2008). Unsaturated fatty acids are effective pharmacologically active components, which are responsible for decreasing harmful lipids in blood and protecting against cardiovascular disease. Therefore it was proved that *C. militaris* is a source of the so-called "good fat" for humans (Reis et al. 2013).

Based on scientific research, it is possible to fetch information on bioactive compounds from *C. militaris*. Content of nucleosides – adenosine and cordycepin in *C. militaris* is more than that in *C. sinensis*. The analysis results showed the content of the following biologically active substances in the fruit bodies of *C. militaris*: ergothioneine and GABA, glycolipids (cerebrosides), glycoproteins (lectins), sterols (ergosterol), statins (lovastatin), xanthophylls including carotenoids (zeaxanthin and lutein), phenolic compounds (flavonoids and phenolic acids), vitamins, and bioelements (e.g., zinc, selenium, magnesium, potassium, and sulfur) (Chan et al. 2015, Chen et al. 2012, Cohen et al. 2014). Importantly, 60°C is the optimal drying temperature for *C. militaris*. A higher temperature each time reduces the content of bioactive compounds, including phenolic compounds and cordycepin (Wu et al. 2019).

Cordycepin (3'-deoxyadenosine) – a structural analog of the nucleoside adenosine is an organic water-insoluble compound. Cordycepin was discovered and isolated from *C. militaris* in 1950. Due to its close chemical structure to adenosine, it is considered a strong bioactive ingredient with nutraceutical potential. On the basis of *in vitro* and *in vivo* experiments performed, this compound has been confirmed to exhibit antitumor, anti-inflammatory, antiviral, immunostimulating, ergogenic, hypoglycemic (type II diabetes), hypolipidemic, and regulation of steroidogenesis and spermatogenesis. A few studies have described the antioxidant activity of cordycepin (Lee et al. 2020, Qin et al. 2019, Tuli et al. 2014). Cordycepin could inhibit cyclooxygenase-2 (COX-2), TNF-α, and iNOS expression in LPS-induced RAW 264.7 macrophage cells indicating its potential anti-inflammatory and anti-diabetic effect. It could regulate type II diabetes especially. Microglia (a kind of macrophage present in the brain), is an important agent in the host defense and tissue repair in the CNS. An impaired immune response might cause severe neurodegenerative ailments, such as Parkinson's disease, Alzheimer's disease, trauma, multiple sclerosis, and cerebral ischemia (Lee et al. 2020, Sacks et al. 2018). Researchers showed that cordycepin could regulate the immune response of BV_2 microglia cells and might be a potential therapeutic agent for neurodegenerative diseases. It could also reduce anxious behavior by increasing the level of IL-4. Cordycepin was reported to protect the integrity of the blood-brain barrier (BBB) in a traumatic brain injury rat model. The results demonstrated that the neurological severity score, brain water content, and brain infarction volume were significantly reduced in the cordycepin-treated group compared to the control group (saline-treated rats). The observed treatment effect was dose-dependent. It has been shown that the two major proteins important for BBB integrity, occludin, and ZO-1 were restored in rats treated (20 mg/kg of cordycepin). Moreover, the study demonstrated that levels of iNOS, myeloperoxidase, IL-1β, nicotinamide adenine dinucleotide phosphate oxidase-1, and matrix metalloproteinases-9 were reduced and those of arginase and IL-10 were increased in treated rats compared to the control group. This study emphasized the potential protective role of cordycepin in maintaining BBB integrity (Ashraf et al. 2020, Lee et al. 2020). One of the other mechanisms of action of cordycepin is ergogenic activity. Cordycepin is an essential energy-producing factor. It is "produced" energy in the form of ATP – adenosine-5'-triphosphate.

Another mechanism of action of this compound may also be the production of nitric oxide (NO) (Qin et al. 2019, Tuli et al. 2014). Several nutraceuticals as well as pharmaceutical preparations made from mushroom dry powder (due to the presence of cordycepin), are marketed and reported to defend hepatic and renal functions, remove toxins from the body, increases oxygenation and natural endurance, improve intracellular energy exchange, delays the aging process, control blood glucose level and lipid profile, stimulates the metabolism of energy and supports the human immune system (Ashraf et al. 2020).

Natural pigments in mushrooms are an important group of bioactive compounds. The presence of carotenoids, including xanthophyll derivatives, has been established in various species of mushrooms, including *C. militaris*. Carotenoids are associated with an intense yellow-orange color of *C. militaris* fruit bodies. The major xanthophylls present in *C. militaris* include zeaxanthin and lutein (Ashraf et al. 2020, Dong et al. 2013, Muszyńska et al. 2016). In addition to studies on the protective effect of zeaxanthin and lutein on eye structures, previous studies have indicated that supplementation with carotenoids improves cognitive functions, and reduces stress symptoms and the level of cortisol in young and adult people. Carotenoids are also responsible for antioxidant activity (Bovier and Hammond 2015, Renzi-Hammond et al. 2017, Stringham et al. 2018).

Other substances characteristic of this species are beauveriolides – cyclodepsipeptides. Beauveriolides exhibited the ability to reduce β-amyloid concentration (neuroprotective effect) and antiatherosclerotic activity (Chen et al. 2020). Militarinones classified as alkaloids are a less-known group of chemical compounds. Structurally, they are pyridine derivatives and have been shown to exhibit cytotoxic and antimicrobial activities (Chen et al. 2020). Another compound – pentostatin is isolated from the fruit bodies of *C. militaris* and is an analog of the purine base hypoxanthine. It demonstrates inhibitory activity for the adenine deaminase enzyme. It also shows immunosuppressive and antitumor activity (Chen et al. 2020). Cordymin, in turn, is a peptide compound with antimicrobial activity (Wong et al. 2011).

The presence of vitamins in *C. militaris* has also been linked to the health-promoting potential of this species. Fruiting bodies of *C. militaris* contain water-soluble vitamins (B_1, B_2, B_3, B_{12}, C) and fat-soluble vitamins (A, E and K). The δ-Tocopherol was one of the quantified isoforms of vitamin E detected (55.86 μg/100 g) (Ashraf et al. 2020, Chan et al. 2015, Reis et al. 2013). Riboflavin (B_2) and niacin (B_3) reduce the feeling of fatigue and tiredness. They are factors responsible for normal energy-yielding metabolism. Ascorbic acid (vitamin C) and vitamin E protect cells from oxidative stress, and ROS, and contributes to maintaining the normal function of the immune system. Vitamin C also increases the absorption of iron from the gastrointestinal tract and supports collagen formation. Vitamin A, in turn, plays a vital role in the process of cell specialization. It is related to the maintenance of the process of correct vision (Commission Regulation (EU) No. 432/2012). Moreover, when we compare the content of vitamins in fruiting bodies and mycelium *C. militaris*, it turns out that mycelial cultures are a better source of vitamins B_2, B_3 and A, while the fruiting bodies are a better source of vitamins C and E (Chan et al. 2015).

Bioelements are necessary to immune regulation and infection prevention. Many bioelements (magnesium, manganese, potassium, copper, selenium, zinc, silver, and sulfur) have been detected in the fruit bodies of *C. militaris*. The increased concentration of minerals in *C. militaris* makes it an alternate source of these bioelements for the human diet. For safety reasons, it is important to test not only for bioelements of therapeutic importance but also for heavy metal content, which causes serious side effects in humans. Other results presented that the content of cadmium, arsenic, lead, copper, silver, and manganese in cultivated fruit bodies of *C. militaris*, respectively, is 0.0041 mg/kg, 0.0047 mg/kg, 0.22 mg/kg, 9.1 mg/kg, 0.024 mg/kg, and 1.6 mg/kg (Cohen et al. 2014, Fenglin et al. 2020).

It is worth emphasizing that selenium, present in *C. militaris* species, is a crucial element for the proper functioning of the human body. Selenium is a trace element, which is present as selenocysteine and constitutes a component of enzymatic proteins – selenoenzymes and selenoproteins. It acts as a cofactor to ensure the proper activity of a few enzymes: thioredoxin reductase (TXNRDs), glutathione peroxidases (GPX), and iodothyronine deiodinases (DIO). Selenium helps in the proper functioning of the human immune system and the thyroid gland. Additionally, it protects cells against oxidative stress and paves the way for the normal spermatogenesis process. In the European Union, the daily reference selenium intake value for young people aged 11–14 years is 55 µg per day, and for people aged 15 years and above is set at 70 µg per day (Agostoni et al. 2014, Rayman 2012). In *C. militaris* species, selenium occurs in an organic form. It is bound to proteins and amino acids. This bioelement is chelated with the amino acid: L-cysteine (selenocysteine) and L-methionine (selenomethionine). Selenium combines with proteins to form methylselenocysteine. On the addition of sodium selenate to *C. militaris* medium, increases the concentration of active compounds in fruit bodies (organic selenium form, polysaccharides, nucleosides, and amino acids) (Dong et al. 2012, Dong et al. 2013). Moreover, enrichment of the *C. militaris* media with organic or inorganic selenium form increases the concentration of adenosine and cordycepin in their fruit bodies. This is due to the ability of the mycelium to accumulate bioelements from the substrate, which in this case affects the metabolic pathway of bioactive compounds (Fenglin et al. 2020, Hu et al. 2019). The experimental work also succeeded in obtaining selenium-fortified polysaccharides from the mycelium of *C. militaris* (SeCSP-I). The antioxidant activity of the obtained seleno-polysaccharide fraction was demonstrated (Zhu et al. 2016a).

3. Biological Activities and Applications

In both *in vivo* and *in vitro* experiments, *C. militaris* or bioactive compounds isolated from these species of mushroom (mainly cordycepin), have been shown to exert immunostimulatory, antitumor, antioxidant, anti-inflammatory, antiviral, neuroprotective and ergogenic activity. Scientific evidence indicated that the biological activity of *C. militaris* is significantly more variable than *C. sinensis* (Tuli et al. 2014, Zhang et al. 2019). The application potential of this species is described in Table 1.

Table 1: Human studies using *C. militaris* species confirming its application potential.

Group	Intake of C. militaris [g/day]	Duration time [week]	Biological activity	Bioactive component	References
28 healthy individuals of both sexes	4	1–3	Ergogenic	Undefined	(Hirsch et al. 2017)
43 healthy individuals of both sexes	1–12	1–4	Ergogenic	Undefined	(Dudgeon et al. 2018)
79 healthy males	1.5	4	Immunostimulation	Cordycepin	(Kang et al. 2015)
100 healthy volunteers	1.5	12	Immunostimulation	Cordycepin	(Jung et al. 2019)
510 patients with chronic bronchitis	Undefined	Undefined	Anti-inflammatory	Undefined	(Cai et al. 2004)
425 patients with chronic bronchitis	3	8	Anti-inflammatory	Undefined	(Gai et al. 2004)
57 patients with mild hepatic impairment aged	1.5	8	Hepatoprotective	Cordycepin	(Heo et al. 2015)

3.1 *Ergogenic Activity*

Ergogenic activity in animal models contributes to the reduction of fatigue, and an increase in tolerance to physical effort. In humans, ergogenic activity in general contributes to the improvement of psychomotor skills (Kerksick et al. 2018, Tokish et al. 2004).

Recommendations and statements for active ingredients used as a dietary ergogenic aid are presented by sports and scientific organizations such as the International Olympic Committee (IOC), the International Society of Sports Nutrition (ISSN) or the Australian Institute of Sport (AIS) (Kerksick et al. 2018, Maughan et al. 2018, https://www.ais.gov.au).

In terms of biological activity, cordycepin is somewhat similar to creatine as an intermediate or indirect precursor of energy in the form of adenosine-5'-triphosphate (ATP). Authorized health claims for creatine informed that increases physical performance in successive bursts of short-term, high-intensity exercise. Supplementation of creatine increases muscle creatine stores and the rate of phosphocreatine resynthesis. Chronic supplementation of creatine correlate with adaptations includes support increases body mass and/or muscle mass, and improvements to muscular strength and power (Kerksick et al. 2018, Kreider and Stout 2021, Maughan et al. 2018).

It has been proven that ATP supplementation can provide ergogenic aid through the improvement in skeletal muscle strength and power development, skeletal muscle mass and changes in lean body mass and recovery from high-intensity training (Freitas et al. 2019, Wilson et al. 2013). In experimental mice, extract from *C. militaris* fruiting bodies showed antifatigue activity correlated with activation of AMPK and AKT/mTOR pathways (Song et al. 2015). Also, cereal grains mixed with *C. militaris* in experimental mice contributed to induced fatigue recovery and prolonged the swimming endurance time (Zhong et al. 2017).

In mice, the extract of *C. militaris* (the content of cordycepin 2.33 mg/g), showed an improvement in exercise performance in a grip strength test. It has been shown to increase the production of ATP, increase the concentration of markers such as AMPK, and PPAR-ɤ, and increase the level of phosphocreatine (Choi et al. 2020a).

Verified the effect of supplementation mushrooms mixture (possessing as primary ingredient *C. militaris*, and other ingredients: *Ganoderma lucidum*, *Hericium erinaceus*, *L. edodes*, *Pleurotus eryngii* and *Trametes versicolor*) on exercise performance during high-intensity exercise. In young recreationally active subjects' consumption of the mushrooms mixture with *C. militaris* showed improved tolerance (Hirsch et al. 2017).

Due to the fact that the exact content of *C. militaris* in the mushroom mixture is not specified, and the co-ingredients mushroom raw materials are also a source of bioactive substances, e.g., hericinen, erinancenin from *H. erinaceus*, the ergogenic activity of *C. militaris* cannot be fully assigned in work Hirsch et al. (2017). The content of cordycepin is also not specified in the study. Further research is required in which volunteers will supplement only *C. militaris*, with a specific dose/portion of mushroom material and a quantified concentration of cordycepin. The mushroom material can also be in different variants: mycelium extract or fruiting body extract, mycelium powdered and fruiting body powdered (Hirsch et al. 2017).

3.2 Immunostimulatory Activity

Extracts from fresh *C. militaris* fruit bodies showed stronger immunostimulatory activity (*in vitro* and *in vivo* experiments) than extracts from dried fruit bodies. There was no significant variation in cordycepin concentration between fresh and dried material, but a significant difference was noticed in the contents of polysaccharides and polyphenols, which were larger in fresh fruit bodies of *C. militaris* (Zhu et al. 2013).

3.2.1 In Vitro Studies

The immunostimulatory activity of *C. militaris* and/or cordycepin correlated with the activation of macrophages for the production of NO and pro-inflammatory cytokines (IL-1β and TNF-α) and prostaglandin-2 (PGE2). It has been shown to increase the activity of enzymes, iNOS and COX-2 (Qin et al. 2019, Tuli et al. 2014).

It has been shown that the polysaccharide signed as CMP-W1 can increase lymphatic spleen cell proliferation (Luo et al. 2017). The novel polysaccharide isolated from cultured *C. militaris* was shown to promote the proliferation of spleen lymphocytes, enhance the cytotoxicity of NK cells and stimulate phagocytosis by macrophages (Bi et al. 2018).

3.2.2 In Vivo Studies

The immunomodulatory activity of cordycepin-enriched *C. militaris*, was associated with the inhibition of tumor proliferation in experimental mice. The study confirmed enhanced IFN-γ production of CD8+ T cells, elevation in the production of IL-4, and a decrease in the concentration of IL-2 and TGF-β (Jeong et al. 2013).

Two studies were conducted with human participants. A group of healthy adult males who used 1.5 g of *C. militaris* daily for 4 weeks, confirmed an increased immunostimulatory activity associated with enhancing NK cells activity and T-cell proliferation, and increased levels of TNF-α and IFN-γ (Kang et al. 2015). Another study demonstrated that 12 weeks of supplementation of *C. militaris* in healthy volunteers (aged 20–70 years), did not have a significant effect on the symptoms of colds. However, it has been shown an enhancement activity of NK cells and increases the concentration of IgA (Jung et al. 2019).

3.3 Antitumor Activity

The mechanism of the antitumor activity of cordycepin has not been fully elucidated, however, it has been recognized that these antitumor activities of cordycepin correllated with disrupting or arresting of the cell cycle, inhibiting proliferation and stimulating apoptosis of neoplastic cells. The above-mentioned effects in cell lines correlate with the biological activity of cordycepin as antimetabolite and inhibition of purine biosynthesis, disturbance in DNA or RNA biosynthesis, interaction with mTOR signal pathway, 5'AMP-activated protein kinase (AMPK), cyclin-dependent kinases (CDK), TNF-α/TNFR1, CXCR4 (Guo et al. 2020, Hu et al. 2021, Tima et al. 2022, Zhang et al. 2022).

Cordycepin has been shown to prevent the growth of colon cancer in a mouse tumor model via immunotropic activity in tumor cells. Cordycepin inhibited the expression of protein-coding gene BNIP3 (BCL2 Interacting Protein 3) and protein thrombospondin-1 (TSP1), which resulted in the inhibition of expression of protein CD47, reduction of phagocytosis CD47 and decrease the apoptosis of T lymphocytes and increase their migration/distribution in tumor cells. *In vitro* experiments on cell lines have shown that cordycepin can stimulate phagocytosis of colon tumor cell No. 26 (CT26) by macrophages (Deng et al. 2022).

3.3.1 In Vitro Studies

Demonstrated that aqueous extract of *C. militaris* inhibits proliferation and induces apoptosis of lung cancer cells in humans. The antitumor activity correlated with an increase in the activity of caspase-3, caspase-8, and caspase-9 and inhibition of the telomerase (Park et al. 2009).

Cordycepin extracted from *C. militaris* inhibits the proliferation and migration of bladder cancer cells in humans. The cytotoxic activity of cordycepin has been associated with various signaling pathways, such as metalloproteinase-9 (MMP-9), TNF-α, NF-κB, and protein activator-1 (AP-1) (Lee et al. 2010).

Antitumor activity has been demonstrated not only for cordycepin but also for the polysaccharides contained in *C. militaris*. CMPS-II polysaccharide was presented to inhibit the growth of lung cancer cells (Liu et al. 2019).

A recent experiment by Guo et al. (2020) showed that cordycepin suppressed the growth of liver cancer cells in humans, via a reduction in the expression of the chemokine CxCR4 (promotor invasiveness of liver cancer cells).

Confirmed that *C. militaris* induced apoptosis in ovarian cancer cells, associated with activation of TNF-α, TNFR1, NF-κB, caspase-3, and caspase-9 and reduction levels of Bcl-2 and BclxL (Jo et al. 2020).

3.3.2 In Vivo Studies

Confirmed that extract of *C. militaris* reduced the volume of tumor and also extended the survival time of mice with lung cancer (Park et al. 2009).

Jeong et al. (2013), demonstrated that *C. militaris* enriched with cordycepin increased the viability of tumor-bearing mice and inhibited tumor growth. This experiment confirmed the immunotropic activity of cordycepin is associated with its antitumor activity.

The antitumor activity of the derivatives of cordycepin such as NUC-7738 was investigated in clinical trials (Plummer et al. 2021, Schwenzer et al. 2021, Symeonides et al. 2020).

3.4 Antioxidant Activity

A significant part of the research confirms the antioxidant activity of the polysaccharide fraction *C. militaris*. However, a few studies have also shown cordycepin's antioxidant activity (He et al. 2013).

Various bioactive compounds in *C. militaris* may influence the antioxidant activity of ergothionine, phenolic compounds, carotenoids and selenium among them (Chan et al. 2015, Chen et al. 2012, Cohen et al. 2014).

3.4.1 In Vitro Studies

The antioxidant capacity of *C. militaris* is higher than *C. sinensis*. The antioxidant activity was related to the of polysaccharides content and phenolic compounds in *C. militaris* (Yu et al. 2006).

In *in vitro* experiments, the antioxidant activity of the polysaccharide fractions was confirmed for the components WCBP50, CMP-1, and SeCSP-I (Chen et al. 2013, Jing et al. 2014, Zhu et al. 2016b). The addition of selenium to the medium of *C. militaris* was found to increase the antioxidant capacity of the polysaccharide fractions (Hu et al. 2019).

3.4.2 In Vivo Studies

Rodents fed *C. militaris* possessing polysaccharides showed an increase in the activity of enzymes such as superoxide dismutase (SOD), catalase, and GPX, and/or a reduction in the level of malondialdehyde (MDA) (Liu et al. 2016, Zhang et al. 2019).

In growing pigs fed on 2 g/kg, *C. militaris* spent mushroom has been shown to increase immunoglobulin secretion and supports antioxidant capacity – increased glutathione peroxidase activity as well as decreased MDA level (Boontiam et al. 2020).

3.5 Neuroprotective Activity

An animal models of vascular dementia caused by cerebral ischemia-reperfusion injury demonstrated the neuroprotection activity of the *C. militaris*. Experimental rats previously exposed to ischemic brain neuronal injury. A post-mortem confirmed that *C. militaris* contributed to significantly increase neuronal cell density in rats' brain tissue in the CA1 region of the hippocampal area. Also this study showed that *C. militaris* treatment improves the short-term memory weakening induced by scopolamine (Kim et al. 2019).

Cordycepin remarkably alleviated long-term potentiation (LTP) impairment induced by cerebral ischemia and excitotoxicity damage in rats. Cordycepin attenuated neuronal loss and protected pyramidal neurons of the hippocampal CA1 region. Cordycepin prevented the reduction of the adenosine A1 receptor level caused by ischemia but did not alter the adenosine A2A receptor level in the hippocampal CA1 area (Dong et al. 2019).

In experimental mice on the model of Alzheimer's disease induced by the β-amyloid protein (Aβ1-42), it was demonstrated that extract *C. militaris* improves functions in the following behavioral tests: water maze, new route and object recognition. *C. militaris* extract has demonstrated the ability to inhibit nitric oxide (NO) production and lipid peroxidation in the brain (He et al. 2019).

In rodents treated with doxorubicin, *C. militaris* extract has been shown to reduce oxidative stress-induced chemotherapy (Veena et al. 2020). In an *in vitro* study, *C. militaris* extract proved to have neuroprotective activity on C6 Glial Cells that had been treated with amyloid-beta peptide (Aβ1-42). Expression enhancement in Brain-Derived Neurotrophic Factor (BDNF) has been confirmed (He et al. 2020). *In vivo* experiments on rodents confirmed that *C. militaris* attenuated Aβ1-42 induced amyloidogenesis and suppressed inflammatory responses, inhibited acetylcholinesterase activity, and enhanced the expression of BDNF (He et al. 2021). Wang et al. (2022), showed that cordycepin modulating activity of the glutamatergic and GABAergic neurotransmissions via a presynaptic mechanism and suppressed the synaptic transmission through the activation of the adenosine A1 receptor (A1AR) (Wang et al. 2022). Ongoing clinical research is carried out to verify the effects of consumption *C. militaris* on the regulation of emotions and mood in humans (https://clinicaltrials.gov/ct2/show/NCT04002219).

3.6 Other Biological Activity

Cordycepin and polysaccharides (isolated or contained in *C. militaris* mushroom material) showed anti-inflammatory activity in *in vitro* and *in vivo* experiments (Smiderle et al. 2014, Tan et al. 2021, Zhang et al. 2019).

In two studies conducted on human subjects, supplementation of *C. militaris* has been shown to support the treatment of bronchitis. However, access to full materials is limited (Cai et al. 2004, Gai et al. 2004).

Several investigations have focused on the effects of *C. militaris* (or its bioactive compounds) on the respiratory system. An experiment in a mouse model of asthma demonstrated that cordycepin alleviates airway hyperresponsiveness, reduces inflammation, decreases IgE and eosinophil levels, and decreases the expression of IL-4, IL-5, IL-13, NF-κB (Qin et al. 2019). In an asthma model in rats, cordycepin was shown to reduce airway remodeling. Cordycepin decrease in the levels of IgE, eosinophils, and neutrophils and a decrease in the expression of TNF-α, TGF-β1, and IL-5, were observed (Fei et al. 2017).

Some evidence indicated the impact of bioactive compounds contained in *C. militaris*, on the cardiovascular system. The cardioprotective activity of the cordycepin confirmed an isolated ischemic rat heart (Park et al. 2014). In *in vitro* experiments, cordycepin showed capacity to inhibited adipogenesis and lipid deposition in adipocytes. These mechanisms of action will be associated with the suppression of the C/EBPβ, PPARγ, and mTORC1 pathways and activation of AMPK (Takahashi et al. 2012). Studies indicate antiplatelet activity and anticoagulant activity of *C. militaris* (Choi et al. 2020b, Lee et al. 2015).

Experiments on rodents confirmed the hypolipidemic activity of the polysaccharide isolated from *C. militaris* (Wang et al. 2015). Also, Huang et al. (2018) showed the hypolipidemic activity of polysaccharides from *C. militaris* in mice fed high-fat food.

Limited studies focused on the effects of *Cordyceps* spp. on the digestive or gastrointestinal system. In *in vitro* studies, *C. sinensis* showed a higher ability than *C. militaris* to affect the gut microbiota by decreasing the pH of the gut environment, stimulating the production of beneficial SCFAs and promoting the growth of beneficial propionic acid-producing bacteria (*Phascolarctobacterium*) and acetic acid- producing bacteria *Bifidobacterium* (Ji et al. 2020). So far, more research indicated that the species *C. sinensis* is useful to support gut microbiota and modulate intestinal immunity (Chen et al. 2021, Ying et al. 2020). In a group of patients with mild liver dysfunction, supplementation of 1.5 g/day *C. militaris* (cordycepin) for eight weeks could protect the liver against lipid accumulation (Heo et al. 2015).

Some scientific evidence indicates that *C. militaris* has a beneficial effect on the locomotor system (Zhang et al. 2015). In experimental rats (with inflammatory-induced osteoporosis model), cordycepin showed an anti-inflammatory action and limited bone loss (Zhang et al. 2014). In rats in an experimental model of osteoarthritis, cordycepin was shown to reduce pain and inflammation in the synovium (Ashraf et al. 2019).

In rodents, cordycepin has been shown to stimulate steroidogenesis as well as increase testosterone concentration (Leu et al. 2011). In rodent experiments, improvement in sperm quality was reported by Kopalli et al. (2019). *In vitro* and *in vivo* experiments (on rodents) confirmed that the *C. militaris* extract (containing cordycepin) regulated the concentration of androgens (Kusama et al. 2021).

In vitro experiments have demonstrated the antiviral activity of cordycepin and its derivatives. Such activity was confirmed against influenza virus, HIV, Epstein-Barr virus, and herpes simplex virus. The mode of antiviral activity of cordycepin is related to the inhibition of reverse transcriptase as well as RNA polymerase of the virus (Montefiori et al. 1989, Mueller et al. 1991, Ryu et al. 2014).

In silico studies on cordycepin created new avenues for the use of this molecule in the treatment of COVID-19. Cordycepin showed interactions with SARS-CoV-2 with receptor-binding domain (RBD) at the spike protein (S) and the main protease

(Mpro). Cordycepin also showed inhibition for RNA-dependent RNA polymerase (RdRp) of SARS-CoV-2 (Bibi et al. 2022, Verma and Aggarwal 2020). Rabie (2022) showed, that cordycepin act as a SARS-CoV-2 replication inhibitor in *in vitro* experiment.

3.7 *Applications in Cosmetology*

The *C. militaris* fruiting bodies are widely used in the pharmaceutical as well as health care industries as medical materials, dietary supplements, and additives in foodstuffs. The bioactive constituents in the fruit bodies are mainly cordycepin, polysaccharides, proteins, and phenolic compounds, which are responsible for multidirectional medicinal effects (Hur 2008).

Recent studies have proven the effectiveness of *C. militaris* bioactive compounds as active ingredients in cosmetic preparations. It has been shown that bioactive compounds – mainly phenolic compounds and carbohydrates contained in crude extracts – are characterized by multidirectional activity, which is effective in cosmetic applications, e.g., antioxidant activity, photoprotective and proliferative activity, showing tyrosinase inhibition ability. Furthermore, it was proved that the extracts were characterized by low cytotoxicity (Pintathong et al. 2021).

The study demonstrated that *C. militaris* broth (CMB) exerts an effect on melanogenesis in B16F0 melanoma cells. This ability was studied by measuring melanin concentration after 3 days of incubation. B16F0 melanoma cells were treated with different concentrations of CMB 10–100 µg/mL and arbutin 200 µM. The phenolic compounds and the antioxidant activity of CMB were also measured. The phenolic compound content of CMB was 3.28 mg/g. The DPPH radical-scavenging and iron ion donor activities were 79.64% and 0.16, respectively. The melanin content and melanoma cell viability under arbutin treatment decreased to 43 and 91%, respectively, compared to the control. The application of CMB application showed a significant inhibitory effect on melanin production by 29, 50 and 56% at concentrations of 50, 80 and 100 µg/mL, respectively, whereas more than 90% of the cells were viable. CMB treatment at concentrations of 50, 80 and 100 µg/mL in culture reduced the extracellular melanin discharge induced by 3-isobutyl-1-methylxanthine (IBMX) treatment by 19%, 38% and 48%, respectively. CMB showed inhibitory effects on intracellular tyrosinase isolated from melanoma cells, while it did not inhibit fungal tyrosinase activity. Cellular glutathione content was increased by CMB in a concentration-dependent manner. These results suggest that CMB inhibited cellular tyrosinase activity and total melanin content in cultured B16F0 melanoma cells without significantly affecting cell proliferation and may be a potential antimelanogenic agent (Cha et al. 2011).

Another research investigates the anti-skin wrinkle properties of *C. militaris* extracts *in vitro* as well as *in vivo*. The anti-skin wrinkle activities of *C. militaris* were investigated by inhibiting matrix metalloproteinase-1 (MMP-1), elastase, and hyaluronidase. Microemulsion and serum formulation containing *C. militaris* extract was developed. The anti-skin wrinkle efficacy and irritation properties of serum formulations were clinically investigated in human volunteers. Cordycepin was identified as a major component of *C. miliaris* extract that was accountable for the inhibition of MMP-1, elastase, and hyaluronidase. The *C. miliaris* water extract (CW) possessed the most potent inhibition on MMP-1 (77.9 ± 5.3%) and elastase (84.4 ± 4.0%). Fascinatingly, the CW was as a potent MMP-1 and elastase inhibitor as oleanolic acid as well as EGCG. The CW was incorporated into the microemulsion with the smallest internal droplet size (146.1 ± 1.5 nm) and further developed as a topical serum formulation. The resulting serum formulation effectively enhanced skin moisture (42.2 ± 14.2%), improved skin elasticity (39.9 ± 7.3%), and did not cause skin irritation in 30 human volunteers. The effectiveness on the skin was detected after a week of applications. Thus, it was suggested as a potent anti-skin wrinkle formulation (Marsup et al. 2021).

C. militaris can be used as an ingredient in cosmetic goods due to its strong antioxidant activity. However, there are limitations to the topical application of these products because the active ingredients are easily degraded. A study was conducted to develop nanoemulsion systems to enhance the stability of cordycepin. In addition, the safety, release profile, and dermal release of cordycepin from nanoemulsions were studied. *C. militaris* was shown to have a free radical scavenging capacity comparable to that of ascorbic acid and a lipid peroxidation inhibition capacity comparable to that of α-tocopherol. Subsequently, ten mg/mL of *C. militaris* was incorporated into nanoemulsions containing sugar squalene, Tween 85 and deionized water. The nanoemulsion, which had the smallest internal droplet size (157.1 ± 2.6 nm), increased the stability of *C. militaris*, showed no cytotoxicity or skin irritation, released the most CW (0.9 ± 0.0% w/w after 24 h) and distributed the highest concentration of *C. militaris* to the skin layer (33.5 ± 0.7% w/w). After the experiment, this mushroom species was proposed as a potent antioxidant in cosmetic products; in order to increase the stability and delivery of *C. militaris* to the skin, a nanoemulsion was proposed (Marsup et al. 2020).

In another experiment, a process was designed and optimized to obtain O/W (oil-in-water) nanoemulsions based on matzo crab meal encapsulated in *Hippophae rhamnoides* (sea buckthorn) fruit oil using an ultrasound process. Tween 80 was used as an emulsifying agent. Chitosan obtained from crab shells acted as a co-surfactant. The stability of the nanoemulsion was tested at different temperatures (4, 25 and 60°C) and storage conditions were tested for a period of three months, where 4°C did not affect stability. The cytotoxicity and anti-inflammatory effects of the obtained nanoemulsion were evaluated *in vitro*. The tested nanoemulsion showed anti-inflammatory activity, proved its antioxidant activity, and its ability to inhibit the growth of *E. coli* as well as *S. aureus*; moreover, the results suggested the absence of toxicity of the obtained product. The tested nanoemulsion was more effective than the free extract and showed greater potential as an ingredient in biomedical and cosmetic products (Rupa et al. 2020).

The antioxidant properties of the extracts from mace allow its antioxidant properties to allow used as an active ingredient in cosmetics. Technological problems that limit the use of *C. militaris* extracts are the fast degradation of active ingredients, first of all, cordycepin after topical application. This is due to the fact that the degradation of the molecule occurs in an acidic environment and the pH of the skin varies between 4.7 and 5.6 pH. In this study, cordycepin applied in alkaline-containing formulations was shown to be stable. To incorporate the nucleoside into acidic pH formulations while protecting it from oxidation, a nanoemulsion system was developed (Marsup et al. 2020).

The solid residues (SBR) that remained after harvesting the fruit bodies of the cultured *C. militaris* mushrooms and the solid-state fermentation residues (SSF) were used to prepare crude extracts. These extracts were examined for their usefulness in cosmetology. The SSFs obtained from a solid medium culture containing defatted rice bran amended with barley, white rice, rice paddy, and wheat were named SBR-B, SBR-R, SBR-Rb and SRB-W, respectively. They were extracted with solvents possessing different polarities. In this study, the total phenolic content (TPC), total flavonoid content (TFC), and total carbohydrate content (TCC) were evaluated. Furthermore, antioxidant activity, tyrosinase inhibition, photoprotection, and cytotoxicity were also evaluated. The results showed that the total bioactive content and biological capacity of the crude SBR extracts were significantly affected by the types of SBR and the extraction solvent (p < 0.05). The SBR-B extracted with hot water showed the highest antioxidant activity (66.62 ± 2.10, 212.00 ± 3.43 and 101.62 ± 4.42 mg TEAC/g extract, respectively) when determined by DPPH methods, ABTS and FRAP, while tyrosinase inhibitory activity (51.13 ± 1.11 mg KAE/g extract) with 90.43 ± 1.96% inhibition at 1 mg/mL concentration was perfectly observed by SBR-Rb extracted with 50% (v/v) ethanol. The correlations between the content of bioactive substances in the crude extracts and their biological activity were found to be mainly high (p < 0.01). The ability of the crude extracts to absorb UV radiation in the 290–330 nm range revealed their potential role as natural UV absorbers and enhancers. Cytotoxicity studies using fibroblast cell lines tested with hot water and 50% ethanolic SBR extracts (v/v) revealed their safety in the concentration range of 0.001–10 mg/mL. Excitingly, their fibroblast proliferation ability, indicating anti-aging properties, was strongly promoted. The chemical composition analyzed by LC-MS/MS techniques showed that seven phenolic acids and four flavonoids were identified in the crude SBR extracts. In addition, nucleosides, nucleobases, amino acids, sugars, phospholipids, alkaloids, organic acids, vitamins, and peptides were among other chemical compounds present. Thus, it is emphasized that SBRs from *C. militaris* can be a promising source for the preparation of crude extracts used in cosmetics. In addition, they can be used as multifunctional ingredients in cosmetics as well as cosmeceuticals. These are very promising results that open up the possibility of using matcha components in cosmeceuticals (Pintathong et al. 2021).

In addition, the anti-photoaging activities of cordycepin have been described. The inhibitory effect of cordycepin on the expression of matrix metalloproteinase-1 (MMP-1) and -3 in human dermis fibroblasts was demonstrated. Western blot and real-time PCR analysis showed that cordycepin inhibited UVB-induced expression of MMP-1 and -3 in a dose-dependent manner. The UVB strongly activated NF-kappaB as determined by IkappaBalpha degradation, nuclear localization of the p50 and p65 subunit, and binding activity of NF-kappaB. However, UVB-induced activation of NF-kappaB and MMP expression was completely blocked by cordycepin administration. These results suggest that cordycepin can prevent UVB-induced MMP expression by inhibiting NF-kappaB activation. In summary, cordycepin can be used as a potential agent to prevent and treat skin photoaging (Lee et al. 2009).

4. Conclusion

Based on scientific data, the *C. militaris* species should be a recognized and valuable food source, not only in Asian countries but also within the European Union. The current lifestyle of stress, nutrient-poor foods, constant overstrain and fatigue is prompting the population to seek sources of supplementation and food that can have a protective and health-promoting effect. An important activity, the strengthening endurance of the human organism includes stimulation of immune system functions as well as reduction of fatigue. Importantly, the species *C. militaris* can fulfill both important functions for the proper functioning of the body, i.e., nutritional and medicinal properties. It is worth emphasizing that not only cordycepin, which is the best-known substance characteristic for the species but also extracts obtained from *C. militaris* species or even powdered mycelia and fruiting bodies are responsible for the medicinal activity. Biotechnological possibilities of mycelium and fruiting body modification contribute to the fact that this species may be a leading component of natural medicines in the future. Many of the studies described are promising and are factors for further work not only with animal models but also with human participants in large-scale clinical trials.

Competing Interest

The authors declare that they have no competing financial interests.

References

Agostoni, C., R. Berni Canani, S. Fairweather-Tait, M. Heinonen, H. Korhonen, S. La Vieille, R. Marchelli, A. Martin, A. Naska, M. Neuhäuser-Berthold, G. Nowicka, Y. Sanz, A. Siani, A. Sjödin, M. Stern, S. Strain, I. Tetens, D. Tomé, D. Turck and H. Verhagen. 2014. EFSA scientific opinion on dietary reference values for selenium. *EFSA Journal*, 12(10): 3846.

Ashraf, S.A., A. Elkhalifa, A.J. Siddiqui, M. Patel, A.M. Awadelkareem, M. Snoussi, M.S. Ashraf, M. Adnan and S. Hadi. 2020. Cordycepin for health and wellbeing: A potent bioactive metabolite of an entomopathogenic cordyceps medicinal fungus and its nutraceutical and therapeutic potential. *Molecules*, 25(12): 2735.

Ashraf, S., M. Radhi, P. Gowler, J.J. Burston, R.D. Gandhi, G.J. Thorn, A.M. Piccinini, D.A. Walsh, V. Chapman and C.H. De Moor. 2019. The polyadenylation inhibitor cordycepin reduces pain, inflammation and joint pathology in rodent models of osteoarthritis. *Sci. Rep.*, 9(1): 1–17.

Australian Institute of Sport. Available from: https://www.ais.gov.au/nutrition/supplements/group_a Accessed: 12.05.2022.

Bi, S., Y. Jing, Q. Zhou, X. Hu, J. Zhu, Z. Guo, L. Song and R. Yu. 2018. Structural elucidation and immunostimulatory activity of a new polysaccharide from *Cordyceps militaris*. *Food Funct.*, 9(1): 279–293.

Bibi, S., M.M. Hasan, Y.B. Wang, S.P. Papadakos and H. Yu. 2022. Cordycepin as a promising inhibitor of SARS-CoV-2 RNA dependent RNA polymerase (RdRp). *Curr. Med. Chem.*, 29(1): 152–162.

Boonstra, E., R. de Kleijn, L.S. Colzato, A. Alkemade, B.U. Forstmann and S. Nieuwenhuis. 2015. Neurotransmitters as food supplements: The effects of GABA on brain and behavior. *Front. Psychol.*, 6: 1520.

Boontiam, W., C. Wachirapakorn and S. Wattanachai. 2020. Growth performance and hematological changes in growing pigs treated with *Cordyceps militaris* spent mushroom substrate. *Vet. World*, 13(4): 768–773.

Bovier, E.R. and B.R. Hammond. 2015. A randomized placebo-controlled study on the effects of lutein and zeaxanthin on visual processing speed in young healthy subjects. *Arch. Biochem. Biophys.*, 572: 54–57.

Breitenbach, J. and F. Krnzlin. (eds.). 1981. Fungi of Switzerland. Vol. 1: Ascomycetes. Edition Mycologia Lucerne, Lucerne.

Cai, H., Y. Wang and J. Pang. 2004. Efficacy of *Cordyceps militaris* capsules in treatment of chronic bronchitis. *Chin. J. New Drugs*, 13: 171–174.

Cha, J.Y., H.J. Yang, M.Y. Park, S.T. Choi, H.I. Moon and Y.S. Cho. 2011. Melanogenesis effect of *Cordyceps militaris* culture broth on the melanin formation of B16F0 melanoma cells. *Immunopharmacol. Immunotoxicol.*, doi: 10.3109/08923973.2011.611810.

Chan, J.S.L., G.S. Barseghyan, M.D.Asatiani and S.P. Wasser. 2015. Chemical composition and medicinal value of fruiting bodies and submerged cultured mycelia of caterpillar medicinal fungus *Cordyceps militaris* CBS-132098 (Ascomycetes). *Int. J. Med. Mushrooms*, 17(7): 649–659.

Chen, B., Y. Sun, F. Luo and C. Wang. 2020. Bioactive metabolites and potential mycotoxins produced by *Cordyceps* fungi: A review of safety. *Toxins*, 12(6): 410.

Chen, S., J. Wang, Q. Fang, N. Dong, Q. Fang, S.W. Cui and S. Nie. 2021. A polysaccharide from natural *Cordyceps sinensis* regulates the intestinal immunity and gut microbiota in mice with cyclophosphamide-induced intestinal injury. *Food Funct.*, 12(14): 6271–6282.

Chen, S.Y., K.J. Ho, Y.J. Hsieh, L.T. Wang and J.L. Mau. 2012. Contents of lovastatin, γ-aminobutyric acid and ergothioneine in mushroom fruiting bodies and mycelia. *LWT—Food Sci. Technol.*, 47(2): 274–278.

Chen, X., G. Wu and Z. Huang. 2013. Structural analysis and antioxidant activities of polysaccharides from cultured *Cordyceps militaris*. *Int. J. Biol. Macromol.*, 58: 18–22.

Choi, E., J. Oh and G.H. Sung. 2020a. Beneficial effect of *Cordyceps militaris* on exercise performance via promoting cellular energy production. *Mycobiology*, 48(6): 512–517.

Choi, E., J. Oh and G.H. Sung. 2020b. Antithrombotic and antiplatelet effects of *Cordyceps militaris*. *Mycobiology*, 48(3): 228–232.

ClinicalTrials.gov. Available from: https://clinicaltrials.gov/ct2/show/NCT04002219 Accessed: 28.06.2019.

Cohen, N., J. Cohen, M.D. Asatiani, V.K. Varshney, H.T. Yu, Y.C. Yang, Y.H. Li, J.L. Mau and S.P. Wasser. 2014. Chemical composition and nutritional and medicinal value of fruit bodies and submerged cultured mycelia of culinary-medicinal higher Basidiomycetes mushrooms. *Int. J. Med. Mushrooms*, 16(3): 273–291.

Davis, M., R. Sommer and J. Menge. (eds.). 2012. Field guide to mushrooms of Western North America. University of California Press, Berkeley, pp. 381–382.

Deng, Q., X. Li, C. Fang, X. Li, J. Zhang, Q. Xi, Y. Li and R. Zhang. 2022. Cordycepin enhances anti-tumor immunity in colon cancer by inhibiting phagocytosis immune checkpoint CD47 expression. *Int. Immunopharmacol.*, 107: 108695.

Dong, J.Z., C. Lei, X.R. Ai and Y. Wang. 2012. Selenium enrichment on *Cordyceps militaris* Link and analysis on its main active components. *Appl. Biochem. Biotechnol.*, 166(5): 1215–1224.

Dong, J.Z., S.H. Wang, X.R. Ai, L. Yao, Z.W. Sun, C. Lei, Y. Wang and Q. Wang. 2013. Composition and characterization of cordyxanthins from *Cordyceps militaris* fruit bodies. *J. Funct. Foods*, 5(3): 1450–1455.

Dong, Z.S., Z.P. Cao, Y.J. Shang, Q.Y. Liu, B.Y. Wu, W.X. Liu and C.H. Li. 2019. Neuroprotection of cordycepin in NMDA-induced excitotoxicity by modulating adenosine A$_1$ receptors. *Eur. J. Pharmacol.*, 853: 325–335.

Dudgeon, W.D., D.D. Thomas, W. Dauch, T.P. Scheett and M.J. Webster. 2018. The effects of high and low-dose *Cordyceps militaris*-containing mushroom blend supplementation after seven and twenty-eight days. *Am. J. Sports Sci.*, 6(1): 1–7.

European Commission RASFF. Available from: https://webgate.ec.europa.eu/rasff-window/screen/notification/434851 Accessed: 24.05.2022.

Fei, X., X. Zhang, G. Zhang, W. Bao, Y. Zhang, M. Zhang and X. Zhou. 2017. Cordycepin inhibits airway remodeling in a rat model of chronic asthma. *Biomed. Pharmacother.*, 88: 335–341.

Fenglin, L., L. Li, J. Xiaokun, L. Yan and L. Hengwei. 2020. Analysis of seven mineral elements in cultivated fruiting bodies of *Cordyceps militaris*. *International Conference on Energy, Environment and Bioengineering (ICEEB 2020), E3S Web Conf.*, 185: 04019.

Freitas, M.C., J.M. Cholewa, J. Gerosa-Neto, D.C. Gonçalves, E.C. Caperuto, F.S. Lira and F.E. Rossi. 2019. A single dose of oral ATP supplementation improves performance and physiological response during lower body resistance exercise in recreational resistance-trained males. *J. Strength Cond. Res.*, 33(12): 3345–3352.

Gai, G., S. Jin, B. Wang, Y. Li and C. Li. 2004. The efficacy of *Cordyceps militaris* capsules in treatment of chronic bronchitis in comparison with Jinshuibao capsules. *Chin. J. New Drugs*, 13: 169–170.

Guo, Z., W. Chen, G. Dai and Y. Huang. 2020. Cordycepin suppresses the migration and invasion of human liver cancer cells by downregulating the expression of CXCR4. *Int. J. Mol. Med.*, 45(1): 141–150.

Hassan, M., R. Rouf, E. Tiralongo, T. May and J. Tiralongo. 2015. Mushroom lectins: Specificity, structure and bioactivity relevant to human disease. *Int. J. Mol. Sci.*, 16(12): 7802–7838.

He, M.T., A.Y. Lee, J.H. Kim, C.H. Park, Y.S. Shin and E.J. Cho. 2019. Protective role of *Cordyceps militaris* in Aβ 1–42-induced Alzheimer's disease *in vivo*. *Food Sci. Biotechnol.*, 28(3): 865–872.

He, M.T., C.H. Park and E.J. Cho. 2021. Caterpillar medicinal mushroom, *Cordyceps militaris* (Ascomycota), attenuates Aβ1-42-induced amyloidogenesis and inflammatory response by suppressing amyloid precursor protein progression and p38 MAPK/JNK activation. *Int. J. Med. Mushrooms*, 23(11): 71–83.

He, M.T., C.H. Park, Y.S. Shin, J.M. Choi and E.J. Cho. 2020. Caterpillar medicinal mushroom, *Cordyceps militaris* (Ascomycetes), protects Aβ1-42-induced neurologic damage in C6 glial cells. *Int. J. Med. Mushrooms*, 22(12): 1203–1213.

He, Y.T., X.L. Zhang, Y.M. Xie, Y.X. Xu and R. Li. 2013. Extraction and antioxidant property *in vitro* of cordycepin in artificially cultivated *Cordyceps militaris*. *Adv. Mat. Res.*, 750-752: 1593–1596.

Heo, J.Y., H.W. Baik, H.J. Kim, J.M. Lee, H.W. Kim, Y.S. Choi, J.H. Won, H.M. Kim, W.I Park and C.Y. Kim. 2015. The efficacy and safety of *Cordyceps militaris* in Korean adults who have mild liver dysfunction. *J. Clin. Nutr.*, 7(3): 81–86.

Hepsomali, P., J.A. Groeger, J. Nishihira and A. Scholey. 2020. Effects of oral gamma-aminobutyric acid (GABA) administration on stress and sleep in humans: A systematic review. *Front. Neurosci.*, 14: 923.

Hirsch, K.R., A.E. Smith-Ryan, E.J. Roelofs, E.T. Trexler and M.G. Mock. 2017. *Cordyceps militaris* improves tolerance to high-intensity exercise after acute and chronic supplementation. *J. Diet. Suppl.*, 14(1): 42–53.

Hu, T., Y. Liang, G. Zhao, W. Wu, H. Li and Y. Guo. 2019. Selenium biofortification and antioxidant activity in *Cordyceps militaris* supplied with selenate, selenite, or selenomethionine. *Biol. Trace Elem. Res.*, 187(2): 553–561.

Hu, Z., Y. Lai, C. Ma, L. Zuo, G. Xiao, H. Gao, B. Xie, X. Huang, H. Gan, D. Huang, N. Yao, B. Feng, J. Ru, Y. Chen and D. Cai. 2021. *Cordyceps militaris* extract induces apoptosis and pyroptosis via caspase-3/PARP/GSDME pathways in A549 cell line. *Food Sci. Nutr.*, 10(1): 21–38.

Huang, Z., M. Zhang, S. Zhang, Y. Wang and X. Jiang. 2018. Structural characterization of polysaccharides from *Cordyceps militaris* and their hypolipidemic effects in high fat diet fed mice. *RSC Adv.*, 8(71): 41012–41022.

Hur, H. 2008. Chemical ingredients of *Cordyceps militaris*. *Mycobiology*, 36(4): 233–235.

Jeong, M.H., C.M. Lee, S.W. Lee, S.Y. Seo, M.J. Seo, B.W. Kang, Y.K. Jeong, Y.J. Choi, K.M. Yang and W.S. Jo. 2013. Cordycepin-enriched *Cordyceps militaris* induces immunomodulation and tumor growth delay in mouse-derived breast cancer. *Oncol. Rep.*, 30(4): 1996–2002.

Ji, Y., A. Su, G. Ma, T. Tao, D. Fang, L. Zhao and Q. Hu. 2020. Comparison of bioactive constituents and effects on gut microbiota by *in vitro* fermentation between *Ophicordyceps sinensis* and *Cordyceps militaris*. *J. Funct. Foods*, 68: 103901.

Jing, Y., X. Cui, Z. Chen, L. Huang, L. Song, T. Liu, W. Lv and R. Yu. 2014. Elucidation and biological activities of a new polysaccharide from cultured *Cordyceps militaris*. *Carbohydr. Polym.*, 102(1): 288–296.

Jo, E., H.J. Jang, K.E. Yang, M.S. Jang, Y.H. Huh, H.S. Yoo, J.S. Park, I.S. Jang and S.J. Park. 2020. *Cordyceps militaris* induces apoptosis in ovarian cancer cells through TNF-α/TNFR1-mediated inhibition of NF-κB phosphorylation. *BMC Complement. Med. Ther.*, 20(1): 1–12.

Jung, S.J., J.H. Hwang, M.R. Oh and S.W. Chae. 2019. Effects of *Cordyceps militaris* supplementation on the immune response and upper respiratory infection in healthy adults: A randomized, double-blind, placebo-controlled study. *J. Nutr. Health*, 52(3): 258–267.

Kang, H.J., H.W. Baik, S.J. Kim, S.G. Lee, H.Y. Ahn, J.S. Park, S.J. Park, E.J. Jang, S.W. Park and J.Y. Choi. 2015. *Cordyceps militaris* enhances cell-mediated immunity in healthy Korean men. *J. Med. Food*, 18(10): 1164–1172.

Kerksick, C.M., C.D. Wilborn, M.D. Roberts, A. Smith-Ryan, S.M. Kleiner, R. Jäger, R. Collins, M. Cooke, J.N. Davis, E. Galvan, M. Greenwood, L.M. Lowery, R. Wildman, J. Antonio and R.B. Kreider. 2018. ISSN exercise & sports nutrition review update: Research & recommendations. *J. Int. Soc. Sports Nutr.*, 15(1): 1–57.

Kim, Y.O., H.J. Kim, G.M. Abu-Taweel, J. Oh and G.H. Sung. 2019. Neuroprotective and therapeutic effect of *Cordyceps militaris* on ischemia-induced neuronal death and cognitive impairments. *Saudi J. Biol. Sci.*, 26(7): 1352–1357.

Kirk, P.M. 2010. Species Fungorum (version 8.0. Sep 2008). *In*: Bisby, F.A., Roskov, Y.R., Orrell, T.M., Nicolson, D., Paglinawan, L.E., Bailly, N., Kirk, P.M., Bourgoin, T., Baillargeon, G. and Ouvrard, D. (eds.). Species 2000 & ITIS Catalogue of Life, DVD, Species 2000: Reading, UK.

Kobayasi, Y. 1941. The genus *Cordyceps* and its allies. pp. 53–260. *In*: Kobayasi, Y. (ed.). Daigaku Science Report of the Tokyo Bunrika Daigaku (Section B, No. 84). Botanical Institute, Tōkyō Bunrika.

Kobayasi, Y. 1982. Keys to the taxa of the genera *Cordyceps* and *Torrubiella*. *Trans. Mycol. Soc. Japan*, 23: 329–364.

Kobayasi, Y. and D. Shimizu. 1982. *Cordyceps* species from Japan. *Bull. Natl. Mus. Nat. Sci. Ser. B.*, 8: 111–123.

Kopalli, S.R., K.M. Cha, S.H. Lee, S.Y. Hwang, Y.J. Lee, S. Koppula and S.K. Kim. 2019. Cordycepin, an active constituent of nutrient powerhouse and potential medicinal mushroom *Cordyceps militaris* Linn., ameliorates age-related testicular dysfunction in rats. *Nutrients*, 11(4): 906.

Kreider, R.B. and J.R. Stout. 2021. Creatine in health and disease. *Nutrients*, 13(2): 447.

Kusama, K., M. Miyagawa, K. Ota, N. Kuwabara, K. Saeki, Y. Ohnishi, Y. Kumaki, T. Aizawa, T. Nakasone and S. Okamatsu. 2021. *Cordyceps militaris* fruit body extract decreases testosterone catabolism and testosterone-stimulated prostate hypertrophy. *Nutrients*, 13(1): 50.

Lee, C.T., K.S. Huang, J.F. Shaw, J.R. Chen, W.S. Kuo, G. Shen, A.M. Grumezescu, A.M. Holban, Y.T. Wang, J.S. Wang, Y.P. Hsiang, Y.M. Lin, H.H. Hsu and C.H. Yang. 2020. Trends in the immunomodulatory effects of *Cordyceps militaris*: Total extracts, polysaccharides and cordycepin. *Front. Pharmacol.*, 11: 575704.

Lee, D.H., H.H. Kim, D.H. Lim, J.L. Kim and H.J. Park,. 2015. Effect of cordycepin-enriched WIB801C from *Cordyceps militaris* suppressing fibrinogen binding to glycoprotein IIb/IIIa. *Biomol. Ther.*, 23(1): 60–70.

Lee, E., W. Kim and S. Moon. 2010. Cordycepin suppresses TNF-alpha-induced invasion, migration and matrix metalloproteinase-9 expression in human bladder cancer cells. *Phytother. Res.*, 24(12): 1755–1761.

Lee, Y.R., E.M. Noh, E.Y. Jeong, S.K. Yun, Y.J. Jeong, J.H. Kim, K.B. Kwon, B.S. Kim, S.H. Lee, C.S. Park and J.S. Kim. 2009. Cordycepin inhibits UVB-induced matrix metalloproteinase expression by suppressing the NF-kappaB pathway in human dermal fibroblasts. *Exp. Mol. Med.*, 41(8): 548–554.

Leu, S.F., S.L. Poon, H.Y. Pao and B.M. Huang. 2011. The *in vivo* and *in vitro* stimulatory effects of cordycepin on mouse leydig cell steroidogenesis. *Biosci. Biotechnol. Biochem.*, 75(4): 723–731.

Li, J., Z. Zhang, X. Liu, W. Wang, F. Mao, J. Mao, X. Lu, D. Jiang, Y. Wan, J.Y. Lv, G. Cao, J. Zhang, N. Zhao, M. Atkinson, D.L. Greiner, G.J. Prud'homme, Z. Jiao, Y. Li and Q. Wang. 2015. Study of GABA in healthy volunteers: Pharmacokinetics and pharmacodynamics. *Front. Pharmacol.*, 6: 260.

Liu, J., C. Feng, X. Li, M. Chang, J. Meng and L. Xu. 2016. Immunomodulatory and antioxidative activity of *Cordyceps militaris* polysaccharides in mice. *Int. J. Biol. Macromol.*, 86: 594–598.

Liu, X.C., Z.Y. Zhu, Y.L. Liu and H.Q. Sun. 2019. Comparisons of the anti-tumor activity of polysaccharides from fermented mycelia and cultivated fruiting bodies of *Cordyceps militaris in vitro*. *Int. J. Biol. Macromol.*, 130: 307–314.

Luo, X., Y. Duan, W. Yang, H. Zhang, C. Li and J. Zhang. 2017. Structural elucidation and immunostimulatory activity of polysaccharide isolated by subcritical water extraction from *Cordyceps militaris*. *Carbohydr. Polym.*, 157: 794–802.

Marsup, P., S. Sirilun, A. Prommaban, S. Sriyab, W. Neimkhum, J. Sirithunyalug, S. Anuchapreeda, C. To-anun and W. Chaiyana. 2021. Potential of topical microemulsion serum formulations to enhance *in vitro* and clinical anti-skin wrinkle benefits of *Cordyceps militaris* extracts. *Res. Sq.*, 10.21203/rs.3.rs-550591/v1.

Marsup, P., K. Yeerong, W. Neimkhum, J. Sirithunyalug, S. Anuchapreeda, C. To-Anun and W. Chaiyana. 2020. Enhancement of chemical stability and dermal delivery of *Cordyceps militaris* extracts by nanoemulsion. *Nanomaterials (Basel)*., 10(8): 1565.

Maughan, R.J., L.M. Burke, J. Dvorak, D.E. Larson-Meyer, P. Peeling, S.M. Phillips, E.S. Rawson, N.P. Walsh, I. Garthe and H. Geyer. 2018. IOC consensus statement: Dietary supplements and the high-performance athlete. *Int. J. Sport Nutr. Exerc. Metab.*, 28(2): 104–125.

Montefiori, D.C., R.W. Sobol, S.W. Li, N.L. Reichenbach, R.J. Suhadolnik, R. Charubala, W. Pfleiderer, A. Modliszewski, W.E. Robinson and W.M. Mitchell. 1989. Phosphorothioate and cordycepin analogues of 2', 5'-oligoadenylate: Inhibition of human immunodeficiency virus type 1 reverse transcriptase and infection *in vitro*. *Proc. Natl. Acad. Sci.*, 86(18): 7191–7194.

Mueller, W.E.G., B.E. Weiler, R. Charubala, W. Pfleiderer, L. Leserman, R.W. Sobol, R.J. Suhadolnik and H.C. Schroeder. 1991. Cordycepin analogs of 2', 5'-oligoadenylate inhibit human immunodeficiency virus infection via inhibition of reverse transcriptase. *Biochemistry*, 30(8): 2027–2033.

Muszyńska, B., M. Mastej and K. Sułkowska–Ziaja. 2016. Biological function of carotenoids and their occurrence in the fruiting bodies of mushrooms. *Medicina Internacia Revuo*, 107(27): 113–122.

Nutraveris. Available from: https://www.nutraveris.com/en/solutions/nol-data-solution/ Accessed: 24.05.2022.

Park, E.S., D.H. Kang, M.K. Yang, J.C. Kang, Y.C. Jang, J.S. Park, S.K. Kim and H.S. Shin. 2014. Cordycepin, 3'-deoxyadenosine, prevents rat hearts from ischemia/reperfusion injury via activation of Akt/GSK-3β/p70S6K signaling pathway and HO-1 expression. *Cardiovasc. Toxicol.*, 14(1): 1–9.

Park, S.E., H.S. Yoo, C.Y. Jin, S.H. Hong, Y.W. Lee, B.W. Kim, S.H. Lee, W.J. Kim, C.K. Cho and Y.H. Choi. 2009. Induction of apoptosis and inhibition of telomerase activity in human lung carcinoma cells by the water extract of *Cordyceps militaris*. *Food Chem. Toxicol.*, 47(7): 1667–1675.

Pintathong, P., P. Chomnunti, S. Sangthong, A. Jirarat and P. Chaiwut. 2021. The feasibility of using cultured *Cordyceps militaris* residues in cosmetics: Assessment of the biological activity of their crude extracts. *J. Fungi (Basel)*., 7(11): 973.

Plummer, R., F. Kazmi, N.M. Haris, T. Ding, F. Aroldi, M. Myers, S. Symeonides and S. Blagden. 2021. Abstract CT136: NUC-7738, a novel ProTide transformation of 3'-deoxyadenosine, in patients with advanced solid tumors. *Cancer Res.*, 81(13): CT136.

Qin, P., X. Li, H. Yang, Z.Y. Wang and D. Lu. 2019. Therapeutic potential and biological applications of cordycepin and metabolic mechanisms in cordycepin-producing fungi. *Molecules*, 24(12): 2231.

Rabie, A.M. 2022. Potent inhibitory activities of the adenosine analogue cordycepin on SARS-CoV-2 replication. *ACS Omega*, 7(3): 2960–2969.

Rayman, M.P. 2012. Selenium and human health. *Lancet*, 379(9822): 1256–1268.

Reis, F.S., L. Barros, R.C. Calhelha, A. Cirić, L.J. van Griensven, M. Soković and I.C. Ferreira. 2013. The methanolic extract of *Cordyceps militaris* (L.) Link fruiting body shows antioxidant, antibacterial, antifungal and antihuman tumor cell lines properties. *Food Chem. Toxicol.*, 62: 91–98.

Renzi-Hammond, L., E. Bovier, L. Fletcher, L. Miller, C. Mewborn, C. Lindbergh, J. Baxter and B. Hammond. 2017. Effects of a lutein and zeaxanthin intervention on cognitive function: A randomized, double-masked, placebo-controlled trial of younger healthy adults. *Nutrients*, 9(11): 1246.

Rupa, E.J., J.F. Li, M.H. Arif, H. Yaxi, A.M. Puja, A.J. Chan, V.A. Hoang, L. Kaliraj, D.C. Yang and S.C. Kang. 2020. *Cordyceps militaris* fungus extracts-mediated nanoemulsion for improvement antioxidant, antimicrobial, and anti-inflammatory activities. *Molecules*, 25(23): 5733.

Ryu, E., M. Son, M. Lee, K. Lee, J.Y. Cho, S. Cho, S.K. Lee, Y.M. Lee, H. Cho and G.H. Sung. 2014. Cordycepin is a novel chemical suppressor of Epstein-Barr virus replication. *Oncoscience*, 1(12): 866–881.

Sacks, D., D. Sacks, B. Baxter, B.C.V. Campbell, J.S. Carpenter, C. Cognard, D. Dippel, M. Eesa, U. Fischer, K. Hausegger, J.A. Hirsch, M.S. Hussain, O. Jansen, M.V. Jayaraman, A.A. Khalessi, B.W.

Kluck, S. Lavine, P.M. Meyers, S. Ramee, D.A. Rüfenacht, C.M. Schirmer and D. Vorwerk. 2018. Multisociety consensus quality improvement revised consensus statement for endovascular therapy of acute ischemic stroke. *Int. J. Stroke*, 13(6): 612–632.

Schwenzer, H., E. De Zan, M. Elshani, R. van Stiphout, M. Kudsy, J. Morris, V. Ferrari, I.H. Um, J. Chettle, F. Kazmi, L. Campo, A. Easton, S. Nijman, M. Serpi, S. Symeonides, R. Plummer, D.J. Harrison, G. Bond and S.P. Blagden. 2021. The novel nucleoside analogue ProTide NUC-7738 overcomes cancer resistance mechanisms *in vitro* and in a first-in-human phase I clinical trial. *Clin. Cancer Res.*, 27(23): 6500–6513.

Singh, R.S., R. Bhari and H.P. Kaur. 2010. Mushroom lectins: Current status and future perspectives. *Crit. Rev. Biotechnol.*, 30(2): 99–126.

Singh, S.S., H. Wang, Y.S. Chan, W. Pan, X. Dan, C.M. Yin, O. Akkouh and T.B. Ng. 2014. Lectins from edible mushrooms. *Molecules*, 20(1): 446–469.

Smiderle, F.R., C.H. Baggio, D.G. Borato, A.P. Santana-Filho, G.L. Sassaki, M. Iacomini and L.J.L.D. Van Griensven. 2014. Anti-inflammatory properties of the medicinal mushroom *Cordyceps militaris* might be related to its linear $(1{\rightarrow}3)$-β-D-glucan. *PLoS One*, 9(10): e110266.

Song, J., Y. Wang, M. Teng, G. Cai, H. Xu, H. Guo, D. Wang and L. Teng. 2015. Studies on the antifatigue activities of *Cordyceps militaris* fruit body extract in mouse model. *Evid. Based Complement. Alternat. Med.*, 2015: 174616.

Stringham, N.T., P.V. Holmes and J.M. Stringham. 2018. Supplementation with macular carotenoids reduces psychological stress, serum cortisol, and sub-optimal symptoms of physical and emotional health in young adults. *Nutr. Neurosci.*, 21(4): 286–296.

Sung, G.H. and J.W. Spatafora. 2004. *Cordyceps cardinalis* sp. nov., a new species of *Cordyceps* with an east Asian-eastern North American distribution. *Mycologia*, 96: 658–666.

Symeonides, S., F. Aroldi, N. Md Harris, S. Kestenbaumum, R. Plummer and S. Blagden. 2020. 600TiP A first-in-human study of, NUC-7738, a 3'-dA phosphoramidate, in patients with advanced solid tumours (NuTide: 701). *Ann. Oncol.*, 31(4): S501.

Takahashi, S., M. Tamai, S. Nakajima, H. Kato, H. Johno, T. Nakamura and M. Kitamura. 2012. Blockade of adipocyte differentiation by cordycepin. *Br. J. Pharmacol.*, 167(3): 561–575.

Tan, L., X. Song, Y. Ren, M. Wang, C. Guo, D. Guo, Y. Gu, Y. Li, Z. Cao and Y. Deng. 2021. Anti-inflammatory effects of cordycepin: A review. *Phytother. Res.*, 35(3): 1284–1297.

Tima, S., T. Tapingkae, C. To-Anun, P. Noireung, P. Intaparn, W. Chaiyana, J. Sirithunyalug, P. Panyajai, N. Viriyaadhammaa, W. Nirachonkul, L. Rueankham, W.L. Aung, F. Chueahongthong, S. Chiampanichayakul and S. Anuchapreeda. 2022. Antileukaemic cell proliferation and cytotoxic activity of edible golden *Cordyceps* (*Cordyceps militaris*) extracts. *Evid. Based. Complement. Alternat. Med.*, 2022: 5347718.

Tokish, J.M., M.S. Kocher and R.J. Hawkins. 2004. Ergogenic aids: A review of basic science, performance, side effects, and status in sports. *Am. J. Sports Med.*, 32(6): 1543–1553.

Tuli, H.S., S.S. Sandhu and A.K. Sharma. 2014. Pharmacological and therapeutic potential of *Cordyceps* with special reference to cordycepin. *3 Biotech.*, 4(1): 1–12.

Varrot, A., S.M. Basheer and A. Imberty. 2013. Fungal lectins: Structure, function and potential applications. *Curr. Opin. Struct. Biol.*, 23(5): 678–685.

Veena, R.K., E.J. Carmel, H. Ramya, T.A. Ajith, S.P. Wasser and K.K. Janardhanan. 2020. Caterpillar medicinal mushroom, *Cordyceps militaris* (Ascomycetes), mycelia attenuates doxorubicin-induced oxidative stress and upregulates krebs cycle dehydrogenases activity and ATP level in rat brain. *Int. J. Med. Mushrooms*, 22(6): 593–604.

Verma, A.K. and R. Aggarwal. 2020. Repurposing potential of FDA approved and investigational drugs for COVID-19 targeting SARS-CoV-2 spike and main protease and validation by machine learning algorithm. *Chem. Biol. Drug Des.*, 97(4): 836–853.

Wang, J., Y. Gong, H. Tan, W. Li, B. Yan, C. Cheng, J. Wan, W. Sun, C. Yuan and L.H. Yao. 2022. Cordycepin suppresses glutamatergic and GABAergic synaptic transmission through activation of A_1 adenosine receptor in rat hippocampal CA_1 pyramidal neurons. *Biomed. Pharmacother.*, 145: 112446.

Wang, L., N. Xu, J. Zhang, H. Zhao, L. Lin, S. Jia and L. Jia. 2015. Antihyperlipidemic and hepatoprotective activities of residue polysaccharide from *Cordyceps militaris* SU-12. *Carbohydr. Polym.*, 131: 355–362.

Wilga, M.S. 2010. New localities of the scarlet caterpillar club *Cordyceps militaris* [L.]. *Przegląd Przyrodniczy XXI*, 4: 11–15. [in polish].

Wilson, J.M., J.M. Joy, R.P. Lowery, M.D. Roberts, C.M. Lockwood, A.H. Manninen, J.C. Fuller, E.O. De Souza, S.M. Baier, S.M. Wilson and J.A. Rathmacher. 2013. Effects of oral adenosine-5'-triphosphate supplementation on athletic performance, skeletal muscle hypertrophy and recovery in resistance-trained men. *Nutr. Metab.*, 10(1): 1–11.

Wong, J.H., T.B. Ng, H. Wang, S.C.W. Sze, K.Y. Zhang, Q. Li and X. Lu. 2011. Cordymin, an antifungal peptide from the medicinal fungus *Cordyceps militaris*. *Phytomedicine*, 18(5): 387–392.

Wu, X., M. Zhang and Z. Li. 2019. Influence of infrared drying on the drying kinetics, bioactive compounds and flavor of *Cordyceps militaris*. *LWT - Food Sci. Technol.*, 111: 790–798.

Yamatsu, A., Y. Yamashita, T. Pandharipande, I. Maru and M. Kim. 2016. Effect of oral γ-aminobutyric acid (GABA) administration on sleep and its absorption in humans. *Food Sci. Biotechnol.*, 25(2): 547–551.

Yang, F.Q., K. Feng, J. Zhao and S.P. Li. 2009. Analysis of sterols and fatty acids in natural and cultured *Cordyceps* by one-step derivatization followed with gas chromatography-mass spectrometry. *J. Pharm. Biomed. Anal.*, 49: 1172–1178.

Ying, M., Q. Yu, B. Zheng, H. Wang, J. Wang, S. Chen, S. Nie and M. Xie. 2020. Cultured *Cordyceps sinensis* polysaccharides modulate intestinal mucosal immunity and gut microbiota in cyclophosphamide-treated mice. *Carbohydr. Polym.*, 235: 115957.

Yu, X.Y., Y. Zou, Q.W. Zheng, F.X. Lu, D.H. Li, L.Q. Guo and J.F. Lin. 2021. Physicochemical, functional and structural properties of the major protein fractions extracted from *Cordyceps militaris* fruit body. *Food Res. Int.*, 142: 110211.

Yu, H.M., B.S. Wang, S.C. Huang and P.D. Duh. 2006. Comparison of protective effects between cultured *Cordyceps militaris* and natural *Cordyceps sinensis* against oxidative damage. *J. Agric. Food Chem.*, 54(8): 3132–3138.

Zhang, D., Z. Wang, W. Qi, W. Lei and G. Zhao. 2014. Cordycepin (3'-deoxyadenosine) down-regulates the proinflammatory cytokines in the inflammation-induced osteoporosis model. *Inflammation*, 37(4): 1044–1049.

Zhang, D.W., H. Deng, W. Qi, G.Y. Zhao and X.R. Cao. 2015. Osteoprotective effect of cordycepin on estrogen deficiency-induced osteoporosis *in vitro* and *in vivo*. *Biomed. Res. Int.*, 2015: 423869.

Zhang, J., C. Wen, Y. Duan, H. Zhang and H. Ma. 2019. Advance in *Cordyceps militaris* (Linn) Link polysaccharides: Isolation, structure, and bioactivities: A review. *Int. J. Biol. Macromol.*, 132: 906–914.

Zhang, Z., K. Li, Z. Zheng and Y. Liu. 2022. Cordycepin inhibits colon cancer proliferation by suppressing MYC expression. BMC Pharmacol. *Toxicol.*, 23(1): 1–8.

Zhong, L., L. Zhao, F. Yang, W. Yang, Y. Sun and Q. Hu. 2017. Evaluation of anti-fatigue property of the extruded product of cereal grains mixed with *Cordyceps militaris* on mice. *J. Int. Soc. Sports Nutr.*, 14(1): 15.

Zhu, S., J. Pan, B. Zhao, J. Liang, J.W. Ze-Yu and J. Yang. 2013. Comparisons on enhancing the immunity of fresh and dry *Cordyceps militaris in vivo* and *in vitro*. *J. Ethnopharmacol.*, 149(3): 713–719.

Zhu, Z.Y., F. Liu, H. Gao, H. Sun, M. Meng and Y.M. Zhang. 2016a. Synthesis, characterization and antioxidant activity of selenium polysaccharide from *Cordyceps militaris*. *Int. J. Biol. Macromol.*, 93: 1090–1099.

Zhu, Z.Y., M.Z. Guo, F. Liu, Y. Luo, L. Chen, M. Meng, X.T. Wang and Y.M. Zhang. 2016b. Preparation and inhibition on α-D-glucosidase of low molecular weight polysaccharide from *Cordyceps militaris*. *Int. J. Biol. Macromol.*, 93: 27–33.

Biofertilizers and Biocontrol

17

Ectomycorrhizal Fungi as Biofertilizers

José Alfonso Domínguez-Núñez

1. Introduction

Biofertilizers can be products based on beneficial soil microorganisms, especially bacteria and/or fungi, living in association or in symbiosis with plants. Biofertilizers naturally help plant nutrition and growth and also improve soil quality. However, in Europe, the term "Biofertilizer" has recently been included within a more general concept, that of "Biostimulant" (EU 2019/1009). In this field, the term biofertilizer would instead be associated with "microbial plant biostimulant".

Ectomycorrhizal fungi (ECM) are higher fungi that develop mutualistic symbioses with several plant species, especially trees (Anderson and Cairney 2007). They are predominantly Basidiomycetes, but we also find some Ascomycetes and a few Zygomycetes. Within mycorrhizae, the Hartig network is the zone where the root and the mycorrhizal fungus exchange water, nutrients and other compounds. The external fungal mantle is connected to the extraradical mycelium that explores the surrounding soil. The ECM fungi are believed to have evolved from multiple saprotrophic species about 300 mya ago (Marcel et al. 2015). About 7,000 fungal species form ectomycorrhizae (Rinaldi et al. 2008). Many of these species form a mycorrhizal association with important commercial tree species (e.g., aspen, birch, oak and pine or spruce) (Wiensczyk 2002). Mushrooms are the carpophores of macromycetes that grow above ground, and truffles are those that develop underground. Many species of ectomycorrhizal mushrooms and truffles are edible

Department of Forest and Environmental Engineering and Management. Universidad Politécnica de Madrid. C/ Jose Antonio Novais, 10. 28040, Madrid, Spain.
Email: josealfonso.dominguez@upm.es

(Savoie and Largeteau 2011). The cultivation of some ECM fungi has been studied in detail, such as boletes (*Boletus edulis*), chanterelles (*Cantharellus* spp.), matsutake (*Tricholoma matsutake*) and black truffle (*Tuber melanosporum*) (Bencivenga 1998, Chevalier 1998).

The functionality and diversity of forest ecosystems depend to a large extent on ECM fungal mycelial networks, which communicate through the soil to a community of plants and trees, as well as promote host plant growth and health (Leake et al. 2004, Smith and Read 2008, Courty et al. 2010a).

Since the 1950s, mycorrhizal fungi have been applied to crops in order to promote and improve plant growth (Feldmann et al. 2009, Koide and Mosse 2004, Miransari 2011). The ECM fungi help to break down the complex minerals and organic matter in the soil and translocate nutrients to the tree. The ECM fungi also appear to increase the tolerance of trees to drought, high temperatures, toxins, and pH extremes.

For example, the ability of ECM fungi as phosphate mobilizers or P solubilizers has been shown in many ECM genera such as *Amanita* sp., *Boletus* sp., *Cenococcum* sp., *Cortinarius* sp., *Elaphomyces* sp., *Laccaria* sp., *Lactarius* sp., *Paxillus* sp., *Piloderma* sp., *Piriformospora* sp., *Pisolithus* sp., *Rhizopogon* sp., *Suillus* sp. and *Tuber* sp. (Lapeyrie et al. 1987, 1991, Arvieu et al. 2003, Courty et al. 2010b, Plassard et al. 2011, Anderson and Cairney 2007).

The ability of ECM fungi as potassium solubilizers (Luciano et al. 2010), or Zinc solubilizers in soil, has also been observed (Naz et al. 2016). A variety of rhizobacterial taxa (e.g., *Bacillus* sp., and *Pseudomonas* sp.), mycorrhizal fungi, or *Saccharomyces* sp. have been reported to increase Zn availability in soil. Zinc solubilization takes place when these microorganisms use chelated ligands and oxidoreductant systems (Kamran et al. 2017).

Much of our knowledge about the functions of ECM fungi in their ecosystems is derived from research aimed at the practical application of ECM fungi in forestry. All these functional capabilities of ECM fungi have motivated the need to "domesticate" their cultivation and their application in the environmental field, as we will discuss below.

2. Molecular Mechanisms of Ectomycorrhizal Fungi as Biofertilizers

Whole genome sequencing of *Laccaria bicolor* and *Tuber Melanosporum* (Martin et al. 2008, 2010) has allowed us to define some variables that regulate mycorrhizal growth and operate at the symbiotic interface of the tree-fungus complex. In *L. bicolor*, 15 genes suspected to encode hexose transporters were recognized, many of them regulated during symbiosis. However, *L. bicolor* is not able to hydrolyze sucrose (abundant in vegetables) into glucose and fructose (since it has no genes encoding invertases), so it depends on its host for glucose uptake. However, *T. melanosporum* has an invertase gene, unlike *L. bicolor*, and can utilize sucrose produced by the host. Other ECM fungi, such as *Amanita muscaria* and *Hebeloma cylindrosporum* are also known to metabolize fructose. This range of mechanisms for accessing plant sugars by ECM fungi appears to reflect some of the differences in nutritional strategies among symbiotic fungi (Bonfante et al. 2010).

Mutualistic interactions between ECM fungi and their host plants (usually forest species) are based on a nutrient exchange; it is expected that fungal genes encoding transporter proteins are well represented in the genomes of ECM fungi. The regulation of transporter genes during symbiosis indicates very intense trafficking across the symbiotic interface of amino acid compounds, oligopeptides and polyamines. Organic and inorganic N and P from soil can be imported by ECM mycelium via ammonium transporters, inorganic P transporters or urea permease. Such compounds then enter the root, the Hartig network (symbiotic interface) and are transferred to crops or trees. The production of cysteine-rich fungal proteins with possible effector function during tree-ECM interaction, such as MISSP7, seems to play an important role, although not for all ECM species (Plett et al. 2011).

3. Inoculation of Ectomycorrhizal Fungi and Formulations

The delivery of biofertilizer products to plants, e.g., products that include ECM fungi, means the incorporation of these fungi into the root or rhizosphere environment of the host plant, i.e., their inoculation.

Traditionally, the three main sources of ECM inoculum used to inoculate plants have been natural soil, spores and fungal mycelium.

Primarily, soil or humus collected from a natural region of interest for the possible presence of ECM fungal propagules was often used. In this case, problems such as the presence of possible soil pathogens, lack of knowledge about possible ECM fungi species in the area, or lack of sufficient inoculum are frequent. Although this method is still applied in low-income countries, since few inputs are needed, it is not a good method of mycorrhizal control. Soil inoculation with chopped and ectomycorrhizal roots has also been applied in nurseries to neighboring seedlings. (Sim and Eom 2006).

Another source of inoculum is spores from ECM carpophores collected in the field. In this case, the spore inoculum is light and an aseptic culture is not necessary. (Marx and Cordell 1989). The most widespread ectomycorrhizal product has been the inoculum (both spores and mycelium) of the ECM species *Pisolithus tinctorius* (Schwartz et al. 2006), on a wide range of forest hosts. Inoculation with spores of ECM species such as *Rhizopogon* spp. has also proved successful; for example, *Pinus radiata* seeds coated with *Rhizopogon luteolus* spores gave rise to *Rhizopogon* mycorrhizal seedlings (Sharma et al. 2017).

However, spore inoculation has traditionally had three major drawbacks: (a) The preservation and viability of spores is a fundamental factor in the success of mycorrhization; (b) significant quantities of fresh carpophores are necessary, and may not be available every year; (c) lack of genetic definition. To preserve the viability of the spores, storage at low temperatures and darkness, as well as freeze-drying, is advisable. There are different formats of spore inoculum depending on the physical carriers of the spores, being applied on the soil in liquid form or as powder, granules, etc.; spores can also be coated or encapsulated on host plant seeds or can be embedded in hydrocolloid chips (Marx and Cordell 1989).

Traditionally, inoculation with ECM mycelium is considered the best procedure, as it allows the selection of specific strains of ECM species (Marx 1980). In general, small portions of carpophores are grown under sterile conditions and give rise to mycelium; sometimes hyphae can be obtained from sclerotia, ectomycorrhizae or sexual spores (Molina and Palmer 1982, Trappe 1969, Fries and Birraux 1980), although the latter is not as effective methods. The mycelium can be carried in liquid or solid medium; the liquid medium is more easily contaminated and involves higher costs in the preparation of the inoculum (Rossi et al. 2007, Cannel and Moo-Young 1980) compared to the solid medium. Genera such as *Amanita, Hebeloma, Laccaria, Pisolithus, Rhizopogon* and *Suillus* are well adapted to mycelial cultures.

The main disadvantage of using mycelial inocula is, that some ECM fungal species are difficult to grow in laboratory conditions, or their growth is very slow. In large-scale plantations, sufficient mycorrhizal seedling production by mycelium inoculation can be complex, as large quantities of mycelium will be needed. The use of mycelium encapsulated in calcium alginate "beads" (Le Tacon et al. 1983) has helped to improve mycelial inocula by increasing the period of mycelial viability under refrigeration. This method may be suitable for the inoculation of seedlings used in the reforestation of large areas, and with the application of approximately 5 g of dry mycelium/plant (Rossi et al. 2007).

As mentioned above, the use of alginate as a protective encapsulating material for microorganisms is quite common. Alginate is a natural polymer, a good choice for formulations with entrapped polymers, and is currently being applied. It is obtained in sustainable quantities from marine macroalgae as well as various bacteria (Yabur et al. 2007, Smidsrød and Skja 1990). Alginate formulations enhance the survival of microbial inoculants, as well as support a high population rate under elevated temperature conditions of 40°C (Viveganandan and Jauhri 2000). Alginate gel allows the generation of a multimicrobial inoculant. Results have demonstrated the superiority of polymer-trapped formulations over all other formulations, such as polymeric formulations of Arbuscular Mycorrhizal and ECM fungi, and many P-solubilizing bacteria (Bashan 1998, Fages 1990, Vassilev et al. 2005).

Although most of the biofertilizers registered worldwide correspond to liquid or solid formulations (powders and granules), new trends in the research and development of this type of product seek to replace them with more innovative formulations, with controlled release matrices or hydrogels based on polymers or polymeric materials (Da Silva et al. 2012). In these new formulations, microorganisms immobilized in matrices could present greater viability under storage conditions, in the soil, or at the target site of action on the plant (Marcelino et al. 2016, Schoebitz et al. 2013).

4. MHB and Co-Inoculation of Microorganisms

The study of "Mycorrhizal Helper Bacteria" (MHB) (Garbaye 1994) has deepened the knowledge of the mycorrhizal fungus-plant interaction (Garbaye and Duponnois 1992, Frey-Klett et al. 2007). The MHB belongs to a wide range of genera, such as *Burkholderia* sp., *Paenibacillus* sp. (Poole et al. 2001), *Pseudomonas* sp. and

Bacillus sp. (Duponnois and Garbaye 1991, Frey-Klett et al. 2007), *Streptomyces* sp. (Maier et al. 2004) and *Rhodococcus* sp. (Heredia-Acuña et al. 2019).

The molecular mechanisms promoting the growth of some ECM fungi through MHBs are still unclear. In some cases, changes in the gene expression of some ECM fungi have been studied at the molecular and functional level when confronted with certain MHBs (Schrey et al. 2005, Riedlinger et al. 2006, Deveau et al. 2007, Zhou et al. 2014). A better knowledge of these mechanisms may facilitate the development of effective combinations of microorganisms as "ad hoc" inoculums (Artursson et al. 2006).

On the other hand, ECM species may show synergistic interactions with other beneficial plant microorganisms; for example, symbiotic N-fixers such as *Bradyrhizobium* sp. (Andre et al. 2005), or synergies between *Glomus mosseae* (VAM), *Pisolithus tinctorius* (ECM) and *Bradyrhizobium* sp. inoculated on *Acacia nilotica* (Saravanan and Natarajan 1996, 2000).

In any case, dual inoculation (bacteria-ECM) has been extensively studied with positive, neutral, or negative responses depending on the affinity between the included microorganisms (Bonfante and Anca 2009, Hrynkiewicz et al. 2010, Wu et al. 2012).

5. ECM Fungi and the Environmental Management

The main applications of ECM fungi in the environmental field are currently related to their use in revegetation, reforestation, restoration of degraded areas, soil phytoremediation, as well as a phytosanitary product (biocontrol). As contemplated in the recent EU regulation on the market of fertilizer products (EU 2019/1009), the application of microorganisms for phytosanitary purposes should be excluded from the concept of "biostimulant" and, therefore, from that of "biofertilizer". We will now comment on the other possible applications mentioned.

5.1 Reforestation Mediated by ECM Fungi

Inoculation of ECM fungi in forest species can have two different objectives: to obtain mushrooms and/or truffles, or to improve the growth and functional quality of their host plant; this has had economic implications (Garbaye 1990) and in the management of the environment. Mycorrhizal plantations have been especially successful in environments that are particularly stressful for seedlings, such as drought or nutritional stress (Duñabeitia et al. 2004). In productive or conservation afforestations, the introduction of dominant ECM species is not desirable to conserve and enhance soil fungal biodiversity. In areas with previous native ECM communities, the introduction of new ECM fungi by planting has not always been successful, perhaps due to poor selection of the ECM species to be introduced; on the other hand, species such as *Pisolithus tinctorius* have been successful, especially in degraded or stressful environments (McAfee and Fortin 1986).

Currently, processes such as erosion, drought, salinization, pollution or lack of soil fertility are increasing soil degradation. Soil microbiota can be very important in maintaining a balanced ecosystem. All biogeochemical processes and the availability of nutrients in the soil depend to a large extent on it. Some ECM fungi can withstand stressful environments for plants, due to temperatures (Tibbett and Cairney 2007, Geml et al. 2011), salinity and metals (Guerrero-Galán et al. 2019, Colpaert et al. 2011), drought (Azul et al. 2010), etc. The application of ECM fungi in environmental management can be very positive, increasing the tolerance of mycorrhizal plants in these environmental conditions. We can talk about bioremediation of contaminated soils, biocontrol of plant diseases and pests, or simply improve the growth of a host plant by using ECM fungi.

The role of ECM fungi on plant aquaporins (AQP) is known; thus, ECM fungi are able to enhance water uptake under drought conditions by modifying plant apoplast and symplast (Lehto and Zwiazek 2011, Maurel and Plassard 2011, Nehls and Dietz 2014). It has also been documented, that symbiosis between plants and ECM fungi helps plants to cope with salt stress (Ishida et al. 2009, Luo et al. 2011, Guerrero-Galán et al. 2019). It has been observed that some ECM fungi are able to maintain the K^+/ Na^+ homeostatic balance (Li et al. 2012) or to change the phytohormonal balance under salt stress conditions (Luo et al. 2009, Szuba 2015). In any case, new pioneer ECM symbionts could be useful in future reforestations (Beaudoin-Nadeau et al. 2016).

5.2 Bioremediation

In the study of plant-microbe interactions, the phytoremediation potential of ECM fungi has been extensively reviewed (Doty 2008, Lebeau et al. 2008, Rajkumar et al. 2012, Khanday et al. 2016).

Trees are useful in phytoremediation as they produce large amounts of biomass (Dickinson and Pulford 2005), can detoxify or be metal tolerant (Bhat et al. 2018), or be associated with bacterial endophytes and mycorrhizal fungi that help recover contaminated soils (van der Lelie et al. 2009, Mrnka et al. 2012). Symbiotic association of trees with ECM fungi can be an advantage in contaminated soils (Kernaghan et al. 2002), enhancing growth, and preventing metal accumulation in their tissues (Adriaensen et al. 2004, 2005, Jourand et al. 2010, Kayama and Yamanaka 2014, Luo et al. 2014), or emitting metallothioneins and glutathione that stabilize metals (Akashi et al. 2004, Khullar and Reddy 2018).

Persistent organic pollutants (POPs) in the soil can also be stabilized by ECM fungi (Meharg and Cairney 2000). ECM fungi can degrade polychlorinated biphenyls (PCBs) and aid in the dissipation of polycyclic aromatic hydrocarbons (Genney et al. 2004, Joner et al. 2006).

The formation of biological barriers (Krupa and Kozdrój 2004), the emission of low molecular weight organic acids (LMWOAc) and metal chelating agents may be other mechanisms of ECM fungi to bind or compartmentalize metal ions and control their transmission to the host plant (Machuca 2011). It is in vacuoles where metal ions are likely forming complexes in chemically inactive forms (Krpata et al. 2009, Colpaert 2011, Luo et al. 2014, Daghino et al. 2016).

6. New European Union Regulation on Fertilizer Products (EU 2019/1009) - Implications

The term "Biofertilizer" at the European level is currently included within the concept of "Biostimulant". Recently the EU has developed a new regulation on the market of fertilizer products, where the concept of "plant biostimulant" is incorporated. According to the EU 2019/1009, "Plant biostimulant" (PIC 6) means an EU fertilizer product whose function is to stimulate plant nutrition processes, irrespective of the nutrient content of the product, with the sole objective of improving one or more characteristics of plants and their rhizosphere as indicated below: (a) nutrient use efficiency, (b) tolerance to abiotic stress, (c) plant quality characteristics, or (d) availability of immobilized nutrients in the soil and rhizosphere.

It also defines "Microbial plant biostimulant" (CFP 6A) as consisting of a microorganism or a group of microorganisms containing *Azospirillum* spp., *Azotobacter* spp., *Rhizobium* spp or Mycorrhizal fungi. As we can see, the bacterial genera that may be incorporated into fertilizer products are limited to the three mentioned above; with respect to mycorrhizal fungi, no genera or species are specified, and the regulation does not even distinguish between the two major mycorrhizal, endomycorrhizal or ECM groups.

A "Microbial Plant Biostimulant" fertilizer product may contain microorganisms, including dead microorganisms or empty cells of microorganisms, and non-harmful residues of the culture media in which they grew, which have not been subjected to any treatment other than drying or freeze-drying. EU 2019/1009 regulation also requires the declaration of all microorganisms intentionally added to the fertilizer product. In case the microorganism has several strains, the added strains should be indicated. Their concentration should be expressed as a number of active units per volume or weight or in any other manner relevant to the microorganism in question (e.g., colony forming units per gram: cfu/g). In the case of mycorrhizal fungi with sporal inoculum, it could be the number of spores/g and the number of spores/ml.

The EU 2019/1009 regulation will probably apply from July 2022. In any case, new materials and microorganisms other than those cited in the EU 2019/1009 regulation, as well as industrial processes that are currently not covered by the EU 2019/1009 regulation, are currently used by several EU member countries to produce fertilizer products. In this regard, the EU technical group GROW/F2 considers the possibility of studying future developments to adapt the annexes of this regulation to technical progress.

7. Conclusion

Research on ECM fungi should address a better knowledge of the functional and molecular mechanisms involved in the mycorrhizal symbiotic interface. From the applicational point of view of these fungi, we should aim to find commercial techniques for large-scale multiplication andinoculation, and formulate "ad hoc" combinations of plant and biostimulant for specific environmental conditions.

The role of ECM fungi as biofertilizers and environmental restorers has been very important so far; ECM fungi can provide stability in ecosystems, increasing plant tolerance against possible biotic and abiotic stresses. The new European regulation on fertilizer products does not provide new information regarding the use of mycorrhizal fungi.

References

Adriaensen, K., D. van der Lelie, A. Van Laere, J. Vangronsveld and J.V. Colpaert. 2004. A zinc-adapted fungus protects pines from zinc stress. *New Phytol.*, 161: 549–555. https://doi.org/10.1046/j.1469-8137.2003.00941.x.

Adriaensen, K., T. Vra°lstad, J.P. Noben, J. Vangronsveld and J.V. Colpaert. 2005. Copper-adapted *Suillus luteus*, a symbiotic solution for pines colonizing Cu mine spoils. *Appl. Environ. Microbiol.*, 71: 7279–7284. https://doi.org/10.1128/AEM.71.11.7279-7284.2005.

Akashi, K., N. Nishimura, Y. Ishida and A. Yokota. 2004. Potent hydroxyl radical-scavenging activity of drought-induced type-2 metallothionein in wild watermelon. *Biochem. Biophys. Res. Commun.*, 323: 72–78.

Anderson, C.I. and W.G.J. Cairney. 2007. Ectomycorrhizal fungi: Exploring the mycelial frontier. *FEMS Microbiol. Rev.*, 31: 388–406.

Andre, S., A. Galiana, C. Le Roux, Y. Prin, M. Neyra and R. Duponnois. 2005. Ectomycorrhizal symbiosis enhanced the efficiency of inoculation with two Bradyrhizobium strains and Acacia holosericea growth. *Mycorrhiza*, 15: 357–364.

Artursson, V., R.D. Finlay and J.K. Jansson. 2006. Interactions between arbuscular mycorrhizal fungi and bacteria and their potential for stimulating plant growth. *Environ. Microbiol.*, 8: 1–10.

Arvieu, J.-C., F. Leprince and C. Plassard. 2003. Release of oxalate and protons by ectomycorrhizal fungi in response to P-deficiency and calcium carbonate in nutrient solution. *Ann. For. Sci.*, 60: 815–821.

Azul, A.M., J.P. Sousa, R. Agerer, M.P. Martín and H. Freitas. 2010. Land use practices and ectomycorrhizal fungal communities from oak woodlands dominated by Quercus suber L. considering drought scenarios. *Mycorrhiza*, 20: 73–88.

Bashan, Y. 1998. Inoculants of plant growth-promoting bacteria for use in agriculture. *Biotechnol. Adv.*, 16(4): 729–770.

Beaudoin-Nadeau, M., A. Gagne, C. Bissonnette, P. Belanger, J. Fortin, S. Roy, C.W. Greer and D.P. Khasa. 2016. Performance of ectomycorrhizal alders exposed to specific Canadian oil sands tailing stressors under *in vivo* bipartite symbiotic conditions. *Can. J. Microbiol.*, 62(7): 543–549. https://doi.org/10.1139/cjm-2015- 0703.

Bencivenga, M. 1998. Ecology and cultivation of *Tuber magnatum* Pico. *In*: Proceedings of the first International Meeting on Ecology, Physiology and Cultivation of Edible Mycorrhizal Mushrooms. Swedish University of Agricultural Sciences, Uppsala, Sweden, 3–4 July.

Bhat, R.A., M.A. Dervash, M.A. Mehmood, B.M. Skinder, A. Rashid, J.I.A. Bhat, D.V. Singh and R. Lone. 2018. Mycorrhizae: A sustainable industry for plant and soil environment. pp. 473–502. *In*: Varma, A., Prasad, R. and Tuteja, N. (eds.). Mycorrhiza—Nutrient Uptake, Biocontrol, Ecorrestoration. Springer.

Bonfante, P. and I.A. Anca. 2009. Plants, mycorrhizal fungi, and bacteria: A network of interactions. *Annu. Rev. Microbiol.*, 63: 363–383.

Bonfante, P. and A. Genre. 2010. Mechanisms underlying beneficial plant–fungus interactions in mycorrhizal symbiosis. *Nat. Commun.*, 1(1): 1–11.

Cannel, E. and M. Moo-Young. 1980. Solidstate fermentation systems. *Process Biochem.*, 15: 24–28.

Chevalier, G. 1998. The truffle cultivation in France: Assessment of the situation after 25 years of intensive use of mycorrhizal seedlings. In: Proceedings of the first International Meeting on Ecology, Physiology and Cultivation of Edible Mycorrhizal Mushrooms. Swedish University of Agricultural Sciences, Uppsala, Sweden, 3–4 July.

Colpaert, J.V., J.H.L. Wevers, E. Krznaric and K. Adriaensen. 2011. How metal-tolerant ecotypes of ectomycorrhizal fungi protect plants from heavy metal pollution. *Ann. For. Sci.*, 68: 17–24.

Courty, P.E., A. Franc and J. Garbaye. 2010. Temporal and functional pattern of secreted enzyme activities in an ectomycorrhizal community. *Soil Biol. Biochem.*, 42(11): 2022–2025.

Courty, P.-E., M. Buée, A.G. Diedhiou, P. Frey-Klett, F. Le Tacon, F. Rineau, M.-P. Turpault, S. Uroz and J. Garbaye. 2010. The role of ectomycorrhizal communities in forest ecosystem processes: New perspectives and emerging concepts. *Soil Biol. Biochem.*, 42: 679–698.

Da Silva, M.F., C. de Souza Antônio, P.J. de Oliveira, G.R. Xavier, N.G. Rumjanek, L.H. de Barros Soares and V.M. Reis. 2012. Survival of endophytic bacteria in polymer-based inoculants and efficiency of their application to sugarcane. *Plant and Soil*, 356(1): 231–243. https://doi.org/10.1007/s11104-012-1242-3.

Daghino, S., E. Martino and S. Perotto. 2016. Model systems to unravel the molecular mechanisms of heavy metal tolerance in the ericoid mycorrhizal symbiosis. *Mycorrhiza*, 26: 263–274. https://doi.org/10.1007/s00572-015-0675-y.

Deveau, A., B. Palin, C. Delaruelle, M. Peter, A. Kohler, J.C. Pierrat, A. Sarniguet, J. Garbaye, F. Martin and P. Frey-Klett. 2007. The mycorrhiza helper Pseudomonas fluorescens BBc6R8 has a specific priming effect on the growth, morphology and gene expression of the ectomycorrhizal fungus *Laccaria bicolor* S238N. *New Phytologist*, 175: 743–755.

Dickinson, N.M. and I.D. Pulford. 2005. Cadmium phytoextraction using short-rotation coppice Salix: The evidence trail. *Environ. Int.*, 31: 609–613.

Doty, S.L. 2008. Enhancing phytoremediation through the use of transgenics and endophytes. *New Phytol.*, 179: 318–333.

Duñabeitia, M., N. Rodriíguez, I. Salcedo and E. Sarrionandia. 2004. Field mycorrhization and its influence on the establishment and development of the seedlings in a broadleaf plantation in the Basque country. *For. Ecol. Manage.*, 195: 129–139.

Duponnois, R. and J. Garbaye. 1991. Mycorrhizal helper bacteria associated with the Douglas fir Laccaria laccata symbiosis: Effects in aseptic and in glasshouse conditions. *Ann. Sci. For.*, 48: 239–251.

EU. 2019/1009. REGULATION (EU) 2019/1009 of the European Parliament and of the Council of 5 June 2019 laying down rules on the making available on the market of EU fertilising products and amending Regulations (EC) No 1069/2009 and (EC) No 1107/2009 and repealing Regulation (EC) No 2003/2003.

Fages, J. 1990. An optimized process for manufacturing an Azospirillum inoculant for crops. *Appl. Microbiol. Biotechnol.*, 32(4): 473–478.

Feldmann, F., I. Hutter and C. Schneider. 2009. Best production practice of arbuscular mycorrhizal inoculum. pp. 319–336. *In*: Varma, A. and Kharkwal, A.C. (eds.). Symbiotic Fungi: Principles and Practice. vol. 18. Springer, Berlin Heidelberg.

Frey-Klett, P., J. Garbaye and M. Tarkka. 2007. The Mycorrhiza helper bacteria revisited. *New Phytol.*, 176: 22–36.

Fries, N. and D. Birraux. 1980. Spore germination in Hebeloma stimulated by living plant roots. *Experimentia*, 36: 1056–1057.

Garbaye, J. 1990. Use of mycorrhizas in forestry. pp. 197–248. *In*: Strullu, D.G. (ed.). Les mycorhizes des arbres et plantes cultivées. Lavoisier, Paris.

Garbaye, J. 1994. Helper bacteria: A new dimesion to the mycorrhizal symbiosis. *New Phytologist*, 128(2): 197–210.

Garbaye, J. and R. Duponnois. 1992. Specificity and function of mycorrhization helper bacteria (MHB) associated with the Pseudotsuga menziesii–Laccaria laccata symbiosis. *Symbiosis*, 14: 335–344.

Geml, J., I. Timling, C.H. Robinson, N. Lennon, H.C. Nusbaum, C. Brochmann, M.E. Noordeloos and D.L. Taylor. 2011. Antarctic community of symbiotic fungi assembled by long-distance dispersers: Phylogenetic diversity of ectomycorrhizal basidiomycetes in Svalbard based on soil and sporocarp DNA. *J. Biogeogr.*, 39: 74–88.

Genney, D.R., I.J. Alexander, K. Killham and A.A. Meharg. 2004. Degradation of the polycyclic aromatic hydrocarbon (PAH) fluorene is retarded in a Scots pine ectomycorrhizosphere. *New Phytol.*, 163: 641–649.

Guerrero-Galan, C., M. Calvo-Polanco and S.D. Zimmermann. 2019. Ectomycorrhizal symbiosis helps plants to challenge salt stress conditions. *Mycorrhiza*, 29: 291–301. https://doi.org/10.1007/s00572-019-00894-2.

Heredia-Acuña, C., J.J. Almaraz-Suarez, R. Arteaga-Garibay, R. Ferrera-Cerrato and D.Y. Pineda-Mendoza. 2019. Isolation, characterization and effect of plant-growth-promoting rhizobacteria on pine seedlings (*Pinus pseudostrobus* Lindl.). *J. For. Res.*, 30: 1727–1734. https://doi.org/10.1007/s11676-018-0723-5.

Hrynkiewicz, K., C. Baum and P. Leinweber. 2010. Density, metabolic activity, and identity of cultivable rhizosphere bacteria on *Salix viminalis* in disturbed arable and landfill soils. *J. Plant Nutr. Soil Sci.*, 173(5): 747–756.

Ishida, T.A., K. Nara, S. Ma, T. Takano and S. Liu. 2009. Ectomycorrhizal fungal community in alkaline-saline soil in northeastern China. *Mycorrhiza*, 19(5): 329–335. https://doi.org/10.1007/s00572-008-0219-9.

Joner, E.J., C. Leyval and J.V. Colpaert. 2006. Ectomycorrhizas impede phytoremediation of polycyclic aromatic hydrocarbons (PAHs) both within and beyond the rhizosphere. *Environ. Pollut.*, 142: 34–38.

Jourand, P., M. Ducousso, R. Reid, C. Majorel, C. Richert, J. Riss and M. Lebrun. 2010. Nickel-tolerant ectomycorrhizal Pisolithus albus ultramafic ecotype isolated from nickel mines in New Caledonia strongly enhance growth of the host plant Eucalyptus globulus at toxic nickel concentrations. *Tree Physiol.*, 30: 1311–1319. https://doi.org/10.1093/treephys/tpq070.

Kamran, S., I. Shahid, D.N. Baig, M. Rizwan, K.A. Malik and S. Mehnaz. 2017. Contribution of zinc solubilizing bacteria in growth promotion and zinc content of wheat. *Front. Microbiol.*, 8: 2593. https://doi.org/10.3389/fmicb.2017.02593.

Kayama, M. and T. Yamanaka. 2014. Growth characteristics of ectomycorrhizal seedlings of *Quercus glauca*, *Quercus salicina*, and *Castanopsis cuspidate* planted on acidic soil. *Trees*, 28: 569–583. https://doi.org/10.1007/s00468-013-0973-y.

Kernaghan, G., B. Hambling, M. Fung and D. Khasa. 2002. *In vitro* selection of boreal ectomycorrhizal fungi for use in reclamation of saline–alkaline habitats. *Restor. Ecol.*, 10: 43–51.

Khanday, M., R.A. Bhat, S. Haq, M.A. Dervash, A.A. Bhatti, M. Nissa and M.R. Mir. 2016. Arbuscular mycorrhizal fungi boon for plant nutrition and soil health. pp. 317–332. *In*: Hakeem, K.R., Akhtar, J. and Sabir, M. (eds.). Soil Science: Agricultural and Environmental Prospective. Springer International Publishing, Switzerland.

Khullar, S. and M.S. Reddy. 2018. Ectomycorrhizal fungi and its role in metal homeostasis through metallothionein and glutathione mechanisms. *Curr. Biotechnol.*, 7(3): 231–241. https://dx.doi.org/1 0.2174/2211550105666160531145544.

Koide, R. and B. Mosse. 2004. A history of research on arbuscular mycorrhiza. *Mycorrhiza*, 14: 145–163.

Krpata, D., W. Fitz, U. Peintner, I. Langer and P. Schweiger. 2009. Bioconcentration of zinc and cadmium in ectomycorrhizal fungi and associated aspen trees as affected by level of pollution. *Environ. Pollut.*, 157: 280–286. https://doi.org/10.1016/j.envpol.2008.06.038.

Krupa, P. and J. Kozdrój. 2004. Accumulation of heavy metals by ectomycorrhizal fungi colonizing birch trees growing in an industrial desert soil. *World J. Microbiol. Biotechnol.*, 20(4): 427–430.

Lapeyrie, F., G.A. Chilvers and C.A. Bhem. 1987. Oxalic acid synthesis by the mycorrhizal fungus Paxillus involutus (Batsch. ex Fr.). Fr. *New Phytologist*, 106: 139–146.

Lapeyrie, F., J. Ranger and D. Vairelles. 1991. Phosphate-solubilizing activity of ectomycorrhizal fungi *in vitro*. *Canad. J. Bot.*, 69: 342–346.

Le Tacon, F., G. Jung, J. Mugnier and P. Michelot. 1983. Efficiency in a forest nursery of an inoculant of an ectomycorrhizal fungus produced in a fermentor and entrapped in polymetric gels. *Ann. of Forest Sci.*, 40: 165–176.

Leake, J., D. Johnson, D. Donnelly, G. Muckle, L. Boddy and D. Read. 2004. Networks of power and influence: The role of mycorrhizal mycelium in controlling plant communities and agroecosystem functioning. *Can. J. Bot.*, 82(8): 1016–1045.

Lebeau, T., A. Braud and K. Jezequel. 2008. Performance of bioaugmentation-assisted phytoextraction applied to metal contaminated soils: A review. *Environ. Pollut.*, 153: 497–522.

Lehto, T. and J.J. Zwiazek. 2011. Ectomycorrhizas and water relations of trees: A review. *Mycorrhiza*, 21(2): 71–90. https://doi.org/10.1007/s00572-010-0348-9.

Luciano, A., V.L. Oliveira, F. Silva and N. Germano. 2010. Utilization of rocks and ectomycorrhizal fungi to promote growth of eucalypt. *Braz. J. Microbiol.*, 41: 676–684.

Luo, Z.B., D. Janz, X. Jiang, C. G€obel, H. Wildhagen, Y. Tan, H. Rennenberg, I. Feussner and A. Polle. 2009. Upgrading root physiology for stress tolerance by ectomycorrhizas: Insights from metabolite and transcriptional profiling into reprogramming for stress anticipation. *Plant Physiol.*, 151: 1902–1917. https://doi.org/10.1104/ pp. 109.143735.

Luo, Z.B., K. Li, Y. Gai, C. Gobel, H. Wildhagen, X. Jiang, I. Feußner, H. Rennenberg, A. Polle. 2011. The ectomycorrhizal fungus (*Paxillus involutus*) modulates leaf physiology of poplar towards improved salt tolerance. *Environ. Exp. Bot.*, 72: 304–311. https://doi.org/10.1016/j.envexpbot.2011.04.008.

Luo, Z.B., C. Wua, C. Zhang, H. Lic, U. Lipkad and A. Polle. 2014. The role of ectomycorrhizas in heavy metal stress tolerance of host plants. *Environ. Exp. Bot.*, 108: 47–62. https://doi.org/10.1016/j.envexpbot.2013.10.018.

Machuca, A. 2011. Metal-chelating agents from ectomycorrhizal fungi and their biotechnological potential. pp. 25. *In*: Rai, M. and Varma, A. (eds.). Diversity and Biotechnology of Ectomycorrhizas. Soil Biology. Springer-Verlag, Berlin, Heidelberg.

Maier, A., J. Riedlinger, H.P. Fiedler and R. Hampp. 2004. Actinomycetales bacteria from a spruce stand: Characterization and effects on growth of root symbiotic and plant parasitic soil fungi in dual culture. *Mycol. Prog.*, 3(2): 129–136.

Marcel, G., A. van der Heijden, F.M. Martin, M.A. Selosse and I.R. Sanders. 2015. Mycorrhizal ecology and evolution: The past, the present, and the future. *New Phytol.*, 205: 1406–1423. https://doi.org/10.1111/nph.13288.

Marcelino, P.R.F., K.M.L. Milani, S. Mali, O.J.A.P. dos Santos and A.L.M. de Oliveira. 2016. Formulations of polymeric biodegradable low-cost foam by melt extrusion to deliver plant growth promoting bacteria in agricultural systems. *Appl. Microbiol. Biotechnol.*, 100(16): 7323–7338. https://doi.org/10.1007/s00253-016-7566-9.

Martin, F., A. Aerts, d. Ahrén, A. Brun, E.G. Danchin, F. Duchaussoy, J. Gibon, A. Kohler, E. Lindquist, V. Pereda, A. Salamov, H.J. Shapiro, J. Wuyts, D. Blaudez, M. Buée, P. Brokstein, B. Canbäck, D. Cohen, P.E. Courty, P.M. Coutinho, C. Delaruelle, J.C. Detter, A. Deveau, S. DiFazio, S. Duplessis, L. Fraissinet-Tachet, E. Lucic, P. Frey-Klett, C. Fourrey, I. Feussner, G. Gay, J. Grimwood, P.J. Hoegger, P. Jain, S. Kilaru, J. Labbé, Y.C. Lin, V. Legué, F. Le Tacón, R. Marmeisse, D. Melayah, B. Montanini, M. Muratet, U. Nehls, H. Niculita-Hirzel, M.P. Oudot-Le Secq, M. Peter, H. Quesneville, B. Rajashekar, M. Reich, N. Rouhier, J. Schmutz, T. Yin, M. Chalot, B. Henrissat, U. Kües, S. Lucas, Y. Van de Peer, G.K. Podila, A. Polle, P.J. Pukkila, P.M. Richardson, P. Rouzé, I.R. Sanders, J.E. Stajich, A.Tunlid, G. Tuskan and I.V. Grigoriev. 2008. The genome of *Laccaria bicolor* provides insights into mycorrhizal symbiosis. *Nature*, 452(7183): 88–92. doi: 10.1038/nature06556.

Martin, F., A. Kohler, C. Murat, R. Balestrini, P.M. Coutinho, O. Jaillon, B. Montanini, E. Morin, B. Noel, R. Percudani, B. Porcel, A. Rubini, A. Amicucci, J. Amselem, V. Anthouard, S. Arcioni, F. Artiguenave, J.M. Aury, P. Ballario, A. Bolchi, A. Brenna, A. Brun, M. Buée, B. Cantarel, G. Chevalier, A. Couloux, C. Da Silva, F. Denoeud, S. Duplessis, S. Ghignone, B. Hilselberger, M. Iotti, B. Marçais, A. Mello, M. Miranda, G. Pacioni, H. Quesneville, C. Riccioni, R. Ruotolo, R. Splivallo, V. Stocchi, E. Tisserant, A.R. Viscomi, A. Zambonelli, E. Zampieri, B. Henrissat, M.H. Lebrun, F. Paolocci, P. Bonfante, S. Ottonello and P. Wincker. 2010. Périgord black truffle genome uncovers evolutionary origins and mechanisms of symbiosis. *Nature*, 464(7291): 1033–1038. doi: 10.1038/nature08867.

Marx, D.H. 1980. Ectomycorrhizal fungus inoculations: A tool for improving forestation practices. pp. 13–71. *In*: Mikola, P. (ed.). Tropical Mycorrhiza Research. Oxford University Press, London.

Marx, D.H. and C.E. Cordell. 1989. The use of specific ectomycorrhizas to improve artificial forestation practices. pp. 1–25. *In*: Whipps, J.M. and Lumsden, R.D. (eds.). Biotechnology of Fungi for Improving Plant Growth: Symposium of the British. Cambridge University Press, Cambridge.

Maurel, C. and C. Plassard. 2011. Aquaporins: For more than water at the plant-fungus interface? *New Phytol.*, 190(4): 815–817. https://doi.org/10.1111/ j.1469-813.2011.03731.x.

McAfee, B.J. and J.A. Fortin. 1986. Competitive interactions of ectomycorrhizal mycobionts under field conditions. *Can. J. Bot.*, 64: 848–852.

Meharg, A.A. and J.W.G. Cairney. 2000. Ectomycorrhizas: Extending the capacities of rhizosphere remediation? *Soil Biol. Biochem.*, 32: 1475–1484.

Miransari, M. 2011. Soil microbes and plant fertilization. *Appl. Microbiol. Biotechnol.*, 92: 875–885.

Molina, R. and J.G. Palmer. 1982. Isolation, maintenance, and pure culture manipulation of ectomycorrhizal fungi. pp. 115–129. *In:* Schenck, N.C. (ed.). Methods and Principles of Mycorrhizal Research. St Paul, Minnesota, USA, The American Phytopathological Society.

Mrnka, L., M. Kuchar, Z. Cieslarova, P. Matejka, J. Szakova, P. Tlustos and M. Vosatka. 2012. Effects of endo- and ectomycorrhizal fungi on physiological parameters and heavy metals accumulation of two species from the family Salicaceae. *Water Air Soil Pollut.*, 223: 399–410.

Naz, I., H. Ahmad, S.N. Khokhar, K. Khan and A.H. Shah. 2016. Impact of zinc solubilizing bacteria on zinc contents of wheat. *American-Eurasian J. Agric. & Environ. Sci.*, 16(3): 449–454. https://www.idosi.org/aejaes/jaes16(3)16/3.pdf.

Nehls, U. and S. Dietz. 2014. Fungal aquaporins: Cellular functions and ecophysiological perspectives. Fungal aquaporins: Cellular functions and ecophysiological perspectives. *Appl. Microbiol. Biotechnol.*, 98(21): 8835–8851. https://doi.org/10.1007/s00253-014-6049-0.

Plassard, C., J. Louche, M.A. Ali, M. Duchemin, E. Legname and B. Cloutier-Hurteau. 2011. Diversity in phosphorus mobilisation and uptake in ectomycorrhizal fungi. *Ann. For. Sci.*, 68: 33–43.

Plett, J.M., M. Kemppainen, S.D. Kale, A. Kohler, V. Legué, A. Brun, B.M. Tyler, A.G. Pardo and F. Martin. 2011. A secreted effector protein of Laccaria bicolor is required for symbiosis development. *Curr. Biol.*, 21(14): 1197–1203.

Poole, E.J., G.D. Bending, J.M. Whipps and D.J. Read. 2001. Bacteria associated with Pinus sylvestris–Lactarius rufus ectomycorrhizas and their effects on mycorrhiza formation *in vitro. New Phytologist*, 151: 743–751.

Rajkumar, M., S. Sandhya, M. Prasad and H. Freitas. 2012. Perspectives of plant-associated microbes in heavy metal phytoremediation. *Biotechnol. Adv.*, 30: 1562–1574.

Riedlinger, J., S.D. Schrey, M.T. Tarkka, R. Hampp, M. Kapur and H.P. Fiedler. 2006. Auxofuran, a novel metabolite that stimulates the growth of fly agaric, is produced by the mycorrhiza helper bacterium Streptomyces strain AcH 505. *Appl. Environ. Microbiol.*, 72: 3550–3557.

Rinaldi, A.C., O. Comandini and T.W. Kuyper. 2008. Ectomycorrhizal fungal diversity: Separating the wheat from the chaff. *Fungal Divers.*, 33: 1–45.

Rossi, M.J., A. Furigo and V.L. Oliveira. 2007. Inoculant production of ectomycorrhizal fungi by solid and submerged fermentations. *Food Technol. Biotechnol.*, 45: 277–286.

Saravanan, R.S. and K. Natarajan. 1996. Effect of *Pisolithus tinctorius* on the nodulation and nitrogen fixing potential of *Acacia nilotica* seedlings. *Kavaka*, 24: 41–49.

Saravanan, R.S. and K. Natarajan. 2000. Effect of ecto and endomycorrhizal fungi along with *Bradyrhizobium* sp. on the growth and nitrogen fixation in *Acacia nilotica* seedlings in the nursery. *J. Trop. Forest Sci.*, 12: 348–356.

Savoie, J.M. and M.L. Largeteau. 2011. Production of edible mushrooms in forests: trends in development of a mycosilviculture. *Appl. Microbiol. Biotechnol.*, 89: 971–979.

Schoebitz, M., M.D. López and A. Roldán. 2013. Bioencapsulation of microbial inoculants for better soil–plant fertilization. A review. *Agron Sustain. Dev. Agro.*, 33(4): 751–765. https://doi.org/10.1007/s13593-013-0142-0.

Schrey, S.D., M. Schellhammer, M. Ecke, R. Hampp and M.T. Tarkka. 2005. Mycorrhiza helper bacterium Streptomyces AcH 505 induces differential gene expression in the ectomycorrhizal fungus *Amanita muscaria. New Phytologist*, 168: 205–216.

Schwartz, M.W., J.D. Horksema, C.A. Gehring, N.C. Johnson, J.N. Klironomos, L.K. Abbot and A. Pringle. 2006. The promise and the potential consequences of the global transport of mycorrhizal fungal inoculum. *Ecol. Lett.*, 9: 501–515.

Sharma, R. 2017. Ectomycorrhizal mushrooms: their diversity, ecology and practical applications. pp. 99–131. *In:* Varma, A., Prasad, R. and Tuteja, N. (eds.). Mycorrhiza—Function, Diversity, State of the Art. Springer International Publishing.

Sim, M.Y. and A.H. Eom. 2006. Effects of ectomycorrhizal fungi on growth of seedlings of *Pinus densiflora. Mycobiology*, 34: 191–195.

Smidsrød, O. and G.J.T.I.B. Skja. 1990. Alginate as immobilization matrix for cells. *Trends Biotechnol.*, 8: 71–78.

Smith, S.E. and D.J. Read. 2008. Mycorrhizal Symbiosis. 3rd Edition, Academic Press, London.

Szuba, A. 2015. Ectomycorrhiza of populus. *For. Ecol. Manage.*, 347: 156–169. https://doi.org/10.1016/j.foreco.2015.03.012.

Tibbett, M. and J.W.G. Cairney. 2007. The cooler side of mycorrhizas: Their occurrence and functioning at low temperatures. *Can. J. Bot.*, 85: 51–62.

Trappe, J.M. 1969. Studies on *Cenococcum graniforme*. An efficient method for isolation from sclerotia. *Can. J. Bot.*, 47: 1389–1390.

Van der Lelie, D., S. Taghavi, S. Monchy, J. Schwender, L. Miller, R. Ferrieri, A. Rogers, X. Wu, W. Zhu, N. Weyens, J. Vangronsveld and L. Newman. 2009. Poplar and its bacterial endophytes: Coexistence and harmony. *Crit. Rev. Plant Sci.*, 28: 346–358.

Vassilev, N., I. Nikolaeva and M. Vassileva. 2005. Polymer-based preparation of soil inoculants: Applications to arbuscular mycorrhizal fungi. *Rev. Environ. Sci. Biotechnol.*, 4(4): 235–243.

Viveganandan, G. and K.S. Jauhri. 2000. Growth and survival of phosphate solubilizing bacteria in calcium alginate. *Microbiol. Res.*, 155(3): 205–207.

Wiensczyk, A.M., D. Gamiet, D.M. Durall, M.D. Jones and S.W. Simard. 2002. Ectomycorrhizas and forestry in British Columbia: A summary of current research and conservation strategies. *Br. Columbia J. Ecosyst. Manage.*, 2: 1–20.

Wu, X.Q., L. Hou, J.M. Sheng, J.H. Ren, L. Zheng, D. Chen and J.R. Ye. 2012. Effects of ectomycorrhizal fungus Boletus edulis and mycorrhiza helper Bacillus cereus on the growth and nutrient uptake by Pinus thunbergii. *Biol. Fertil. Soils*, 48(4): 385–391.

Yabur, R., Y. Bashan and G. Hernández-Carmona. 2007. Alginate from the macroalgae Sargassum sinicola as a novel source for microbial immobilization material in wastewater treatment and plant growth promotion. *J. Appl. Phycol.*, 19(1): 43–53.

Zhou, A.D., X.Q. Wu, L. Shen, X.L. Xu, L. Huang and J.R. Ye. 2014. Profiling of differentially expressed genes in ectomycorrhizal fungus *Pisolithus tinctorius* responding to mycorrhiza helper Brevibacillus reuszeri MPt17. *Biologia.* Section *Cellular and Molecular Biology*, 69(4): 435–442.

18

Edible Mushroom Bioproducts for Controlling Parasitic Nematodes

Liliana Aguilar-Marcelino,[1,*] *Filippe Elias de Freitas Soares,*[2] *Juliana Marques Ferreira,*[3] *Jesús Antonio Pineda-Alegría,*[1,4] *Gloria Sarahi Castañeda-Ramírez,*[1] *Julio Cruz-Arévalo,*[4,5] *Susan Yaracet Páez-León*[1,4] *and Benjamín Mendoza-Ortega*[1]

1. Introduction

Parasitic nematodes of livestock and plants cause a loss of livestock and various crops. Within the wide range of parasitic nematodes that affect ruminants and plants, gastrointestinal and root-knot nematodes stand out, which have a negative impact on animal and plant health and are considered one of the groups of parasites that cause serious losses in the agricultural sector.

[1] Centro Nacional de Investigación Disciplinaria en Salud Animal e Inocuidad, INIFAP, Jiutepec, Morelos, Mexico.
[2] Department of Chemistry, Universidade Federal de Lavras, Lavras, Minas Gerais, Brazil.
[3] Institute of Tropical Pathology and Health, Universidade Federal de Goiás, Goiânia, Brazil.
[4] Biotechnology Research Center, UAEM, Cuernavaca, Morelos, Mexico.
[5] Universidad Tecnológica de la Selva. Entronque Toniná Km. 0.5 Carretera Ocosingo-Altamirano, Ocosingo, Chiapas, México.
* Corresponding author: aguilar.liliana@inifap.gob.mx

The economic impact caused by these parasitic nematodes is considerable. The main technique of control gastrointestinal nematodes of ruminants and plants is the use of anthelmintics. But the use of these chemicals in a non-recommended way has led to anthelmintic resistance in gastrointestinal and phytopathogenic nematodes at a global and national level. Therefore, the continuous search for sustainable alternative control methods with less impact on the environment where they are used is necessary. The use of edible fungi (macromycetes) worldwide and in Latin America has a significant ethnomedicinal tradition because of their medicinal effects and great potential to produce different molecules such as enzymes, proteins, peptides and bioactive secondary metabolites with therapeutic properties. Additionally, they have been proposed as nutraceutical food because of their nutritional value. This chapter presents an overview of edible mushrooms as an emerging source of bioproducts for the control of livestock nematodes as well as plant parasites.

2. Nematodes

Nematodes or roundworms have cylindrical, thin, elongated bodies and ends that gradually taper. They are organisms with bilateral symmetry and vary in size from 100 μm to 1 mm, as in the case of *Caenorhabditis* spp., but can reach more than 1 m, which is the case for *Drancunculus* spp. (Blumenthal and Davis 2004). They are covered by a cuticle secreted by the hypodermis, consisting mainly of collagen. On the outside, it can be covered by lipids, carbohydrates (which allow the evasion of host defense in parasitic species) and proteins, impermeable to solutes (Navone et al. 2011). The main functions of the cuticle are protection and provision of rigidity; however, it has limited flexibility and is therefore frequently changed to allow growth due to cell elongation. Internally, the body of the nematodes has a cavity or pseudocoelom that houses the different organs, such as a complete digestive system containing a mouth, oesophagus, intestine and anus (Barnes and Ruppert 1996). The mouth is made up of three to six lips, and there is a cavity between the mouth and the oesophagus that can have thick walls and forms the buccal capsule, where there are accessory structures such as plates, teeth, spines and stylets. All of them specialised to obtain and process food, such as unicellular organisms, tissues, and vascular fluids of plants, insects and mammals. Some species prey on other nematodes (Hodda 2022b).

Regarding reproduction, nematodes are dioecious, with a notorious sexual dimorphism due to larger females; males have the posterior end curved ventrally. There are species where the male is absent, and the females reproduce by parthenogenesis. In some cases, the eggs of nematodes are oviposited with only a couple of cells inside, whereas others contain the already developed larvae some others present oviparity (Barnes and Ruppert 1996). The development of the larva (L) or juvenile is mediated by conditions such as temperature and food. They present four ecdyses during their growth, which determines the different larval stages (or juveniles): L1, L2, L3 and L4. In stages L1 and L2, there are bacteriophages and saprophytes. Stage L3 does not feed and uses the energy reserves stored in the previous phases; generally, this is the infective stage in parasitic species, but in certain cases, it is also the L2. The L4 precedes the adult phase and may present hypobiosis, which provides resistance to unfavourable environmental conditions (Navone et al. 2011). Parasitic nematodes can reach their hosts in three different ways: passive penetration, given by the ingestion of eggs or larvae that remain in the soil or in an intermediate host; active penetration, where the larvae dynamically seek out and force their way through the body walls of their host; vector transmission, where some hematophagous insect transmits the infective larva.

Parasitic species of vertebrates can be found mainly in the colon and intestine. In species that infect invertebrates, they can be found in the intestine and haemocoel. Phytoparasitic nematodes are classified as ectoparasites that feed on plant fluids through their stylet, the semi-endoparasites that partially enter their body and endoparasites that penetrate different plant tissues (Hodda 2022b, Navone et al. 2011). Just as there are parasitic and free-living nematodes, there are others that can be facultative, alternating between both lifestyles. Thanks to this versatility, it is a highly successful phylum with around 30,000 described species, constituting less than 5% of said taxa (Hodda 2022). Regarding their ecological importance, they serve as effective decomposers of organic matter, accelerating biogeochemical cycles since they are also highly abundant. This also makes them bioindicators in multiple environments (Franco et al. 2022, Yeates and Bongers 1999).

3. Free-Living Nematodes

Nematodes are the most abundant organisms on Earth (Perry and Moens 2013) and are present in all regions, including aquatic and terrestrial environments, such as deserts, and Antarctica. According to their mode of life, three categories are distinguished: animal parasites, plant parasites and free-living nematodes (FLN). Most nematode species are free-living ones (APS 2022). In line with Iqbal and Jones (2017), more than half of the nematode species exist as free-living forms, which have important ecological functions. They can feed on bacteria, fungi, algae, other nematodes and decaying matter. According to Sochová et al. (2006), The FLNs are closely related to other soil organisms and contribute to the decomposition of organic matter and the flow of nutrients. They are pest controllers by feeding on pathogens such as bacteria, fungi or other N-rich nematodes; as they consume more N than needed, the excess is released into the soil as NH_4.

Yadav et al. (2018) indicate that nematode excretions account for up to 19% of the N present in the soil. Furthermore, they are a food source for other organisms and facilitate the transport and dispersal of bacteria in the soil (Ingham 2010). They also contribute to the decomposition of dead matter, allowing plants to have better accessibility to nutrients (Iqbal and Jones 2017) (Table 1).

Many researchers value FLNs because they are used as study models to assess the efficacy in nematicides of agricultural and livestock interests (Castañeda-Ramírez et al. 2020, Li and Zhang 2014). In addition, they are used as bioindicators of toxicity in environmental samples or to assess the possible negative impacts of natural products (Rivas-Morales et al. 2016). The most studied species are *C. elegans* and *P. redivivus* (Castañeda-Ramírez et al. 2020, Pica-Granados 2008), of which the latter

Table 1: Examples of some free-living nematodes.

Group	Examples	References
Fungivores	- *Aphelenchus avenae, Aphelenchoides* spp.	(Lagerlöf et al. 2011)
Bacteriophages	- *Panagrellus redivivus, Caenorhabditis elegans*	(Aguilar-Marcelino et al. 2020)
Carnivores	- *Mononchus* spp., *Parazercon radiates*	(Yadav et al. 2018)
	- *Butlerius butleri*	(Aguilar-Marcelino et al. 2020)
Omnivores	- *Dorylaimus* spp., *Lysigamasus lapponicus*	(Yadav et al. 2018)

is currently farmed in large volumes to feed fish larvae in the aquaculture industry (Brüggemann 2012, Kolesnik et al. 2019). Edible mushroom metabolites also have been tested in *P. revivivus* and *C. elegans* to assess their efficacy as nematicides. Castañeda-Ramírez et al. (2020) published a complete revision of the nematicidal effect of edible mushrooms, providing evidence that *P. redivivus* is widely used as a model. Li and Zhang (2014) list several basidiomycetes tested against different kinds of nematodes, highlighting *P. redivivus* and *C. elegans*. Soares et al. (2019) also used *P. redivivus* to assess the nematicidal effect of an aqueous extract of spent mushroom compost derived from the edible mushroom *Hypsizygus marmoreus*. Marques et al. (2019) revealed the nematicidal activity of spent mushroom compost of *Flammulina velutipes* on *Panagrellus* sp. larvae.

Based on the above, FLNs are important due to their ecological functions. In addition, the characteristics of their cuticle are useful to test nematicidal compounds and serve as bioindicators for pollution monitoring. The ease of cultivation and high prolificacy of nematodes, such as *P. redivivus*, allows for fast and accurate experiments. Generally, experiments on nematodes are designed to assess the predatory, toxicity, immobilisation and detrimental effects of nematicidal products.

4. Phytoparasitic Nematodes

Phytoparasitic nematodes are responsible for huge economical losses in agriculture worldwide. There are around 4,300 species of plant-parasitic nematodes (Lionetti et al. 2017), with several ways of feeding. Phytoparasitic nematodes contain a stylet and can be ectoparasitic, endoparasitic and semi-endoparasitic (Figure 1). Ectoparasitic nematodes spend their complete life cycle outside of the host, whereas semi-endoparasitic nematodes have the posterior part out of the root; endoparasitic nematodes enter the root and feed on internal tissues (Sato et al. 2019). Endoparasitic nematodes are considered economically important because of the damage they cause to the plant. Although there are more than 4,000 nematode species, *Meloidogyne* spp., *Globodera* spp. and *Heterodera* spp., are the two major groups and the most economically important ones (Jones et al. 2013). The use of chemical nematicides is the main strategy used to control phytoparasitic nematodes; however, this frequently results in environmental pollution and anthelmintic resistance (Laith et al. 2020). Several alternative control strategies have been proposed, such as biological control, cultural control and the use of botanical and fungal secondary metabolites (Aguilar-Marcelino et al. 2020) (Figure 1).

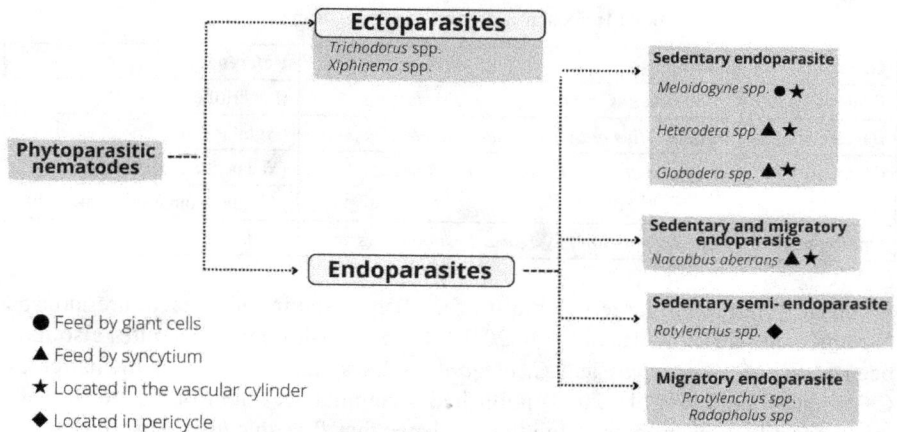

Figure 1: Lifestyles of some plant parasitic nematodes. Source: Own elaboration with information taken from Eves-van den Akker et al. (2021), and Sato et al. (2019).

According to Li and Zhang (2014), numerous basidiomycetes have nematicidal substances with potential activity against plant-parasitic nematodes. Some wood-rotting mushrooms prey upon and consume nematodes to obtain N and other nutrients (Armas-Tizapantzi et al. 2019, Barron and Thorn 1987, Luo et al. 2007). There are several reports about the nematicidal effect of edible mushrooms; specifically, for the genus *Pleurotus*, effects have been observed on free-living, gastrointestinal and phytoparasitic nematodes, including *Meloidogyne* spp. (Heydari et al. 2006), *Heterodera* sp. (Palizi et al. 2009) and *Globodera* sp. (Arteaga et al. 2020). Degenkolb and Vilcinskas (2016) report the nematicidal potential of several basidiomycetes, of which *Pleurotus* spp. are the most cited ones. Some studies have confirmed the presence of nematicidal compounds imbibed in spherical knobs attached to hyphae of *Pleurotus* spp. and *Coprinus comatus* (Armas-Tizapantzi et al. 2019, Luo et al. 2007, Soares et al. 2018); however, nemato-toxic substances are also present in fruiting body extracts and broth culture filtrates of several edible mushrooms. Table 2 shows some examples of nematicidal effects of edible mushrooms on plant parasitic nematodes. Most studies focus on *Meloidogyne* spp. due to the worldwide economic importance.

The use of *Pleurotus* spp. is an interesting alternative because of the easy cultivation and its use in a variety of agricultural residues. Thus, they have some advantages in terms of yield, worldwide cultivation, short time lifecycle and nematicidal effects in both stages (vegetative and reproductive). In addition, mushroom by-products have metabolites with antagonistic properties on plant parasitic nematodes.

5. Gastrointestinal Nematodes in Ruminants

Gastrointestinal nematodes represent a health problem for livestock worldwide because animals infected with these parasites can present diarrhoea, changes in their behaviour, lack of appetite and growth reduction. This is reflected in poor weight

Table 2: Edible mushrooms with nematicidal effects on phytoparasitic nematodes.

Source	Mushroom	Nematode tested	Effect	Reference
Fruiting bodies (extract)	*P. ostreatus*	*M. incognita*	Reproduction	(Youssef and El-Nagdi 2021)
	Cordyceps militaris, Dictyophora indusiata and *Hericium erinaceous*	*M. incognita*	Egg hatchability and mortality	(Soliman et al. 2022)
	P. ostreatus, P. citrinopileatus, P. pulmonarius and *Boletus* sp.	*M. incognita*	Egg hatchability, mortality and reproduction	(Wille et al. 2019)
Broth culture	*P. ostreatus*	*G. pallida*	Mortality and nematostatic effect	(Arteaga et al. 2020)
	Lentinula edodes and *P. eryngii*	*M. javanica*	Mortality and nematostatic effect	(Hahn et al. 2019)
	Pleurotus spp.	*M. javanica*	Mortality	(Heydari et al. 2006)
	Pleurotus spp.	*Heterodera schachtii*	Mortality and paralysis	(Palizi et al. 2009)
Mycelium	*Coprinus comatus*	*M. incognita*	Immobilisation	(Luo et al. 2007)
	L. edodes and *P. eryngii*	*M. javanica*	Mortality and nematostatic effect	(Hahn et al. 2019)
	Pleurotus spp.	*M. javanica*	Mortality	(Heydari et al. 2006)
	Pleurotus spp. and *Hypsizygus ulmarius*	*Heterodera schachtii*	Mortality and paralysis	(Palizi et al. 2009)

gain and a decreased productive efficiency of cattle (Pinto et al. 2021). The adverse effects caused by these nematodes on their hosts are due to the low absorption of nutrients, hormonal changes, tissue damage and immunopathological effects. The symptomatology depends on the environmental conditions, the susceptibility and the production system in which the host is located (Charlier et al. 2020). Infections by these nematodes are more frequent in ruminant species raised in extensive grazing systems, mainly in tropical areas, where animals can acquire infections throughout the year (Almeida et al. 2018). Gastrointestinal nematode (GIN) infections are usually mixed, but there are species that have been found to be more prevalent in ruminants such as *Haemonchus, Cooperia, Trichostrongylus, Ostertargia,* among others (Barone et al., 2020).

All ruminant GINs have a direct life cycle. The eggs are shed from the host along with the faeces and hatch in the soil (Figure 2). Subsequently, the development of the larva consists of moulting from stage 1 (L1) to stage 3 (L3; infective phase) within the first 2 weeks, as long as the environmental conditions are optimal. The L3 has a protective sheath that makes it more resistant to environmental conditions and allows it to survive waiting for the host. Once the L3 is ingested, it is unsheathed in the rumen and reaches the abomasum or the intestine (depending on the species), where it will have two more moults until it reaches the adult larvae stage. In this

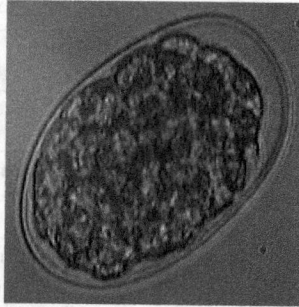

Figure 2: Photograph of an egg of the sheep parasitic nematode *Haemonchus contortus* under an optical microscope (40X) (photograph by Dr. Liliana Aguilar-Marcelino CENID-SAI, INIFAP).

stage, the larvae sexually differentiate, and the females produce eggs to start a new cycle (Zajac 2006, Charlier et al. 2020).

To reduce the impacts of GINs on cattle, anthelmintics are widely used. Nevertheless, the continuous application of these anthelmintics has generated populations of nematodes resistant to anthelmintics around the world (Dimunová et al. 2022). For this reason, it is necessary to continue the search for alternative control methods. The need for the search for new control treatments and knowledge of the lifecycle of GINs has led to the development of *in vitro* techniques that allow evaluation of the efficacy of extracts of natural products isolated from plants and edible fungi against different stages of the disease. Some of these techniques are inhibition of egg hatching, inhibition of larval motility, inhibition of larval shedding, inhibition of larval development and mortality in L3 and L4 larvae (Jackson and Hoste 2010, Pineda-Alegría et al. 2017, González-Cortazar et al. 2020).

These techniques allow the identification of potential compounds for the control of GINs at the *in vitro* level, which should subsequently be directly evaluated in ruminants to measure their effectiveness. This can be considered as an alternative or complementary technique to the control methods that currently exist.

6. Bioproducts of Edible Mushrooms

The differences between edible and medicinal mushrooms are difficult to demonstrate because many edible mushrooms are medicinal and vice versa (Guillamón et al. 2010, Ergönül et al. 2013). The global production of edible mushrooms has increased over 30-fold since 1978; by 2012, more than 31 million tons were recorded, generating US$20 billion, with the per capita mushroom consumption exceeding 4.70 kg annually (Chang and Wasser 2012). In 2020, global production of 42,792,893 tons was recorded (FAO 2022). More than 50 species of mushrooms are considered fit for human consumption, including *Agaricus, Coprinus, Ganoderma, Hericium, Lentinula, Pleurocybella, Pleurotus* and *Trametesvar* (Castañeda-Ramírez et al. 2020).

Royse and Sanchez (2017) mention that only five genera comprise about 85% of the world's mushroom supply, with *L. edodes* (shiitake) being the most cultivated, accounting for 22% of the global production, followed by *Pleurotus* spp. (19%), *Auricularia* spp. (18%), *A. bisporus* (15%), *F. velutipes* (11%) and *Volvariella* (5%).

Edible mushrooms have a high nutritional value as they are rich in protein, with high contents of indispensable amino acids and fibre, a poor fat content, and large amounts of important fatty acids, providing nutritionally significant contents of vitamins (B1, B2, B12, C, D and E). This makes them a superb source of various different nutraceuticals, used directly in the human diet as well as promoting health because of the synergistic effects of all the bioactive compounds present (Valverde et al. 2015). Besides the nutritional components of edible mushrooms, there are also antioxidant, anticancer, antidiabetic, antiallergic, immunomodulatory, cardiovascular protective, anticholesterol, antiviral, antibacterial, antiparasitic, antifungal, detoxifying and hepatoprotective effects, in addition to preventing tumours and the development of inflammatory processes (Valverde et al. 2015, Castañeda-Ramírez et al. 2020). These functions are based on the synthesis of biologically active compounds that can be classified into secondary metabolites (acids, terpenes, polyphenols, sesquiterpenes, alkaloids, lactones, sterols, metal chelators, nucleotide analogues and vitamins), glycoproteins and polysaccharides (mainly beta-glucans) (Table 3). New biologically active proteins have also been discovered for use in biotechnological processes and the development of new drugs and lignocellulose-degrading enzymes, lectins, proteases and protease inhibitors, ribosome-inactivating proteins and hydrophobins (Mattila et al. 2001, Barros et al. 2007).

The contents and types of biologically active compounds in edible mushrooms can vary considerably; the concentrations of these substances vary by differences in strain, substrate, culture, developmental stage, age, storage conditions, processing and cooking practices (Erjavec et al. 2012).

Mushroom production of bioactive metabolites is a key aspect of the development of effective biotechnological techniques to obtain these metabolites. However, based on extensive research, fungi contain ingredients with remarkable properties to treat different types of diseases, and future studies on the mechanism of action of fungal bioproducts are needed to better characterise the valuable functions and properties of different fungi for the prevention and treatment of diseases or against different organisms such as bacteria, viruses and nematodes.

7. Nematotoxic Activity of Secondary Metabolites of Edible Mushrooms

Edible mushrooms have begun to be studied more extensively for the properties they possess and their biological activity. Recent studies have indicated a direct correlation between the culture medium and the metabolites obtained from the fungus; in particular, the selection of the carbon source depends on the type of metabolites desired. According to Chegwin (2013), the use of wheat flour as a carbon source increases the production of fatty acids, whereas cereal and legume flour favour other types of materials with biological activities (Chegwin and Nieto 2013). Therefore, several studies have been conducted on the nematicidal activity and the characteristics of the fungus used. One of the first reports of nematicidal activity

Table 3: Compounds are identified in different edible mushrooms (Muszyńska et al. 2018).

Compounds	Mushroom species
Saccharides: - trehalose, - β-glucans - chitos	*Agaricus bisporus,* *Lentinula edodes,* *Trametes versicolor,* *Inonotus obliquus,* *Pleurotus ostreatus* and *Ganoderma lucidum*
Proteins - lectins	*Lentinula edodes,* *Agaricus bisporus,* *Pleurotus ostreatus* and *Taraxacum mongolicum*
Fatty acids: - oleic acid - palmitic, - linoleic acid and α-linoleic	*Agaricus bisporus,* *Cantharellus cibarius* and *Imleria badia*
Phenolic compounds: - gallic acid - protocatechuic acid - p-hydroxybenzoic acid - p-coumaric acid - cinnamic acid - caffeic acid	*Imleria badia,* *Agaricus bisporus,* *Boletus edulis,* *Calocybe gambosa,* *Hygrophorus marzuolus,* *Lactarius deliciosus* and *Elaphomyces granulatus*
Indole compounds: - melatonin - L-trytptophan - 5-hydroxy-L-tryptophan	*Cantharellus cibarius,* *Agaricus bisporus,* *Imleria badia,* *Lactarius deliciosus,* *Leccinum scabrum,* *Suillus bovinus,* *Suillus luteus,* *Pleurotus ostreatus,* *Tricholoma equestre* and *Armillaria mellea*
Vitamins: - vitamins of B group (thiamine, riboflavin, biotin, pyridoxine) - viamine C - tocopherols	*Cantharellus cibarius,* *Agaricus bisporus,* *Termitomyces microcarpus,* *Termitomyces tyleranus,* *Termitomyces clypeatus,* *Volvariella speciosa* and *Polyporus tenuiculus*
Bioelements - zinc - selenium - copper	*Agaricus bisporus,* *Cantharellus cibarius* and *Imleria badia* *Lycoperdon perlatum*

was published in 1984, when mycelia of the fungi *Pleurotus spp* were assessed *in vitro* against a free-living rhabditoid nematode, showing a mortality effect (Thorn and Barron 1984). Later, the mycelium of the edible mushroom *P. pulmonarius* was evaluated against different nematodes; in this study, the larvae were immobilised

(Larsen and Nansen 1991). Later, in 1992, a possible compound with nematicidal activity was identified, for which an *in vitro* assessment of the mycelium of the edible mushroom *P. ostreatus* against the nematode *Panagrellus* sp. was carried out; the reported activity was attributed to trans 2-decenedioic acid (Kwok et al. 1992).

The edible mushroom *P. pulmonarius* was also evaluated against *C. elegans* larvae, and the following compounds with nematostatic effect were observed in the active fraction: S-coriolic acid, linoleic acid, p-anisaldehyde, p-anisyl alcohol, 1-(4-methoxyphenyl)-1,2-propanediol and 2-hydroxy-(4′-methoxy)-propiophenone (Stadler et al. 1994). For its part, another study evaluated the extract of *P. ferulae* against a nematode that affects wood (*Bursaphelenchus xylophilus*), and activity was found at 72 h at 54.7 mg L^{-1}. Subsequently, three nematicidal compounds, cheimonophyllon E, 5α,8α-epidioxyergosta-6,22-dien-3β-ol and 5-hydroxymethyl-2-furancarbaldehyde, were isolated from a bio-directed chemical fractionation of the *P. ferulae* extract (Li et al. 2007). In 2016, a bio-directed fractionation of the hydroalcoholic extract of mycelium of *P. ostreatus* mushroom mycelium against *H. contortus* eggs was performed. Meeting "I" showed 100% inhibition of hatching at a concentration of 1.25 mg mL^{-1} after 72 hours. The metabolites reported were four fatty acids, namely hexadecanoic acid, octadecanoic acid, ethanol 2-butoxy phosphate and ethanol 2-butoxy phosphate (3:1), as well as a sugar named "xylitol" (Cedillo 2016). Subsequently, a bio-directed study was conducted where the hydroalcoholic extract of *P. djamor* fructifications was assessed against *H. contortus*. In another study, the highest activity observed was against *H. contortus* eggs, with up to 100% inhibition in the E1 fraction at 10 mg mL^{-1} after 72 h. The compounds present in the E1 fraction were four fatty acids (pentadecanoic, hexadecanoic, octadecadienoic and octadecanoic) as well as a terpene identified as β-sitosterol (Pineda-Alegria et al. 2017).

Likewise, another study evaluated *in vitro* the ethanolic extract of seven strains of *P. eryngii* species against eggs and larvae (L3) of *H. contortus*. The extract with the highest activity was fractionated (F1 to F5), and F5 showed an activity of 88.7% and 91.87% at 20 and 40 mg/mL against eggs and larvae, respectively. The GC-MS analysis of F5 revealed the presence of trehalose, polyols (L-iditol, galactitol, D-mannitol, D-glucitol and myoinositol), adipic acid, stearic acid, squalene and β-sitosterol (Cruz-Arévalo et al. 2020). Regarding reports on the anthelmintic activity of the degraded substrate, it was decided to evaluate the different degraded substrates of the edible mushroom *P. djamor* and the time of colonisation of the mycelium and after harvesting against *H. contortus* larvae. The extract that showed the highest activity was the extract of the first harvest on which fractionation was performed, and the fraction with the highest potential was obtained, as well as the components of this fraction, which were alcohol, 4-hydroxy-3,5,5-trimethyl-4-[3-oxo-1-butenyl]-2-cyclohexen-1-one, caffeine and 5,6-dimethoxy-1(3 H) isobenzofuranone (Colmenares-Cruz et al. 2021).

8. Nematotoxic Activity Macromolecules of Edible Mushrooms

Macromolecules have a molecular mass in the range of thousands to many millions of Daltons. In biochemical terms, they are large organic molecules present in cells and can be classified as carbohydrates, lipids, proteins and nucleic acids (Nelson and Cox 2017). Edible mushrooms produce a series of macromolecules with several effects of interest: anticancer, antioxidant, anti-allergic, anti-diabetic, immunomodulatory, antimicrobial and nematotoxic effects (Dubey et al. 2015, Mishra et al. 2021). Among the classes of macromolecules produced by mushrooms, carbohydrates stand out as those with pronounced nutraceutical effects. Edible mushrooms produce a number of polysaccharides: heteropolysaccharides rich in fucose, galactose, mannose, and xylose, glycogen-like glucans and structural cell walls polysaccharides (Smiderle et al. 2017). A number of linear/branched glucans and heteroglycans have been isolated from edible mushrooms and showed a wide range of biological functions (Maity et al. 2021). However, although there are numerous studies on carbohydrates produced by edible mushrooms, only a few studies have reported macromolecules with nematotoxic activity derived from edible mushrooms. Cruz-Arévalo et al. (2020) evaluated the *in vitro* effects of acetone (AE) and hydroalcoholic (HA) extracts from *Pleurotus eryngii* against the eggs and larvae (L3) of *H. contortus*. After chromatography, one of the fractions acquired (F5) was the most effective against eggs, with hatching inhibition percentages (88.77% and 91.87% at 20 and 40 mg mL^{-1}, respectively). This fraction contained some monosaccharides (e.g., trehalose) and polyols.

Cedillo (2016) evaluated the action of the hydroalcoholic extract of the mushroom *P. ostreatus* against *H. contortus*. The extract showed nematicidal activity and contained xylitol. González-Cortázar et al. (2021) identified that the mushroom *P. djamor* produces the monosaccharide allitol, which acted on *H. contortus* larvae, with a reduction of 92.56%. On the other hand, in relation to the class of lipids, the production of lipid toxins by edible mushrooms has been demonstrated (Thorn and Barron 1984). Fatty acids are long-chain carboxylic acids found in fats, oils and as components of membrane lipids. The species of mushrooms of the genus *Pleurotus* produce some fatty acids with demonstrated nematotoxic action.

The edible mushroom *P. ostreatus* produces trans-2-decenoic acid, derived from linoleic acid. This fatty acid is a toxin with nematotoxic action and immobilised 95% of test nematodes (*Panagrellus redivivus*) within 1 hr (Kwok et al. 1992). *P. djamor*, an edible mushroom that has medicinal values, produces an extract containing four fatty acids: pentadecanoic, hexadecanoic, octadecadienoic and octadecanoic acid, as well as one terpene, β-sitosterol. It demonstrated a high nematotoxic *in vitro* effect on *H. contortus* exsheathed infective larvae (L3) (Pineda-Alegría et al. 2017). Subsequently, the same research group carried out a study using these pure lipid compounds on *H. contortus* infective eggs and larvae; palmitic acid and stearic acid inhibited *H. contortus* egg hatching up to 100% (Pineda-Alegría et al. 2017). On the contrary, in the larval mortality assessment, the consortium of the five compounds showed a dose-dependent impact and 100% mortality was evident 24 h post-incubation (Pineda-Alegría et al. 2020). Stadler et al. (1994) studied fatty acids and other compounds with nematicidal activity from cultures particularly *P. pulmonaryius* and *Hericium coralloides*. These mushrooms produced fatty acids with toxic effects towards the saprophytic nematode *Caenorhabditis elegans*; *P. pulmonarius* produced linoleic acid and 1-(4-methoxyphenyl)-1,2-propanediol, in addition to other metabolites. *Hericum coralloides* produced a nematicidal fatty acid mixture containing linoleic acid, oleic acid and palmitic acid as its main components, which showed both repellent and nematicidal effects. Based on these authors, linoleic acid works according to the nematicidal principle of several nematophagous fungi, including edible mushrooms (Stadler et al. 1993).

Proteins are macromolecules with diverse biological functions, and edible mushrooms produce a myriad of proteins with diverse biological activities interesting for human use. Among the protein functions, the nematotoxic action is highlighted here. Lectins (previously called agglutinins) are a diverse group of carbohydrate-binding proteins present in all organisms (Žurga et al. 2014). In this sense, several fruiting body lectins from edible mushrooms have been demonstrated to be toxic to insects and nematodes (Sabotic et al. 2016). The mechanism of lectin nematotoxicity occurs by means of the binding of the lectin to specific glycans in the nematode gut (Butschi et al. 2010). Sabotic et al. (2016) reviewed and listed the entomotoxic and nematotoxic lectins from fungal fruit bodies. However, this line of research is ongoing, with some new studies (Bleuler-Martinez et al. 2017, Tayyrov et al. 2018, Adedoyin et al. 2021).

Enzymes are proteins whose biological function is to catalyse chemical reactions. Some of these proteins catalyse the hydrolysis of constituents of nematodes, which implies toxicity and/or lethality. Proteases (EC 3.4) catalyse the hydrolysis of peptide bonds, thus leading to impairment of the nematode cuticle and digestion of its body contents. Based on the literature, edible mushrooms produce proteases with nematicidal action. Ferreira et al. (2019) suggested that the nematicidal action of *Flammulina velutipes* extract is due to protease activity. Genier et al. (2015) also demonstrated that proteases of *P. ostreatus* have nematicidal action. Moreover, Sufiate et al. (2017) indicated that, in addition to proteases, chitinases (EC 3.2.1.14) produced by *P. eryngii* affect *M. javanica* egg and juvenile development.

Recently, Tayyrov et al. (2019) suggested that cytoplasmic lipases (EC 3.1.1.3) from the edible mushroom *Coprinopsis cinerea* constitute a unique class of fungal defense proteins against nematode predators. Ribonucleases (EC 3.1) from edible mushrooms, known as ribotoxins, are enzymes that have also recently been linked to toxic action. These enzymes, the prototype of which is ageritin from *Cyclocybe aegerita* (syn. *Agrocybe aegerita*), catalyse the hydrolysis of a single phosphodiester bond within the universally conserved sarcin-ricin loop of ribosomes and prevent protein biosynthesis (Ragucci et al. 2021). However, recombinant ageritin was not active against any of the tested nematode species (*Caenorhabditis briggsae*, *C. elegans*, *C. tropicalis*, *Distolabrellus veechi* and *Pristionchus pacificus*), which indicates its specificity for insects (Tayyrov et al. 2019).

9. Use of Mushroom Bioproducts for Parasitic Nematodes Control

In the industrial-scale production of edible mushrooms, an enormous amount of waste is generated, known as spent mushroom compost (SMC). For the production of 1 kg of fresh mushrooms, approximately 5 kg of SMC, and it has been estimated that more than 50 million tons of SMC are generated each year worldwide (Rajavat et al. 2022). Therefore, SMC recycling is a necessary and urgent approach.

Spent mushroom compost contains residual fungal mycelium, a wide variety of degenerated lignocellulosic biomass types, amendments, nutrients, and a high level of organic matter and enzymes. The composition of SMC will vary greatly depending on the type of mushroom grown, the climate, and the composition, among other factors (Najafi et al. 2019, Leong et al. 2022).

Some uses for SMC are bioremediation, organic fertiliser, soil conditioner, animal feed, pest and disease management, energy feedstock, vermiculture, and source of degradative enzymes (Cunha Zied et al. 2020), such as α-amylase (EC 3.2.1.1), cellulase (EC 3.2.1.4), β-glucosidase (EC 3.2.1.21), xylanase (EC 3.2.1.8), laccase (EC 1.10.3.2) and protease (Ko et al. 2005, Ferreira et al. 2019, Soares et al. 2019).

It has long been known that some mushrooms, including edible mushrooms, can feed on nematodes (Drechsler 1941). However, it is only recently that SMC has been used for nematotoxic purposes. As highlighted and discussed above, edible mushrooms produce a series of macromolecules (in addition to other metabolites smaller in size) with action on nematodes. Therefore, it is to be expected that such substances will be found and recovered from SMC.

In this context, Ferreira et al. (2019) demonstrated that the extract from the SMC from *F. velutipes* comprises proteases and metabolites which showed nematicidal activity. In that paper, the active SMC crude extract obtained displayed a reduction (43% and 57% for 24 and 48 h of incubation, respectively), compared to the control group (water). In addition, exploring the proteolytic activity of the crude extract obtained from the SMC of an edible mushroom (*H. marmoreus*), Soares et al. (2019) observed significant nematicidal action ($p < 0.01$) on *P. redivivus* and bovine infective larvae, with reduced levels (52% and 26%, respectively). Hydroalcoholic extracts from the SMC of edible mushrooms are the subjects of some studies regarding their nematotoxic activities. Valdez-Uriostegui et al. (2021) assessed the *in vitro* anthelmintic activity of hydroalcoholic extracts of *P. ostreatus* SMC against eggs and unsheathed larvae of *H. contortus*. However, that extract showed no activity against *H. contortus* larvae (Cuevas-Padilla et al. 2018). The hydroalcoholic extract of the SMC of *P. djamor* showed the highest nematocidal activities evaluated by gas chromatography coupled to mass spectrometry (GC–MS) in another study. The following compounds were identified: veratryl alcohol, 4-hydroxy-3,5,5 trimethyl-4-[3-oxo-1-butenyl]-2- cyclohexen-1-one, caffeine and 5,6-dimethoxy-1(3 H) isobenzofuranone (Colmenares-Cruz et al. 2021). These authors point out that the substrate is a possible candidate for nematocidal food-inducing activity against the L3 larvae of *H. contortus*.

10. Conclusions and Future Perspectives

Livestock nematodes and phytoparasitic nematodes cause huge economic losses around the world, and new products to control these different nematode species need to be developed. Nowadays, chemical nematicides are the most popular and effective products used to manage nematodes, although their long-term use leads to an increasingly resistant population (Chen et al. 2020). In this context, the use of nematophagous fungi as biocontrollers is extremely promising as they can reduce resistant nematode populations and represent a sustainable control method, without residues in the environment. According to Soares, Sufiate and Queiroz (2018), there are several fungal species that commonly present saprophytic behaviour and prey on nematodes. Most of these species are found in the soil, but just a few are studied under laboratory conditions. There is growing evidence that some species of edible mushrooms can act as nematophagous fungi, enabling a new perspective on the use of nematophagous fungi in bionematicides since they are cultivated, commercialised and studied worldwide. Such fungi use different strategies to prey upon nematodes, such as the generation of traps and the secretion of toxins, secondary metabolites and macromolecules with nematotoxic activity. They also produce different enzymes, such as proteases, chitinases and lipases, to degrade nematode cuticles. These enzymes act on eggs as well as on juveniles and adults, which allows the control of all phases of the nematode life cycle.

Besides the use of the edible mushroom itself, the residue of its commercial production, SMC, can also be employed in the biological control of nematodes. This residue is rich in organic matter as well as fungal mycelium. The application of SMC on the soil creates a favourable microenvironment for the development of the fungus in the new soil and facilitates the predation of the nematodes present. In this context, edible mushrooms are likely candidates to be used in the biological control of nematodes. Different fungal structures can be used in the formulation of biopesticides, such as mycelium, SMC, enzymes, secondary metabolites or macromolecules with nematotoxic activity. Biopesticides can be formulated using just the fungus itself or combined with other molecules. Thus, this chapter demonstrates that edible mushrooms have a great potential to be used in a clean and efficient approach to nematode control in the future.

References

Adedoyin, I.O., T.S. Adewole, T.O. Agunbiade, F.B. Adewoyin and A. Kuku. 2021. A purified lectin with larvicidal activity from a woodland mushroom, *Agaricus semotus* Fr. *Acta Biologica Szegediensis*, 65(1): 65–73.

Aguilar-Marcelino, L., L.K. Tawfeeq Al-Ani, G.S. Castañeda-Ramírez, V. Garcia-Rubio and J. Ojeda-Carrasco. 2020. Microbial technologies to enhance crop production for future needs. pp. 29–47. *In:* Ali Asghar Rasteri, Ajar Nath Yadav and Neemal Yadav (eds.). New and Future Developments in Microbial Biotechnology and Bioengineering. Elsevier Inc.

Aguilar-Marcelino, L., P. Mendoza-de-Gives, G. Torres-Hernández, M.E. López-Arellano and R. González-Garduño. 2020. *Butlerius butleri* (Nematoda: Diplogasteridae) feeds on *Haemonchus contortus* (Nematoda: Trichostrongylidae) infective larvae and free-living nematodes in sheep faecal cultures under laboratory conditions: Preliminary report. *Acta Parasitologica*, 0123456789: 1–9.

Almeida, F.A., M.L.S.T. Piza, C.C. Basseto, R.Z.C. Starling, A.C.A. Albuquerque, V.M. Protes, C.M. Pariz, A.M. Castilhos, C. Costa and A.F.T. Amarante. 2018. Infection with gastrointestinal nematodes in lambs in different integrated crop-livestock systems (ICL). *Small Ruminant Research*, 166(1): 66–72. https://doi.org/10.1016/j.smallrumres.2018.07.009.

APS. 2022. Exercise: Free-Living and Plant-Parasitic Nematodes (Roundworms). https://www.apsnet.org/edcenter/disandpath/nematode/intro/Nematode/Pages/Background.aspx.

Armas-Tizapantzi, A., G. Mata, L.V. Hernández-Cuevas and A.M. Montiel-González. 2019. Estructuras tipo toxocistos en *Pleurotus ostreatus* y *P. pulmonarius*. *Scientia Fungorum*, 49(May), e1250: 1–5.

Arteaga, M.B., C.A. Soria and M.E. Ordoñez. 2020. Determinación del potencial nematicida y nematostático *in vitro* de *Pleurotus ostreatus* (Jacq. ex Fr.) sobre larvas J2 de *Globodera pallida* (Stone). *Revista Ecuatoriana de Medicina y Ciencias Biológicas*, 41(1): 45–50.

Barnes, R.D. and E.E. Ruppert. 1996. Zoología de Los Invertebrados. McGraw-Hill Interamericana Editores S.A de C.V. 1114.

Barone, C.D., J. Wit, E.P. Hoberg, J.S. Gilleard and D.S. Zarlenga. 2020. Wild ruminants as reservoirs of domestic livestock gastrointestinal nematodes. *Veterinary Parasitology*, 279: 109041. https://doi.org/10.1016/j.vetpar.2020.109041.

Barron, G.L. and R.G. Thorn. 1987. Destruction of nematodes by species of *Pleurotus*. *Canadian Journal of Botany*, 774–778.

Barros, L., P. Baptista, D.M. Correia, S. Casal, B. Oliveira and I.C. Ferreira. 2007. Fatty acid and sugar compositions, and nutritional value of five wild edible mushrooms from Northeast Portugal. *Food Chemistry*, 105(1): 140–145.

Bleuler-Martinez, S., K. Stutz, R. Sieber, M. Collot, J.M. Mallet, M. Hengartner and M. Künzler. 2017. Dimerization of the fungal defense lectin CCL2 is essential for its toxicity against nematodes. *Glycobiology*, 27(5): 486–500.

Blumenthal, T. and R.E. Davis. 2004. Exploring nematode diversity. *Nature Genetics*, 36(12): 1246–47.

Brüggemann, J. 2012. Nematodes as live food in larviculture—A review. *Journal of the World Aquaculture Society*, 43(6): 739–763.

Butschi, A., A. Titz, M.A. Walti, V. Olieric, K. Paschinger, K. Nobauer, X. Guo, P.H. Seeberger, I.B. Wilson, M. Aebi, M.O. Hengartner and M. Künzler. 2010. *Caenorhabditis elegans* N-glycan core beta-galactoside confers sensitivity towards nematotoxic fungal galectin CGL2. *PLoS Pathogens*, 6(1): e1000717. doi:10.1371/journal.ppat.1000717.

Castañeda-Ramírez, G.S., J.F.D.J. Torres-Acosta, J.E. Sánchez, P. Mendoza-de-Gives, M. González-Cortázar, A. Zamilpa, L.K.T. Al-ani, C. Sandoval-Castro, F.E. de Freitas Soares and L. Aguilar-Marcelino. 2020. The possible biotechnological use of edible mushroom bioproducts for controlling plant and animal parasitic nematodes. *BioMed Research International*, 1–12.

Cedillo, C. 2016. Estudio químico biodirigido del extracto hidroalcohólico del hongo *Pleurotus ostreatus* con actividad nematicida contra *Haemonchus contortus*. Tesis de Licenciatura Universidad Politécnica del Estado de Morelos. Ingeniería en Biotecnología. Jiutepec, Morelos, México, 72 pp.

Chang, S.T. and S.P. Wasser. 2012. The role of culinary medicinal mushrooms on human welfare with a pyramid model for human health. *International Journal of Medicinal Mushrooms*, 1: 95–134.

Charlier, J., J. Höglund, E.R. Morgan, P. Geldhof, J. Vercruysse and E. Claerebout. 2020. Biology and epidemiology of gastrointestinal nematodes in cattle. The veterinary clinics of North America. *Food Animal Practice*, 36(1): 1–15. https://doi.org/10.1016/j.cvfa.2019.11.001.

Chegwin, A.C. and R.I.J. Nieto. 2013. Influence of culture medium in the production of secondary metabolites of edible fungus *Pleurotus ostreatus* cultivated by liquid state fermentation using grain flour as a carbon source. *Revista Mexicana de Micología*, 37.

Chen, J., Q.X. Li and B. Song. 2020. Chemical nematicides: Recent research progress and outlook. *Journal of Agricultural and Food Chemistry*, 68(44): 12175–12188.

Colmenares-Cruz, S., M. González-Cortazar, G.S. Castañeda-Ramírez, R.H. Andrade-Gallegos, J.E. Sánchez and L. Aguilar-Marcelino. 2021. Nematocidal activity of hydroalcoholic extracts of spent substrate of *Pleurotus djamor* on L3 larvae of *Haemonchus contortus*. *Veterinary Pasitology*, 300: 109608. https://doi.org/10.1016/j.vetpar.2021.109608.

Cruz-Arévalo, J., J.E. Sánchez, M. González-Cortázar, A. Zamilpa, R.H. Andrade-Gallegos, P. Mendoza-de-Gives and L. Aguilar-Marcelino. 2020. Chemical composition of an anthelmintic fraction of *Pleurotus eryngii* against eggs and infective larvae (L3) of *Haemonchus contortus*. *BioMed. Research International*, 2020.

Cruz-Arévalo, J., J.E. Sánchez, M. González-Cortázar, A. Zamilpa, H. Andrade-Gallegos, P. Mendoza de Gives and L. Aguilar-Marcelino. 2020. Chemical composition of an anthelmintic fraction of *Pleurotus eryngii* against eggs and infective larvae (L3) of *Haemonchus contortus*. *BioMed. Research International*. https://doi.org/10.1155/2020/4138950.

Cuevas-Padilla, E.J., L. Aguilar-Marcelino, J.E. Sánchez, M. González-Cortázar, A. Zamilpa-Álvarez, M. Huicochea-Medina and R. González-Garduño. 2018. A *Pleurotus* spp. Hydroalcoholic fraction possess a potent *in vitro* ovicidal activity against the sheep parasitic nematode *Haemonchus contortus*. Updates on tropical Mushrooms. *Basic and Applied Research*, 193.

Cunha Zied, D., J.E. Sánchez, R. Noble and A. Pardo-Giménez. 2020. Use of spent mushroom substrate in new mushroom crops to promote the transition towards a circular economy. Agronomy, 10(9): 1239.

Degenkolb, T. and A. Vilcinskas. 2016. Metabolites from nematophagous fungi and nematicidal natural products from fungi as alternatives for biological control. Part II: metabolites from nematophagous basidiomycetes and non-nematophagous fungi. *Applied Microbiology Biotechnology*, 1(1), 1–12.

Dimunová, D., P. Matoušková, M. Navrátilová, L.T. Nguyen, M. Ambrož, I. Vokřál, B. Szotáková, and L. Skálová. 2022. Environmental circulation of the anthelmintic drug albendazole affects expression and activity of resistance-related genes in the parasitic nematode *Haemonchus contortus*. *Science of the Total Environment*, 822: 153527. https://doi.org/10.1016/j.scitotenv.2022.153527.

Drechsler, C. 1941. Predaceous fungi. *Biological reviews. Cambridge Philosophical Society*, 16(4): 265–290. https://doi.org/10.1111/j.1469-185X.1941.tb01104.x.

Dubey, S.K., V.K. Chaturvedi, D. Mishra, A. Bajpeyee, A. Tiwari and M.P. Singh. 2019. Role of edible mushroom as a potent therapeutics for the diabetes and obesity. *3 Biotech.*, 9(12): 1–12.

Erjavec, J., J. Kos, M. Ravnikar, T. Dreo and J. Sabotič. 2012. Proteins of higher fungi–from forest to application. *Trends in Biotechnology*, 30(5): 259–273.

Eves-van den Akker, S., B. Stojilković and G. Gheysen. 2021. Recent applications of biotechnological approaches to elucidate the biology of plant-nematode interactions. *Current Opinion in Biotechnology*, 70: 122–130.

FAO (Food Agriculture Organization of the United Nations) 2022. Available online: https://www.fao.org/faostat/es/#data/QCL.

Ferreira, J.M., D.N. Carreira, F.R. Braga and F.E.D.F. Soares. 2019. First report of the nematicidal activity of *Flammulina velutipes*, its spent mushroom compost and metabolites. *3 Biotech.*, 9(11): 1–6.

Franco, André L.C., P. Guan, S. Cui, C.M. de Tomasel, L.A. Gherardi, O.E. Sala and D.H. Wall. 2022. Precipitation effects on nematode diversity and carbon footprint across grasslands. *Global Change Biology*, 28(6): 2124–32.

Genier, H.L.A., F.E. de Freitas Soares, J.H. de Queiroz, A. de Souza Gouveia, J.V. Araújo, F.R. Braga and M.C.M. Kasuya. 2015. Activity of the fungus *Pleurotus ostreatus* and of its proteases on *Panagrellus* sp. larvae. *African Journal of Biotechnology*, 14(17): 1496–1503.

González-Cortázar, M., J.E. Sánchez, M. Huicochea-Medina, V.M. Hernández-Velázquez, P. Mendoza-de-Gives, A. Zamilpa, M.E. López-Arellano, J.A. Pineda-Alegría and L. Aguilar-Marcelino. 2021. *In vitro* and *in vivo* nematicide effect of extract fractions of *Pleurotus djamor* against *Haemonchus contortus*. *Journal of Medicinal Food*, 24(3): 310–318. https://doi.org/10.1089/jmf.2020.0054.

Guillamón, E., A. García-Lafuente, M. Lozano, M.A. Rostagno, A. Villares and J.A. Martínez. 2010. Edible mushrooms: Role in the prevention of cardiovascular diseases. *Fitoterapia*, 81(7): 715–723.

Günç Ergönül, P., I. Akata, F. Kalyoncu and B. Ergönül. 2013. Fatty acid compositions of six wild edible mushroom species. *The Scientific World Journal*, 2013: 163964.

Hahn, M.H., L.L. May De Mio, O.J. Kuhn and H. da S. Silveira Duarte. 2019. Nematophagous mushrooms can be an alternative to control *Meloidogyne javanica*. *Biological Control*, 138: 1–9.

Heydari, R., E. Pourjam and E.M. Goltapeh. 2006. Antagonistic effect of some species of *Pleurotus* on the root-knot nematode, *Meloidogyne javanica in vitro*. *Plant Pathology Journal*, 5(2): 173–177.

Hodda, M. 2022a. Phylum nematoda: A classification, catalogue and index of valid genera, with a census of valid species. *Zootaxa*, 5114(1): 1–289.

Hodda, M. 2022b. Phylum nematoda: Feeding habits for all valid genera using a new, universal scheme encompassing the entire phylum, with descriptions of morphological characteristics of the stoma, a key, and discussion of the evidence for trophic relationships. *Zootaxa*, 5114(1): 318–451.

Ingham, E.R. 2010. Nematodes. pp. 20–22. *In:* Tugel, A.J., Lewandowsky, A.M. and Happe-vonArb, D. (eds.). The Soil Biology Primer. USDA.

Iqbal, S. and M.G.K. Jones. 2017. Crop diseases and pests. *In: Encyclopedia of Applied Plant Sciences*, 3: 113–119.

Jackson, F. and H. Hoste. 2010. *In vitro* methods for the primary screening of plant products for direct activity against ruminant gastrointestinal nematodes. pp. 25–45. *In:* Vercoe, P.E., Makkar, H.P.S.

and Schlink, A.C. (eds.). *In vitro* Screening of Plant Resources for Extra-nutritional Attributes in Ruminants: Nuclear and Related Methodologies. Springer.

Jones, J.T., A. Haegeman, E.G.J. Danchin, H.S. Gaur, J. Helder, M.G.K. Jones, T. Kikuchi, R. Manzanilla-López, J.E. Palomares-Rius, W.M.L. Wesemael and R.N. Perry. 2013. Top 10 plant-parasitic nematodes in molecular plant pathology. *Molecular Plant Pathology*, 14(9): 946–961.

Kolesnik, N., M. Simon, O. Marenkov and O. Nesterenko. 2019. The cultivation of the nematodes *Panagrellus*, *Turbatrix* (Anguillula) and *Rhabditis* for using in fish feeding. *Ribogospodars'ka Nauka Ukraïni*, 3(49): 16–31.

Kwok, O.C.H., R. Plattner, D. Weisleder and D.T. Wicklow. 1992. A nematicidal toxin from *Pleurotus ostreatus* NRRL 3526. *Journal of Chemical Ecology*, 18(2): 127–136.

Lagerlöf, J., V. Insunza, B. Lundegårdh and B. Rämert. 2011. Interaction between a fungal plant disease, fungivorous nematodes and compost suppressiveness. *Acta Agriculturae Scandinavica Section B: Soil and Plant Science*, 61(4): 372–377.

Laith, T.A.-A., L. Aguilar-Marcelino, J. Fiorotti, V. Sharma, M. Sharif and A. Herrera-estrella. 2020. Biological control agents and their importance for plant health. pp. 13–36. *In*: Microbial Services in Restoration Ecology. Elsevier Inc.

Larsen, M. and P. Nansen. 1991. Ability of the fungus *Pleurotus pulmonarius* to immobilise preparasitic nematode larvae. *Research in Veterinary Science*, 51: 246–249. http://dx.doi.org/10.15226/2381-2907/1/1/00104.

Leong, Y.K., T.W. Ma, J.S. Chang and F.C. Yang. 2022. Recent advances and future directions on the valorization of spent mushroom substrate (SMS): A review. *Bioresource Technology*, 344: 126157.

Li, G.-H. and K.-Q. Zhang. 2014. Nematode-toxic fungi and their nematicidal metabolites. pp. 348–357. *In*: Zhang, K.-Q. and Hyde, K.D (eds.). Nematode-Trapping Fungi. Springer Science+Business Media B.V., Dordrecht, The Netherlands and Kevin D. Hyde, Chiang Rai, Thailand.

Li, G., X. Wang, L. Zheng, L. Li, R. Huang and K. Zhang. 2007. Nematicidal metabolites from the fungus *Pleurotus ferulae* Lenzi. *Annals of Microbiology*, 57: 527. https://doi.org/10.1007/BF031753.

Lionetti, V., E. Venter, V. Pastor, M.A. Ali, F. Azeem, H. Li and H. Bohlmann. 2017. Smart parasitic nematodes use multifaceted strategies to parasitize plants. *Frontiers in Plant Science*, 8: 1699.

Luo, H., Y. Liu, L. Fang, X. Li, N. Tang and K. Zhang. 2007. *Coprinus comatus* damages nematode cuticles mechanically with spiny balls and produces potent toxins to immobilize nematodes. *Applied and Environmental Microbiology*, 73(12): 3916–3923.

Maity, P., I.K. Sen, I. Chakraborty, S. Mondal, H. Bar, S.K. Bhanja, S. Mandal and G.N. Maity. 2021. Biologically active polysaccharide from edible mushrooms: A review. *International Journal of Biological Macromolecules*, 172: 408–417.

Marques, J., F. Dhiogo, N. Carreira, F. Ribeiro, B. Filippe and E. de Freitas. 2019. First report of the nematicidal activity of *Flammulina velutipes*, its spent mushroom compost and metabolites. *3 Biotech.*, 1–6.

Mattila, P., K. Könkö, M. Eurola, P. Juha-Matti, J. Astola, L. Vahteristo, V. Hietaniemi, J. Kumpulainen, M. Valtonen and V. Piironen. 2001. Contents of vitamins, mineral elements, and some phenolic compounds in cultivated mushrooms. *Journal of Agricultural and Food Chemistry*, 49(5): 2343–2348.

Mishra, V., S. Tomar, P. Yadav and M.P. Singh. 2021. Promising anticancer activity of polysaccharides and other macromolecules derived from oyster mushroom (*Pleurotus* sp.): An updated review. *International Journal of Biological Macromolecules*, 182: 1628–1637.

Muszyńska, B., A. Grzywacz-Kisielewska, K. Kała and J. Gdula-Argasińska. 2018. Anti-inflammatory properties of edible mushrooms: A review. *Food Chemistry*, 243: 373–381.

Najafi, B., S. Faizollahzadeh Ardabili, S. Shamshirband and K.W. Chau. 2019. Spent mushroom compost (SMC) as a source for biogas production in Iran. *Engineering Applications of Computational Fluid Mechanics*, 13(1): 967–982.

Navone, G., A. Fernanda, J. Notarnicola and L. Zonta. 2011. CAPÍTULO 9 Phylum Nematoda. *Digital. Cic. Gba. Gob. A*, 9(0): 128–56.

Nelson, D.L. and M.M. Cox. 2017. Lehninger Principles of Biochemistry (7th ed.). W.H. Freeman.

Palizi, P., E.M. Goltapeh, E. Pourjam and N. Safaie. 2009. Potential of oyster mushrooms for the biocontrol of sugar beet nematode (*Heterodera schachtii*). *Journal of Plant Protection Research*, 49(1): 27–33.

Perry, R.N. and M. Moens. 2013. Plant nematology. pp. 4. *In*: Perry, R.N. and Moens, M. (eds.). 2nd ed. CABI International.

Pica-Granados, Y. 2008. Ensayo de toxicidad con el nemátodo *Panagrellus redivivus*. pp. 139–154. *In*: Ensayos toxicológicos para la evaluación de sustancias químicas en agua y suelo. La experiencia en México (Primera). SEMARNAT.

Pineda-Alegría, J.A., J.E. Sánchez-Vázquez, M. Gonzalez-Cortazar, A. Zamilpa, M.E. López-Arellano, E.J. Cuevas-Padilla and L. Aguilar-Marcelino. 2017. The edible mushroom *Pleurotus djamor* produces metabolites with lethal activity against the parasitic nematode *Haemonchus contortus*. *Journal of Medicinal Food*, 20(12): 1184–1192.

Pineda-Alegría, J.A., J.E. Sánchez-Vázquez, M. González-Cortazar, A. Zamilpa, M.E. López-Arellano, E.J. Cuevas-Padilla, P. Mendoza-de-Gives and L. Aguilar-Marcelino. 2017. The edible mushroom *Pleurotus djamor* produces metabolites with lethal activity against the parasitic nematode *Haemonchus contortus*. *Journal of Medicinal Food*, 20(12): 1184–1192. https://doi.org/10.1089/jmf.2017.0031.

Pineda-Alegría, J.A., J.E. Sánchez, M. González-Cortazar, E. von Son-de Fernex, R. González-Garduño, P. Mendoza-de Gives, A. Zamilpa and L. Aguilar-Marcelino. 2020. *In vitro* nematocidal activity of commercial fatty acids and β-sitosterol against *Haemonchus contortus*. *Journal of Helminthology*, 94.

Pinto, A., K. May, T. Yin, M. Reichenbach, P.K. Malik, R. Roessler, E. Schlecht and S. König. 2021. Gastrointestinal nematode and *Eimeria* spp. infections in dairy cattle along a rural-urban gradient. *Veterinary Parasitology, Regional Studies and Reports*, 25: 100600. https://doi.org/10.1016/j.vprsr.2021.100600.

Ragucci, S., N. Landi, R. Russo, M. Valletta, P.V. Pedone, A. Chambery and A. Di Maro. 2021. Ageritin from pioppino mushroom: The prototype of ribotoxin-like proteins, a novel family of specific ribonucleases in edible mushrooms. *Toxins*, 13(4): 263.

Rajavat, A.S., V. Mageshwaran, A. Bharadwaj, S. Tripathi and K. Pandiyan. 2022. Spent mushroom waste: An emerging bio-fertilizer for improving soil health and plant productivity. *In*: New and Future Developments in Microbial Biotechnology and Bioengineering (pp. 345–354). Elsevier.

Rivas-Morales, C., M.A. Oranday-Cárdenas and M.J. Verde-Star. 2016. Actividad antioxidante y toxicidad. *In*: Investigación en plantas de importancia médica (Issue May 2017, pp. 351–410). *OmniaScience*.

Royse, D.J. and J.E. Sánchez. 2017. Producción mundial de setas *Pleurotus* spp. con énfasis en países iberoamericanos. pp. 17–25. *In*: Sánchez, J.E. and Royse, D. (eds.). La biología, el cultivo y las propiedades nutricionales y medicinales de las setas *Pleurotus* spp. El Colegio de la Frontera Sur, Tapachula.

Sabotic, J., R.A. Ohm and M. Kunzler. 2016. Entomotoxic and nematotoxic lectins and protease inhibitors from fungal fruiting bodies. *Applied Microbiology and Biotechnology*, 100: 91–111.

Sato, K., Y. Kadota and K. Shirasu. 2019. Plant immune responses to parasitic nematodes. *Frontiers in Plant Science*, 10(September): 1–14.

Smiderle, F.R., D. Morales, A. Gil-Ramírez, L.I. de Jesus, B. Gilbert-López, M. Iacomini and C. Soler-Rivas. 2017. Evaluation of microwave-assisted and pressurized liquid extractions to obtain β-d-glucans from mushrooms. *Carbohydrate Polymers*, 156: 165–174.

Soares, de F.F., M.V. Nakajima, L.B. Sufiate, S.L.A. Satiro, G.E. Honorato, F.F. Vieira, S.F. Porto, F.R. Braga and J.H. de Queiroz. 2019. Proteolytic and nematicidal potential of the compost colonized by *Hypsizygus marmoreus*. *Experimental Parasitology*, 197(December 2018): 16–19.

Soares, F.E. de F., B.L. Sufiate and J.H. de Queiroz. 2018. Nematophagous fungi: Far beyond the endoparasite, predator and ovicidal groups. *Agriculture and Natural Resources*, 52(1): 1–8.

Soares, F.E.S., V.M. Nakajima, B.L. Sufiate, L.A.S. Satiro, E.H. Gomes, F.V. Fróes, F.P. Sena, F.R. Braga and J.H. de Queiroz. 2019. Proteolytic and nematicidal potential of the compost colonized by *Hypsizygus marmoreus*. *Experimental Parasitology*, 197: 16–19.

Soares, F.E.F., B.L. Sufiate and J.H. Queiroz. 2018. Nematophagous fungi: Far beyond the endoparasite, predator and ovicidal groups. *Agriculture and Natural Resources*, 52(1): 1–8.

Sochová, I., J. Hofman and I. Holoubek. 2006. Using nematodes in soil ecotoxicology. *Environment International*, 32(3): 374–383.

Soliman, G., W. Elkhateeb, T.-C. Wen and G. Daba. 2022. Mushrooms as efficient biocontrol agents against the root-knot nematode, Meloidogyne incognita. *Egyptian Pharmaceutical Journal*, 0(0): 2–8.

Stadler, M., H. Anke and O. Sterner. 1993. Linoleic acid—The nematicidal principle of several nematophagous fungi and its production in trap-forming submerged cultures. *Archives of Microbiology*, 160(5): 401–405.

Stadler, M., A. Mayer, H. Anke and O. Sterner. 1994. Fatty acids and other compounds with nematicidal activity from cultures of Basidiomycetes. *Planta Medica*, 60: 128–132. https://doi.org/10.1055/s-2006-959433.

Sufiate, B.L., F.E. de Freitas Soares, S.S. Moreira, A. de Souza Gouveia, T.S.A. Monteiro, L.G. de Freitas and J.H. de Queiroz. 2017. Nematicidal action of *Pleurotus eryngii* metabolites. *Biocatalysis and Agricultural Biotechnology*, 12: 216–219.

Tayyrov, A., S. Azevedo, R. Herzog, E. Vogt, S. Arzt, P. Lüthy and M. Künzler. 2019. Heterologous production and functional characterization of ageritin, a novel type of ribotoxin highly expressed during fruiting of the edible mushroom *Agrocybe aegerita*. *Applied and Environmental Microbiology*, 85(21): e01549–19.

Tayyrov, A., S.S. Schmieder, S. Bleuler-Martinez, D.F. Plaza and M. Künzler. 2018. Toxicity of potential fungal defense proteins towards the fungivorous nematodes *Aphelenchus avenae* and *Bursaphelenchus okinawaensis*. *Applied and Environmental Microbiology*, 84(23): e02051–18.

Thorn, R.G. and G.L. Barron. 1984. Carnivorous mushrooms. *Science (New York, N.Y.)*, 24(4644): 76–78. https://doi.org/10.1126/science.224.4644.76.

Thorn, R.G. and G.L. Barron. 1984. Carnivorous mushrooms. *Science*, 224(4644): 76–78.

Valdez-Uriostegui, L.A., A.D. Sánchez-García, A. Zamilpa, J.E. Sánchez, R. González-Garduño, P. Mendoza-de-Gives and L. Aguilar-Marcelino. 2021. *In vitro* evaluation of hydroalcoholic extracts of mycelium, basidiomata and spent substrate of *Pleurotus ostreatus* against *Haemonchus contortus*. *Tropical and Subtropical Agroecosystems*, 24(2).

Valverde, M.E., T. Hernández-Pérez and O. Paredes-López. 2015. Edible Mushrooms: Improving human health and promoting quality life. *International Journal of Microbiology*. DOI http://dx.doi.org/10.1155/2015/376387.

Wille, C.N., C.B. Gomes, E. Minotto and J.S. Nascimento. 2019. Potential of aqueous extracts of basidiomycetes to control root-knot nematodes on lettuce. *Horticultura Brasileira*, 37(1): 54–59.

Yadav, S., J. Patil and R.S. Kanwar. 2018. The role of free-living nematode population in the organic matter recycling. *International Journal of Current Microbiology and Applied Sciences*, 7(06): 2726–2734.

Yeates, G.W. and T. Bongers. 1999. Nematode diversity in agroecosystems. *Agriculture, Ecosystems and Environment*, 74(1-3): 113–35.

Youssef, M.A. and W.M.A. El-Nagdi. 2021. New approach for biocontrolling root-knot nematode, *Meloidogyne incognita* on cowpea by commercial fresh oyster mushroom (*Pleurotus ostreatus*) Jordan. *Journal of Biological Sciences*, 14(1): 173–177.

Zajac, A.M. 2006. Gastrointestinal nematodes of small ruminants: Life cycle, anthelmintics, and diagnosis. The Veterinary clinics of North America. *Food Animal Practice*, 22(3): 529–541. https://doi.org/10.1016/j.cvfa.2006.07.006.

Žurga, S., J. Pohleven, M. Renko, S. Bleuler-Martinez, P. Sosnowski, D. Turk, M. Künzler, J. Kos and J. Sabotič. 2014. A novel β-trefoil lectin from the parasol mushroom (Macrolepiota procera) is nematotoxic. *The FEBS Journal*, 281(15): 3489–3506.

19

Mushroom Products in Malaysia

Shin Yee Fung[1],* and Chon-Seng Tan[2]

1. Introduction

Increasing demand for food that is dense in nutrients, but low in fat and cholesterol has boosted products of "superfood" origin. Superfood is not a term of choice for experts, dieticians and nutrition scientists; but has been used arbitrarily amongst marketers to refer to food that offers maximum nutritional benefits for minimal calories. Fortune Business Insights (https://www.fortunebusinessinsights.com/industry-reports/mushroom-market-100197) (accessed 25th March 2022) in a recent report (Report ID: FBI100197) revealed that the globally mushroom was valued at 12.74 million tonnes (MT) in 2018 and is projected to attain 20.84 million ton (MT) by 2026, showing a cumulative annual growth rate (CAGR) of 6.41% in the forecasted period. Mushrooms are commercially grown at a large scale in the United States, China and many European countries. Malaysia's mushroom market which saw cultivation initiated in the 1960s has been relatively small. 49% of mushrooms are in exported (fresh button mushroom (*Agaricus bisporus*)) whilst 51% are in processed form (Ling Zhi (*Ganoderma lucidum*), wood ear (*Auricularia auricula-judae*) and straw mushroom (*Volvariella volvacea*)) (Dasar Agro-Makanan Negara (2011–2020)). The main destination for export is the United States, Brazil and Singapore.

[1] Medicinal Mushroom Research Group (MMRG), Department of Molecular Medicine, Universiti Malaya, 50603 Kuala Lumpur, Malaysia.
[2] LiGNO Research Initiative, LiGNO Biotech Sdn. Bhd., Balakong Jaya 43300, Selangor, Malaysia.
* Corresponding author: syfung@um.edu.my

Being a superfood, mushrooms are marketed in current years as more than just a fresh, culinary ingredient, but has been made into other value-added products such as soup powders, biscuits, nuggets, sauce, candies, chips and other ready-to-eat products such as curries (Wakchaure 2011). They have also been made into cosmetics, pet foods and supplements for therapeutic applications. The increased awareness of the dietary benefits offered by mushrooms is gaining traction and is driving the growth of the market.

2. Mushroom Products and Preservation Techniques

Certain exquisite mushroom species are delicacies with significant nutritional and functional values. Other more commonly known mushrooms are part of the daily diet for some consumers. Industries have also considered mushrooms to be nutraceutical foods with organoleptic value, medicinal potential and economic significance (Chang and Miles 2008, Ergonul et al. 2013). They are packed with optimal content consisting of indispensable amino acids (protein), fibre, and meagre fat but with excellent essential fatty acids contents. In addition, edible mushrooms provide a nutritionally significant content of vitamins (B_1, B_2, B_{12}, C, D, and E) (Heleno et al. 2010, Mattila et al. 2001). They also possess secondary metabolites (e.g., terpenes, steroids, anthraquinones, benzoic acid derivatives and quinolones) and primary metabolites (e.g., oxalic acid). These are characteristics of immense interest for the health-conscious and lay the basis of product development.

Being a perishable ingredient, many of the mushroom products in the market have adopted different processing technologies to retain bioactive components in their marketed products. Postharvest preservation methods to maintain the quality and extension of the shelf life of mushrooms are classified into three groups: thermal (cooling and drying), physical (pulsed electrical field, irradiation, packaging, plasma), and chemical (washing, coating, adopting the use of ozone and electrolyzed water) (Zhang et al. 2018). These strategies aim to slow down or halt chemical deterioration, and microbial growth and avoid recontamination. These postharvest preservation techniques are not without drawbacks. Zhang et al. (2018) described concerns of safety along with decreases in nutritional value, alterations in colour and texture as well as possible contamination by pathogens that cause bad flavours. The high cost of capital involved in setting up these preservation techniques and high energy consumption are also cause for concern.

3. Challenges of a Thriving Mushroom Industry

In Malaysia, mushrooms have been cultivated since the late 1960s. In recent years and with increasing awareness, mushrooms have now been declared an industrial crop under the 4th National Agricultural Policy (2010–2015) by the Ministry of Agriculture, and as one of the seven high-valued industrial crops in Malaysia that will be commercially grown under the Dasar Agro-Makanan Negara (2011–2020).

The export of mushrooms in Malaysia has grown at a 19% rate from RM12 million in the year 2000 to RM67 million in 2010. The main destination for export is the United States, Brazil and Singapore. Grey oyster mushroom (*Pleurotus ostreatus*) is the dominant commercial variety due to its ease of growth and the capability of this species to be cultivated in lowlands. Most producers are operating under a small-scale agriculture model and produce 50–500 kg with no more than 10 farms producing tons of fresh mushrooms per day according to Vikineswary (2007). Other mushrooms such as abalone (*Pleurotus cystidiosus*), Ling Zhi (*Ganoderma lucidum*), button (*Agaricus bisporus*), shiitake (*Lentinus edodes*), wood ear (*Auricularia polytricha*) and monkey head (*Hericium erinaceus*) mushrooms are also cultivated for commercial reasons (Vikineswary et al. 2007), but sustainability challenges remain.

Issues and challenges in the Malaysian mushroom industry are attributed to two main factors namely, (1) production or cultivation technology and (2) marketing. Production of mushrooms depends largely on environmental temperature, and mushrooms that thrive in a cooler environment such as shiitake and button mushrooms will find the tropical climate of Malaysia challenging. The hot Malaysian weather can cause the cultivation medium to dry easily leading to a general decline of the mushroom species in cultivation plants unless a cultivation plant with a controlled environment is built. Advancement in technology enabling a controlled environment for mushroom cultivation has been proven effective in some countries. Pest contamination (in the form of mold, insects and rodents) is yet another major problem the mushroom growers need to reconcile with, especially in a hot and humid local environment.

The quality of the seed culture and raw materials used are also a hurdle for many attempting cultivators in Malaysia. Most of the cultivators/growers depend on private suppliers for the seeds at exorbitant prices. The cost of raw materials, for instance, rubber wood sawdust has been increasing annually and small-scale cultivators/growers find it economically exasperating. The limited availability of rubber wood sawdust has also caused conspicuous competition and met with an unrealistic price.

The obstacles listed have universally increased the production cost for mushroom cultivators/cultivators and indirectly handicapped Malaysian mushroom growers from competing with other countries that are able to produce mushrooms at a significantly lower cost. Hence, mushroom products in Malaysia which are largely dependent on the fungi-farming industry to thrive are indirectly impacted.

A brief SWOT analysis of the challenges of the Malaysian mushroom industry is depicted in Figure 1.

4. Successful Mushroom Product Companies in Malaysia

The commercial scenario for the mushroom industry in Malaysia is largely divided into two segments: companies specializing in farming/cultivating edible mushrooms

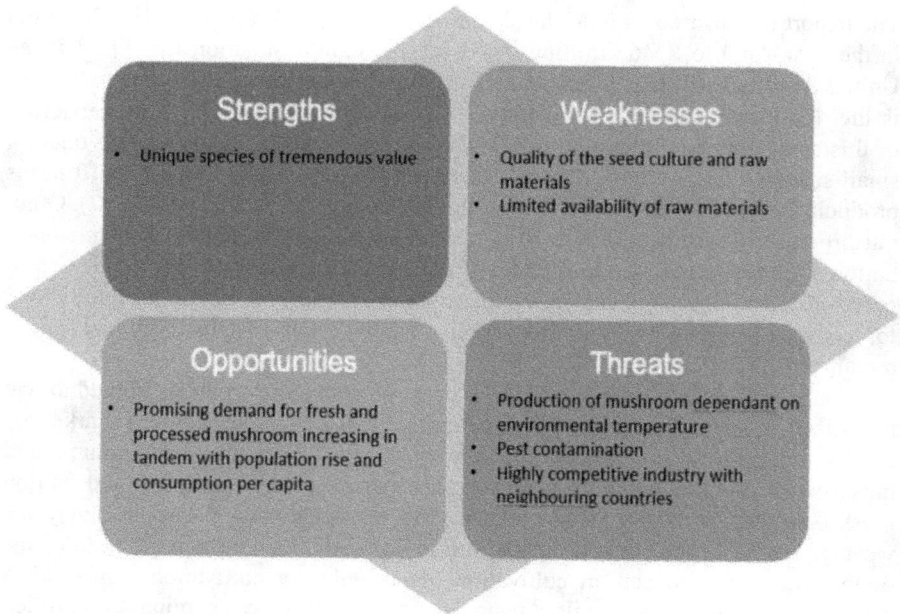

Figure 1: SWOT analysis of the mushroom industry in Malaysia.

and medicinal mushrooms. Examples of companies that farm edible mushrooms are: C&C Mushroom Cultivation Farm Sdn. Bhd. (oyster mushroom), ChampFungi Sdn. Bhd. (button mushroom), Agro Bio-Future Gunung Jerai (shiitake) and Hokto Malaysia Sdn. Bhd. (bunashimeji, bunapi, maitake); whilst medicinal mushroom-centred companies are DXN Holdings Bhd. (*Ganoderma lucidum*), BioFact Life Sdn. Bhd. (Timo Cordyceps) and LiGNO™ Biotech Sdn. Bhd. (Tiger Milk Mushroom, *Ophiocordyceps sinensis*, Chaga, Antrodia, Sanghuang).

4.1 ChampFungi Sdn. Bhd.

ChampFungi Sdn. Bhd. located in Telok Gong, Port Klang, Malaysia was established in 2002 and its initial business activity was button mushroom farming. They have since diversified to also grow Shiitake and Eryngii. They make their own substrates as a base material for their mushrooms, pack and transport their cultivated variety of fresh produce products. They sell directly to supermarkets or restaurants via a wide sales and distribution network across South East Asia. On their website (champfungi. com) they revealed that they cultivate the mushrooms in a "controlled environment" (enclosed growing houses/greenhouses) to sustain a stable supply of mushrooms. To ensure the freshness of their produce, they ensure a time frame of no longer than 24 hours from farm to shelf. A short video depicting their cultivation facility can be found at https://champfungi.com/pages/farm-tours.html.

4.2 *BioFact Life Sdn. Bhd.*

Incorporated in 2005 to specialise in the cultivation of *Cordyceps* sp. (冬虫草). Their cordyceps species are cultured in a cleanroom and the airborne particle is controlled to specified limits, free from chemical, yeast, mould and bacterial contaminations. Their registered products are in the form of capsules, liquid, powder, tablets and ointment under the brand 'TIMO' (http://timo.com.my/) and are sold in retail pharmacies and drug houses locally and internationally.

4.3 *LiGNO™ Biotech Sdn. Bhd.*

Established in the year 2009, this small-medium enterprise was founded to mass cultivate a local Malaysia medicinal mushroom – the tiger milk mushroom (*Lignosus rhinocerus*) registered with the Plant Varieties Board of the Malaysian Government bearing the Registration number: PBR0182 as TM02® cultivar protected under Section 28 of the Protection of New Plant Varieties (PNPV) Act 2004 (Act 634). Numerous attempts were made to cultivate the sclerotium of this rare medicinal mushroom to obtain high nutritional and bioactive values.

Various methods of cultivating mushrooms have been tried ie. (a) submerged fermentation and (b) solid-state fermentation (Elisashvili 2012, Letti et al. 2018). The solid-state fermentation technology employed provides a higher yield of production and allows the formation of sclerotium (Letti et al. 2018). Developing a suitable environment and condition of harvest for this specific species is crucial for this species as a controlled environment allows for the production of nutrient and bioactive-filled sclerotia.

The advent of LiGNO™ Biotech Sdn. Bhd. has taken Malaysia one step ahead as this is the first local Malaysian cultivator who has achieved successful cultivation on a large scale of a medicinal mushroom of local origin and made it into a commodity widely sought after for various uses in the nutraceutical industry. They have also taken the same cultivation technology and expanded to cultivate other prized mushrooms such as the *Ophiocordyceps sinensis*, *Antrodia camphorata* and *Inonotus obliquus*. These mushrooms have been made into functional food products in capsules, beverage sachets, noodles and even skincare products. Examples of a wide range of products from LiGNO™ Biotech Sdn. Bhd. is shown in Figure 2(A-D) and Figure 3(A-H).

5. Future Prospects

The National Agro-Food Policy (National Agro-Food Policy (Dasar Agro-Makanan Negara) 2011–2020) predicted that the export value of Malaysian mushrooms had reached RM300 million in 2020 with an annual increase of 16%. While there is no concrete data to show if this had come to pass, it is undeniable that the prospect of the mushroom industry in Malaysia is promising as the demand for both fresh and processed mushroom production has increased steadily. This demand increased in

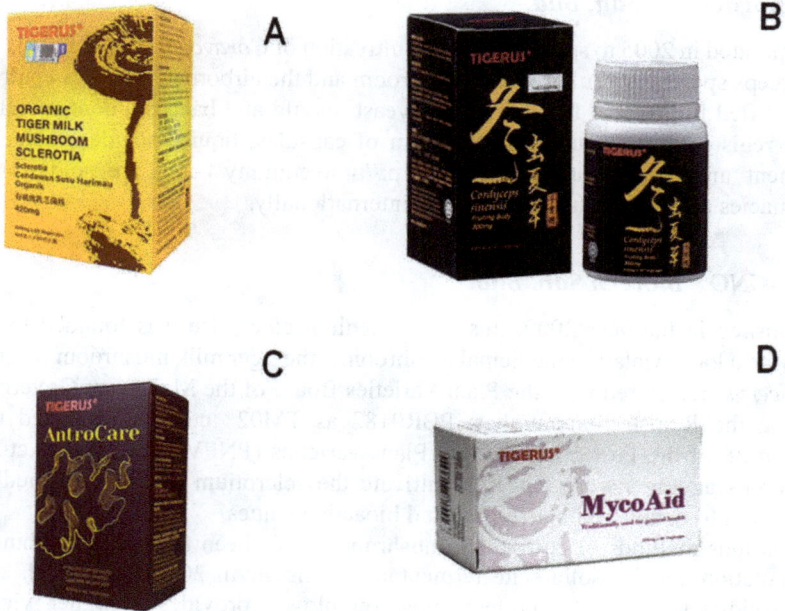

Figure 2: Mushroom Products – (A) TIGERUS Tiger Milk Mushroom (*Lignosus rhinocerus* TM02™); (B) TIGERUS *Ophiocordyceps sinensis* (C) TIGERUS Antrocare: combination of *Antrodia camphorata*, *Ophiocordyceps sinensis* and *Lignosus rhinocerus* TM02™ (D) TIGERUS MycoAid: a combination of *Antrodia camphorata*, *Ophiocordyceps sinensis*, Chaga and *Lignosus rhinocerus* TM02™.

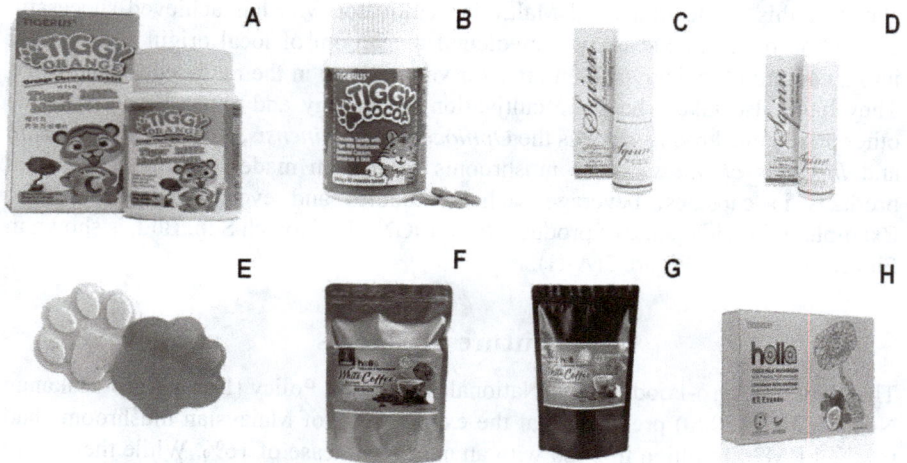

Figure 3: Mushroom Enhanced Products – (A) TIGERUS Tiggy Orange and (B) TIGERUS Tiggy Cocoa (both A and B are vitamin C with *Lignosus rhinocerus* TM02TM) (C) SQINN Skin cream with *Lignosus rhinocerus* TM02™ (D) SQINN Skin spray with *Lignosus rhinocerus* TM02™ (E) SQINN Soap with *Lignosus rhinocerus* TM02™ (F) Inulin white coffee with *Lignosus rhinocerus* TM02™ (G) Sugar-free white coffee with *Lignosus rhinocerus* TM02™ (H) HOLLA Passion fruit with *Lignosus rhinocerus* TM02™.

tandem with the population rise and consumption per capita (1 kg as of 2008 to 2.4 kg in 2020). Production of mushrooms has also increased by as much as 16%, from 15 000 metric tonnes in 2010 to 67000 metric tonnes in 2020. Several innovations have been reported to enhance the growth of the cultivation industry (Teik et al. 2021).

Due to the pandemic that hit the world in 2020, a new report revealed that the scarcity of wood supply has impacted mushroom cultivation, which relies on sawdust as its main source of raw material (https://www.nst.com.my/business/2021/07/707517/ wood-supply-scarcity-adversely-affecting-mushroom-cultivation-sector). In a highly competitive industry with neighbouring countries such as Thailand and Vietnam, local cultivators are encountering many difficulties to sustain the industry and to subsequently recover from output slump.

For further improvement and expansion of the Malaysia mushroom industry, growers, stakeholders and the Malaysian Government can concentrate more on increasing markets; broadening consumer awareness; developing novel products and technology; and building up a network with other agro-food-based industries (Rosmiza et al. 2016).

References

Castellanos-Reyes, K., R. Villalobos-Carvajal and T. Beldarrain-Iznaga. 2021. Fresh Mushroom preservation techniques. *Foods*, 10: 2126. https://doi.org/10.3390/ foods10092126.

Chang, S.T. and P.G. Miles. 2008. Mushrooms: Cultivation, Nutritional Value, Medicinal Effect, and Environmental Impact. CRC Press, USA.

Dasar Agro-Makanan Negara 2011–2020 (2011). Kementerian Pertanian dan Industri Asas Tani. Percetakan Watan Sdn. Bhd.

Elisashvili, V. 2012. Submerged cultivation of medicinal mushrooms: Bioprocesses and products. *Int. J. Med. Mushr.*, 14: 211–239. 10.1615/IntJMedMushr.v14.i3.10.

Ergonul, P.G., I. Akata, F. Kalyoncu and B. Ergonul. 2013. Fatty acid compositions of six wild edible mushroom species, *The Scientific World Journal*, 2013: Article ID 163964.

Heleno, S.A., L. Barros, M.J. Sousa, A. Martins and I.C.F.R. Ferreira. 2010. Tocopherols composition of Portuguese wild mushrooms with antioxidant capacity. *Food Chemistry*, 119(4): 1443–1450.

http://timo.com.my/ (accessed 27th of March 2022).

https://champfungi.com (accessed 27th of March 2022).

https://www.fortunebusinessinsights.com/industry-reports/mushroom-market-100197 (accessed 25th March 2022).

https://www.nst.com.my/business/2021/07/707517/wood-supply-scarcity-adversely-affecting-mushroom-cultivation-sector (accessed 5th April 2022).

Letti, L., F. Vítola, G. Pereira, S. Karp, A.B.P. Medeiros, E.S.F. da Costa, L. Bissoqui and C.R. Soccol 2018. Solid-state fermentation for the production of mushrooms. pp. 285–318. *In*: Pandey, A., Larroche, C. and Soccol, C.R. (eds.). Current Developments in Biotechnology and Bioengineering. Current Advances in Solid-state Fermentation, Elsevier, B.V.

Mattila, P., K. Könkö, M. Eurola, J. Pihlava, J. Astola, L. Vahteristo, V. Hietaniemi, J. Kumpulainen, M. Valtonen and V. Piironen. 2001. Contents of vitamins, mineral elements, and some phenolic compounds in cultivated mushrooms. *J. Agric. Food Chem.*, 49(5): 2343–2348.

Rosmiza, M.Z., W.P. Davies, C.R. Rosniza Aznie, M.J. Jabil and M. Mazdi. 2016. Prospects for increasing commercial mushroom production in Malaysia: Challenges and opportunities. *Med. J. Social Sci.*, 7(1): 406–415. doi:10.5901/mjss.2016.v7n1s1p406.

Teik, T.S., G. Krishnen, K.A. Khulidin, M.A. Mohamad Tahir, M.H. Hashim and S. Khairudin. 2021. Automated controlled environment mushroom house. *Adv. Agri. Food Res. J.*, 2(2): a0000230. https://doi.org/10.36877/aafrj.a0000230.

Vikineswary, S., A. Norlidah, I. Normah, Y.H. Tan, D. Fauzi and E.B.G. Jones. 2007. Edible and medicinal mushroom. In. Malaysian Fungal Diversity. NRE Malaysia, 287–305.

Wakchaure, G.C. 2011. Mushrooms-value added products. pp. 233–238. *In*: Singh, M., Vijay, B., Kamal, S. and Wakchaure, G.C. (eds.). Mushrooms-cultivation, Marketing and Consumption. Directorate of Mushroom Research, Solan, Himachal Pradesh 173213, India.

Zhang, K., Y.Y. Pu and D.W. Sun. 2018. Recent advances in quality preservation of postharvest mushrooms (*Agaricus bisporus*): A review. *Tr. Food Sci. Technol.*, 78: 72–82.

Index

About the Editors

Dr. Sunil Kumar Deshmukh is Scientific Advisor to Greenvention Biotech, Uruli-Kanchan, Pune, India and Agpharm Bioinnovations LLP, Patiala, Punjab, India. Veteran industrial mycologist, spent a substantial part of his career in drug discovery at Hoechst Marion Roussel Limited [now Sanofi India Ltd.], Mumbai, and Piramal Enterprises Limited, Mumbai. He has also served TERI-Deaken Nano Biotechnology Centre, TERI, New Delhi, and as an Adjunct Associate Professor at Deakin University, Australia. He has to his credit 8 patents, 145 publications, and 18 books on various aspects of fungi and natural products of microbial origin. He is a president of the Association of Fungal Biologists (AFB) and a past president of the Mycological Society of India (MSI). He is a fellow of Mycological Society of India, the Association of Biotechnology and Pharmacy, the Society for Applied Biotechnology, and Maharashtra Academy of Sciences. Dr. Deshmukh is Associate editor of Frontiers in Microbiology, series editor of Progress in Mycological Research published by CRC press. Dr. Deshmukh serves as a referee for more than 20 national and international journals. He has approximately four decades of research experience in getting bioactives from fungi and keratinophilic fungi.

Prof. Kandikere R. Sridhar is an adjunct faculty at Mangalore University, India. His primary area of study is aquatic fungi in freshwaters and marine waters. He has research collaborations in the USA, Canada, Germany, Switzerland and Portugal. He was president of the Mycological Society of India (2018), a distinguished Asian Mycologist (2015) and one of the world's top 2% scientists in the field of mycology (2019–21). He has published over 500 research articles and edited 10 books. He is in the editorial board of several national and international journals and reviewed over 150 research papers.

Prof. Dr. rer. Nat. Hesham Ali El Enshasy (Dr. rer. Nat.; M.Sc., M.Sc.) is the current Director of Institute of Bioproduct Development (IBD), and professor in bioprocess engineering, Facultyl of Chemical and Energy Engineering, Universiti Teknologi Malaysia (UTM). Since 2020, he is recognized as one of the top 2% researchers in the world in the field of Biotechnology, according Stanford University research study for both career and single year lists. Prof. El Enshasy received his B.Sc. and M.Sc. in Microbiology (Ain Shams University, Egypt), Dr. rer. Nat. in Industrial Biotechnology (TU Braunschweig, Germany 1998), M.Sc. Technology

Management (UTM, Malaysia). He has more than 350 publications in form of peer reviewed international journals, book Chapters, and books. He was also invited as keynote, plenary, and guest speaker in more than 100 international conferences in field of industrial biotechnology and BioEconomy. He serves as editor, associate editor, and editorial board member in several international journals.